Reconfigurable Circuits and Technologies for Smart Millimeter-Wave Systems

Get up to speed on the modeling, design, technologies, and applications of tunable circuits and reconfigurable millimeter-wave (mm-wave) systems. Coverage includes smart antennas and frequency-agile radio frequency (RF) components, as well as a detailed comparison of three key technologies for the design of tunable mm-wave circuits: complementary metal–oxide–semiconductor (CMOS), RF microelectromechanical systems (MEMS), and microwave liquid crystals. Measurement results of state-of-the-art prototypes are also considered. Numerous examples of tunable circuits and systems are included that can be practically implemented for the reader's own needs. This book is ideal for graduate students studying RF/microwave engineering and researchers and engineers involved in circuit and system design for new communication platforms such as mm-wave 5G and beyond, high-throughput satellites in geostationary orbit (GSO), and future satellite constellations in medium Earth orbit (MEO) and low Earth orbit (LEO), as well as for automotive radars, security, and biomedical mm-wave systems.

Philippe Ferrari is a professor at the RFIC-Lab at the University of Grenoble.

Rolf Jakoby is a professor and the managing director of the Institute for Microwave Engineering and Photonics at the Technical University of Darmstadt.

Onur Hamza Karabey is the chief executive officer at ALCAN Systems GmbH, Darmstadt.

Holger Maune is Privatdozent and leads the tunable microwave devices group at the Technical University of Darmstadt.

Gustavo P. Rehder is an associate professor at the Department of Electronic Systems at the University of São Paulo.

EuMA High Frequency Technologies Series

Series Editor
Peter Russer, Technical University of Munich

Homayoun Nikookar, *Wavelet Radio*
Thomas Zwick, Werner Wiesbeck, Jens Timmermann, and Grzegorz Adamiuk (Eds),
 Ultra-wideband RF System Engineering
Er-Ping Li and Hong-Son Chu, *Plasmonic Nanoelectronics and Sensing*
Luca Roselli (Ed), *Green RFID Systems*
Vesna Crnojević-Bengin, *Advances in Multi-band Microstrip Filters*
Natalia Nikolova, *Introduction to Microwave Imaging*
Karl F. Warnick, Rob Maaskant, Marianna V. Ivashina, David B. Davidson, and Brian
 D. Jeffs, *Phased Arrays for Radio Astronomy, Remote Sensing, and Satellite
 Communications*
Apostolos Georgiadis, Ana Collado, and Manos M. Tentzeris, *Energy Harvesting:
 Technologies, Systems, and Challenges*

Reconfigurable Circuits and Technologies for Smart Millimeter-Wave Systems

Edited by

PHILIPPE FERRARI
University of Grenoble

ROLF JAKOBY
Technical University of Darmstadt

ONUR HAMZA KARABEY
ALCAN Systems GmbH, Darmstadt

HOLGER MAUNE
Technical University of Darmstadt

GUSTAVO P. REHDER
Polytechnic School of the University of São Paulo

CAMBRIDGE
UNIVERSITY PRESS

University Printing House, Cambridge CB2 8BS, United Kingdom

One Liberty Plaza, 20th Floor, New York, NY 10006, USA

477 Williamstown Road, Port Melbourne, VIC 3207, Australia

314–321, 3rd Floor, Plot 3, Splendor Forum, Jasola District Centre, New Delhi – 110025, India

103 Penang Road, #05–06/07, Visioncrest Commercial, Singapore 238467

Cambridge University Press is part of the University of Cambridge.

It furthers the University's mission by disseminating knowledge in the pursuit of education, learning, and research at the highest international levels of excellence.

www.cambridge.org
Information on this title: www.cambridge.org/9781107102477
DOI: 10.1017/9781316212479

First published 2022

A catalogue record for this publication is available from the British Library.

ISBN 978-1-107-10247-7 Hardback

Contents

Contributors

Alfredo Bautista
Advanced Silicon, Lausanne, Switzerland

Sylvain Bourdel
University of Grenoble, Grenoble, France

Marcelo N. P. Carreño
Polytechnic School of the University of São Paulo, São Paulo, Brazil

Philippe Ferrari
University of Grenoble, Grenoble, France

Alexander Göritz
Innovations for High Performance Microelectronics, Frankfurt (Oder), Germany

Rolf Jakoby
Technical University of Darmstadt, Darmstadt, Germany

Matthias Jost
Merck KGaA, Darmstadt, Germany

Onur Hamza Karabey
ALCAN Systems GmbH, Darmstadt, Germany

Mehmet Kaynak
Innovations for High Performance Microelectronics, Frankfurt (Oder), Germany

Holger Maune
Technical University of Darmstadt, Darmstadt, Germany

Matthias Nickel
Technical University of Darmstadt, Darmstadt, Germany

Ersin Polat
Technical University of Darmstadt, Darmstadt, Germany

Thomas Quémerais
STMicroelectronics, Grenoble, France

Roland Reese
Deutsche Telekom Technik GmbH, Bonn

Gustavo P. Rehder
Polytechnic School of the University of São Paulo, São Paulo, Brazil

Ariana L. C. Serrano
Polytechnic School of the University of São Paulo, São Paulo, Brazil

Henning Tesmer
Technical University of Darmstadt, Darmstadt, Germany

Christian Weickhmann
ALCAN Systems GmbH, *Darmstadt, Germany*

Matthias Wietstruck
Innovations for High Performance Microelectronics, Frankfurt (Oder), Germany

Selin Tolunay Wipf
Innovations for High Performance Microelectronics, Frankfurt (Oder), Germany

Preface

Information and communication technologies (ICTs) are the foundation of growth and development in the modern global economy, striving to bring robust connectivity to all corners of the globe, driving both innovation and ways in which technologies can be used to improve economic and social development to build a "smart society." New means of connectivity plus enhanced architectures promise improved coverage, greater capacity, higher data rates, more efficient use of spectrum resources, much quicker round-trip times or lower latency, higher system reliability, and more flexibility for effective delivery of ICT services, including, e.g., bandwidth-on-demand and guarantee of a chosen quality of service. Innovative wireless technologies will be a critical component in this development, which opens up new applications such as those composing Mobile Internet (MI) and the Internet of Things (IoT), including machine-to-machine (M2M) communications, triggering a massive number of use cases such as ultra-high density (UHD) and 3D video, virtual reality, eHealth, online gaming, mobile cloud, remote computing, collaborative robots, smart offices, smart cities, smart manufacturing, critical infrastructure monitoring, intelligent transportation systems, and self-driving and connected cars, transforming everything.

The prospect of constant connectivity and complex interconnection of devices will also shape private-sector business models. Providers of communication services as well as manufacturers of hardware components and systems are facing a world that is hungry for more data, more speed, lower latency, and competitive pricing. Some of them are moving quickly, investing in future systems and enhanced and new components, and exploring new commercial opportunities with other industry sectors as they try to find their place in a new ecosystem that demands flexibility to meet changing demands. Increasingly, new classes of companies are developing new capabilities and developing innovative products and services that rely on technological innovations, existing wired and wireless terrestrial and satellite-based systems, but in particular on enhanced and new platforms and network architectures. In this frame, ICTs are advancing in the following key areas:

1. Enhanced and new *platforms* such as future 5G (IMT-2020), 6G systems, and beyond, high-throughput satellites (HTSs) in geostationary, medium Earth orbits (MEO), or low Earth orbit (LEO) satellite constellations; high-altitude platforms (HAPs); and other wireless network technologies that will enable new forms of connectivity for the delivery of broadband connectivity, enabling new services and

the possibility of being always connected, boosting user experience and empowering industries with ICT. To reach their full potential will require a new spectrum of resources and spectrum management. Moreover, the volume and expected growth of these deployments are forcing policymakers to consider how to address a world increasingly powered by ICTs.

2. Enhanced and new network *architectures*, software advances, and other complementary technologies that increase the flexibility and efficiency of services, such as cloud radio access network (RAN), heterogeneous networks, network function virtualization (NFV), network slicing, multiple input–multiple output (MIMO), beam forming, and beam steering are increasing the capabilities of platforms and Internet-based services.

3. New *applications (use cases)* of these technologies, primarily driven by MI and IoT, such as connected cars, and other applications for infrastructure, manufacturing, and health.

Moreover, enhanced and new *hardware concepts and technologies* for smart user devices and terminals as well as for the satellites, relay and base stations are crucial for the deployment of the new platforms and services mentioned earlier, following to some extent the software-defined radio (SDR) or more precisely, the software-controlled radio (SCR) approach. A bottleneck in this approach are the components in the radio frequency (RF) stage of the transceiver hardware and the antennas, to provide smart system functionalities, particularly for smart millimeter-wave (mm-wave) systems. Therefore, this book addresses the evolutionary process in wireless communications toward *mm-wave systems*, focusing particularly on *reconfigurable circuits and technologies* such as frequency-agile RF components and smart antennas to make these systems smart and efficient.

Chapter 1 of this book aims to give first an overview of recent developments in new platforms and technologies and explores their implications in communications, including future mobile traffic, the 5G vision, trends in satellite communication platforms, spectrum allocation, key technology drivers, markets, and perspectives.

These are mainly driven by an ever-growing demand for higher data rates to meet the needs for increasing capacity per subscriber and the increasing number of subscribers, i.e., to keep up with the remarkable speed-up of fiber optic networks. This demand on extremely high bandwidth leads inevitably to increasing operating frequencies of wireless communication and access systems up to the mm-wave range, where still large frequency resources are available. However, in particular, power link budget considerations at these high frequencies require highly directive, high-gain antennas to focus the narrow beam toward the desired hub. Traditionally, this is done with a static beam formed by a simple parabolic dish. However, when the hub and user terminal are moving relative to each other, these antennas must be steered by heavy and bulky mechanical systems, which are impractical for deploying wide-scale applications. Therefore, smart antennas are aimed toward the aforementioned platforms, dynamically creating a desired antenna pattern. The smart antenna concept covers a wide range of techniques with different complexities, including MIMO, beam-forming, and beam-steering techniques. However, implementing these techniques is challenging at mm-wave frequencies, where complexity, technological constraints, and costs increase. Hence, there are different approaches to enable cost-efficient, low-profile electronically steerable antennas (ESAs), using different technologies to improve the performance while reducing the manufacturing cost to an economical price point.

One of the largest impediments to realize an SCR at mm-wave frequencies is the lack of tunable RF bandpass filters, in particular with a high-quality factor (Q-factor). Filters can be tuned mechanically but have limited practical use. Selecting one filter from a bank of available filters is another technique. However, this is impractical for mobile devices. Another issue is adaptive matching, for example, of an antenna or even more challenging, of a wideband high-power amplifier in the transmitter stage.

Therefore, in many applications, it would be a significant advantage to dynamically change/tune the characteristics of mm-wave components such as filters and antennas electronically (instead of mechanically) to enable adaptive matching, polarization agility, frequency agility, and beam-forming capabilities (space agility). However, implementing these smart system functionalities means a considerable technological challenge to enable suitable low-cost, robust, and reliable hardware solutions, especially at mm-wave frequencies. This will be part of the book, where Chapter 2 starts with some link-budget considerations of mm-wave communications (Section 2.1) and some common basics of reconfigurable circuits and technologies for smart mm-wave systems, focusing particularly on various building blocks for smart antenna systems (Section 2.2) and tunable filters (Section 2.3), and promising technologies for tunable mm-wave components (Section 2.4).

Chapters 3–5 of this book then deal with three efficient technologies to realize reconfigurable and tunable components and circuits for smart mm-wave systems: silicon semiconductor technology (complementary metal–oxide–semiconductor, CMOS and bipolar CMOS, BiCMOS), RF microelectromechanical systems (RF-MEMS), and microwave liquid crystal (MLC) technology. In these chapters, an overview of the corresponding technology is given first, along with the characteristics and performance that can be potentially achieved, followed by examples of reconfigurable components, circuits, and systems.

Abbreviations

A/D	analog-to-digital
ADC	analog-to-digital converter
AESA	active electronically scanned array
AF	array factor
ASIC	application specific integrated circuits
AWGN	additive white Gaussian noise
BB	baseband (signal)
BEOL	back-end-of-line
BER	bit error rate
BST	barium strontium titanate
C_0	speed of light (299,792,458 m/s)
CAGR	compounded annual growth rate
CCP	cross coupled pair
CDMA	code division multiple access
CMOS	complementary metal–oxide–semiconductor
CPM	cavity perturbation method
CPW	coplanar waveguide
CR	cognitive radio
CRLH	composite left-right-handed (transmission line)
D2D	device-to-device communications
DC	direct current
DK	Design Kit
DRM	Design Rule Manual
DVB	digital video broadcasting
DW	dielectric waveguide
EDGE	enhanced data rate for gsm evolution
EF	element factor
EIRP	equivalent isotropic radiated power
EM	electromagnetic
eMBB	enhanced/extreme mobile broadband
ESA	electronically steerable antenna
ETSI	European Telecommunications Standards Institute
FCC	Federal Communications Commission
FDD	frequency division duplex

FDFD	finite difference frequency domain
FDMA	frequency division multiple access
FDTD	finite difference time domain
FEM	finite element method
FET	field-effect transistor
FoM	figure of merit
FPGA	field-programmable gate arrays
FSS	frequency-selective surface
FTR	frequency tuning range
gCPW	grounded coplanar waveguide
GEO	geostationary (satellite)
GPRS	general packet radio services
GSM	Global Systems for Mobile communications
H2M	human-to-machine (communication)
HAPS	high-altitude platform station
HBT	heterostructure bipolar transistor
HDPE	high-density polyethylene
HPBW	half-power beam width
HSPA	High Speed uplink/downlink Packet Access
HTS	high-throughput satellite
IC	integrated circuit
IF	intermediate frequency
IMT	International Mobile Telecommunications
IMUX	input multiplexer section
IoT	Internet of Things
IP	Internet Protocol
ITO	indium tin oxide
ITU	International Telecommunication Union
LC	liquid crystal
LCD	liquid crystal displays
LCP	liquid crystal polymer
LEO	low Earth orbit (satellites)
LH	left-handed (transmission line)
LNA	low-noise amplifier
LO	local oscillator
LOS	line-of-sight (communication)
LTCC	low-temperature co-fired ceramic
LTE	Long-Term Evolution technology
LTE-A	LTE-Advanced
M2M	machine-to-machine (communications)
MEMS	microelectromechanical system
MEO	medium Earth orbit (satellite)
MI	mobile Internet

MIM capacitor	metal insulator metal capacitor
MIMO	multiple input–multiple output
MLC	microwave liquid crystal
mm wave	millimeter wave
MMI	multimode interference
MMIC	monolithic microwave integrated circuit
MMS	multimedia messaging service
mMTC	massive machine-type communications
MSL	microstrip line
MuT	material under test
NaM	nanowire membrane
NF	noise figure
NGEO	non-geostationary orbit
nLOS	non-line-of-sight (communication)
NRD	non-radiative dielectric (waveguide)
NRW	Nicolson–Ross–Weir
OFDM	orthogonal frequency division multiplexing
OSI	open systems interconnection
PACS	personal access communications system
PCB	printed circuit board
PMP	polymethylpentene (TPX)
PPDW	parallel-plate dielectric waveguide
PTFE	polytetrafluoroethylene (Teflon)
QoS	quality-of-service
RAN	radio access network
RF	radio frequency
RH	right-handed (transmission line)
RTPS	reflection type phase shifters
RX	receiver
S-CPS	slow-wave coplanar stripline
S-CPW	slow-wave coplanar waveguide
SCR	software-controlled radio
SDR	software-defined radio
SIW	substrate integrated waveguide
SLL	side lobe level
SNR	signal-to-noise ratio
SPDT	single-pole double-throw
SRR	split-ring resonators
SWO	standing-wave oscillators
TDD	time division duplex
TDMA	time division multiple access
TDS	time-domain spectroscopy
TEM	transverse electromagnetic (wave)

TRL	thru reflect line (calibration)
TWO	traveling-wave oscillator
TX	transmitter
UMTS	universal mobile telecommunications systems
URLLC	ultra-reliable and low-latency communications
UWB	ultra-wide band radio
VCO	voltage-controlled oscillator
VGA	variable gain amplifiers
VNA	vector network analyzer
VSAT	very small aperture terminal
WiMAX	worldwide interoperability for microwave access
WLAN	wireless local area network
WMAN	wireless metropolitan area network
WPAN	wireless personal area network
WSatN	wireless satellite network
YIG	yttrium iron garnet

1 Introduction and Motivation

Rolf Jakoby and Holger Maune

The rapidly evolving world of wireless communications is driven by the ever-ongoing enhancement of already existing services and the emergence of new services. This includes the usage of video-on-demand services such as ultra-high definition (UHD) television and 3D video (*enhanced mobile broadband*), which will continue to grow and account for a large portion of all mobile data traffic. Moreover, Mobile Internet (MI) and Internet of Things (IoT), including machine-to-machine (M2M), device-to-device (D2D) and human-to-machine (H2M) connectivity, will be major market drivers beyond 2020 and are expected to trigger a large range of use cases. Some of these emerging use cases will significantly contribute to increasing mobile data traffic. The rate at which applications are being adopted is thereby accelerating (*application uptake*). Another driving factor is the massive growing number of connections (connected devices) and the spread of smart mobile devices (*device proliferation*), smartphones, and tablets, but also other kinds of devices such as wearable and machine-type communication devices. This is running alongside a growing number of users/subscribers with an ever-increasing demand for *flexibility, mobility,* and *higher user data rates.* All these driving factors will lead to an explosion of data traffic in mobile communications [1]. To enable higher data rates and spectrum efficiency to cope with this data traffic explosion as well as to meet users' demands for flexibility and mobility, various enabling techniques, technologies, and smart hardware solutions have to be developed and devised such as advanced modulation, coding and multiple access schemes, flexible backhaul and dynamic radio access configurations, flexible spectrum usage, flexible downlink/uplink resource allocation schemes, flexible (*frequency-agile* and *environmental-agile*) radio interfaces, as well as *polarization- and space-agile* antenna systems, including analog beam-steering and beam-forming antenna arrays as well as active and massive multiple-input–multiple-output (MIMO) antenna configurations.

Because of the huge spectrum availability in the millimeter-wave (mm-wave) frequency range from 30 GHz to 300 GHz, and connected with it, the feasible large bandwidths to meet the requirements for ever higher user data rates, mm-wave communication will be the next frontier in wireless technology, in particular for future fifth generation (5G) and beyond terrestrial as well as future satellite-based hardware platforms and infrastructures. The latter will include intrinsic multicast/broadcast and high-throughput satellites (HTSs) in the K_u-, K_a-, and V-bands, but also future non-geostationary orbit (NGEO) satellite fleets for real-time communication services.

Consequently, one of the next evolutionary steps will be to form hybrid networks, combining satellite and terrestrial Internet Protocol version 6 (IPv6) connectivity, for example for connected, self-driving cars (Internet of Vehicles) and IPv6 on robotics, and later in the next generations of mobile communications systems, 6–8G [2]. Some of the HTSs and NGEOs operate or will operate below the mm-wave frequency range, for example from about 10 GHz to 30 GHz, that is, belonging to the microwave[1] frequency range. This part of the spectrum is also considered in this book.

Sections 1.1 and 1.2 give an overview of the different communication platforms and infrastructure, depicting the evolution and trends in terrestrial and satellite-based mobile communications and technologies impacting emerging and future applications. These sections will also provide details about the driving factors behind these evolutionary processes, with facts and figures, including the major drivers influencing the growth of future mobile traffic; the 5G activities, vision, and objectives; the spectrum allocation schemes; the key technology drivers; the markets and perspectives; the evolving high-altitude platform stations (HAPSs); and trends of HTSs in geostationary and medium Earth orbits (GEOs and MEOs), as well as of low Earth orbit (LEO) HTS constellations, picturing also future interoperable satellite and hybrid terrestrial 5G satellite networks.

To integrate more and more users, services, and standards, and related to this, more and more frequency bands over a wide frequency range, as well as to enable broadband access, for example to internet, and at the same time, high energy efficiency, future generations of mobile communication systems require much higher flexibility for multiservice, multistandard, and multiband operation than in state-of-the-art systems. While reconfigurable architecture in the digital baseband already allows this flexibility for multiservice and multistandard operation, using powerful field-programmable gate arrays (FPGAs), controllers, and analog to digital converters with high dynamic and high speed, it has not yet been achieved in the analog radio frequency (RF) front end. There, up to now, multiband operation is often realized by individual transceivers for each frequency band. However, parallel implementation of transceivers means higher cost as well as larger space and volume, and in addition, higher overall energy consumption and hence less efficiency, in particular for the high-power amplifiers in the transmitter branches. Hence, frequency-agile RF front-end components such as filters, antennas, or impedance matching networks, tuning its center frequency over a wide frequency range or changing its bandwidth and/or impedance on demand and controlled by software would enable compact and smart *frequency-agile, software-controlled radios*. However, these approaches require new system concepts and corresponding innovative technological solutions for hardware with high integration density and low power consumption, in particular for hand-held devices. Another technological challenge in mm-wave mobile communications, in particular for relatively long distances, are electronic beam-steering or even

---

[1] Microwaves are a form of electromagnetic radiation with wavelengths ranging from about 1 meter to 1 millimeter, with frequencies between 300 MHz and 300 GHz [https://en.wikipedia.org/wiki/Microwave]. They include millimeter waves, ranging from 30 GHz to 300 GHz.

beam-forming, high-gain antennas to enable flexible and reliable line-of-sight (LOS) communications.

High-gain, narrow-beam antennas are desired because of the following physical imperative: as the operational frequency increases, the free-space path loss increases at least quadratically, resulting in a stringent link budget. Moreover, utilizing a larger amount of bandwidth results in a higher amount of received noise power. Both effects lead to a significantly reduced signal-to-noise ratio (SNR) for a given communication distance, making mm-wave communication scenarios quite different from existing approaches in the radio-frequency and microwave ranges. The reduced SNR can be circumvented by increased equivalent isotropic radiated power (EIRP), which can be achieved by large output power of the transmitter's power amplifier and by high-gain antennas. However, it is a challenge to obtain high transmit power in the mm-wave range no matter which device technology is chosen; therefore, such systems are typically built upon high-gain, narrow-beam antenna array architectures, utilizing electronic polarization agility, beam-steering, beam-forming, or adaptive spot-beam capabilities. This would enable flexible and reliable LOS mm-wave communications for wireless backbones and small cells such as micro-, pico-, and femtocells in cellular networks or for mobile users of satellite systems.

A first introduction and overview of concepts and technologies to implement these smart system functionalities is provided in Section 1.3, including the software-defined and software-controlled radio approaches; examples of reconfigurable/tunable RF components and circuits; various technologies to enable suitable low-cost, robust, and reliable hardware solutions suited for smart mm-wave systems; and different electronically steerable antenna (ESA) concepts, introducing also approaches of various spin-off companies for producing and commercializing low-profile, cost-efficient large-scale ESAs.

1.1   Evolution of Mobile Communications

Since the 1990s, mobile communication has become a most significant platform, leading to a revolutionary transformation in the way to communicate, experience entertainment, and make use of the Internet, driven by users' demands for mobility and tremendous amount of growth in connectivity and data traffic volume. Since these trends are evolving dramatically, we are now at the dawn of a new era in mobile communications, aiming to overcome limits of state-of-art systems toward innovative mobile broadband applications and to meet new and unexpected demands beyond the capability of previous generations of systems to provide a mobile and everywhere-connected society [3, 4]. Since the sociotechnical evolution in the past decades has been significantly driven by this evolution of mobile communications, which has been closely integrated in the daily life of all of society, it is expected that both will remain tightly coupled, forming an even stronger foundation for society beyond 2020 [4]. Figure 1.1 illustrates the evolution in mobile communications, that is, all evolving generations that use a licensed spectrum, including the future 5G. Wireless

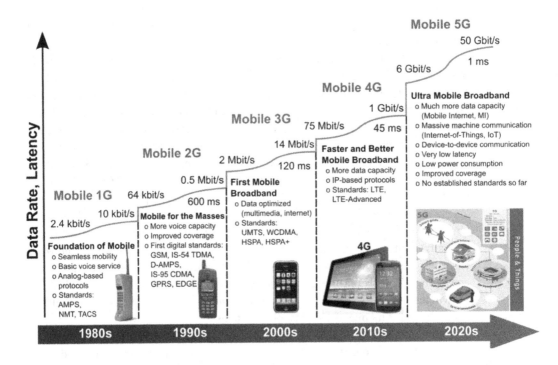

Figure 1.1 Evolution in mobile communications with some characteristics of the various generations.

technologies that are using an unlicensed spectrum (nomadic) such as WiFi, Bluetooth, and WiMAX will not be considered. During the process of evolution in mobile communications, the data rate, mobility, coverage, and spectral efficiency generally increase.

It all started in 1979 with the foundation of the first cellular mobile communication systems by NTT in Japan. These 1G cellular mobile communication systems were focused on basic voice service with analog-based protocols, using most often frequency modulation (FM) and frequency division multiple access (FDMA) schemes. A unique feature was the use of first cellular networks built up of hexagonal cells. Data rates were up to 2.4 kbit/s, but often accompanied with some drawbacks such as unreliable handoff and poor voice links. Major standards were the Advanced Mobile Phone System (AMPS), Nordic Mobile Telephone (NMT), and Total Access Communication System (TACS) [2, 5].

The first digital standards were developed in the 1990s in the second generation of mobile communications with additional voice capacity and improved coverage and mobility. Standards were Global Systems for Mobile communications (GSM), IS-54 TDMA (Time Division Multiple Access), IS-95 CDMA (Code Division Multiple Access), and the Personal Access Communications System (PACS). Beyond voice communication, 2G also provided services such as Short Message Service (SMS) and email. Data rates were up to 64 kbit/s. 2G-mobile handset battery lasted longer, because radio signals have low power. 1G and 2G used a circuit-switched, connection-based technology, where the end systems are devoted for the entire call period, having low efficiency in the usage of bandwidth and resources and inability to handle complex data such as video. In 2.5G, packet switching was applied along with circuit switching, supporting high data rates up to 144 kbit/s. The main technologies were General Packet Radio Services (GPRS) and Enhanced Data rate for GSM Evolution (EDGE) [2, 5]. While 1G and 2G technologies used a circuit-switched, connection-based technology, where the end systems are devoted for the entire call period, 2.5G and 3G used both circuit and packet switching, and the next generations from 3.5G to 5G use packet switching.

3G was established with the first commercial network of NTT DoCoMo in October 2001 in Japan, merging high-speed mobile access to services based on Internet Protocol (IP). It allowed transmission rates of up to 2 Mbit/s for fixed applications in local coverage areas, enabling video calls and video conferences, and 144 kbit/s to 384 kbit/s for mobile applications in wide coverage areas. Beside improved voice quality, it provided global roaming for the first time. However, 3G handsets often required more power than most 2G models. Since 3G utilized Wideband Code Division Multiple Access (WCDMA), Universal Mobile Telecommunications Systems (UMTS) and Code Division Multiple Access (CDMA) 2000 technologies, evolving technologies such as High Speed uplink/downlink Packet Access (HSPA) and Evolution-Data Optimized (EVDO) made an intermediate wireless generation 3.5G with improved data rates of 5–30 Mbit/s, allowing downloading of huge files in less time. The peak upload rate was 5 Mbit/s and the peak download rate is 100 Mbit/s [2, 5, 6]. Thus, 3G was the first mobile broadband standard.

After the introduction of Long-Term Evolution (LTE) technology and Fixed Worldwide Interoperability for Microwave Access (WiMAX) already during the 3G evolutionary period, subsequent LTE-Advanced (LTE-A) as a forthcoming 4G standard along with mobile WiMAX, developed in the 2010s, provided a substantial number of users with faster and better mobile broadband services such as video-on-demand, peer-to-peer file sharing and composite web services, Multimedia Messaging Service (MMS), Digital Video Broadcasting (DVB), and video chat as well as high-definition TV and mobile TV. Thus, 4G truly constitutes mobile broadband. All this is possible because a supplementary spectrum is accessible and operators manage their network more efficiently, offering better coverage with improved performance such as higher throughput and spectral efficiency for lower cost, and most decisively, the prevailing communication networks could be improved by imparting a complete and reliable solution based on IP, where the 128-bit IP version (IPv6) replaces the 32-bit IP version (IPv4), supporting a high quality of services, security, and mobility [5, 6].

4G uses orthogonal frequency division multiplexing (OFDM) and ultra-wide band radio (UWB). The frequency range is between 2 and 8 GHz with a bandwidth of 100 MHz and data rates of 20 Mbit/s up to 1 Gbit/s, where the peak upload and download speed may be up to 500 Mbit/s and 1 Gbit/s, respectively. Mobile speed can be up to 200 km/h. The high performance is achieved by the use of long-term channel prediction, in both time and frequency, scheduling among users, and multiple antennas combined with adaptive modulation and power control. First efficient software-defined radios (SDRs) are introduced, where the radio can be configured or defined by software. Primarily, SDR benefits from the high processing power, enabling multistandard and multiband operation, and where terminals will adapt the air interface to the available radio access technology [2]. Although 4G connections represented only 26% of all mobile connections, 4G traffic already accounted for 69% of all mobile traffic in 2016, while 3G connections represented 33% and only 24% of all mobile traffic. Thus, a 4G connection generated four times more traffic on average than a 3G connection [7].

To cope with this exponential increase in mobile data traffic and smart device connections in the future, but also to meet the network operators' and users' demands for flexibility and mobility as well as for enhanced and new services and applications, often requiring significant higher data rates, and hence, bandwidth, the next major phase in the evolution of mobile communications is upcoming, named 5G. 5G is considered as beyond 2020 mobile communications technology, supporting IPv6 and flat IP, where it is commonly assumed that 5G cellular networks face some challenges that are not effectively addressed by 4G: higher capacity, higher data rate, lower end-to-end latency, massive device connectivity, reduced cost, and consistent quality service (QoS) provisioning. Recently introduced IEEE 802.11ac, 802.11ad, and 802.11af standards are acting as building blocks on the road toward 5G [5].

1.1.1   Major Drivers Influencing the Growth of Future Mobile Traffic

The evolution of wireless communications is driven primarily by the anticipated data traffic explosion and its economical perspective. Many white papers and reports present some of the major global data traffic projections and growth trends. According to the CISCO White Paper [7], the forecast for global mobile data traffic per month ranges from about 7 EB to 49 EB between 2016 and 2021, which means a compounded annual growth rate (CAGR) in the overall mobile data traffic of 47% from 2016 to 2021. The key figures in [8] are quite similar: the total mobile data traffic is estimated as 8.5 and 69 EB/month in 2016 and 2022, respectively, with a CAGR of 45% compared to 70 and 170 EB/month for the total fixed data traffic in 2016 and 2022, respectively, with a CAGR of 20%. Better user experience and lower prices of "smart" devices will accelerate its proliferation. Thus, the projected IPv6 mobile traffic forecast will increase over the same time period from 1.03 EB in 2016 to 27.4 EB in 2021, which means an increase from 14.7% to 56% of the total mobile-traffic noted earlier. The CAGR in IPv6 network traffic is then 92% from 2016 to 2021 [7].

Some drivers and trends that impact traffic growth are described in detail in Report ITU-R M.2370 [1], containing global International Mobile Telecommunications (IMT) traffic estimates beyond 2020 from several sources, where the global mobile traffic estimates per month beyond 2020 with and without M2M traffic are illustrated in Figure 1.2. These traffic estimates are significantly higher than those of the aforementioned White Paper [7] for 2020 and 2021. They reach 62 EB/57 EB,

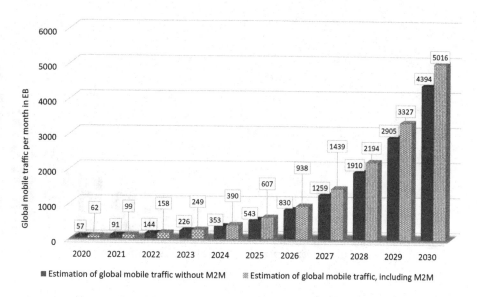

Figure 1.2 Mobile traffic estimates per month in EB between 2020 and 2030 without and with M2M traffic. All data values are taken from [1]. The network capacity unit ExaByte is 10^{18} bytes.

607 EB/543 EB, and 5016 EB/4394 EB with and without M2M traffic in 2020, 2025, and 2030, respectively. This means a CAGR of around 55% and 54% in 2020–2030 with and without M2M traffic, respectively. These estimates anticipate that global IMT traffic will grow in the range of 10–100 times from 2020 to 2030 [1].

Among many characteristics and trends that are expected to impact the anticipated traffic growth and the overall traffic demand in 2020 and beyond, there are three *major drivers* influencing the growth of future mobile traffic [1]:

- *Enhanced mobile broadband*, in particular by video usage
- *Application uptake*, that is, the rate at which applications are being adopted
- *Device proliferation*, accompanied with an evolution toward ever smarter mobile devices in different form factors and with continuously enhanced capabilities and intelligence, which require increasing bit-rates and bandwidth

All are expected to evolve over time.

Within the different service types behind this anticipated traffic growth is M2M with a traffic volume of about 7% and 12% of the total in 2020 and 2030, respectively [1]. By far the largest traffic volume will be consumed by the usage of video-on-demand services (*video usage*), which will continue to grow, and resolution of these videos will continue to increase. Beyond 2020, people will watch more ultra-high-resolution audio-visual content, regardless of the way the content is delivered. Video streaming with an *enhanced mobile broadband* is expected to account for almost 75% of all global mobile traffic in 2021, being 4.2 times and 6 times higher than non-video in 2025 and 2030, respectively [1]. The popularity of ultra-high resolution and large

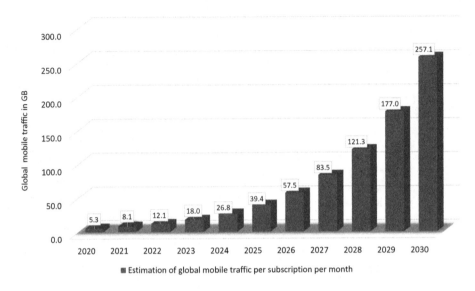

Figure 1.3 Estimations of global mobile data traffic per subscriptions per month from 2020 to 2030 without M2M traffic. All data values are taken from figure 10 in [1]

screens as well as the increasing number of high-performance smart devices such as smartphones and tablets, widely used around the world, together with the growth of mobile subscriptions will dramatically increase the mobile traffic volume consumed by each user according to Figure 1.3. The figure indicates that each subscriber will consume 5.3 GB, 39.4 GB, and 257.1 GB of data traffic per month on average in 2020, 2025, and 2030, respectively.

Moreover, the rate at which applications are being adopted by users is accelerating. The annual global downloading of applications was 102 billion apps in 2013 and grew to about 270 billion in 2017. This mobile *application uptake* and usage of applications will contribute to increased mobile broadband traffic [1] and the broadband access growth rates given by various laws.

Nielsen's Law of Users' Internet Bandwidth, that is, users' internet connectivity in bits per second, is based on data from 1983 to 2016. It states that high-end users' connection speed grows by 50% per year, reaching 120 Mbit/s and 240 Mbit/s in 2014 and 2016, respectively [9]. This means it doubles roughly every 21 months. This 50% CAGR is close to the more established Moore's Law, which states that computers double in capabilities every 18 months, which corresponds to about 60% annual growth. An observation similar to Nielsen's has been made by Ulm et al. in [10]. It reflects a 50% CAGR for the downstream and about 30% CAGR for the upstream capacity, respectively. Similar exponential curves, that is, fitted straight lines of data rates plotted logarithmically against time, can be found for the wireline, wireless (cellular), and nomadic (WLAN) technologies according to Edholm's Law of Bandwidth. Some trends can be observed from the figures of data rates plotted logarithmically against time [11–13]:

- Backhaul network bandwidth (wireline) is larger than the ones for cellular (wireless) and WLAN (nomadic) technologies. For example, in 2010, data rates of cellular connections were around 10 Mbit/s, for WLAN connections at about 1 Gbit/s, and for Ethernet connections of office desktops about 10 Gbit/s.
- Backhaul network bandwidth grows annually at 20–30%, while the bandwidth of wireless and nomadic technologies exhibits a 50% CAGR according to Nielsen's Law.
- Since wireless and nomadic data rates are growing much faster than wireline ones, the curves will converge sometime in the future.
- Wireless and nomadic bandwidth lines are almost parallel with a gap in between of about 100 times, however, gradually converging and cutting at around 2030. This is because use cases for mobile data on devices will converge. Consumers only care about applications and services, not about the underlying network; thus a truly heterogeneous network is needed, leading to a new inflection point of technology [11].

Because the highest data speed offered is a determining factor for sizing the network, some open questions arise [10]: will this broadband demand for wireless and nomadic technologies end, that is, Nielsen's Law break as Moore's Law or will service providers offer a residential 10+ and 100+ Gbit/s Internet service by 2025 and 2030, respectively, according to Nielsen's Law?

As mentioned earlier, another driving factor is the *device proliferation* accompanied with an evolution toward ever smarter mobile devices in different form factors and with continuously enhanced capabilities and intelligence, which require increasing bitrates and bandwidth. Globally, the growth of wireless devices, accessing mobile networks worldwide, grew to 8.0 billion in 2016, and it will grow to 11.6 billion by 2021 at a CAGR of 8%, where 8.3 billion will be handheld or personal mobile-ready devices and 3.3 billion M2M connections according to the 2017 CISCO White Paper [7]. Again, the values for the personal mobile-ready devices are close to the one in [8]. The growth of global mobile subscriptions between 2020 and 2030 was estimated in 2015 by the International Telecommunication Union [1], where the values in 2020 and 2021 are higher than the ones of the aforementioned CISCO Report. It estimated the following number of global mobile subscriptions: 10.7 billion in 2020, 13.8 billion in 2025, and 17.1 billion in 2030, respectively. This is illustrated in Figure 1.4, together with the estimated number of mobile connected M2M devices, which would be around 7 billion in 2020, 34 billion in 2025, and 97 billion in 2030, respectively. Hence, it is expected that the total number of devices connected by global mobile communications networks will surpass 100 billion by 2030, which will be more than 10 times greater than the world population at that time, which was estimated at 7.52 billion in July 2017, 7.76 billion in July 2020, 8.14 billion in July 2025, and 8.5 billion in July 2030 [14].

Better user experience and lower prices of smart devices will accelerate the growth rate of the penetration of smart devices. Thus, concerning the share of total devices, there will be a rapid decline of non-smartphones or feature phones from about 40%

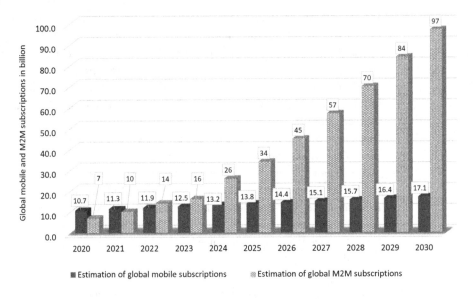

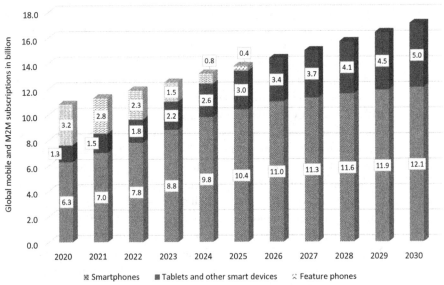

Figure 1.4 Estimations of global mobile and global M2M subscriptions in billions between 2020 and 2030 (top). In addition, the global mobile subscriptions are separated into three categories (bottom). All data values are taken from [1]

(3.3 billion) in 2016 down to 13% (1.5 billion) in 2021. At the same time, the share of smartphones, including phablets, will increase from about 45% (3.7 billion) in 2016 up to 53% (6.1 billion) in 2021. The shares of tablets, PCs, and other smart devices will smoothly go up from 4% in 2016 to 5% in 2021, while a noticeable growth is expected in M2M connections with shares of 10% in 2016 to 29% in 2021 [7]. Again, the estimation of global mobile subscriptions, separated for feature phones, smartphones, tablets, and other smart devices between 2020 and 2030 from [1], is given in Figure 1.4.

1.1.2 5G Activities, Vision, and Objectives

The framework, the overall objectives, and timelines of the development of IMT, including user and applications trends, usage scenarios and capabilities, technology trends and technical feasibility, traffic growth estimation, and spectrum implications as well as the expected evolution is regularly defined and released in Recommendations of the International Telecommunication Union (ITU), offering at the time of its publication some visions of future systems and networks. The most important ones for mobile communications are taken as basic references for this book chapter.

The Recommendation for IMT-2000 in 2000 [15] provides a single set of standards for 3G, which marks a major change in cellular mobile communications evolution. Based on IMT-2000 (3G), the ITU has released Recommendation ITU-R M.2012 in the terrestrial radio interface of IMT-Advanced in 2012, after nine years for its development, where 4G specifications are based on the IMT-Advanced requirements. Along with the enhancement of IMT-2000 (3G) and IMT-Advanced (4G) systems, a new IMT vision was released in 2015 [4], defining the framework and overall objectives of the development of future radio interfaces with far more enhanced and new capabilities of IMT in 2020 and beyond, called IMT-2020 (5G).

In parallel, different activities, involving academic research groups and telecommunication companies, have worked on or are still working on 5G scenarios, test cases, new system concepts, network architectures, and technologies, influencing the standardization and regulatory processes. A brief overview over some activities is given in [6]. An example is the largest 5G project METIS (Mobile and wireless communications Enablers for the Twenty–twenty Information Society) within the Framework Program 7 (FP7) of the European Union, where top telecommunication companies and top academic institutions have been evolved. Their comprehensive outcomes are summarized in the METIS final project report in 2015 [16] (Deliverable D8.4), based on in total 29 deliverables, including scenarios, requirements, and key performance indicators for 5G mobile and wireless system (D1.1), channel models (D1.2, D1.4), test-bed/demonstration results (D1.3), requirements and general design principles for new air interface (D2.1), novel radio link concepts and state of the art analysis (D2.2), components of a new air interface-building blocks and performance (D2.3), proposed solutions for new radio access (D2.4), multi-node/multi-antenna transmission technologies (D3.1, D3.2, D3.3), network-level solutions (D4.1, D4.2, D4.3), spectrum needs and usage principles (D5.1, D5.3, D5.4), METIS system concept and technology roadmap (D6.6), and architecture and system evaluation results (D6.4, D6.3). In the subsequent METIS-II project (https://5g-ppp.eu/metis-ii/), the key objectives have been

to develop an overall 5G radio access network design and to provide the technical enablers needed for an efficient integration and use of the various 5G technologies and components currently developed. On the strategic level, METIS-II provided the 5G collaboration framework with the 5G-PPP (5G Infrastructure Public–Private Partnership) for a common evaluation of 5G radio access network concepts and prepare concerted action towards regulatory and standardization bodies.

Stimulated by the METIS outcome and on the aforementioned 5G/IMT – 2020 Report [4], Wei Xiang et al. published a comprehensive book in 2017 [3], in which a large number of authors have been involved, revealing the enabling techniques for 5G networks and addressing the challenges and opportunities of 5G mobile communications.

Based on these three major literatures on 5G, its vision is briefly summarized in the text that follows, starting with the major market drivers beyond 2020, which is expected to be MI and IoT, including M2M or D2D connectivity [3, 4]. Both MI and IoT will trigger a massive number of *use cases*, for example

- Mobile Internet (MI): Ultra-high density (UHD) and 3D video, augmented reality, virtual reality, online gaming, mobile cloud, remote computing, tactile Internet, 3D connectivity to aircrafts and drones, collaborative robots, smart office, as well as
- Internet of Things (IoT): Smart grid and critical infrastructure monitoring, mobile surveillance, environmental monitoring, industrial automation, eHealth services, smart wearables and smart body area networks, sensor networks, smart homes/ buildings, smart cities, smart transportation, self-driving and connected cars (Internet of Vehicles).

All these use cases can be grouped into three usage scenarios [3, 4] or generic 5G services [16], addressing different use case characteristics according to Figure 1.5:

- Enhanced/extreme mobile broadband (eMBB) for improved performance and an increasingly seamless user experience, offering human-centric use cases for access to multimedia services at least with IMT capabilities (see Table 1.1)
- Massive machine-type communications (mMTC) characterizes a large number of connected devices, typically with relatively low-volume, non-delay sensitive data at low cost and very long battery lifetime
- Ultra-reliable and low-latency communications (URLLC) with stringent requirements for capabilities such as throughput, latency, and availability for applications such as industrial process monitoring and manufacturing, remote medical diagnosis and surgery, as well as safety and automation in smart grids.

While MI is focused on people-oriented communications, IoT provides communications between things and between things and people. Some of its use cases are indicated in Figure 1.5. UHD and 3D videos, over-the-top services such as social networking, as well as desktop cloud, augmented reality, and online gaming will significantly drive up data rates, even at high mobility in high-speed trains and connected cars (Internet of Vehicles), where massive data activities of a vast number of users will require a high system capacity. For services such as smart homes, smart cities, as well as environmental monitoring, future networks have to support massive

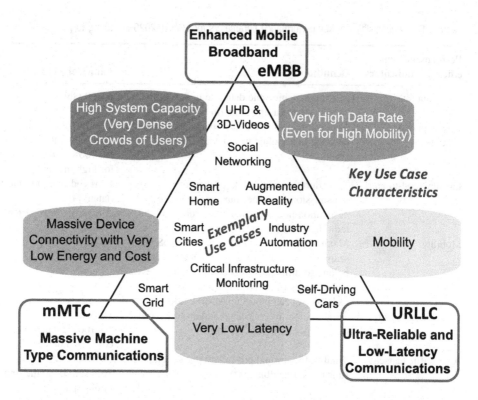

Figure 1.5 Usage scenarios (generic 5G services) for mobile communications beyond 2020 with key use case characteristics and some exemplary use cases according to [4, 16].

connectivity of devices with low energy and cost. While some use cases such as desktop cloud, augmented reality, and online gaming will additionally need "imperceptible" latency, "safety-critical" use cases such as e-banking, eHealth, self-driving cars, smart grid, and critical infrastructure monitoring face extreme security requirements, that is, in fast response with very low latency in the lower millisecond range and with nearly absolute reliability and perception of 99.999% availability [2–4, 16].

It is expected that 5G/IMT-2020 will provide far more enhanced capabilities than 4G/IMT-Advanced, defining nine key performance and efficiency indicators summarized in Table 1.1.

Among these key capabilities, the user experienced data rate, connection density, and latency might be the most fundamental ones. While all key capabilities may to some extent be important for most of the aforementioned use cases, the relevance of certain key capabilities may be significantly different for the individual use cases or usage scenarios [3, 4, 16]. In these references, the importance of key capabilities for the three usage scenarios are given in form of a "spider-web" and "blooming flower" pictures together with some explanations, and in addition, the enhancement of key capabilities from 4G/IMT-Advanced to 5G/IMT-2020.

Table 1.1 Key capabilities or key performance parameters for 5G/IMT-2020 according to [2–4, 16]

Performance and efficiency indicators	Definition	Values/objectives
Peak data rate	Maximum achievable data rate under ideal conditions per user/device (in Gbit/s)	10–20 Gbit/s
User experienced data rate	Achievable data rate that is available in a real network to a mobile user/device (in Mbit/s or Gbit/s)	Typical 0.1–1 Gbit/s, up to 10 Gbit/s 10–100× higher data rates (even for high mobility)
Latency (end-to-end)	Time duration of the data packet for transmission from the source node to the destination node, i.e., the contribution by the radio network (in ms)	1 ms end-to-end round-trip delay (latency) at least 5× lower latency
Mobility	Maximum speed at which a defined QoS and seamless transfer between radio nodes that may belong to different layers and/or radio access technologies (multilayer/radio access technology) can be achieved (in km/h)	500 km/h
Connection density	Total number of connected and/or accessible devices per unit area (per km^2)	10^6/km^2 10–100× higher number of connected devices
Area traffic capacity	Total traffic throughput served per geographic (unit) area (in Mbit/s/m^2)	10 Mbit/s/m^2 100–1,000× higher traffic throughput
Spectrum efficiency	Average data throughput per unit of spectrum resource and per cell (bit/s/Hz)	3–5× higher spectrum efficiency
Energy efficiency	Number of information bits per unit energy consumption (in bit/Joule) o On the network side, refers to the quantity of information bits transmitted to/ received from users, per unit of energy consumption of the radio access network o On the device side, refers to quantity of information bits per unit of energy consumption of the communication module	100+ 90% reduction in network energy usage 10× longer battery life for low-power devices (up to 10 years)
Cost efficiency	Number of information bits that can be transmitted per unit cost (in bit/€)	100+

The third column indicates values/objectives and improvements compared to 4G/IMT-Advanced.

Compared to 4G, 5G should have three to five times higher spectrum efficiency and more than 100 times improvement on energy and cost efficiency to outperform previous generations of mobile communications systems. Moreover, to keep up with the rapid traffic growth, some core design objectives of 5G wireless networks are [2–4, 16, 17]:

- Implementation of massive (up to 1,000-fold) area traffic capacity (see Table 1.1) and massive device connectivity for at least up to 100 billion devices according to the predictions in Figure 1.4.

- Providing a fiber-like user experienced data rate by 10–100 times (see Table 1.1), that is, of up to 10 Gbit/s, for example, to support mobile cloud service, capable of extremely low latency and response times.
- Support for an increasingly diverse set of services, applications, and users – all with extremely diverging requirements as discussed earlier for the usage scenarios.
- Flexible and efficient use of all available noncontiguous spectrum for wildly different network deployment scenarios, which will be described in the next section.

1.1.3 Spectrum Allocation

To cope with the anticipated 1,000 times higher traffic capacity and the up to 100 times higher typical user data rate in 5G wireless networks, considerably more spectrum is required than currently available for mobile and wireless communication systems. Hence, in the 5G vision, another significant driver is the vast amount of spectrum available in the mm-wave range [3, 16, 18–20], for example, for cellular hot-spot coverage to satisfy consumer demand for high-speed wireless access with ultra-low latency. Technical feasibility of radio interface technology and systems operating in frequency bands between 6 GHz and 100 GHz, considering propagation characteristics, antenna technology, active and passive components, physical layer, and medium access control design as well as deployment architectures, is carried out by simulations and performance tests and trials and are published in [18, 19, 21].

Spectrum has been in the past and will also be in future one of the most valuable resource for mobile communications. Therefore, agencies and standardization organizations worldwide aim for international harmonized spectrum and full-spectrum access, especially above 6 GHz. Hence, beyond the sub-6 GHz bands for 5G in Europe 3.4–3.8 GHz, USA 3.1–3.55 GHz and 3.7–4.2 GHz, Japan 3.6–4.2 GHz and 4.4–4.9 GHz and China 3.3–3.6 GHz, 4.4–4.5 GHz and 4.8–4.99 GHz for 5G phase I, frequency bands in the mm-wave range are already foreseen for 5G phase II: in Europe 24.25–27.5 GHz and 31.8–33.4 GHz, United States 27.5–28.35 GHz and 37–40 GHz, Japan 27.5–29.5 GHz, China 24.75–27.5 GHz, South Korea 26.5–29.5 GHz. Moreover, some studies and tests are carried out by Samsung at 13.4–14 GHz, 18.1–18.6 GHz, 27.0–29.5 GHz, and 38.0–39.5 GHz. China allocated 40.5–42.5 GHz and 48.4–50.2 GHz for fixed point-to-point wireless access systems as well as 42.3–47.0 GHz and 47.2–48.4 GHz for mobile point-to-point wireless access systems [3].

In the V-band, a continuous spectrum is allocated for wireless communications in unlicensed mode: 57–64 GHz in the United States, 57–66 GHz in Europe, 59–64 GHz in China, and 59–66 GHz in Japan [3, 19, 20]. In Europe, the European Telecommunications Standards Institute (ETSI) is working to facilitate the use of the E-band from 71.0 to 76.0 GHz and 81.0 to 86.0 GHz, and in the future, on the channelization of the W-band from 92.0 to 114.5 GHz and the D-band from 130.0 to 174.8 GHz for large-volume (high capacity) backhaul and front-haul systems as well as for innovative solutions for fixed broadband access [20].

Most of the GEO- and MEO-HTS make efficient use of both, K_u-band and K_a-band (e.g., O3b downlink 17.7–20.2 GHz, uplink 27.5–30 GHz). LEO high-throughput satellite constellations also aiming to operate in the K_u-, K_a- and V-bands.

International harmonized spectrum and spectrum regulation will support future mobile network implementation and other services, enabling large-scale usage of the spectrum. Parts of the aforementioned bands do not have global mobile allocation, and since several other services such as different satellite services, radiolocation, and radionavigation will use portions of these bands, some compatibility studies to determine the feasibility of using these bands must be discussed and managed on a global level, for example at World Radiocommunication Conferences [22].

1.1.4   Key Technology Drivers

The deployment of 5G networks is expected to emerge between 2020 and 2030. In [2-4, 16, 17] different deployment scenarios and environments are considered in detail such as various indoor hotspots (offices, shopping malls, stadiums) and various urban and rural, high-speed, and highway environments, as well as detailed requirements on the key performance indicators and many more capabilities, including also the requirements for the next generation of radio access network (RAN), which should support a wide range of intercell coordination schemes and accommodate new mobile access technologies for massive capacity, huge numbers of connections, and ultra-fast network speeds. Figure 1.6 illustrates some deployment scenarios for 5G and beyond cellular systems, indicating some usage scenarios such as cellular 5G, LTE-A, HSPA or GSM, D2D, mm-wave- and intervehicular communications in macro-, micro-, pico- and femtocells, where external communication flow is provided by a core network, which might consist of hybrid radio backhaul and/or fiber access networks, including satellites. Hence, 5G will realize networks capable of providing connectivity between people and machines/devices (M2M and H2M) over very small as well as long distances. Moreover, 5G radio access will be built upon both, evolved existing wireless technologies (GSM, HSPA, LTE, LTE-A, M-WiMAX, and WiFi) and new radio access technologies, that is, 5G core networks will also be equipped to seamlessly integrate current mobile core networks.

While previous generations of wireless networks were characterized by fixed radio parameters and spectrum blocks, 5G network nodes will be flexible/configurable based on cloud, software-defined networking, and network function virtualization technologies [4].

Integration of mass-scale cloud architectures or cloud-based RAN will infuse mobile networks with capabilities for delivering services in a very flexible way, available to the customer on demand, which might leverage new services and applications. It will provide on-demand resource processing, storage, and network capacity (in a centralized server) wherever needed in a rapid speed, in order to cope with the tremendous growth in mobile data traffic, the diversification of mobile app innovation, IoT connectivity, and security, but also to facilitate on-demand customization. Thus, it will ensure QoS, and at the same time, reduce network cost and energy consumption.

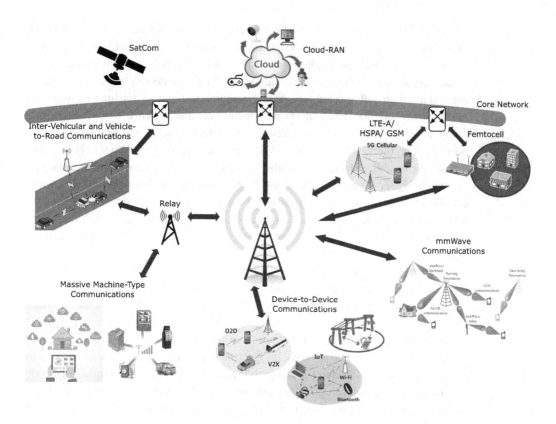

Figure 1.6 Deployment scenarios envisioned for 5G and beyond cellular systems.

Software-defined air interface technologies will be seamlessly integrated into 5G wireless access network architectures toward a "hyper transceiver" approach to mobile access, which will help to realize the joint-layer optimization of how radio resources are efficiently utilized [17].

In addition, new integrated access node and backhaul design will enable ultra-dense networking of radio nodes. Plug-and-play will become essential to deployment, where such nodes will need to access and self-organize available spectrum blocks for both access and backhauling. This capability will be a key for enabling high-frequency spectrum radio access. All these adaptive network solutions will allow a much higher flexibility to utilize "any" spectrum and "any" access technology for the best delivery of services and to speed up the creation of massive-scale services and applications [17].

Moreover, if cognitive radio technology can be used for 5G cellular systems, then, it would enable the user equipment/handset to look at the radio landscape in which it is located and choose the optimum RAN, modulation scheme, and other parameters to configure itself to gain the best connection and optimum performance [2]. For this, reconfigurable baseband and RF hardware architectures are required to enable computationally intensive and adaptive air interfaces.

Hence, 5G will use advanced heterogeneous networks and new kinds of network deployments, in order to create an integrated system that can provide and support a variety of services and applications with flexibility, including ubiquitous ultra-broadband network infrastructure, mass-scale cloud architectures, ultra-dense radio networking with self-backhauling, D2D communications, and dynamic radio access infrastructure sharing. Thus, in 5G and beyond cellular systems, users will concurrently be connected to several wireless access technologies and seamlessly move between them (pervasive networks). With different access schemes, it will be possible to link, for example, devices to nearby devices to provide ad-hoc wireless networks for much higher data flow and low-latency connectivity. Furthermore, in current cellular networks, data rates fall toward the cell edge, where interference levels are higher and signal levels lower. To overcome this, that is, to make high data rates available over a wider area of the cell, group cooperative relay will be applied in 5G networks [2].

As enabler for designing programmable/flexible air interfaces, various advanced waveform technologies combined with advances in modulation and coding as well as advances in multiple access schemes such as Filtered OFDM, Filter Bank Multi Carrier, Pattern Division Multiple Access, Sparse Code Multiple Access, Interleave Division Multiple Access, and Beam Division Multiple Access are essential to achieve continuing improvements in spectral efficiency. In the latter concept, an orthogonal beam is allocated to each mobile station, that is, this technique will divide the antenna beam according to locations of the mobile stations for giving multiple accesses to the mobile stations, which correspondingly increases the capacity of the system [4, 5, 16, 23].

Other techniques such as flexible backhaul and dynamic radio access configurations can also enable enhancements to the radio interface. Moreover, flexible spectrum usage, joint management of multiple radio access technologies, and flexible uplink/downlink resource allocation such as Time Division Duplex (TDD) and Frequency Division Duplex (FDD) joint operation and dynamic TDD provide technical solutions, which address the growing traffic demand and allow more efficient and flexible use of radio resources. The latter one could also be attained by advanced RF-domain processing, for example using single-frequency full-duplex radio technologies, where simultaneous transmission and reception on the same frequency with self-interference cancellation could increase spectrum efficiency significantly. Improvements in these areas will drive overall network costs down while achieving improved energy efficiency [4, 17, 23].

Another major element of any 5G cellular system in the future are smart antennas, to alter the narrow beam direction to enable more direct communications and suppressing interference effectively, thus increasing the overall cell capacity and spectrum efficiency. This can be done by using advanced antenna technologies such as analog or digital 3D-beam-forming, active antenna systems, massive MIMO, and network MIMO. However, a significantly more advanced baseband computation and RF and antenna hardware are required to meet the complex requirements of digital 3D-beam-forming and mass-scale MIMO [23].

Finally, to cite [17]:

> breakthroughs in radio technologies for mobile devices are required to support a vast range of capabilities, from ultra-low energy sensors to ultra-fast devices with long-lasting battery life. Miniaturized multi-antenna technologies will be critical for enabling Gbit/s-level access speeds with less spectrum and lower power consumption. Extending the capability of mobile devices could support certain base station functionalities. This will allow device-based, on-demand mobile networking for services like instant device-to-device communications.

1.2   Evolution and Trends in Satellite Communications Technology

Similar to terrestrial cellular mobile communications, satellite communications technology evolution can also be scaled in different generations of development in terms of significant increase in throughput [24]. However, compared to terrestrial mobile communication evolution, it starts much earlier (see Appendix 1, but in much slower development cycles. This is because satellite solutions are characterized by quite long design, development, implementation, and testing phases before deployment, which is rather different from terrestrial solutions, where technologies and solutions are easier to be tested in situ. In addition, the time to fix a larger LEO or MEO satellite constellation in orbit might also take a long time period. Moreover, to manage and continuously replenish these constellations can be a rather complex task, especially for the recently proposed large LEO satellite constellations of up to several hundred satellites. Moreover, GEO satellites are planned for a long lifetime of up to about 15 or even 20 years (usually limited only on fuel), while LEO/MEO satellites last only about 5–8 years because of atmospheric contamination and eventually radiation from the van Allen belts. Therefore, critical systems on board of satellites are implemented with redundancy to ensure communications over their expected lifetimes in a rather harsh environment. This also means that satellite solutions tend to rely on proven and very robust technology, and historically on-board capabilities of satellites tend to be rather modest in comparison to their terrestrial counterparts. Although hardware updates of satellite on-board systems are not possible, they must, therefore, be reconfigurable; this means the RF front end, and in some cases, the multibeam and steerable beam antenna arrays. All these various aspects result in completely different business models to recap the development costs and investment in deployment over the long operational lifetime. However, once in orbit, a satellite infrastructure offers a quick and flexible roll-out of services and provisioning connectivity [25].

In the text that follows, the generations of satellite communications technology are briefly characterized in terms of significant increase in throughput according to chapter 15 in [24]:

- First-generation satellites primarily in geostationary orbit were characterized by C-band links. A typical communications satellite payload had 12–24 transponders, each with 70–120 MHz bandwidth, operating with fixed beam full earth coverage antennas. Transmissions were primarily analog with limited flexibility in adjustments to modulation or processing to enhance communication rates.

- Second-generation satellites in the early 1980s, including GEO broadcast and data relay satellites and LEO/MEO mobile and radio navigation satellites, are characterized by K_u-band extension, the implementation of digital communications technology, first introduction of beam-steering antennas, and on-board processing transponders. In this time period, satellite capacity improved significantly with the use of high-power solid-state transmitters, shaped beam antenna coverage, and steerable spot beams for higher data rate users. Typical data rates were in-between a fraction and several hundred Mbit/s.

- Many of the third-generation, fully digital communications satellites in the mid-1990s operated with a larger number of K_u-band transponders, some even of higher data rate K_a-band transponders. Moreover, few satellites employed frequency reuse the first time by using multibeam technology. All this leads to a considerable increase in satellite capacity at the global level.

- Current fourth-generation communications satellites expanded capacity even more significantly by a factor of 20 or more compared to earlier geostationary orbit networks by the application of enhanced solar power systems, though with increased flexibility and on-board processing to maximize the spectrum use. Moreover, hybrid terrestrial/satellite innovations can divert latency-sensitive traffic over shorter terrestrial routes. By 2020–2025, it is assumed there will be more than 100 HTS systems in orbit delivering terabytes of connectivity across the world with reduced unit bandwidth costs by an estimated factor of 10 [22].

- To ensure global coverage, mobility with low latency, high throughput and reasonable signal power, huge LEO satellite constellations are planned to be deployed for connectivity anywhere, anytime to any device. These future LEO constellations, briefly considered in Section 1.2.2, might be defined as fifth-generation communications satellites.

1.2.1   High-Throughput Satellites in Geostationary and Medium Earth Orbits

These HTS in the geostationary orbit make efficient use of the allocated frequency spectrum, both at K_u-band and K_a-band, use much larger capacity payloads with several hundred transponders and extensive multibeam and steerable beam antenna arrays. Moreover, adjustable spot beams enable greater flexibility for the operator to direct capacity according to the customers' demands, considering an average lifespan of a GEO satellite of about 15–20 years during which market requirements can change significantly.

The use of multibeam antennas to provide coverage of small service areas, typically with 100–250 km diameter footprints (see Appendix 1), rather than the individual wide beams or spot beams of earlier satellites, has provided a significant increase in data handling capacity and throughput for all primary satellite communications services fixed satellite service broadcasting satellite service and mobile satellite service, because of frequency reuse, and higher EIRP on the transmit side and higher

G/T on the receiver side. The design and principles of multibeam antennas are given in chapter 15.2 of [24, 26, 27], where the total available bandwidth in Hz and the capacity or overall throughput of the multibeam satellite in bit/s can be determined by

$$B_{\text{tot}} = \frac{P \cdot N_b}{N_c} B_w \quad \text{and} \quad C_{\text{sat}} = \eta_s B_{\text{tot}} \tag{1.1}$$

where $P = 1$ if polarization diversity is not employed and $P = 2$ for dual-polarization frequency reuse and N_b is the number of beams of the satellite, N_c is the cluster size, $F_r = P \cdot N_b/N_c$ is the frequency reuse factor, B_w is the bandwidth allocate to the satellite, and η_s is the spectral efficiency in bit/s/Hz of the modulation and coding scheme.

By software-defined payloads, these GEO-HTS will provide flexible coverage, power, and connectivity [28, 29]. Moreover, advanced modulation and coding schemes in the next-generation modems allow users to increase channel throughput, and adaptive coding enables more efficient use of higher frequency bands that are intrinsically susceptible to rain attenuation. Typical services that GEO-HTS provides are high data rate very small aperture terminal (VSAT), transmission control protocol/Internet protocol (TCP/IP) over satellite, IP-TV, HD-TV, and UHD-TV [24, 26].

Few providers plan to build up global networks with GEO-HTS, such as Intelsat, for applications such as maritime and aeronautical, cellular backhaul, consumer broadband, and enterprise networks. Moreover, some providers started with K_u-band transponders to develop the market, then moved to the K_a-band as demand increased [28, 29]. One of the earliest broadband satellites in geostationary orbit was Thaicom 4 (IPSTAR), launched in August 2005, operating in the K_u-band and providing multibeam coverage with 84 two-way spot beams, 3 shaped broad area beams, and 7 one-way broadcast beams. Examples of other broadband, high-capacity satellites are the following, with brief descriptions of their capabilities according to [24, 30]:

- **K_a-Sat 9A**, built by Eutelsat, launched December 2010, location 9° east, lifetime 15 years, 82 K_a-band spot beams, 250 km footprint, 10 gateways, 70 Gbit/s total throughput capacity
- **ViaSat-1**, built by Space Systems Loral, launched October 2011, position 115° west, lifetime 12 years, 56 K_a-band transponders, 72 spot beams, coverage continental United States, Alaska, Hawaii, and Canada, 140 Gbit/s total throughput capacity
- **EchoStar 17**, built by Space Systems Loral, launched July 2012, position 107° west, lifetime 15 years, 60 K_a-band downlink beams
- **Intelsat 29e Epic**, built by Boeing Satellite Systems, launched January 2016, global coverage, transponders: 14 C-band, 56 K_u-band, 1 K_a-band, wide beams, area beams, 25–60 Gbit/s total throughput capacity per satellite of the Intelsat Epic^{NG} platform
- **Intelsat 33e Epic**, built by Boeing Satellite Systems, launched August 2016, position 60° east longitude, lifetime >15 years, coverage: Europe, Africa, Asia, transponders: 20 C-band, 249 K_u-band, 1 K_a-band, wide beams, area beams

- **ViaSat-2**, built by Space Systems Loral, launched June 2017, position 69.9° west, lifetime >15 years, K_a-band transponders, coverage North and South America, 300 Gbit/s total throughput capacity, hybrid (chemical and electric) propulsion

To reduce latency, HTS can also be placed in a MEO, which can result in three to four times lower delay compared with GEO. However, in these MEO-HTS, users require steerable antennas to track the satellite and retain signal connectivity by moving from one satellite to another one within the MEO constellation [24, 26]. An example of an MEO-HTS system is the O3b satellite constellation, operated by O3b Networks, founded in 2007 and that is now a wholly owned subsidiary of SES S.A. It is designed for telecommunications and data backhaul from remote locations, providing high-speed connectivity to Internet service providers and phone companies. The first four satellites were launched in June 2013, and eight more in 2014. However, only nine are used operationally. There are plans to extend this to 20 satellites. The satellites were deployed in an MEO along the equator at an altitude of 8062 km (4.8 times closer to Earth, 14 dB less path loss, that is, 20 times lower power requirement than GEO), each making 4 orbits a day. Its service area can be covered continuously by a relatively small constellation compared to LEO. Each satellite is equipped with 12 fully steerable K_a-band (downlink 17.7–20.2 GHz, uplink 27.5–30 GHz) antennas (2 beams for gateways, 10 beams for remotes) that use 4.3 GHz of spectrum (2×216 MHz per beam) with a proposed throughput of 1.6 Gbit/s per beam (800 Mbit/s per direction), resulting in a total capacity of 16 Gbit/s per satellite. Each beam's footprint measures 700 km in diameter. O3b claims a one-way latency of 179 ms for voice communication, and an end-to-end round-trip latency of 140 ms for data services compared to about 500 ms for GEO. The satellites weigh approximately 700 kg each [31–34]. More examples and details about GEO- and MEO-HTS are given in chapter 15 in [24] and chapter 3.12 in [26].

The growth in HTS has been accelerated in recent years as costs for ground terminals and satellite transponders have been coming down. Globally, the number of potential subscribers for satellite services is expected to rise quickly from about 1.5 and 3.1 million subscribers in 2010 and 2015, respectively, up to about 6 million in 2020 according to the projected growth in satellite broadband subscribers by ITU assessments. Thus, the number doubles every 5 years. This rapid anticipated growth highlights the need for continued development and launching of HTS networks, which promise a more competitive market and a key enabler to meet universal broadband targets than previous iterations of satellite technologies [22, 24].

1.2.2   Low Earth Orbit High-Throughput Satellite Constellations

The demand for connectivity anywhere, anytime and to any device, will continue to grow in traditional sectors such as wireless connectivity and enterprise networking, but it will also open up new opportunities in providing services to the fast-growing mobility sector, that is, people/vehicles-on-the-move[2], aiming in particular for MI (including ultra-high density and 3D video, augmented reality, virtual reality, online gaming, mobile cloud, remote computing, tactile Internet, and 3D connectivity to aircrafts and drones and smart office) and to extend IoT (including M2M and H2M) to broadly distributed entities [24, 35]. This could include access to people's smart-phones, tablets, and laptops, as well as to various smart devices in recreational vehicles, buses, trucks, in future autonomous driving and connected automobiles, trains, boats, ships, and airplanes. Advances in smaller, less costly low-orbit satellites, capable of delivering high-speed, bidirectional services will make this possible. By positioning in low orbit rather than geostationary, low-power processors and wireless transceivers for the device back on the ground have similar technology gains [36].

To ensure global coverage, mobility with low latency, high throughput, and reasonable signal power will require to deploy huge satellite constellations in LEO and/or hybrid communications solutions combining wireless terrestrial and satellite networks that work together seamlessly, since no single technology or company might reach all the possible markets and customers that will be available in the future [24, 35]. Thus, satellite connectivity will play a critical role.

According to [37], nearly 4.4 billion people around the world have never been online, at least are "unconnected or underconnected" [38–40]. For example, in India, 80% lack access to the Internet. Even in the United States and in Europe, full coverage is not achieved, and a certain percentage of the population are lacking high-speed broadband. This is because of both economic and technological factors. For example,

---

[2] Depending on the speed of motion, the corresponding applications could be divided into three categories:

A. *Portable applications*, where stand-alone portable radio access terminals or radio access modules in caravans, cars, or ships automatically steer the antenna beam toward the hub (satellite, access point, relay, or base station) after positioning them. In terms of the antenna's dynamic beam-scanning capabilities, we define it as antenna calibration, since electronically steerable antennas (ESAs) have to be precisely aligned toward the hub to avoid additional alignment losses, which is not an easy task for high-gain, narrow beam antennas, particularly at mm-waves. However, the response times for automatic alignment must not be that fast (could be in few seconds or even more). Another example might be an ESA integrated into the cover of a portable device such as a tablet or laptop for connecting them to wireless local or personal area networks in indoor as well as outdoor environments.

B. *Slow-moving applications*, where slow-moving devices, for example, of a pedestrian, have to be connected to the hub constantly. Another example might be an intersatellite connection, where the fixed geostationary relay satellite has to be connected constantly to a slow-moving LEO satellite. In this case, antenna tracking is required for high gain, narrow-beam antennas, where the ESA has to direct its beam constantly toward the target of interest, not to lose track and hence gain. However, the response time of the ESA could still be relatively low (in hundreds of milliseconds or few seconds) depending on the specific application, that is, the technological challenge for its implementation could also be less.

C. *On-the-move applications*, where fast-moving vehicles or airplanes have to be connected very dynamically, in order not to lose the track toward the hub. Examples might be direct links of airplanes to ground stations or airplanes connecting satellites as well as mobile user terminals at an automobile or ship to satellites or base stations of terrestrial mobile systems, for example for onboard broadband Internet access. In this case, the requirements on the high gain, narrow-beam ESA are most challenging, to ensure constant broadband access to the hub. Response times must be less than 30 ms.

rural areas have much lower average revenue per user than urban ones, since it would entail significantly large backhaul infrastructure [32].

This motivates new ventures to provide reliable and fast Internet to the masses via new satellite constellations. By December 2019, at least 20 systems are announced for Earth observation and at least 6 new LEO systems for telecommunications [40–43]. Google, SpaceX (Space Exploration Technologies Corporation), and Boeing recently announced large investments in LEO constellations around the globe to provide Internet to rural and developing areas of the world. Google announced plans to deploy 180 small LEO satellites to provide Internet access to underserved regions of the globe. SpaceX's proposed constellation Starlink would comprise a network of 4,425 K_a- and K_u-band small operational satellites in the category between 100 kg and 500 kg to make global Internet service a reality. Launching with SpaceX's own rockets, the full constellation would take about 10–15 years, starting in 2019. Latency is assumed under 20–30 ms with speeds up to 1 Gbit/s through ground terminals [32, 38, 44]. Moreover, SpaceX proposes an even larger constellation of 7,500 V-band satellites, which would circle in an orbit below the first constellation [44]. Also, Boeing applied for a license of the US Federal Communications Commission (FCC) to operate between 1,396 and 2,956 satellites in LEO to provide Internet access in the V-band around the world [39, 41]. With all these satellite constellations, price per bit is aimed to approach terrestrial cellular and fiber offerings.

Because of first-mover advantage, another entity, OneWeb (founded by Greg Wyler of O3b), seems to be ahead of the schedule of all the aforementioned campaigns [45]. Among many investors are SoftBank Group, Qualcomm, Virgin Group, Hughes, Intelsat, and Airbus Group. However, in March 2020, OneWeb faced a liquidity crisis and considered bankruptcy. Then, in July/August 2020, the UK Government announced that it had acquired a 45% stake in OneWeb Global for US$500 million in a joint venture with Sunil Mittal's Bharti Global of India who would hold 55% (formerly a partner of OneWeb). It was reported that the UK would repurpose the satellites for its own Global Navigation Satellite System. It was also announced that Hughes Network Systems would invest US$50 million in the consortium [46].

The original schedule of OneWeb was to design and build 648 (+250 spare) small communication satellites in a constellation, which will offer high-speed Internet with global coverage, connecting both Internet gateways (cellular backhaul, seamlessly extend the networks of mobile operators) and end users for satellite broadband, emergency services, and mobiles, connecting people/vehicles-on-the-move. The 648 satellites in the 150 kg category will operate in circular LEOs (in 18 orbital planes of 36 satellites each) at approximately 1200 km altitude, transmitting and receiving in the K_a (20/30 GHz) and Ku (11/14 GHz) bands. Their mission life time is expected to be at least 5 years. OneWeb aims to be the first fully global, pole-to-pole HTS system with a total throughput capacity of each satellite of more than 7.5 Gbit/s and low latency less than 30 ms round trip delay and a downlink data rate to the user terminals of 50 Mbit/s, making interactive applications such as 5G mobile telephony, game playing, and web surfing possible [29, 32, 44, 46, 47]. According to [36], each satellite is designed to provide an aggregate downlink capacity of 17–23 Gbit/s. The first 74 satellites of OneWeb had already been launched in 2019. Generally, to launch a large number of satellites for LEO constellations is challenging the satellite launch services industry, since even each rocket launch might carry about 35 satellites, it requires several tens of launches [48, 49]. For example, the establishment of the aimed OneWeb constellation would have required the greatest rocket campaign in the history of spaceflight. Arianespace, OneWeb's primary launch partner, booked already 20 Soyuz rockets to launch clusters of 32–36 satellites into orbit. But OneWeb also had contracts with Virgin Orbit for 39 launches, using its still-in development LauncherOne dedicated smallsat vehicle as well as BlueOrigin for 5 launches with a future New Glenn rocket [44, 45]. The high-cost structure of OneWeb's contracted launches and competition, particularly with SpaceX's self-launched Starlink constellation, might have accelerated OneWeb's failure [46]. However, this means not the end of LEO constellations. At the moment, the race is led by Elon Musk's Starlink, which has already launched more than 600 satellites till August 2020, but whose target is 42000 satellites. Moreover, Amazon just got approval from the Federal Communications Commission (FCC) of the United States to launch its Kuiper project with its 3326 satellites (784 at a height of 590 km, 1296 to 610 km and 1156 are located at 630 km) into space, but it is not the only one with this goal in mind [49].

1.2.3   Evolving High-Altitude Platform Stations

Beside terrestrial and satellite platforms, other high-altitude platform stations (HAPSs) are evolving due to the growing demand of broadband services, using solar-powered air balloons and space glider or drones. The ideas are not all new, but emerging technologies and innovative Internet entrepreneurs might leverage them. In the project Loon, Google aims to build up an inexpensive ground-based infrastructure with solar-powered, remote-controlled air balloons within Earth's atmosphere to provide 3G/LTE connectivity directly to the users on the ground. These HAPSs shall be placed above 20 km height to provide a wide area below compared to the very small ones with cell towers, but with latency-like terrestrial technologies [22, 32, 38]. Facebook works on a small solar-powered space glider or drone called "Aquila," with a wingspan roughly the same as that of a Boeing 737, but weighs only about 450 kg. The upper surface of the wing is covered by solar cells to power the aircraft's four electric motors and to charge the batteries as power storage for night flights. Aquila will be able to stay at an altitude between 18 and 27 km for months at a time. It uses lasers to transmit data between planes and to terrestrial stations within 50–80 km, which can then provide Internet service locally [22, 50–52]. Also, Google has shown interest in high-altitude drones to deliver wireless Internet access around the world, using mm-waves, which could offer up to 40 times higher datarate than today's 4G LTE systems [22, 32].

The advantages of using HAPSs as alternatives to satellites are their ease of deployment and ability to move easily to new locations in order to meet the demand

and the changing requirements of the operator or service provider's business plan. These concepts of HAPSs are realizable now because of the improvements in composite materials of the lightweight aircraft technology, low-power computing, battery technology, and solar panels. With respect to spectrum resources, the ITU Radio Regulations currently designated several frequency bands for HAPSs in 2 GHz, 6.5 GHz, 27–31 GHz, and 47–48 GHz ranges. As part of its agenda, WRC-19 will consider additional spectrum requirements for gateway and fixed terminal links for HAPS in the 21.4–22 GHz, 24.25–27.5 GHz, and 38–39.5 GHz bands allocated to the fixed service and could take a decision on designation of some additional bands for HAPSs [22].

1.2.4   Satellite Markets and Perspective

Revenues in the satellite industry are a relatively small percentage of global telecommunications revenues. In 2015, they comprised about 4% according to the Satellite Industry Association (SIA). But in terms of money, it is still a large market with an overall worldwide satellite industry revenue of $260.5 billion in 2016 with a 2% growth rate [43]. Figure 1.7 exhibits the global satellite industry revenues and growth rates from 2007 to 2016, indicating a 10-year global industry growth of two times.

The satellite industry revenue in [43] is subdivided into four major segments:

- Satellite Services is the largest industry segment, with revenues reaching $127.7 billion, but remained flat with 0.2% growth. It includes broadcast, radio, and

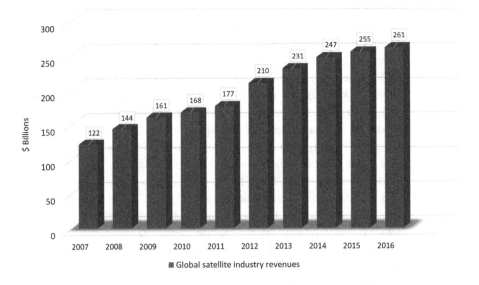

Figure 1.7 Global satellite industry revenues and growth rates from 2007 to 2016 according to [43].

broadband among the consumer communication services with $104.7 billion, but also revenues from fixed satellite services with $17.4 billion, comprising transponder agreements and managed network services, mobile satellite services with $3.6 billion as well as Earth observation with $2.0 billion.

- Satellite Manufacturing revenues dropped down from $16.0 billion in 2015 to $13.9 billion in 2016, suffering the greatest revenue loss of 13% globally.
- Satellite Launch Services Industry revenues, which include revenues for all commercially competitive launches that occurred in 2016, increased by 2% to $5.5 billion after decreasing by 9% from 2014 to 2015. There were 64 worldwide commercially procured launches in 2016, subdivided in orbits: 28 GEO, 2 beyond GEO, 6 MEO, and 28 LEO.
- Satellite Ground Equipment includes network equipment (gateways, control stations, very small aperture terminals) and consumer equipment (satellite TV dishes, satellite radio equipment, satellite broadband dishes, satellite phones and mobile satellite terminals, and satellite navigation stand-alone hardware). The revenues rose by 7% over 2015 to reach $113.4 billion.

By the end of 2016, the total number of operational satellites was 1,459, with an increase of 47% over the preceding 5 years, where 35%, 14%, and 1% are for commercial, government, and nonprofit communications, respectively; 19% and 1% for Earth and space observation, respectively; 12% and 5% for R&D and scientific purposes, respectively; 7% for navigation; 2% for meteorology; and 6% for military surveillance [43]. Currently, there are 1,738 satellites orbiting Earth [40].

The growing importance of the industry is underlined by both the continued increase in the sheer number of operational satellites in orbit and by the announced plans for new satellites with increased capabilities as well as multiple constellations with very small, lightweight satellites, having very low production costs to make them economically viable. The challenge is enormous, since satellites have never been mass produced before. A requirement to produce several small satellites a day has inspired engineers and researchers to develop innovative designs and processes that will dramatically lower the cost in large volumes for high-performance space applications [45]. These smallsats will take advantage of recent advances in consumer electronics, incorporating the latest technologies and many sophisticated functions from smartphones.

Hence, constellations of inexpensive smallsats will be a major market driver (of the manufacturing and launch markets) for the next decade. The shape of the next decade is strongly reliant on the deployment of constellations with inexpensive smallsats. For the future LEO satellite market, it is expected that more than 65% of the total 3,832 satellites will be launched between 2017 and 2025 (2,067 until 2020 and 1,765 from 2021 to 2025) will operate in constellation, experiencing a 385% growth as compared to the previous decade, where the peak of satellites per year to be launched will be around 2021 [42]. These new classes of commercial non-geostationary orbit (NGSO) systems have received a wave of investment from key players inside and outside the traditional satellite industry.

Future LEO satellite constellations aim to provide direct broadband capacity to the user all around the globe, extending terrestrial broadband connectivity and providing direct-to-customer Internet connectivity in remote areas. These emerging networks will have a global impact on humanity

- By delivering ubiquitous high-bandwidth communications to underserved but hard-to wire regions of the world, which contain nearly 60% of the world's population, including remote and rural (difficult-to-reach) areas, for example to obviate the need for expensive fiber infrastructure
- To support maritime, aviation, and other vertical markets
- To enable signaling offload or creating an emergency network, for example an emergency 5G slice in the future, during times of disaster when terrestrial networks are damaged [22, 25]

The use of electric propulsion for keeping a satellite in its operating position has already changed the global satellite industry. Although it takes months with this type of propulsion technology (because electric thrusters produce very small flow), rather than weeks with traditional chemical propulsion (which ejects large amount of propellant) to reach the final operating position or to maneuver the satellite into another slot during the satellite's lifetime (which could be up to 25% of an operator's fleet), economic advantages make it attractive to replace chemical propulsion by all-electric propulsion in a satellite. This approach aims to reduce the spacecraft weight, which could be only 50% of the full-chemical propellant one, and hence it could reduce the launch cost significantly, and possibly extend the spacecraft life [26].

Moreover, new launch platforms such as the Falcon 9 rockets operated by SpaceX, LauncherOne by Virgin Orbit, and New Glenn rocket by BlueOrigin are approaching the market. Competition and modular design with reusable rocket technology will lower the cost of launching satellites [36, 44, 45].

1.2.5   Interoperable Satellite Networks

In the past, there had already been schemes of interoperable satellite systems in geostationary and low orbits, to combine their strengths such as low latency, low loss, small terminals, high look angle, and pole coverage for LEO as well as wide beams for global coverage (e.g., for broadcast) or spot beams for high-density areas (e.g., for broadband services) for GEO, using its fixed positions. With the emerging new LEO constellations, schemes are going ahead, using GEO and LEO interoperable networks [28, 29], relying on the core wireless standard 3GPP to seamlessly integrate with future networks such as 5G phase II and 6G. According to [28], "Intelsat is working closely with players outside of the satellite industry such as QUALCOMM to integrate their chip technology into terminals. This will transform satellite communications from last mile solution to become an integral part of the Telecommunications network," where the future architecture and different path to HTS are shown in some illustrations there.

1.2.6　Toward Hybrid Terrestrial 5G/6G–Satellite Networks

Besides direct broadcasting, satellite communication technology has been associated with expensive proprietary solutions, addressing primarily niche markets and applications. In the past years, satellite technology has improved both its technical performance and capabilities as well as becoming much more competitive with terrestrial solutions and infrastructure. However, in the long run, satellite networks, even LEO constellations, might not exist as stand-alone islands in the sea of connectivity. Up to now, satellites have been used in terrestrial cellular mobile networks primarily for backhaul services for long-distance and international connections only, but not to any great extent in terrestrial cellular mobile networks. However, this begins to change as fourth-generation mobile systems move into the fifth and sixth generation, where IP-based services will be widely deployed, allowing true explicit end-to-end device addressability. The integration of satellite communication promises to provide a powerful networking infrastructure. Hence, the emerging 5G and 6G visions opens up a new chapter in communications, since in the move toward 5G phase II and 6G there will be a convergence of service delivery via multiple networks and systems by aiming to merge several technologies, integrating terrestrial cellular networks alongside and in combination with satellite systems and possibly high-altitude stratospheric platforms, operating on global distribution levels. The satellite will require very high capacity links, extending the performance of current HTS systems and that might operate in expanded and new frequency allocation bands to support increasing traffic capacity [24–26].

Obviously, there are some characteristic constraints of satellite communications technology compared to terrestrial solutions: (1) the low signal level; (2) the comparatively large propagation delay, that is, the large latency, even for LEO constellations; and (3) the relatively low capacity, even for GEO-HTS, since satellites are difficult to build and launch and power is limited even considering powerful satellites, relying on solar power or batteries.

All these factors might be improved with the new LEO-HTS systems in future. At least for long-distance communications, latencies can be better than fiber, since speed of light is around 40% higher in space than in silica glass and satellites need fewer hops with less repeaters [32]. But can LEO-HTS systems bypass existing providers? Most probably not, since these new networks will be reliant on high-bandwidth interconnects. LEO constellations are undoubtedly an efficient solution for delivering service with an aggregate downlink capacity of several Gbit/s into large expanses as well as inaccessible or expensive to cable or terrestrial network areas. Here, satellite technology offers the potential for lower capital expenditure. But its downlink capacity is a fraction of modern domestic fiber interconnects higher than 100 Gbit/s.

In opposite, there are also some features and capabilities of satellite communication that distinguish it from terrestrial solutions: (1) the capability to address large geographic regions, even continents, using a minimum amount of infrastructure on the ground; (2) its flexibility; and (3) its intrinsic broadcast capability, which enables satellite communication to deliver the same content to a very large number of network nodes and user devices with unparalleled efficiency. Because of these capabilities of satellites, it should be considered as complementary, rather than a replacement for fixed or terrestrial mobile infrastructure, extending coverage into areas where it is not cost effective to do so terrestrially or supporting terrestrial solutions in delivering on the 5G and 6G promises, some of which are very challenging [25, 32, 36].

1.3   Technologies

As described in the previous Sections, mobile traffic will increase dramatically in the next years and decade. The proliferation of smartphones, tablets, laptops, and other portable devices is placing greater demand for new services, increasing the capacity or data rates significantly. This requires improved spectrum efficiency methods as well as higher frequencies with larger bandwidth, which is aimed for in the future 5G mm-wave systems as well as geostationary and particularly non-geostationary satellite and high-altitude platforms. With different frequency allocations for different services, mobile devices are required to operate in multiple bands simultaneously. One approach is to use frequency-agile components in the RF front end such as frequency-agile multiband antennas, tunable filters and adaptive matching networks, for example integrated into wideband high-power amplifiers. This would reduce parallel trunks, and hence space, since many portable devices have constraints on the dimensions. Another key component, particularly for on-the-move applications, is the "smart antenna," creating dynamically a desired antenna pattern toward the target of interest, for example a satellite, by constantly tracking the antenna beam toward it, during driving or flying. Figure 1.8 exhibits a scenario of various applications and platforms with some important characteristics, which will be discussed in the text that follows, following to some extent the software-defined radio (SDR) approach.

An SDR is a radio in which some or all physical layer functions are software defined, where the term physical layer is defined by the Open Systems Interconnection (OSI) model. The broad implication of the term software defined is that different waveforms can be supported by modifying the software or firmware but not changing the hardware, where the term waveform refers to a signal with specific values for all the parameters such as carrier frequency, bandwidth or data rate, modulation, coding, and so forth [53]. According to the strictest interpretation of this definition, most radios are not software defined but rather software controlled. While a software-controlled radio is limited to functionality explicitly included by the designers, an SDR may be reprogrammed for functionality that was never anticipated. In the ideal SDR, the user data are mapped to the desired waveform in the microprocessor. The digital samples are then converted directly into an RF signal and sent to the antenna.

The ideal SDR hardware should support any waveform at any carrier frequency and any bandwidth. Assuming a 60 GHz RF signal, without an RF front end to select the band of interest the entire band must be digitized. According to Nyquist's criterion, the signal must be sampled twice at the maximum frequency, that is, 2×60 GHz = 120 GHz. Capabilities of currently available A/D converters are nowhere

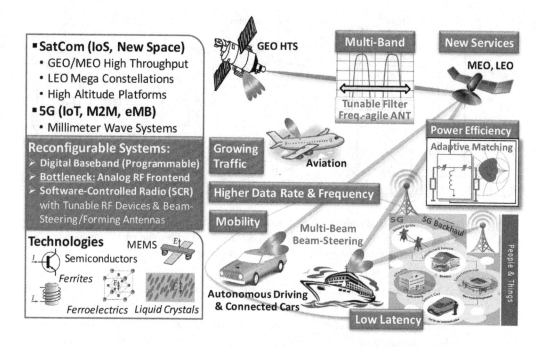

SatCom (IoS, New Space)
- GEO/MEO High Throughput
- LEO Mega Constellations
- High Altitude Platforms

5G (IoT, M2M, eMB)
- Millimeter Wave Systems

Reconfigurable Systems:
➤ Digital Baseband (Programmable)
➤ Bottleneck: Analog RF Frontend
➤ Software-Controlled Radio (SCR)
  with Tunable RF Devices & Beam-
  Steering/Forming Antennas

Technologies   MEMS
  Semiconductors
  Ferrites
  Ferroelectrics   Liquid Crystals

GEO HTS   Multi-Band   New Services
                         MEO, LEO
Tunable Filter
Freq.-agile ANT

Power Efficiency
Adaptive Matching

Growing
Traffic   Aviation

Higher Data Rate & Frequency

Mobility   Multi-Beam
            Beam-Steering

            5G   5G Backhaul

Autonomous Driving
& Connected Cars   Low Latency

People & Things

Figure 1.8 Scenario of various applications and platforms with some important characteristics, including beam-steering antennas and agile components, following the software-controlled radio approach.

close to 120×10^9 samples per second. Furthermore, interfering signals can be much stronger than the wanted signal, for example having a large power difference of 120 dB. The digitizer must have sufficient dynamic range to process both the strong and the weak signals. An ideal digitizer provides about 6 dB of dynamic range per bit of resolution. Thus, the digitizer would then have to provide well over 20 bits of resolution. However, an A/D converter with these specifications violates the Heisenberg uncertainty principle and is therefore not realizable. The maximum A/D precision at 120×10^9 samples per second is limited to 14 bits [54]. Thus, this is beyond the capabilities of modern processors and is likely to remain so in the foreseeable future. Moreover, the digitizer must be very linear. Nonlinearity causes intermodulation between all the signals in the digitized band. Even a high-order intermodulation component of a strong signal can swamp a much weaker signal. Furthermore, physical size and power consumption are other issues, in particular for handheld devices. Hence, an ideal SDR may be perfect for a research laboratory, but not suitable for a given application, considering these technical problems [53].

In contrast, a software-controlled radio (SCR) has an RF front end, which should be reconfigurable or frequency-agile, controlled by software in order to tune carrier frequency and/or the bandwidth of the waveform dynamically, at least over a wide frequency range, including several frequency bands. Some of the SDR functionalities are then implemented smartly into the RF front end, which could save physical size

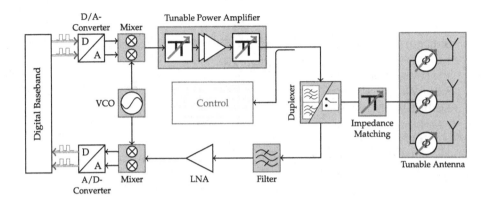

Figure 1.9 Example of a frequency-agile transceiver with baseband (software) reconfiguration for multistandard operation and a reconfigurable RF front end with some reconfigurable/tunable analog RF components for multiband operation and beam-steering.

and power consumption and overcome the technical problems of an ideal SDR. Of course, at a first sight, an SCR, similar to an ideal SDR, faces the problem of higher complexity and higher cost compared to conventional radios, because the frequency-agile components of the front end are more complex and costly than fixed ones. But when several parallel trunks for different frequency bands can be reduced with *frequency-agile* technologies, it might be the opposite. Moreover, it reduces chip area and complex power management and it provides higher energy efficiency through adaptive matching. However, it needs control for the adaption and baseband processing. An example is illustrated in Figure 1.9.

Independent of the specific approach, SDR or SCR both address the following characteristics of a communication system, which are briefly summarized below [53]:

- Interoperability to seamlessly communicate with multiple incompatible radios or act as a bridge between them.
- Spectrum efficiency, that is, efficient use of resources under varying conditions to adapt the waveform for maximizing the key metric. For example, a low-power waveform can be selected if the radio is running low on battery or a high-throughput waveform can be selected to quickly download a file. Delay can be related to throughput. By choosing the appropriate waveform for every scenario, the radios can provide a better user experience.
- Opportunistic frequency reuse (cognitive radio) to take advantage of an underutilized spectrum. If the owner of the spectrum is not using it, an SDR/SCR can "borrow" the spectrum until the owner comes back. This technique has the potential to dramatically increase the amount of available spectrum. This requires "intelligent" spectrum sensing capabilities.
- Power efficiency, that is, a tradeoff of bandwidth or data rate and power, which are key constraints in a wireless system.
- Upgrading to support the latest communications standards. This capability is especially important for radios with long life cycles. Thus, a new standard or

new allocated frequency band can be rolled out by remotely loading new software into a base station or user terminal, saving the cost of new hardware and the installation labor.

- Lower cost by adapted its use in multiple markets and for multiple applications, for example for 5G and satellite communication systems or in hand-held devices as well as in automobiles.

The last two features affect the deployment efficiency, that is the network throughput per unit deployment cost, consisting of both the hardware and operational expenses.

The lack of tunable RF bandpass filters, in particular with a high quality factor (Q-factor) and at mm-wave frequencies, is one of the largest impediments to realize a SCR. Of course, filters can be tuned mechanically, resulting in limited practical use. Selecting one filter from a bank of available filters is a technique often used currently. However, this takes up a great deal of space, might cause some additional losses, depending on the switching level and might pass some interference via the "off" switches, since switches are not ideal. Moreover, in both cases, bandwidth cannot be adjusted. Another issue is *adaptive matching*, for example of the antenna or even more challenging, of the high-power amplifier in the transmitter stage.

A key component of all wireless systems is the *antenna*, where four characteristics are of specific interest: (1) its input impedance, which should be matched to the following RF circuit; (2) its frequency response, since an antenna inherently acts as a filter; (3) its space response by designing its pattern, which can be used as a space filter; and (4) its polarization, which can be used to increase the capacity.

In many applications, particularly in the SDR/SCR approach, it would be a significant advantage to dynamically change/tune these characteristics of an antenna electronically (instead mechanically) to enable adaptive matching, polarization-agility, frequency-agility, and beam-forming or beam-steering (space-agility). That could mean in the case of a frequency-agile antenna, instead of using a wideband antenna with low performance over the entire frequency range of interest with different frequency bands, for example for different services, we could use a tunable narrowband-antenna with good performance, and then tune it over the entire frequency range, thus selecting the individual band of interest. A tunable narrowband antenna typically provides higher gain than a similar sized wideband antenna, or vice versa saves space for similar gain. At the same time, these frequency-agile multiband antennas could be adaptively matched to optimize dynamically any detuning due to proximity effects or a changing environment, for example of handsets [55, 56]. This kind of antenna can provide an inexpensive solution and might better meet the constraints on the form factor, particularly size and shape, of handsets and portable devices.

In the approach of the next generation of terrestrial mobile and satellite systems, including 5G mm-wave and 6G systems, and constellations in all orbits (GEO, MEO, and LEO) as well as future high-altitude platforms, higher frequency bands, in particular K_a-band, V-band, and W-band, are exploited in order to provide high bandwidth and hence higher data rates. To compensate some of the huge path losses (mainly free-space propagation losses) in the power link budget at these high frequencies, large (in terms of wavelength), highly directive, high-gain antenna configurations are required to focus the narrow beam toward the desired hub (satellite, access point, relay, or base station). Conventionally this can be done with a static beam formed by a simple parabolic dish, or more complex, by double-reflector configurations such as Cassegrain or Gregorian antennas. However, when the hub, for example an LEO satellite, and/or the customer or user terminal, for example in an automobile, are moving, then these antennas must be mechanically steered to maintain the direction of the static beam accurately toward the target of interest. This requires a gyroscopic-controlled mechanical system that moves the whole antenna to scan the desired space. These antenna systems with mechanical motorized gimbals for steering are large, heavy, and bulky, and hence lack aerodynamic qualities. They consume a great deal of power for steering, are expensive, and can require additional high maintenance. This is usually also the case for hybrid electronic-mechanical steered antennas, making both impractical to deploy for large-scale applications.

Therefore, smart antennas with dynamic beam-forming capabilities are aimed for the aforementioned platforms, creating dynamically a desired antenna pattern. An example might be a mobile (RV, bus, truck, self-driving cars, train, boat, ship, or airplane), where its antenna constantly tracks the desired hub during moving by positioning the antenna's main beam continuously toward the hub, and at the same time might put pattern nulls in directions of strong interferers.

The smart antenna concept covers a wide range of techniques with different complexity and performance/capabilities (degree of parameters to be changed), including multiple input–multiple output (MIMO), beam-forming and beam-steering techniques. Beam-forming is a technique that combines appropriately scaled and delayed signals of individual antenna elements to create the desired shape of the beam. While MIMO is fully implemented in the digital domain, beam-forming and beam-steering could be realized in the analog RF or IF band, in the digital baseband, or even in a hybrid analog/digital implementation, depending on the system requirements and cost. The various working principles and schemes as well as advantages and disadvantages of its hardware and software implementations are given in Section 2.2.

One focus of the book at hand is on beam-forming and beam-steering techniques. While beam-forming requires electronically controlled amplitude tapering and phase shifting across the individual antenna elements, beam-steering needs only electronically controlled phase-shifting to track the desired hub by positioning the narrow beam constantly toward it. Because of the lower complexity of beam-steering, it is potentially more cost effective and less power consuming, in particular for large arrays with a great number of antenna elements, for example 64×64, but of course, enables less performance than beam-forming and MIMO. However, in many applications, it would be already a breakthrough to enable cost-efficient electronically steerable antennas (ESAs), in particular for mm-wave systems, where complexity, technological constraints, and cost increase, and where implementing ESAs is still very challenging because of the conflict in finding robust, reliable, low-cost hardware solutions in particular for large antennas and wide-scale applications that match, at the same time, the most important system's requirements or at least provide antenna performance sufficient for the desired applications.

There are different ESA concepts that had been investigated worldwide such as transmit arrays, reflectarrays, digital active and analog passive, as well as metamaterial-based phased arrays, using different technological approaches to steer the beam. Phased arrays are most common, where the first phased array transmission was originally shown in 1905 by Nobel laureate Karl Ferdinand Braun, who demonstrated enhanced transmission of radio waves in one direction. It creates beam-steering from a large array of small antenna elements for high gain, which are most often printed microstrip patches with a half wavelength squared dimension. By electronically changing the relative phase of up to 360° for the signal that each antenna element transmits, the combination of all of these small signals creates a larger focused beam in a particular direction, which can be directed instantaneously in any direction by fully electronic control, and hence track the movement, for example of any satellite, no matter how or where you move, without the need for any mechanical moving parts.

The challenge is to miniaturize the phase shifter and to improve the performance, while reducing the manufacturing cost to an economical price point, in particular for large arrays. Phase shifters (see Section 2.2) can be realized in different technologies: in semiconductor technologies, in particular silicon technologies (complementary metal–oxide–semiconductor, CMOS and bipolar CMOS, BiCMOS) that offer much lower cost as compared to InP or GaAs technologies, and can address consumer applications, with RF microelectromechanical systems (RF-MEMS), and by using functional materials such as ferrites, ferroelectrics (in particular barium strontium titanate [BST]) and microwave liquid crystals (MLCs). Advantages and drawbacks of the different technologies for higher frequency bands are given in Section 2.4.

Concerning system's requirements, appropriate response times had already been mentioned in Section 1.2, in the context of different categories of applications – portable, slow-moving, and on-the-move – where for the last one, they must be at least in the range of few tens of milliseconds. Other important characteristics of ESAs are the scanning range and insertion loss; for example, in cases in which analog phase shifters are being implemented, the phase shift should be 360°, accompanied with low losses. This is expressed in the phase shifter figure-of-merit (FoM) in [°/dB]. Moreover, in the transmit mode, high power handling is a key figure for many systems, where antennas must manage tens up to hundreds of watts, for example in satellites, with limited power supply. Hence, insertion loss and power consumption to steer these antennas must be as low as possible to avoid massive power losses and power dissipation in the antenna and to operate the ESA terminals, particularly for vehicles, ships, mobile terminals, and autarkic M2M systems, with battery, renewable energy, and off-grid power solutions. Moreover, ESAs have to be manufactured to withstand the vibration and thermal environment on the aircrafts, vehicles, trains, ships, and so forth. Hence, a robust and reliable technology has to be used for manufacturing, where the key figure will be the cost per terminal to enable wide-scale applications.

Currently, there are different approaches by some spin-off companies for developing, producing, and commercializing cost-efficient large-scale ESAs for satellite systems, but most probably also for future 5G mm-wave systems. All of them aim for low-profile and light ESAs with a modular concept to be scaled to any requirement and a thickness of few centimeters only (maximum 10 cm at K_u-band). Moreover, all work on the form factor of their ESAs, for compact stand-alone terminals with aesthetic appearance and to integrate them smartly or nearly seamlessly into the structure or body of a carrier object such as into the skin of an airplane, or into the rooftop of a vehicle or the cover of a notebook-like terminal. Four spin-off companies will be briefly introduced (status Dec. 2019).

Phasor (www.phasorsolutions.com) builds up a modular, digital phased array antenna based on patented innovations in the architecture and design of custom RF & Digital Application Specific Integrated Circuits (ASICs or microchips). Digital beam-steering is done by using a large number of ASICs, each directly connected to the small patch antenna elements of the array. The embedded microprocessors are able to dynamically control the signal phases of each element to combine and steer the beams in any direction. This active array consists of a number of smaller repeated modules. Each module fits onto two printed circuit boards (PCBs). The upper PCB hosts the array of patch antenna elements on the front and the ASIC microchip on the back. A second back-board provides the power, control, and communications for the system. All this stacks up to be less than about 2.5 cm thick. These modules can be combined to create a larger antenna to meet the required performance. Moreover, modules can also conform and be shaped to any curved surface, for example to host the ESA in the fuselage of an aircraft or in the superstructure of a yacht. Because of the active circuitry, price per terminal and power consumption will still be quite high, particularly for arrays with a large number of antenna elements, for example 64 × 64, to achieve high gain.

Isotropic Systems (www.isotropicsystems.com) uses a passive "optical" beamformer (specifically designed lens) to control and concentrate radio waves with a high degree of accuracy onto a number of active feeds and associated circuitry. With this hybrid approach, Isotropic Systems claims to "reduce the required circuitry of conventional active phased arrays and flat panels by 70–95%" with corresponding results of lower cost and lower DC power consumption. Moreover, because its solution implements true time delay, it provides significantly more bandwidth. The scanning range of their terminal is driven by the number of feeds within each module, which can be optimized to match the scanning requirements of a specific application. The number, arrangement, and orientation of modules set the gain of the terminal, which is more than 10 cm thick for the K_u-band. Each module can be tilted and conformed to fit any curved surface. This modular concept allows combined Tx/Rx in the same aperture but offers the option to have separate Tx/Rx lenses, or lenses with combined Tx/Rx in the same lens.

Kymeta Corporation (www.kymetacorp.com) provides a flat-panel passive ESA by using metamaterial structures to form "holographic" beams that could link to satellites while the antenna is in motion. Kymeta's Metamaterials Surface Antenna Technology (MSA-T) is going back to the fundamental research on metamaterials of Prof. Smith and his team at Duke University [57, 58]: and Dr. Kundtz's pioneering use of metamaterials technologies in electronic beam-forming applications, which ultimately led to the spin-out of Kymeta Corporation in August 2012. The principle is described in one of his patents [59]: "scattering elements are made adjustable by disposing an electrically adjustable material, such as a liquid crystal, in proximity to the scattering elements. Methods and systems provide control and adjustment of surface scattering antennas for various applications." The entire passive array could then build up by a stack flat panels, reducing the complexity of the $N \times N$ phase shifter, biasing and feeding network of a conventional passive phased array, ultimately reducing the manufacturing costs for the same technology. Moreover, owing to the MLC used, there is potential for another dramatic cost reduction and for a lower power consumption compared to semiconductor based phased arrays. Kymeta flat-panel antenna can steer the beam and to control its polarization. However, metamaterial structures have inherently frequency-selective characteristics, limiting the antenna's operating bandwidth.

ALCAN Systems (www.alcansystems.com) is a German-based smart antenna technology start-up developing LC-based phased array antennas. It is a multilayer approach on different stacks, which is similar to Phasor's, with a specially designed MLC layer inside a conventional passive phased-array antenna structure with a number of radiating elements, and at the back side, a semiconductor integrated circuit to control LC-based phase shifters. This MLC technology has been researched and developed by Prof. Jakoby's group at Technische Universität Darmstadt in close cooperation with the research department of the Performance Materials Division of Merck KGaA, Darmstadt since 2002. The viability of broadband MLC-based phased array antennas was proved by Dr. Karabey in 2012 and published in [60, 61]. One of those early prototypes is shown in Figure 1.10. Following the modular concept, one

Figure 1.10 A photo of one of the first LC-based phased array antenna fabricated in an LC TV production line (2012, courtesy of ALCAN Systems GmbH). The photo shows the backside of the 1.5 mm thick antenna with fully integrated 8 × 8 spiraled microstrip delay line phase shifters, feeding network, and thin bias lines (not visible).

module consists of 8×8 microstrip patch antenna elements. Then, several modules build up the ESA with an appropriate size to meet the application's specific requirements. By using the loaded-line concept with periodic tunable LC varactors of about 4 μm thickness [62–64], the response time is less than 30 ms. Moreover, polarization degradations can be compensated by using LC polarization-agile schemes [60, 65].

In conclusion of this section, Figure 1.11 summarizes all the aforementioned RF smart system functionalities in the upper line below the schematic pictures and some examples of tunable components below as well as some materials and technologies to implement them, where both are described in more detail in Chapter 2.

Beyond the reconfigurable circuits and technologies, there are other fixed (not reconfigurable or smart) key components for mm-wave systems such as power amplifier, low-noise amplifier, mixer, and analog-to-digital and digital-to-analog converter. These devices are described in many books in the recent literature, and therefore will not be discussed in this book.

In this book, three technologies that are suitable for the design of mm-wave tunable/reconfigurable components, circuits, and devices are described, namely silicon technology (CMOS and BiCMOS) in Chapter 3, RF-MEMS in Chapter 4, and MLC

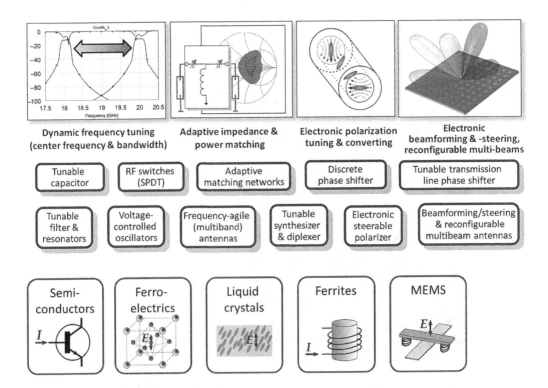

Figure 1.11 Schematic pictures of RF smart system functionalities and some examples of reconfigurable and tunable components as well as some related materials and technologies for hardware implementation.

technology in Chapter 5. In Chapter 2, basic concepts of tunable components and circuits are presented, to introduce the components and circuits described in the next three chapters.

References

1. International Telecommunication Union (ITU), "IMT traffic estimates for the years 2020 to 2030," Report ITU-R M.2370-0 2015.
2. S. R. Tripathi and S. Khaparde, "Analysis and survey on past, present, and future generation in mobile communication," in *National Conference on Recent Trends in Computer Science and Information Technology (NCRTCSIT)*, 2016.
3. W. Xiang, X. Shen, and K. Zheng, *5G Mobile Communications*. Cham, Switzerland: Springer International.
4. International Telecommunication Union (ITU), "MT vision: Framework and overall objectives of the future development of IMT for 2020 and beyond," Recommendation ITU-R M.2083-0 I, 2015.
5. A. Rafay. (2016). "Evolution of wireless technologies," Available at: www.linkedin.com/pulse/evolution-wireless-technologies-arqam-rafay?articleId=7418986240225826962
6. R. N. Mitra and D. P. Agrawal, "5G mobile technology: A survey," *ICT Express*, vol. 1, no. 3, pp. 132–137, 2015.
7. Cisco. "Cisco Visual Networking Index: Global mobile data traffic forecast update, 2016–2021 White Paper." Available at: www.cisco.com/c/en/us/solutions/collateral/service-provider/visual-networking-index-vni/mobile-white-paper-c11-520862.html
8. Ericsson, "The Ericsson Mobility Report." Available at: www.ericsson.com/en/mobility-report
9. J. Nielsen. (1998). Nielsen's law of internet bandwidth, 1998. Available at: www.nngroup.com/articles/law-of-bandwidth/
10. J. Ulm, J., T. Cloonan, M. Emmendorfer, J. Finkelstein, and J. Fioroni, "Is Nielsen ready to retire? Latest developments in bandwidth capacity planning," 2014.
11. G. Fettweis and S. Alamouti, "5G: Personal mobile internet beyond what cellular did to telephony," *IEEE Communications Magazine*, vol. 52, no. 2, pp. 140–145, 2014.
12. G. Kim. (2014). "Bandwidth growth: Nearly what one would expect from Moore's law," Available at: https://ipcarrier.blogspot.de/2014/02/bandwidth-growth-nearly-what-one-would.html
13. S. Cherry, "Edholm's law of bandwidth," *IEEE Spectrum*, vol. 41, no. 7, pp. 58–60, 2004.
14. *"Current world population."* Available at: www.worldometers.info/world-population/
15. International Telecommunication Union (ITU), "Framework and overall objectives of the future development of IMT-2000 and systems beyond IMT-2000." Recommendation ITU-R M.1645, 2003.
16. "FP7 Integrating Project METIS (ICT 317669)." Available at: https://metis2020.com/documents/deliverables/
17. HUAWEI, "5G: A technology vision." Available at: www.huawei.com/ilink/en/download/HW_314849
18. International Telecommunication Union (ITU), "Technical feasibility of IMT in bands above 6 GHz," Recommendation ITU-R M.2376-0, 2015.

19. T. S. Rappaport, R. W. Heath, R. C. Daniels, and J. N. Murdock, *Millimeter Wave Wireless Communications.* Prentice Hall Communications Engineering and Emerging Technologies Series. Upper Saddle River, NJ: Pearson Education.

20. ETSI, "Work Programme 2016–2017." Available at: www.etsi.org/images/files/WorkProgramme/etsi-work-programme-2016-2017.pdf

21. T. Rappaport, "Spectrum frontiers: The new world of millimeter-wave mobile communication." Available at: https://transition.fcc.gov/oet/5G/Workshop/Keynote%20Rappaport%20NYU.pdf

22. K. Martin, K. O'Keefe, and L. Finucan, "Emerging technologies and the global regulatory agenda." Available at: www.itu.int/en/ITU-D/Conferences/GSR/Documents/ITU_EmergingTech_GSR16.pdf

23. International Telecommunication Union (ITU), "Future technology trends of terrestrial IMT systems." Recommendation ITU-R M.2320-0, 2014.

24. L. J. Ippolito, *Satellite Communications Systems Engineering: Atmospheric Effects on Satellite Link Design and Performance*, 2nd ed. John Wiley & Sons, 2017.

25. A. Kapovits. (2016). Satellite communications and 5G – An overview Available at: https://m.eurescom.eu/news-and-events/eurescommessage/eurescom-message-summer-2016/satellite-communications-and-5g-an-overview.html

26. D. Minoli, *Innovations in Satellite Communication and Satellite Technology.* Hoboken, NJ: John Wiley & Sons, 2015.

27. R. Swinford and B. Grau. "High throughput satellites: Delivering future capacity needs." Available at: www.adlittle.com/en/insights/viewpoints/high-throughput-satellites

28. R. Suber, "Next generation satellites: The path for the Pacific Islands." Available at: www.apt.int/sites/default/files/2017/04/4_Next_Generation_Satellites_The_Path_for_the_Pacific_Islands.pdf

29. G. d. Dios, "Satellite technology trends: A perspective from Intelsat." In *ITU International Satellite Symposium*, 2017.

30. Intelsat. "The Intelsat EpicNG platform: High throughput, high performance to support next-generation requirements." Available at: www.intelsat.com/wp-content/uploads/2016/03/Intelsat-Epic-Positioning-6493-wp.pdf

31. ArianeSpace, "A batch launch for the O3b constellation." Available at: www.arianespace.com/wp-content/uploads/2015/10/VS05-O3b-launchkit-EN.pdf

32. V. Velivela, "Small satellite constellations: The promise of 'Internet for All'." Available at: www.orfonline.org/wp-content/uploads/2015/12/IBrief1071.pdf

33. Wikipedia. (2019). "O3b (satellite)." Available at: https://en.wikipedia.org/wiki/O3b_

34. R. Barnett, "O3b – A different approach to Ka-band satellite system design and spectrum sharing," in *ITU Regional Seminar*, 2012.

35. K. Emery. (2017). "Trends in Broadband." Available at: www.intelsat.com/news/blog/2017-trends-in-broadband

36. M. Kelly. (2017). "Data from above: Supporting the next generation of satellite broadband." Available at: www.itproportal.com/features/data-from-above-supporting-the-next-generation-of-satellite-broadband/

37. T. Fitzsimons. (2014). "Why 4.4 billion people still don't have Internet access." Available at: www.npr.org/sections/alltechconsidered/2014/10/02/353288711/why-4-4-billion-people-still-dont-have-internet-access

38. C. Kelby. (2017). "SpaceX, OneWeb, Google and the Battle for Accessible Internet." Available at: www.dunmore.com/functional-films/spacex-oneweb-google-and-the-battle-for-accessible-internet

39. M. Alleven. (2017). "2018 Preview: Satellite industry 2.0 tees up bevy of satellites in name of broadband for all." Available at: www.fiercewireless.com/wireless/2018-preview-satel lite-industry-2-0-tees-up-bevy-satellites-for-broadband-for-all
40. D. Grossman. (2018). "The race for space-based Internet is on." Available at: www .popularmechanics.com/technology/infrastructure/a14539476/the-race-for-space-based-inter net-is-on
41. D. Werner. (2017). "Boeing's LEO constellation hinges on V-band's viability." Available at: https://spacenews.com/boeings-leo-constellation-hinges-on-v-bands-viability/
42. M. Puteaux, "Defining the LEO market compliant to SDM solutions," ed, 2016. https://indico. esa.int/event/128/attachments/729/799/04_CleanSpace_Industrial_days_-_23_05_2016_pre- sentation_new.pdf
43. Bryce, "State of the satellite industry," 2017. Available at: www.sia.org
44. C. Henry. (2017). "SpaceX, OneWeb detail constellation plans to Congress." Available at: http://spacenews.com/spacex-oneweb-detail-constellation-plans-to-congress
45. BBC. (2017). "Satellite mega-constellation production begins." Available at: www.bbc .com/news/science-environment-40422011
46. Wikipedia. (2020). *OneWeb* Available: https://en.wikipedia.org/wiki/OneWeb
47. T. Azzarelli, "OneWeb global access." Available at: www.itu.int/en/ITU-R/space/work shops/SISS-2016/Documents/OneWeb%20.pdf
48. Paul Hastings LLC. "Building the world's largest satellite constellation." Available at: www.paulhastings.com/area/internet-of-things/internet-of-things-case-study
49. Rafi Pratt. (2020). *The race for internet satellites: Amazon, OneWeb and Starlink.* Available: : https://techbriefly.com/2020/08/12/the-race-for-internet-satellites-amazon-oneweb-and-starlink/
50. *The Guardian.* (2017). "Facebook drone that could bring global internet access completes test flight." Available at: www.theguardian.com/technology/2017/jul/02/facebook-drone- aquila-internet-test-flight-arizona
51. Futurism. (2017). "Facebook's drone is one step closer to beaming Internet to the world. Available at: https://futurism.com/facebooks-drone-is-one-step-closer-to-beaming-internet- to-the-world/
52. Wikipedia. (2018). "Facebook Aquila." Available at: https://en.wikipedia.org/wiki/ Facebook_Aquila
53. E. Grayver, *Implementing Software Defined Radio.* New York: Springer, 2013.
54. R. H. Walden, "Analog-to-digital conversion in the early twenty-first century," in *Wiley Encyclopedia of Computer Science and Engineering.*
55. A. Petosa, *Frequency-Agile Antennas for Wireless Communications.* Norwood, MA: Artech House.
56. Y. Zheng, "Tunable multiband ferroelectric devices for reconfigurable RF-frontends," PhD- Thesis, Technische Universität Darmstadt.
57. D. R.. Smith, W. J. Padilla, D. C. Vier, S. C. Nemat-Nasser, and. S. Schultz., "Composite medium with simultaneously negative permeability and permittivity," Phys. Rev. Lett. 84, 4184–4187 (2000).
58. R. A. Shelby, D. R. Smith, and S. Schultz, "Experimental verification of a negative index of refraction," Science 292, 77–79 (2001).
59. A. Bily, A. K. Boardman, R. J. Hannigan, J. Hunt, N. Kundtz, D. R. Nash, R. A. Stevenson, P. A. Sullivan, "Surface scattering antennas" Patent US 20120194399, 2011.
60. O. H. Karabey, "Electronic beam steering and polarization agile planar antennas in liquid crystal technology," PhD thesis, Technische Universität Darmstadt.

61. O. H. Karabey, A. Gaebler, S. Strunck, and R. Jakoby, "A 2-D electronically steered phased-array antenna with 2 × 2 elements in LC display technology," *IEEE Transactions on Microwave Theory and Techniques,* vol. 60, no. 5, pp. 1297–1306, 2012.

62. F. Goelden, A. Gaebler, S. Mueller, A. Lapanik, W. Haase, and R. Jakoby, "Liquid-crystal varactors with fast switching times for microwave applications," *Electronics Letters,* vol. 44, no. 7, pp. 480–481.

63. Goelden, F., A. Gaebler, M. Goebel, A. Manabe, S. Mueller, and R. Jakoby, "Tunable liquid crystal phase shifter for microwave frequencies," *Electronics Letters,* vol. 45, no. 13, pp. 686–687, 2009.

64. R. Jakoby, A. Gäbler, C. Weickhmann, "Microwave liquid crystal enabling technology for electronically steerable antennas in SATCOM and 5G millimeter-wave systems", *Crystals,* 2020, vol. 10(6), pp. 514, doi:10.3390/cryst10060514.

65. O. H. Karabey, S. Bildik, S. Bausch, S. Strunck, A. Gaebler, and R. Jakoby, "Continuously polarization agile antenna by using liquid crystal-based tunable variable delay lines," *IEEE Transactions on Antennas and Propagation,* vol. 61, no. 1, pp. 70–76, 2013.

2 Reconfigurable Devices and Smart Antennas

Holger Maune and Matthias Nickel

As presented in Chapter 1, modern communication systems for the second phase of the fifth generation (5.1G) will include millimeter (mm)-wave communication links to achieve high throughput with low latency even for communication on-the-move. These three properties (mobility, throughput, and latency) seem contradictory. From deployed systems, we know that, for example, the throughput is decreasing with increasing speed of the mobile terminal. So, having a high-speed Internet connection in a high-speed train cannot be realized with a single communication system. Today, the data are aggregated at a base station in the train and multiplexed over a single or few communication links. In the future, this backhauling will gain more and more importance by the implementation of small nano or femto cells. This process cannot be stopped and requires fast and dynamic adaptation of the underlying network structure. Hence, concepts such as software-defined radio (SDR) and cognitive radio (CR) will gain much more influence for future communication systems. These ideas are basically developed for the lower frequency bands, below 10 GHz, where most of the modern communication systems are established. The fifth generation (5G) of mobile communication will introduce software-defined network function on the carrier level. Software-defined radio frequency (RF) functionality will not be part of the first 5G version. Here, the access network will be based on the installations of 4G communication systems such as Long-Term Evolution (LTE) and LTE-Advanced.

Software-defined functionality requires also that the hardware back end of the communication system can be adapted to meet the requirements. The great vision of having a seamless comprehensive system allowing every band and standard combination or at the limit working without any standard cannot be fulfilled with state-of-the-art technologies. Many technologies and realizations are available at lower frequencies, but as complementary metal–oxide–semiconductor (CMOS) integration can handle many different parallel front-end implementations at the same time at comparatively low cost, there is no real need for reconfigurable/tunable components in business models today.

At mm-wave and sub-mm-wave frequencies, the story is different. With upcoming new mm-wave communication systemss different scenarios can be addressed. Let's consider three main categories: vehicular, indoor, and wire replacement scenarios. (1) *Vehicular scenarios* are particularly challenging because nodes often move at high speed. Thus, access points must perform beam-steering continuously to track nodes in point-to-point links in order to minimize power consumption. Vehicular scenarios can be further subdivided. In info-fueling scenarios, the network exploits a transient halt of a vehicle at a certain location to deliver a large amount of data in a very short time. Since the vehicle is not moving during that time, continuous tracking of nodes is not needed. A car stopping at a traffic light remains still for at least a few tens of seconds, enough to download megabytes or even gigabytes of data. The location of the car waiting at the traffic light is well known. Thus, only limited steering capability is needed to deliver data to each car. Since the car's waiting phase is in the order of tens of seconds if not minutes, a steering delay of a few milliseconds is negligible. A very similar scenario concerns airplanes or trains arriving at a gate or at the station and establishing a high bandwidth connection, for example to update the content of the entertainment and information system.

The key advantage of (2) *indoor deployments* is that nodes are typically carried by humans and thus the speed at which they move is very limited compared to many vehicular applications. As a result, the requirements on beam steering are less stringent, except concerning power consumption. The environment is much more complicated, resulting in a complex nonstationary radio channel with quickly varying shadowing and attenuation. For instance, if only a few people are sitting at a meeting table in the room, each user can be served by a dedicated narrow beam. However, if the room is crowded, access points would have to switch to wider beams to ensure that all users in the room are within the coverage range. For the Internet of Things (IoT), the position of nodes might vary throughout the day. The mm-wave communication system compensates for different channel requirements by aligning the beam.

Finally, we focus on cases where mm-wave is envisioned to (3) *replace a traditionally wired connection*. Such scenarios inherently feature low mobility with static channel properties. The throughput achievable is large enough to substitute wired access with wireless links. Companies envision the deployment of access points on lamp posts. In principle, such a deployment would be fully static, since both the lamp posts and the user equipment do not move. However, the use of reconfigurable systems and antennas enables flexibility. For instance, the operator may reconfigure last mile links depending on the requirements of each household at different times of the day.

Each of the scenarios discussed poses a different set of requirements on the communication system. An initial estimation on those requirements is summarized in Table 2.1.

In the following, we motivate why and at which places reconfigurable components are important for mm- and sub-mm-wave systems. After this motivation from the system level, we present the different building blocks and how the desired functionality can be implemented. After that, different technologies for the realization of tunable components at mm-waves are described. First, a general overview is given, followed by a detailed description in the upcoming dedicated chapters. Within these chapters, realization examples of different components are given.

Table 2.1 Different mm-wave communication scenarios and their requirements for reconfigurable antennas

	Beam steering				
	Horizontal	Vertical	Beam width	Beam shape	Steering speed
Traffic light	±60°	±30°	20°	No control	ms
Airport gate	±30°	±45°	20°	No control	Irrelevant
Conference room	±45°	±30°	10°–60°	Beam width	ms
IoT	±180°	±45°	20°	No control	s
Last mile	±90°	±45°	10°	Beam-width nulling	Irrelevant

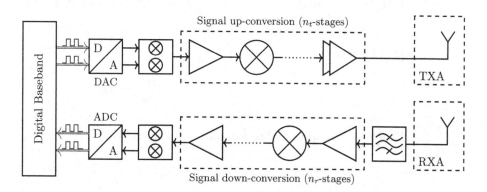

Figure 2.1 Simplified schematic of a general RF communication transceiver consisting of analog-to-digital (ADC) and digital-to-analog (DAC) converters, IQ mixers, up conversion and down conversion, filters, amplifiers, and antennas.

2.1 Millimeter-Wave Communication

Figure 2.1 shows the general schematic of a heterodyne transmitter and receiver structure as it is used in most communication systems. In contrast to a direct-conversion (homodyne) receiver, this structure allows a much easier channel selection as the filter requirements in terms of quality factor and slope steepness are much more relaxed at the intermediate frequency stages. Another very important reason for using heterodyne mixing is the limited feasibility to realize mixers with low leakage as the local oscillator signal will directly interfere with the receiving signal.

The requirements of the transmitter (TX) and receiver (RX) are derived mainly from the specific channel between TX and RX. There are some fundamental differences between low-frequency well-established communication systems and mm-wave systems. Communication channels at mm-waves are challenged mainly by two well-known phenomena: (1) the huge path loss between transmitter and receiver and (2) by the atmospheric absorption due to water vapor and other chemicals, especially oxygen molecules. In the following the channel properties are discussed in detail. Based on this discussion, the requirements for the interface between channel and RX/TX circuit, namely the antenna, are derived.

Starting from the transmitting antenna (TXA), a spherical wave is generated, which results in the power density at the receiver antenna:

$$p\left(\vec{r}\right) = p(\Phi,\Theta,r) = \frac{P_{\text{EIRP}}}{4\pi r^2} \tag{2.1}$$

with the spherical coordinate (Φ,Θ,r) and the equivalent isotropic radiated power (EIRP) P_{EIRP}, respectively. In the far field of the antenna, the wave has a plane wave front and can be described by a transverse electromagnetic (TEM) wave, meaning that the received power can be calculated easily from the integral of the power density and the effective aperture of the receiving antenna. The effective aperture of the antenna A_e is directly linked to the gain of the antenna G_0 by $A_e = \frac{\lambda^2}{4\pi}G_0$, resulting in a received power that can be expressed as

$$P_{RX} = \int_{A_e} p(\vec{r})\,d\vec{A} = \frac{P_{\text{EIRP}}\cdot\lambda^2}{(4\pi r)^2}\cdot G_0. \tag{2.2}$$

Hence, the free space propagation loss α_{fs} is expressed as

$$\alpha_{fs} = 20\log_{10}\left(\frac{4\pi r}{\lambda}\right) = 20\log_{10}\left(\frac{4\pi r f_0}{c_0}\right), \tag{2.3}$$

which is a function of operation frequency f_0. The second part of the channel loss is related to atmospheric attenuation, especially resonances of oxygen and water molecules. To consider these losses, a linear range dependent term is added for the overall channel loss α_C:

$$\alpha_C = 20\log_{10}\left(\frac{4\pi r f_0}{c_0}\right) + \alpha(f_0)\cdot r^*, \tag{2.4}$$

with $\alpha(f)$ the frequency dependent attenuation. For most satellite systems, the signal travels only a very short distance in atmosphere (r^*), so that $r^* \ll r$. In contrast to terrestrial communications, where $r^* = r$. Figure 2.2 shows the atmospheric attenuation over frequency according to ITU-R P.676-11 [1]. For applications above 15 GHz, the atmospheric attenuation becomes an issue. For 60 GHz and 100 GHz systems, $\alpha(f)$ reaches values of 15 dB/km and 0.5 dB/km, respectively. It shows a strong dependency on the water vapor content especially at 100 GHz. The high loss at 60 GHz occur due to a resonance of the oxygen molecule and are independent of the water vapor content.

Figure 2.3 shows the channel loss for different frequencies used for communication systems, including the influence of the atmospheric attenuation in addition to the free space propagation loss. For comparison, the link between a geostationary satellite ($r \approx$ 36,000 km) at 12 GHz results in a channel loss of 210 dB. At 60 GHz or 100 GHz this

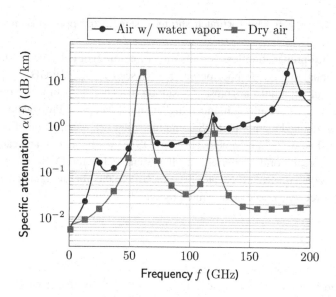

Figure 2.2 Specific atmospheric attenuation versus frequency according to ITU-R P.676-11 (09/2016) for dry air and air with water vapor (air temperature 15°C, pressure 1,013.25 hPa, water-vapor density 7.5 g/m³).

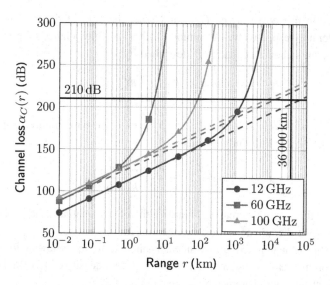

Figure 2.3 Channel loss for different frequencies as the sum of free space propagation loss (dashed lines) and the atmospheric attenuation (ITU standard atmosphere; compare Figure 2.2) for homogeneous medium between transmitter and receiver. Note that for satellite communication $r^* \ll r$, so that the free space propagation loss dominates over the atmospheric attenuation.

attenuation is already present for much shorter distances of approximately 4 km or 80 km, respectively.

The free space propagation loss for this direct line-of-sight (LOS) communication link is proportional to r^2. In mobile communication scenarios at low-GHz frequencies, often two- or multipath links are present, resulting in a channel path loss proportional to r^n with $n = [3 \ldots 4]$, resulting in much higher signal damping. Propagation of mm-wave signals is defined mainly by LOS communication [2]. Attenuation of mm-wave signals for non-line-of-sight communication (nLOS) is much too high and can usually be neglected in real-world scenarios. In all systems, also for LOS systems, some amount of the radiated wave propagates off-axis (not on the line-of-sight path between transmitter and receiver). The area defined by the ellipsoid:

$$F(r_{TX}, r_{RX}) = \sqrt{\lambda \frac{r_{TX} \cdot r_{RX}}{r_{TX} + r_{RX}}} \qquad (2.5)$$

with radius F and distance r_{RX} and r_{TX} from TX antenna and RX antenna, respectively, is called Fresnel zone. This zone must be kept largely free from obstructions. However, some obstruction of the Fresnel zones can often be tolerated.

The realized output power of the transmitter is a second parameter. With different active technologies, the output power is usually of a few milliwatts up to 1 watt, as illustrated in Tables 2.2 and 2.3, respectively.

Combining the results for the link budget, we get the receiver power level L_{RX}:

$$L_{RX} = L_{TX} + G_{TXA} - \alpha_C + G_{RXA}, \qquad (2.6)$$

with L_{TX} the transmit power level, G_{TXA} and G_{RXA} the gain of the transmitting and receiving antenna, respectively. Assuming further a white Gaussian noise of the channel, Shannon's channel capacity is described by

$$C = B \cdot \log_2 \left(1 + \frac{P_{RX}}{k \vartheta_{sys} B}\right), \qquad (2.7)$$

with B the channel bandwidth and ϑ_{sys} the system noise temperature, respectively. Adding now the power link budget leads to

$$C = B \cdot \log_2 \left(1 + \frac{P_{TX} \cdot g_{TXA} \cdot g_{RXA}}{\alpha_C \cdot k \vartheta_{sys} B}\right). \qquad (2.8)$$

It is obvious that with increasing bandwidth of the signal, the overall gain $g_\Sigma = g_{TXA} \cdot g_{RXA}$ of the antennas must be maximized.

Figure 2.4 shows the required receive antenna gain g_{RXA} for the following four different scenarios of communication systems with a power spectral density of 13 dBm/MHz. The corresponding channel loss parameters are given in Table 2.4.

1. *Wireless personal area networks* (WPANs) are used for communication in an environment very close to the user. The path loss might be treated as a multipath channel with a maximum distance of a few meters. Besides free space path loss, only atmospheric loss is considered (see Figure 2.4a).

Table 2.2 Comparison of different technologies for mm-wave transmitters

Freq. GHz	BW GHz	Pout dBm	Conversion gain dB	Power cons. mW	PAE %	Supply V	Size mm²	Technology	Process	Ref	Year
60	57–65	5.7	8.6	76	N/A	N/A	2.1	Super-heterodyne mixer	90-nm CMOS	[3]	2009
60	57–65	8.6	12.4	112	N/A	N/A	1.95	Super-heterodyne mixer	90-nm CMOS	[3]	2009
60	57–66	15.6	26 dB	217	25	1	0.33	Direct conversion outphasing modulator	40-nm CMOS	[4]	2012
61.56	57.24–65.88	10.3	15	251	N/A	1.2	3.855	Direct conversion IQ-modulator with 4-channel bonding	65-nm CMOS	[5]	2014
94	N/A	10	N/A	2,100	N/A	N/A	64	Direct conversion IQ-modulator	IBM 12SOI CMOS	[6]	2014
45	N/A	28.6	N/A	2,022	3.1	5.5	16.75	Direct conversion IQ-modulator with digital pre-distortion, 2x2 Array	PA: 45-nm SOI CMOS, IQ-mod: 120 nm SiGe BiCMOS	[7]	2015
78.5	71–86	12	11	102	25	0.9	0.225	Direct conversion IQ-modulator	40-nm CMOS	[8]	2015
60	3.1	10.8	24.3	40.2	N/A	0.9	0.18	Direct conversion polar IQ-modulator	40-nm SOI CMOS	[9]	2016
28	24–30	13	16.5	100	25	2.5	3	N/A	45-nm CMOS	[10]	2018

BiCMOS, bipolar CMOS; CMOS, complementary metal–oxide semiconductor; IQ, in-phase and quadrature-phase; PAE, power added efficiency; SOI, silicon on insulator

Table 2.3 State of the art technologies for mm-wave power amplification

Freq.	BW	Psat	Gain	OP1	PAE	Supply	Topology	Class	Size	Technology	Ref	Year
GHz	GHz	dBm	dB	dBm	%	V			mm²			
42	9	28.4	18.5	N/A	10	Cascode driver stage: 4, Output stage: 2.4	16-way zero-degree combiner of cascode driver with CE output stage	N/A	5.55	130-nm BiCMOS	[11]	2013
91	15	19.2	12.4	N/A	14	3.4	Multi-drive three-stack PA	A	0.228	45-nm SOI CMOS	[12]	2014
73	7.6	22.6	25.3	18.9	19.3	1.8	2-stage NBCA with 4-way diff. parallel-series combiner	AB	0.25	40-nm CMOS	[13]	2015
58	9.7	17.9	21.5	14.9	20.5	1	3-stage CS pseudo-differential with a class E/F2 output stage	E/F_2	0.25	40-nm CMOS	[14]	2015
30	0.25	15.3	16.3	14.3	36.6	1.15	2-stage	N/A	0.16	28-nm CMOS	[15]	2016
24	8.5	30.8	16.1	24.6	17.6	5.8	4 parallel combined PAs of CE stage followed by cascode stage	N/A	5.37	350-nm SiGe	[16]	2016
24	24–28	25.3	13	23.8	20	10.8	Three stacked dynamically-biased triple cascode cells	AB	0.28	45-nm SOI CMOS	[17]	2016
46	42 – 54	22.4	17.4	18.6	46	6	Two stacked triple-cascode cells	N/A	0.281	45-nm SOI CMOS	[17]	2016
78	27	19.6	12	N/A	18	N/A	Single stage, three-stacked PA	A	0.12	32-nm SOI CMOS	[18]	2016
78	27	18.7	11	N/A	24	3.6 – 4.5	Single stage, three-stacked PA	AB	0.12	32 nm SOI CMOS	[18]	2016
85	75 – 105	22	17	N/A	19.1	4.4	5-stage with double-stacked output stage HBT	E	1.425	90 nm SiGe BiCMOS	[19]	2016
83	75 -105	23.3	18.7	N/A	17.1	6.5	6-stage with triple-stacked output stage HBT	E	1.95	90 nm SiGe BiCMOS	[19]	2016
88	75 – 105	19.5	15	N/A	15.9	2.2	Non-stacked HBT	E	1.425	90 nm SiGe BiCMOS	[19]	2016
40.5	39–43	18	15	16	43	First stage: 1.5, second stage: 2.4	2-stage	F^{-1}	0.57	130 nm SiGe BiCMOS	[20]	2016

BiCMOS, bipolar CMOS; CE, common emitter; CMOS, complementary metal–oxide semiconductor; CS, common source; HBT, heterostructure bipolar transistor; NBCA, neutralized bootstrapped cascode amplifier; PAE, power added efficiency; SOI, silicon on insulator.

(a)

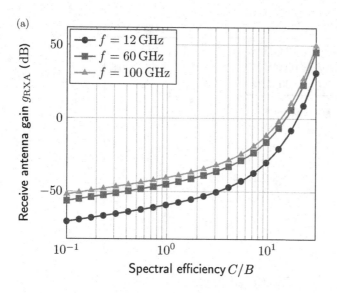

(b)

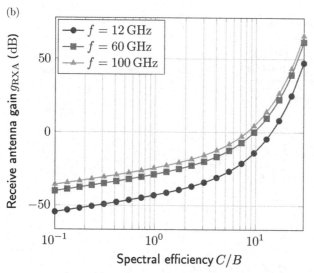

Figure 2.4 Required antenna gain for different wireless communication scenarios. (a) WPAN: Close distance (5 m) LOS indoor communication. (b) WLAN: Long-range (10 m) nLOS indoor communication. (c) WMAN: Long-range (10 km) LOS outdoor communication. (d) WSatN: Long-range (300 km) LOS earth to LEO satellite communication assuming 10 km atmospheric propagation. (e) WSatN: Long-range (300 km) LOS intersatellite communication.

2. *Wireless local area networks* (WLANs) for communication within a room or between a few rooms on the same floor. For communication within a single room the channel path loss was chosen corresponding to a distance of up to 10 m. Again, only atmospheric damping is considered as an additional channel loss parameter. For communication between different rooms an additional damping

(c)

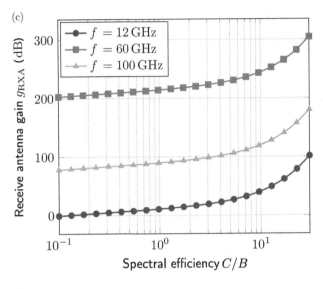

(d)

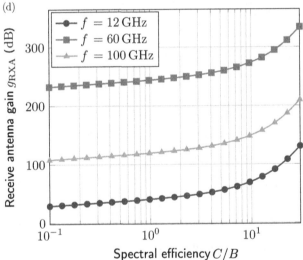

Figure 2.4 (cont.)

term of 10 dB for the walls was considered. Commonly accepted values for standard walls are in the range between 5 dB and 18 dB [2]. Multipath propagation is reduced compared to the WPAN scenario (see Figure 2.4b).

3. *Wireless metropolitan area networks* (WMANs) for communication over a larger area, such as a street or in a small village. The distances here can easily reach 10 km. Additional attenuation due to the effects of atmosphere, rain, and fog has to be accounted for. This scenario is a purely LOS scenario, as the second path is already attenuated to the extent that the link relies completely on a line of sight. The mobility in those systems is usually much smaller than in WPAN or WLAN

(e)

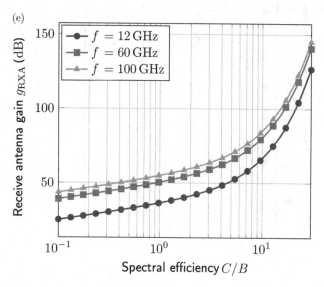

Figure 2.4 (cont.)

scenarios, as these networks are used mainly for backhauling applications (see Figure 2.4c).

4. *Wireless satellite networks* (WSatNs) deliver content from low Earth orbit (LEO) satellites to the user on Earth (see Figure 2.4d), from a satellite to a flighting platform, or between satellites as a relay service (see Figure 2.4e). These systems come with the highest path loss, but also the possibility to install large antennas on the satellite. The mobility in such systems has to be described as moderate, as the satellites are always moving relative to the users.

Beside the required gain, there are strict limitations in the allowed equivalent isotropic radiated power (EIRP) for human safety. For mm-wave communication systems the allowed power levels per occupied bandwidth are up to 13 dBm/MHz. For distance evaluation, compare Figure 2.5, users transmit with maximum power, so

$$\text{EIRP}_{\text{max}} = \frac{P_{TX} \cdot g_{TXA}}{B} = 13 \text{ dBm/MHz},\tag{2.9}$$

where the second parameter of the receiving antenna is the antenna gain over system noise temperature:

$$\frac{G}{T} = \frac{g_{RXA}}{\vartheta_{sys}}.\tag{2.10}$$

Taking these system parameters into account, the channel capacity per bandwidth becomes

$$C/B = \log_2\left(1 + \frac{\text{EIRP}_{\text{max}} \cdot G/T}{\alpha_C \cdot k}\right),\tag{2.11}$$

Table 2.4 Summary of loss mechanisms for the discussed communication link scenarios

	Frequency in GHz	α_{fs} in dB			α_{atm} in dB			α_{rain} in dB			α_{fog} in dB			α_{pen} in dB	α_c in dB		
		12	60	100	12	60	100	12	60	100	12	60	100	12, 60, 100	12	60	100
Link	5 m indoor	68.01	81.99	86.43	0	0.07	0	0	0	0	0	0	0	0	68.01	82.06	86.43
	10 m indoor	74.03	88.01	92.45	0	0.15	0.01	0	0	0	0	0	0	10	84.03	98.16	102.45
	50 m indoor	88.01	101.99	106.43	0	0.74	0.03	0	0	0	0	0	0	10	98.01	112.73	116.45
	5 km outdoor	128.01	141.99	146.43	0.1	74	2.58	2.02	25.38	32.6	0.02	0.48	1.1	0	130.15	241.85	182.71
	10 km outdoor	134.03	148.01	152.45	0.19	147.99	5.16	3.53	44.26	56.84	0.04	0.96	2.2	0	137.79	341.22	216.65
	300 km atmosphere	163.57	177.55	181.99	0.19	147.99	5.16	3.52	44.26	56.84	0.04	0.96	2.2	0	167.33	370.76	246.19
	300 km, no atmosphere	163.57	177.55	181.99	0	0	0	0	0	0	0	0	0	0	163.57	177.55	181.99

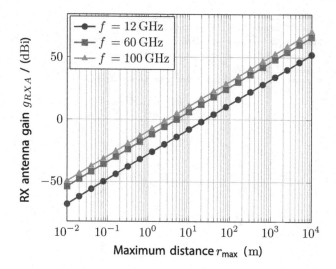

Figure 2.5 Required antenna gain versus distances for a 13 dBm/MHz PSD EIRP and a spectral efficiency of 15 bit/s/Hz.

and hence the maximum distance defined by the Shannon theorem is limited to

$$r_{max} = \frac{c_0}{4\pi f} \sqrt{\frac{EIRP_{max} \cdot G/T}{\left(2^{C/B} - 1\right) \cdot k}}. \tag{2.12}$$

It is obvious that the only remaining possibility to increase the communication distance is an increase of G/T, which can easily be achieved by high-gain and low-loss antennas.

For the first two scenarios, it can be read that the antenna gain must be in the range of 0–55 dBi for a spectral efficiency up to 30 bit/s/Hz. This holds for the case in which the EIRP can be achieved with the antenna system. Otherwise the gain of the antenna must be increased to reach the desired distances. In addition, a suitable margin must be implemented into the system because of environmental conditions such as rain or snowfall and system imperfections. At mm-waves especially, rain becomes a major issue for communication systems, as shown in Figure 2.6. Depending on the region on Earth, an additional margin of up to 30 dB/km for tropical rain with 100 mm/h rain must be considered.

With increasing frequency, the effect of wave polarization compared to the rain's falling direction must be considered as well. Figure 2.7 shows the additional attenuation introduced by traveling of a polarized wave across rain fall. Even in the extreme case of monsoon, a 100 GHz signal is additionally attenuated by approximately 1 dB/km due to the polarization effect.

Also fog and other weather effects might disturb the communication link. Figure 2.8 shows the influence of fog on the communication channel according to ITU-R P.840-7 [22]. Thick fog with a visibility of 50 m has a liquid water

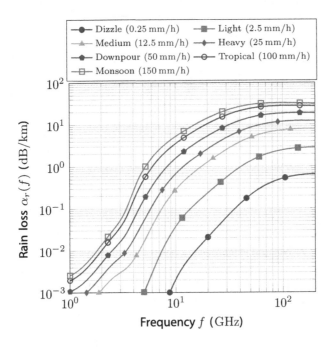

Figure 2.6 Attenuation by rain in dB/km for different strength of rainfall at microwave and mm-waves according to ITU-R P.838-3 [21].

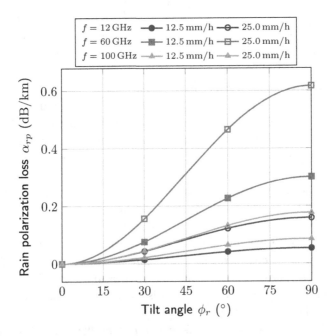

Figure 2.7 Additional attenuation due to polarization of wave. The tilt angle is the difference in main rain falling direction and polarization angle according to ITU-R P.838-3.

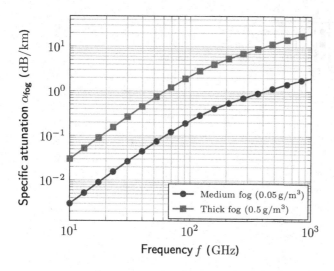

Figure 2.8 Additional attenuation due to water content of the atmosphere (fog) versus frequency. Two different scenarios are calculated, one with medium and one with thick fog compared to a visibility of the order of 300 m and 50 m, respectively.

density of about 0.5 g/m³, which yields to almost negligible attenuation below 1 dB/km at 60 GHz.

With these additional losses the worst-case channel loss $\alpha_C^{(WC)}$ sums up to

$$\alpha_C^{(WC)} = \alpha_{fs}(r) + \left[\alpha_{atm} + \alpha_{rain} + \alpha_{rp} + \alpha_{fog}\right] \cdot r^*. \qquad (2.13)$$

Figure 2.9 shows the difference between worst-case channel and free space propagation loss for three different rain scenarios. It is obvious that mm-wave communication channels are varying largely versus environmental conditions. Rain attenuation especially shows a significant impact already at short distances of less than 1 km.

Astonishingly, the 60 GHz channel shows less environmental impact, as the already very high attenuation is impacted only to a small fraction by rain and fog. On the contrary, the rain attenuation at 100 GHz shows a much stronger impact. Already at short distances of less than 1 km the attenuation in addition to the free space attenuation loss of strong rain can be in the range of 20–30 dB. Figure 2.10 shows the influence of rain attenuation versus the transmission distance for different rain scenarios and operating frequencies. The saturation of the rain attenuation versus distance is in the range of 15–35 dB for 12 GHz and above 160 dB for 60 GHz and 100 GHz communication frequencies, respectively. From Figure 2.3 it might be possibe to establish a 100 GHz communication link over a distance of 10 km, but taking the results of the signal's rain attenuation into account, these values seem not to be very conclusive. From a system engineer's perspective the bridgeable distance correlates not only to the static losses, which can be balanced with high gain antennas or better receivers, but also the dynamic channel characteristics. Taking all these effects into account and allowing a certain outage of the system, the antenna gain

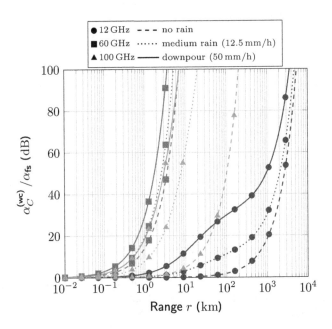

Figure 2.9 Additional attenuation due to weather conditions versus link distance for three different frequencies for worst-case polarization.

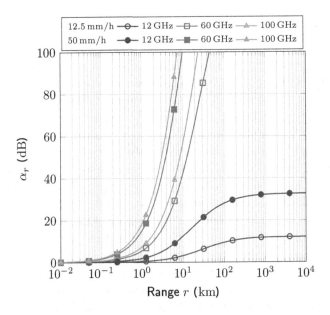

Figure 2.10 Channel attenuation by rain versus communication distance for different operating frequencies and rain scenarios (medium rain of 12.5 mm/h and downpour of 50 mm/h). The polarization represents the worst-case scenario.

(a)

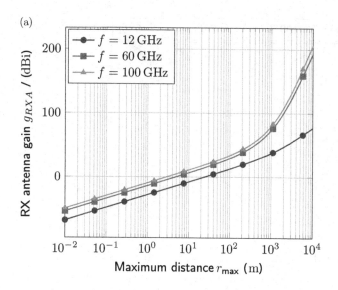

(b)

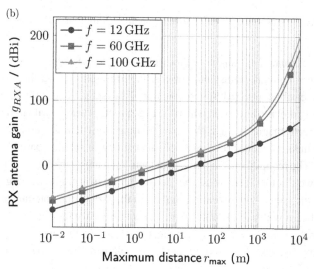

Figure 2.11 Required antenna gain for different distances considering a 99.99% system availability of a terrestrial communication system for the worst-case rain rate of $R = 110$ mm/h and a medium of $R = 60$ mm/h. (a) Indonesia ($R_{p=0.01\%}$=110 mm/h). (b) Europe ($R_{p=0.01\%}$ = 60 mm/h).

must be increased to keep the same reach. Figure 2.11 is an extension of Figure 2.5 for a terrestrial communication system considering the attenuation caused by rainfall for a 99.99% system availability [23].

As can be seen, mm-wave communication systems are suitable for short-distance and/or intersatellite communication in space. The range is directly determined by the allowed EIRP, which must compensate the channel loss. Satellite as well as stationary

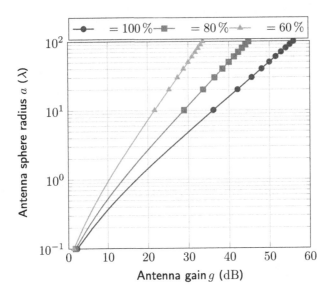

Figure 2.12 Required antenna size to realize a certain gain for different antenna efficiencies.

systems must be evaluated in a slightly different way, as the EIRP of such systems can be much higher, since humans are usually not near the transmitting antenna, or the area of exposure can be easily controlled by the operator. This is a general difference from, for example, handheld mobile devices, in which the user directly interacts with the antenna and the transmitted electromagnetic wave. Without the stringent EIRP restrictions, much longer communication links as needed for terrestrial backhauling or satellite-based services can also be realized. The requirements in terms of antenna gain are again higher in these cases.

There are two factors that must be considered when implementing high-gain antennas. First, the gain of the antenna is linked to the size of the antenna, and second, the half-power beam width (HPBW) becomes very narrow for high-gain antennas. The first one is somehow balanced by the frequency scaling of the antenna by the small wavelength at mm waves. The maximum antenna gain can be approximated [24] by

$$g_{max} = (ka)^2 + 2ka = \left(2\pi \cdot \frac{a}{\lambda}\right)^2 + 4\pi \cdot \frac{a}{\lambda}, \tag{2.14}$$

with k the wave number and a the radius of the sphere enclosing the antenna. Figure 2.12 shows the required antenna size to realize a certain gain level for different efficiencies of the antenna.

For the different application scenarios, the maximum antenna size is usually the limiting factor. As an instance for mobile applications, as in the WPAN and WLAN scenarios, a size of max. 16 cm^2 would be acceptable to still fit into mobile devices. For stationary links, for example backhauling, as in the WMAN scenario, the size could be increased up to 1 m^2. Hence, in practice, fabrication cost and power consumption are the factors limiting the maximum antenna size. Weight and mobility are additional factors when it comes to WSatN scenarios. Here, only fixed uplink stations are flexible in terms of size and power consumption. But mobile satellite users as well as the satellites themselves are limited in space, weight, and available power, allowing antennas of up to $1\,m^2$ in size in the worst case. Table 2.5 summarizes the requirements to the antenna and the achievable system performance for the different use cases.

These antennas can easily fit into the system in terms of space restrictions, as large antennas are required for stationary (terrestrial and satellite) communications, in contrast to the small antennas required for mobile WPAN and WLAN application scenarios.

The second parameter is much more severe, as the HPBW does not scale with increasing frequency. The well-known relation between the antenna's gain and HPBW is given by Kraus' equation [25]:

$$g \approx \eta \cdot D = \eta \cdot \frac{4\pi}{\text{HPBW}_{az} \cdot \text{HPBW}_{el}}, \tag{2.15}$$

with HPBW_{az} and HPBW_{el} being the half-power beam width in azimuth and elevation in radiant, respectively, and η the antennas efficiency.

For a symmetrical beam pattern ($\text{HPBW}_{az} = \text{HPBW}_{el} = \text{HPBW}$) the HPBW in degree is given by

$$\text{HPBW} \approx \eta \cdot \frac{360°}{\sqrt{\pi \cdot g}}. \tag{2.16}$$

For visualization, Figure 2.13 shows the HPBW versus gain and Figure 2.14 shows the relation of beam width in azimuth and elevation for an idealized antenna with certain gain.

It becomes obvious that the required gain values can be achieved only with narrow pencil beam antennas with a HPBW well below 10° or even 1°. This implies three important aspects for mm-wave communication systems:

1. As the HPBW must be very narrow, the alignment of the antennas is a critical part of the communication system, as the user must align both antennas within a few degrees of precision.
2. Serving multiple users with a single fixed antenna configuration is very difficult, as the users must be in very close vicinity.
3. In the case of mobile users, the antenna beams must be kept aligned to each other all the time. So, the communication systems must enable user tracking mechanisms.

A possible solution for these problems are smart antennas, which can automatically steer the beam in the direction of the communication partner or switch between different beam patterns.

2.2 Smart Antenna Systems

To achieve tunability/reconfigurability of the antenna, one has different possibilities as summarized in Figure 2.15. Choosing a particular realization results in limitation of the functionality. Beam steering antennas can steer a fixed beam pattern into different

Table 2.5 Summary of different scenarios and the resulting system parameters (for all scenarios, an antenna efficiency of 60% was assumed)

Scenario	Distance	Mobility	PSD_{EIRP} dBm/MHz	$a_c(f)$ dB	$g_{RXA,\,max}$ dBi	$A_{ant,\,max}$	$HPBW_{min}$ °	C/B_{max} bit/s/Hz	Possible links GHz
WPAN	<5 m	++	− (13)	+ indoor (68–86)	12–31	16 cm^2	6–48	24	12, 60, 100
WLAN I	<10 m	++	− (13)	+ indoor (84–102)	12–31	16 cm^2	6–50	18	12, 60, 100
WLAN II	<50 m	++	− (13)	+ indoor (98–116)	13–31	16 cm^2	6–47	14	12, 60, 100
WMAN	<5 km	−	+ (50)	− outdoor (128–146)	40–59	1 m^2	0.24–1.97	13–25	12, 100
WMAN	<10 km	−	+ (50)	− outdoor (138–217)	41–58	1 m^2	0.25–1.87	2–22	12, 100
WSatN	<300 km	+	++ (50)	− outdoor 10 km (167–246)	41	1 m^2	2	12	12
WSatN	<300 km	+	++ (50)	+ only FSPL (164–182)	40–59	1 m^2	0.2–2	14	12, 60, 100

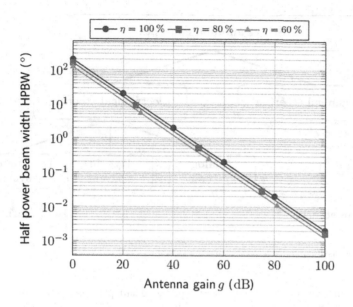

Figure 2.13 Half-power beam width of an antenna versus certain gain for symmetrical (conical) beams and different antenna efficiencies.

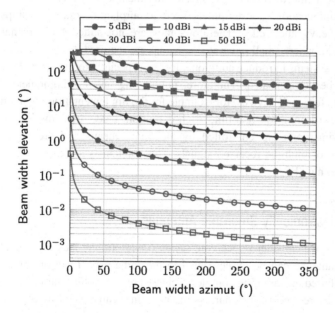

Figure 2.14 Relation of beam width in azimuth and elevation for an idealized antenna with certain gain.

directions, while the resolution of the steering can be very high. Beam switching antennas can be included in this class, although they can only switch between fixed beam configurations, resulting in a much lower resolution, for example 15°. In contrast to beam steering antennas, beam forming antennas are not only able to steer

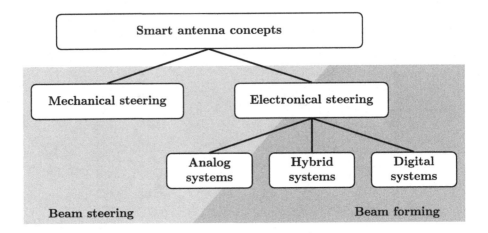

Figure 2.15 Classification of adaptive antenna systems.

the beam in any direction but also adapt the beam pattern at the same time, for example to null interferers or to serve multiple users at different directions.

The most basic and well-established possibility is a mechanical steering of a fixed beam antenna in the direction of interest. This method is widely used for mobile satellite communications for terminals such as airplanes, trains, or ships. These systems are heavy and bulky. Therefore, they are not suited for personal mobile communication, where small integrated antennas are required. For backhauling applications, these systems introduce additional forces, especially torque, into the structure, which must be compensated. Therefore, nonmoving antenna implementations are highly desired for small devices, for example for the implementation on an unmanned aerial vehicle (drone) for real-time high-definition video transmission. Antennas without moving parts are controlled by electronic means and are based on the weighted sum of multiple received signals. These antennas are called electronically steerable antennas (ESAs).

2.2.1   Basics of Electronically Steerable Antennas

In this section, we consider a receiving scenario without any lack of generality. The transmitting antenna can be implemented in the same way. Electronically steerable antennas are based on summing complex weighted replicas of the same input signal to generate the desired received signal. Assuming a linear antenna array consisting of N antenna elements, as shown in Figure 2.16, the received signal $s_r(t)$ is described by

$$s_r(t) = \sum_{n=1}^{N} w_n \cdot p(\vec{r}) \cdot s(t),　\quad (2.17)$$

with w_n the complex weighting functions, $p(\vec{r})$ the antenna positioning, and $s(t)$ the wanted signal as shown in Figure 2.17.

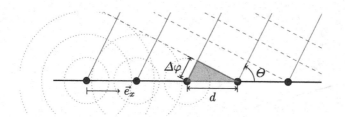

Figure 2.16 Signal flow graph for an electronically steerable antenna.

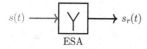

Figure 2.17 Linear antenna array configuration with plane wave excitation at angle Θ resulting in a phase offset $\Delta\varphi$ between neighboring antenna elements.

For the linear antenna array with element spacing d, the antenna position results in a phase offset of $\Delta\angle s = \Delta\varphi = kd\cos\Theta$ between neighboring elements and approximately constant amplitude $\Delta|s| = 0$ over the array:

$$s_r(t) = \sum_{n=1}^{N} w_n \cdot e^{j \cdot n \cdot k \cdot d \cdot \cos\Theta} \cdot s(t) \qquad (2.18)$$

To compensate the phase offsets of the single array elements $\Delta\varphi$, the weighting function must shift the phase in such a way that all elements of the sum are in phase, meaning

$$w_n \cdot e^{j \cdot n \cdot k \cdot d \cdot \cos\Theta} = \text{const} \qquad (2.19)$$

In this case, the signal coming from direction Θ is summed up in phase and can be treated as input signal. Signals from other directions are out of phase and interfere destructively. Based on this simplified example, beam forming antennas can be built by solving an equation system for N weighting functions, allowing the building of the received signal and suppression of $N-1$ interfering signals from other directions. This effect is called space filtering. Up to now we only considered a linear array with equal spacing of the isotropic antenna elements. First, we want to extend the linear array into a 2D array by adding elements in a second dimension as depicted in Figure 2.18.

For an $N \times M$ antenna array the weighted sum results in

$$s_r(t) = \sum_{n=1}^{N}\sum_{m=1}^{M} w_{nm} \cdot e^{jk(d_n \sin\Theta\cos\Phi + d_m \sin\Theta\sin\Phi)} \cdot s(t), \qquad (2.20)$$

with Θ, Φ the incident angle of the signal. Equation (2.20) is still based on the assumption of isotropic radiators as antenna elements. The angular dependency of

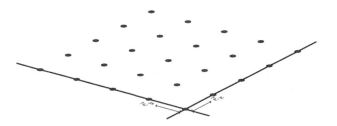

Figure 2.18 Two-dimensional antenna array configuration with plane wave excitation at angle Θ, Φ resulting in a phase offset $\Delta\varphi_{nm}$ at each antenna element.

the antenna elements must be considered. This so-called element factor $EF(\Theta, \Phi)$ is independent of the arrangement of the individual elements and is constant for a certain element type, such as isotropic radiator ($EF(\Theta, \Phi) \propto 1$), dipole ($EF(\Theta, \Phi) \propto \sin\Phi$), patch antenna, or any other antenna type. For example, the element factor of a patch antenna, being widely used in today's antenna arrays, can be approximated by the equation proposed in [25]:

$$EF(\Theta, \Phi) \approx$$
$$\frac{jk_0 h W E_0 e^{-jk_0 r}}{\pi r} \left\{ \sin\Theta \frac{\sin\left(X(\Theta, \Phi)\right)}{X(\Theta, \Phi)} \frac{\sin\left(Z(\Theta, \Phi)\right)}{Z(\Theta, \Phi)} \right\} \cos\left(\frac{k_0 L_e}{2} \sin\Theta \sin\Phi\right) \tag{2.21}$$

with

$$X(\Theta, \Phi) = \frac{k_0 h}{2} \sin\Theta \cos\Phi \tag{2.22}$$

and

$$Z(\Theta, \Phi) = \frac{k_0 W}{2} \cos\Theta, \tag{2.23}$$

where h is the antennas substrate thickness and W the physical width of the patch, whereas L_e represents its electrical length (typical half a wavelength), respectively.

The element factor relates $s(t)$ to the plane wave. For arrays with identical elements it is constant. So, the received signal is calculated as

$$s_r(t) = s(t) \cdot EF(\Theta, \Phi) \cdot \sum_{n=1}^{N} \sum_{m=1}^{M} w_{nm} \cdot e^{jk(d_n \sin\Theta \cos\Phi + d_m \sin\Theta \sin\Phi)}$$
$$\propto EF(\Theta, \Phi) \cdot AF(\Theta, \Phi). \tag{2.24}$$

The double sum with the weighting factors w_{mn} and the geometrical arrangement of the antenna elements can be summarized to the array factor $AF(\Theta, \Phi)$, which is independent on the individual element properties. It is important to note that the nulls of the array and of the element factor will be included in the final antenna pattern defined by $EF(\Theta, \Phi) \cdot AF(\Theta, \Phi)$. So, if an antenna element has a null in a certain direction Θ', Φ', the antenna array cannot receive any signal from this direction. This must be considered in the final array design. One chance to overcome this limitation are conformal arrays, which are built on a curved surface rather than in a planar configuration [26–28].

As described earlier, the beam forming of the antenna can be realized only by changing the array factor by introducing different weighting functions to solve the system equation:

$$\sum_{n=1}^{N}\sum_{m=1}^{M} w_{nm} \cdot e^{jk(d_n \sin\Theta \cos\Phi + d_m \sin\Theta \sin\Phi)} = \begin{cases} \max & \text{for } \Theta_S, \Phi_S \\ 0 & \text{for } (\Theta,\Phi)_{li} \end{cases} \qquad (2.25)$$

for the wanted signal with incident direction Θ_S, Φ_S and the unwanted signals from directions $(\Theta, \Phi)_{li}$. The system can be used to generate a maximum of $N \cdot M$ linear independent terms. If all terms are not needed to null interference signals, the overdetermined system can be used to optimize the gain of the antenna array, which is proportional to $AF(\Theta_S, \Phi_S)$. Many different algorithms can be implemented for the determination of the weighting functions. The most common ones are the Howells–Applebaum [29] and the Widrow–Hoff LMS [30] algorithms and their derivatives.

2.2.2   Implementation of Electronically Steerable Antennas

Let's come back to the hardware implementation of antenna arrays. As shown in Figure 2.15 the weighting functions can be implemented in many different flavors. In the first approach, the weighting functions are realized in the analog domain by means of phase shifters and variable gain amplifiers (VGAs); see Figure 2.19a. The weighted signal is then combined with an analog summing network (feeding network) before it is mixed down and converted to the digital domain by means of A/D conversion.

The opposite design is presented in Figure 2.19b. Here, the signals are converted as soon as possible to the digital domain. The weighting and summing of the individual elements take place in the digital domain. There are other concepts to enable a signal weighting in antenna arrays. Beside a direct weighting after the antenna elements, this can also be carried out at every single intermediate frequency (IF) in heterodyne receivers or transmitters, as shown in Figure 2.20a. This relaxes the requirements to the phase shifters tremendously, as the operating frequency can be chosen much lower than in the first case. The main drawback of this IF analog weighting is the number of mixers, which is as high as in the digital case (one per antenna element). The local oscillator (LO) signal to be distributed in addition to the signals results in a much more complicated feeding network. In case of transmit and receive antennas, four feeding networks must be implemented without changing the element's distance, resulting in a very complex and complicated structure for more than few elements. The pros and cons are similar for weighting the LO signal as shown in Figure 2.20b. Here the LO signal is phase shifted and distributed with different phase offsets to the mixers.

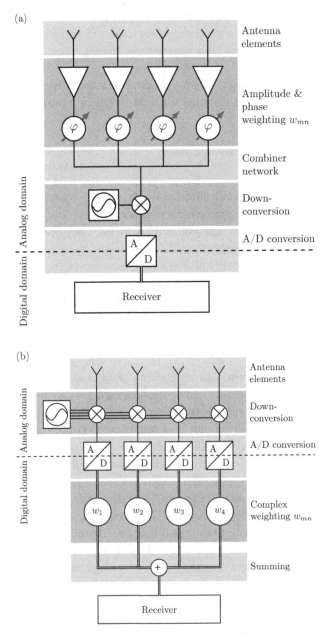

Figure 2.19 Possible implementation of the weighting functions for antenna array. The block diagrams are simplified in such a way that all components and stages not essential for signal weighting are omitted. (a) Analog RF weighting of array signals. (b) Digital weighting of array signals.

(a)

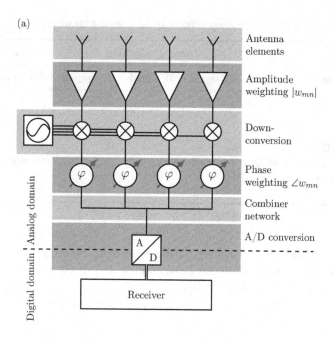

(b)

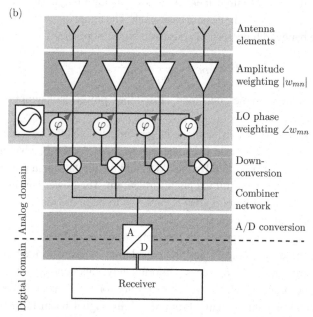

Figure 2.20 Different realizations of analog weighting for antenna arrays. The block diagrams are simplified in such a way that all components and stages not essential for signal weighting are omitted. Also, the position of the amplitude weighting with the first (low noise) amplifier is arbitrarily chosen for simplification. (a) Analog IF weighting of array signals. (b) Analog LO weighting of array signals.

Table 2.6 Comparison of different weighting technologies for antenna array with *N* elements

	Analog weighting			Digital weighting
Phase shifting of	**RF signal**	**IF signal**	**LO signal**	**BB signal**
No. of ADC/DAC	1	1	1	N
No. of mixers	1	N	N	N
No. of feeding networks	1	2	2	1
No. of digital I/Os	1	1	1	N
Amplitude and phase weighting	Separated	Separated	Separated	Combined
Power consumption	+	+/−	+/−	−
Flexibility	−	+/−	+/−	+

IF intermediate frequency; LO, local oscillator; RF, radiofrequency.

Hybrid implementations combining all these concepts, for example analog RF beam steering in elevation and digital beam forming in azimuth are also possible. Even subarrays with different technologies can be implemented depending on the system's design. Table 2.6 summarizes the properties and advantages of the different topologies.

Today, the bottleneck of all implementations is the analog-to-digital converter (ADC), which converts the signal with the desired bandwidth. The figure of merit (FoM) for ADC is the achievable bandwidth per watt of consumed power normalized to one bit:

$$\text{FoM}_{ADC} = \frac{SR}{P_{DC}} = \frac{n \cdot B}{P_{DC}}, \tag{2.26}$$

with $SR = n \cdot B$ the sample rate and P_{DC} the power consumption of the ADC, respectively.

A major improvement as in semiconductor industry cannot be seen in ADC development over the last decade [31]; see Figure 2.21. As the bandwidth in mm-wave systems will be at least 1–2 GHz, the ADC and the digital signal processor must work at least with 5 GS/s to fulfill the Nyquist theorem. The power consumption of a 12-bit ADC with a bandwidth of 5 GHz is expected to be in the range of up to 10 W for commercially available products.

The digital beam former comes with the highest degree of freedom, as all signals are available in the digital domain and the calculation of beam former weights and received signal are very easy to access. Unfortunately, this concept comes with the highest power consumption. Hence, it is suited only for small-sized arrays with few elements or special antenna systems. Prominent fully digital beam forming antenna arrays are used today in military installations, where power consumption and cost do not play a role.

Analog beam forming is usually implemented as phase weighting only, resulting in a phased array that can only steer the beam and is not able to force nulling of interferers. The implementation of the amplitude weighting is neglected in that case. With this simplification, the antenna array is only able to steer the beam and not to shape the beam freely. Major drawbacks are the constant amplitude tapering,

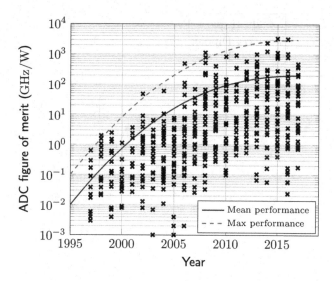

Figure 2.21 Development of ADC performance over the last decade based on [31] for published analog-to-digital converter at International Solid-State Circuits Conference (ISSCC) and Symposia on VLSI Technology and Circuits.

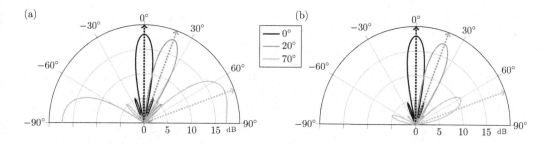

Figure 2.22 Beam steering by means of phase weighting only for different steering angles and for an 8-element linear array with $\lambda_0/2$ spacing. (a) Isotropic elements. (b) Dipole elements.

resulting in a si $x = \dfrac{\sin x}{x}$ shaped antenna pattern and the creation of unwanted side lobes, especially at large tilting angles, as shown in Figure 2.22.

In Figure 2.22 the growing side lobe level (SLL) is obvious. As the antenna elements are exchanged by dipoles, the growth in the side lobe level cannot be observed anymore, but here the gain drops significantly down to zero for 90° steering angle. A comparison of different antenna types usually used for antenna arrays is shown in Figure 2.23.

For most practical applications, only patch antennas can be used, as they can be implemented easily.

The element spacing d is another important parameter. For optimum operation, $d \leq \lambda_0$ has to be fullfilled. If this condition is violated, ambiguities in phase relation

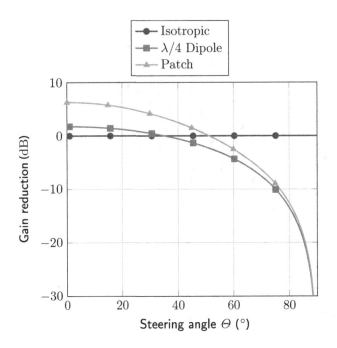

Figure 2.23 Inherent gain reduction of the element factor *EF* of different antenna types.

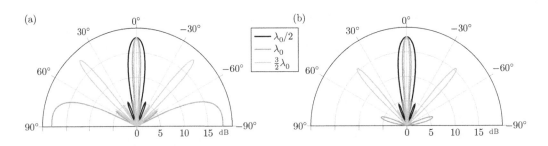

Figure 2.24 Generation of grating lobes for an 8-element linear array steered to 0° with different element spacing. (a) Isotropic elements. (b) Dipole elements.

lead to grating lobes appearing in the pattern. Figure 2.24 shows the generation of grating lobes for two different antenna arrays.

Antenna spacing can get crucial, as the components for each antenna element must fit inside a space of less than $\lambda_0/2 \times \lambda_0/2$. This limitation is independent of the chosen weighting technology, as it originates directly from array theory with Eq. (2.20). If the elements cannot be placed within this area, a more sophisticated 3D integration technology has to be chosen. Especially in active electronically scanned arrays (AESAs), which implement the fully digital weighting in receive and transmit, plank-based or tile-based achritectures are considered [32, 33].

2.2.3   Feeding Networks

The feeding or summing network of analog steered antenna arrays can be implemented in many different ways. For planar arrays, there exist different possibilities to
feed the individual antenna elements. Not all of them are suited for the realization of
steerable antenna systems.

At first, there is the linear fed antenna array as shown in Figure 2.25a. In this case
the antennas are connected to a single feeding line by means of coupling or power
splitting circuits. The feeding point of the antenna can vary. Figure 2.25a shows an
end-fed antenna array, but center-fed arrays can also be built. The main drawback of
this approach is the nonconstant phase offset between different antenna elements,
resulting in an additional phase compensation by the phase shifters. In addition, the
power splitting is of crucial importance, as all elements should be fed with the same
power level. The feeding structure can easily be implemented in a multilayer substrate
stack, for example by aperture coupling of patch elements. Series feed arrays are
frequency sensitive. When the frequency of operation is changed, the phase at the
radiating elements changes proportionately to the electrical length of the feed line so

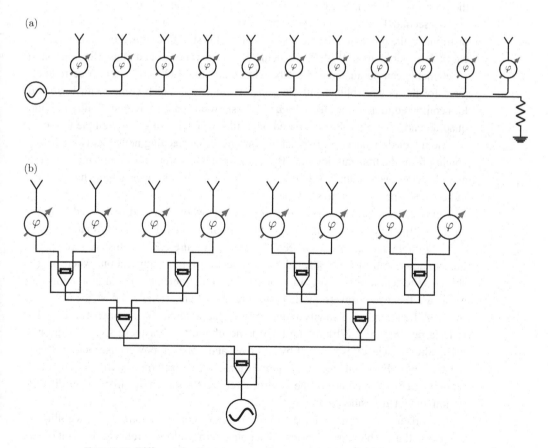

Figure 2.25 Different feeding structures for antenna arrays. (a) Linear feed. (b) Corporate feed.

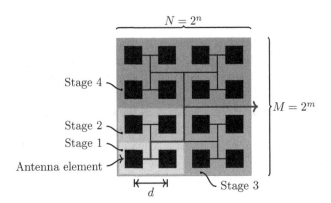

Figure 2.26 Design of a corporate feeding network for quadratic arrays of arbitrary size.

that the phase at the aperture tilts in a linear manner and the beam is scanned. This effect can be compensated in electronic scanning series–fed antenna arrays, allowing the operation over a larger bandwidth than for a static series fed array.

The second feeding structure is based on a power splitting of the input before distributing the power to the antenna elements. This structure shown in Figure 2.25b is called a corporate feed. With this structure, a constant power and phase distribution to all elements can be guaranteed easily. The main drawback is a larger space requirement for the power splitter and the transmission lines, which comes with increased loss compared to the series feed. Figure 2.26 shows a space saving possibility to build quadratic array of arbitrary size based on a mirroring technology of a single unit cell.

One major drawback of all transmission line–based feeding networks is the losses coming from the transmission line. By revoking the requirements of a planar structure and allowing also voluminous structures with a thickness much larger than several millimeters, space–fed arrays can be built.

Figure 2.27 shows two different configurations where the feed is realized by open air [34–39]. The phase shifters are now implemented in a layer that forms a reconfigurable lens in front of a feeding antenna (Figure 2.27a). This concept, called transmitarray, can also be adapted for a reflection type configuration, where phase shifters shorted at one port are connected to an antenna. By illuminating the antennas, the wave travels through the phase shifters, is reflected at the short, is phase shifted again and finally radiated from the antenna. This concept is called a reflectarray (Figure 2.27b). While there is no ohmic loss for these space-fed arrays, the feeding efficiency is reduced by spillover and noneven power distribution across the elements. For high-frequency applications, the reflectarray configuration offers many degrees of freedom, as the control circuitry for the phase shifters can directly be implemented beside them.

The offset feed is the main drawback of these configuration. One possibility to overcome this problem is to implement a polarization-selective reflector in front of the array and to put the feed in the same plane as the tunable phase shifters. This concept

(a)

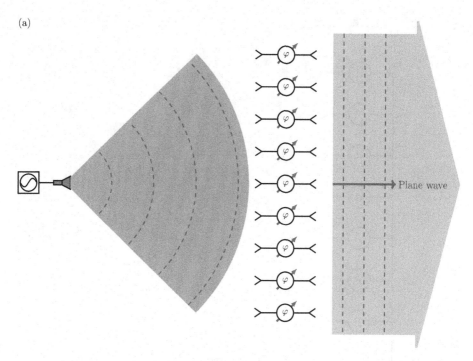

Figure 2.27 Array configurations with space feed. (a) Transmitarray. (b) Reflectarray.

is called folded reflectarray. Figure 2.28 shows the basic principle. This approach was implemented in a slightly modified version in [38].

2.2.4 *G/T* and Signal-to-Noise Ratio

The efficiency of the antenna is of major concern for mm-wave communication systems. To evaluate different antenna scenarios, the G/T ratio has been identified as figure-of-merit (FoM) for the receiving antenna [40, 41] while the allowed EIRP is connected to the transmitting antenna. For communication engineers the signal-to-noise ratio (SNR) at the receiver is used for evaluating the performance. The SNR depends on the equivalent isotropic received power P_{RX} and the bandwidth B of the signal according to

$$\text{SNR} = P_{RX} + \frac{G}{T} - 10 \log_{10}(k_B B). \tag{2.27}$$

The equivalent isotropic received power is obtained from the EIRP of the transmitter and the channel loss α_C, which comprises free space propagation loss α_{fs} and frequency-dependent atmospheric path loss $\alpha(f_0)$ according to Eq. (2.4). The SNR can therefore be derived from

(b)

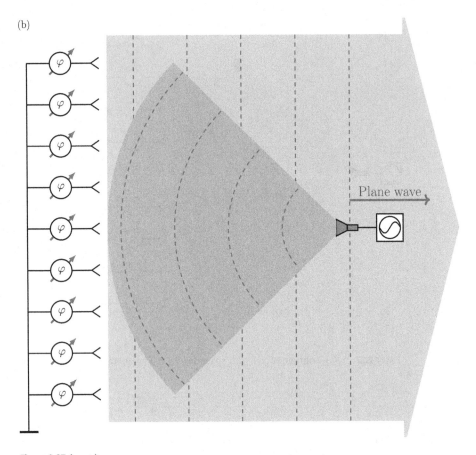

Plane wave

Figure 2.27 (cont.)

$$\text{SNR} = \underbrace{\text{EIRP}}_{\text{TX}} \underbrace{-20 \log_{10} \left(\frac{4\pi r f}{c_0} \right) - \alpha(f) r^*}_{\text{Channel}} \underbrace{-10 \log_{10}(k_{\text{B}} B)}_{\text{RX}} + \frac{G}{T} \qquad (2.28)$$

and is therefore linearly dependent on the G/T ratio and on channel properties. Depending on the chosen modulation scheme, a certain SNR ratio is required for error-free transmission. Figure 2.29 shows the theoretical bit error rate (BER) for different modulation schemes. Signals with high spectral efficiency (many states in in phase and quadrature phase (IQ) plane) require higher signal strength for identical BER performance.

With the required SNR the required G/T ratio can be determined with

$$\frac{G}{T} > \text{SNR} - \text{EIRP} + 20 \log_{10} \left(\frac{4\pi r f}{c_0} \right) - \alpha(f) r^* + 10 \log_{10}(k_{\text{B}} B). \qquad (2.29)$$

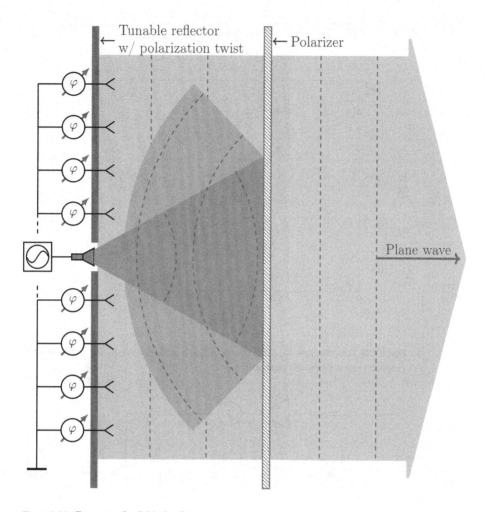

Figure 2.28 Concept of a folded reflectarray.

2.2.5   Antenna Efficiency

For the evaluation of planar antennas, the block diagrams in Figures 2.19 and 2.25 have to be combined and extended to reflect practical implementations.

Figure 2.30 shows a segment of a hybrid corporate fed antenna array with low-noise amplifiers (LNAs) and feeding structures. All elements are represented by their characteristic values, such as the losses of the feeding line ($\alpha_f, \alpha_{l,1}, \ldots$), the phase shifters ($\alpha_P$) and the combiners ($\alpha_C$), the noise figure, and gain of the amplifiers (F, g). The position of the amplifiers in Figure 2.30 is at the first stage, resulting in the highest amount of $M \cdot N$ amplifiers needed for the realization of an $M \times N$ antenna array. From the theoretical point of system noise temperature, this is the optimum solution. But this comes with the highest cost and many active elements distributed over the whole array. If the amplifiers are placed at a later stage, for example at stage $n = 2$ or

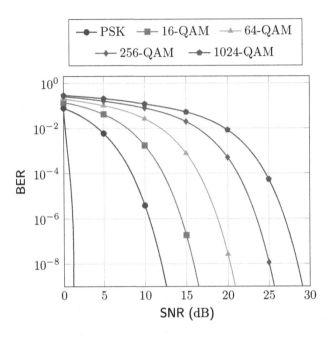

Figure 2.29 Bit error rate for different modulation schemes versus the SNR for an additive white Gaussian noise (AWGN) channel.

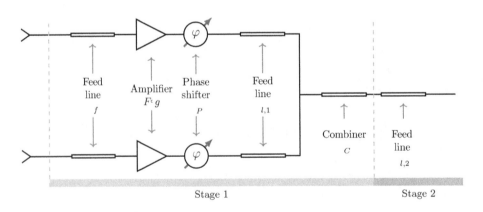

Figure 2.30 Design of a corporate feeding network for quadratic arrays of arbitrary size.

3 in the corporate feed network, the number of amplifiers can be reduced by a factor of 2^{n-1}. The price to pay is a lower noise figure, since the insertion loss before the LNAs directly impacts the system's noise figure. The overall system noise temperature of such a corporate feed network with variable amplifier positioning is given by [42]. For an amplifier placement at first stage ($n = 1$), the system noise temperature is evaluated by

$$T_{sys} = T_1 + T_p \left(\left(a_c^{\log_2(MN)} - 1 \right) \frac{a_f a_p}{g} + \left(e^{2\alpha l} - 1 \right) \frac{a_f a_p a_c^{\log_2(MN)}}{g} \right), \qquad (2.30)$$

where

$$T_1 = T_b + T_p\left((a_f - 1) + (F - 1)a_f + (a_p - 1)\frac{a_f}{g}\right).$$ (2.31)

For a placement at later stages, the system noise temperature has to be modified according to

$$T_{sys} = T_3 + T_p\left(\left(a_c^{\log_2(MN)-(n-1)} - 1\right)\frac{a_f a_p a_c^{(n-1)} e^{2\alpha l_1}}{g} + \left(e^{2\alpha(l-l_1)} - 1\right)\frac{a_f a_p e^{2\alpha l_1} a_c^{\log_2(MN)}}{g}\right),$$ (2.32)

with

$$T_3 = T_2 + T_p(F - 1)a_f a_p a_c^{n-1} e^{2\alpha l_1},$$ (2.33)

$$T_2 = T_1 + T_p\left(\left(a_c^{n-1} - 1\right)a_f a_p + \left(e^{2\alpha l_1} - 1\right)a_f a_p a_c^{n-1}\right),$$ (2.34)

and

$$T_1 = T_b + T_p\left((a_f - 1) + (a_p - 1)a_f\right).$$ (2.35)

Equations (2.31)–(2.35) permit to estimate the equivalent system temperature at a virtual point located between the antenna elements and the rest of the active array. From this, the G/T ratio at the system output is calculated by the array directivity in main lobe direction $G/T = D/T_{Sys}$, whereby all losses and the amplifier gain are already incorporated in T_{Sys}. The gain is calculated considering isotropic radiators for a steering angle of up to 60°. Using patch antennas, the gain of the radiating element is increased by approximately 3 dB, and hence, the G/T value is biased by this amount. For a systematic evaluation, this linear offset is of minor importance and can easily be calculated out. The feed line length l has been calculated assuming a corporate configuration as depicted in Figure 2.26. However, the presented approach neglects several effects such as mismatch, mutual coupling, frequency dependency, and phase state dependent insertion loss. Nevertheless, important design functions can be extracted. Figure 2.31 shows example results for a 60 GHz antenna array with the amplifiers positioned at different stages of the corporate feed network. All G/T results are normalized by multiplying with 1K before converting to decibels, indicated by the term re(1/1K) as recommended by NIST [43]. The assumed network parameters are given in Table 2.7. The phase shifter loss and the feed line loss are estimated in a full-wave simulation. The LNA parameters are taken from a commercially available product. It can be clearly seen that positioning the amplifiers at the first stage $n = 1$ gives the best results in terms of G/T. For a -15 dB G/T, equivalent to an SNR of 10 dB for an 8.5-km communication link with an PSD$_{EIRP}$ of 13 dBm/MHz, the size of the array has to be minimum 8×8 elements. By moving the amplifiers more into direction of the receiver the size must be fourfold (16×16 elements), but the number of amplifiers can be reduced to 32 instead of 64. Without any active parts, the antenna array size has to be 32×32 elements. It is interesting to note that the G/T cannot be

Table 2.7 Parameters used for antenna efficiency analysis

Parameter	A_f (dB)	A_P (dB)	A_L (dB/m)	A_C (dB)	F (dB)	g (dB)
Value	0.1	5	50	1	4	13

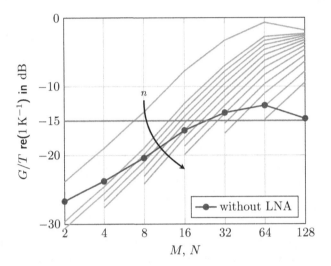

Figure 2.31 Influence of LNA position on G/T in dB related to 1/1K ranging from $n = 1$ to $n = 13$ in descending order as depicted by the arrow. As an example, a design goal of 15 dB is depicted showing possible implementations at the intersection points.

improved beyond -12.5 dB for a purely passive antenna array, based on the values given in Table 2.7.

With this model, parameter studies of antenna arrays with different parameter sets can be carried out. Figure 2.32 shows the analytical results for swept feed line loss from the theoretical lossless case up to 170 dB/m of line loss.

The results reveal that the maximum practical size of corporate fed antenna arrays is limited by the feed line losses. The practical size of this 60 GHz antenna array is limited to maximum 32×32 or 64×64 elements. Even by putting LNAs at every single antenna element does not allow to build array of larger size. The only way to cope with this circumstance is by using lower loss transmission lines or amplifiers with higher gain, as shown in Figure 2.33.

From the model, the following general correlations between amplifier requirements and properties of the lossy passive components can be drawn. For a placement at stage $n = 1$:

$$\frac{1 - 1/g}{F - 1} = \frac{1}{a_P \cdot (a_C)^{\log_2(MN)} \cdot e^{2al} - 1}; \qquad (2.36)$$

otherwise, there will be no improvement in terms of G/T ratio. Equivalently, for a placement at $1 < n \leq \mathrm{ld}(MN) + 1$ the condition becomes

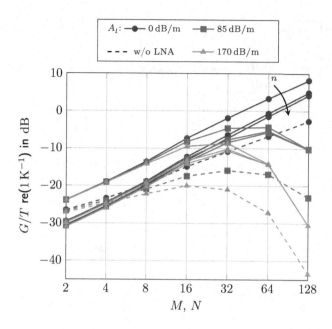

Figure 2.32 *G/T* in dB related to 1/1K vs. feed line loss A_L for different array configurations. The curve sets of the same line style show an LNA placement at stages from $n = 1$ to $n = 3$ as depicted by the arrow.

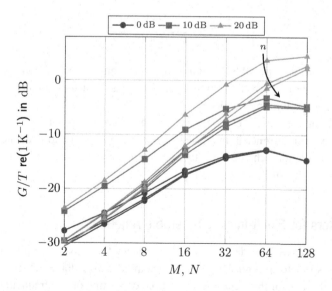

Figure 2.33 *G/T* in dB related to 1/1K for different amplifier gain *G* versus array configurations. The curve sets of the same line style show an LNA placement at stages from $n = 1$ to $n = 3$ as depicted by the arrow.

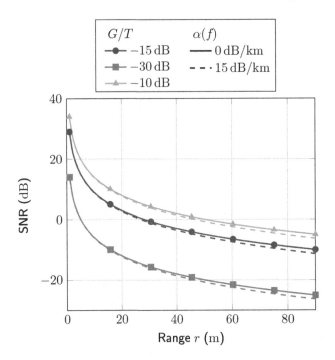

Figure 2.34 SNR values for different *G/T* ratios versus the range for 60 GHz communication. Solid lines represent free space propagation loss only, while dashed lines represent a channel loss taking atmospheric attenuation with 15 dB/km into account.

$$\frac{1 - 1/g}{F - 1/g} > \frac{(a_C)^{n-1} \cdot e^{2al_1}}{(a_C)^{\log_2(MN)} \cdot e^{2al}}. \tag{2.37}$$

With this method, it is possible to find an optimum antenna array configuration in terms of array size and LNA positioning, reaching a defined minimum *G/T* ratio under a given set of limiting factors.

As the *G/T* ratio is a channel independent measure for the quality of the receiving antenna, the system designer is interested in the achieved SNR, to evaluate suitable coding and modulation schemes and to calculate coverage and interference ranges. These results are shown in Figure 2.34.

2.2.6   Phase Shifters for Electronically Steerable Antennas

A phase shifter is a two-port device, which can change the phase of a radio frequency signal with respect to a control signal, for example bias voltage. There are many different possibilities for the technical realization, depending on the frequency. At low frequencies, for example, parts of reactances are used, which occur at inductances and capacitances, while at high frequencies, time differences due to delay lines are used. There are two different ways of phase shifting:

1. In non–frequency-dependent phase shifters, the spectrum is shifted uniformly by a certain angle. Since those phase shifters make use of the Hilbert transformation, they are also called Hilbert transformer. Each of them causes a rotation of the spectrum by 90° [44].
2. For frequency-dependent phase shifters, a phase shift of a specific frequency is accomplished by delaying the input signal. The phase shift becomes frequency dependent because different frequencies have different periodic times. Those kinds of phase shifters can be realized by many circuits, as instance delay lines, reflection-type phase shifters, or all-pass filters.

The class of frequency-dependent phase shifters is widely used. The earliest generations of phase shifters have all been mechanical phase shifters based on variable length transmission lines, which met nearly all phase shifting requirements until the 1960s [45–47]. The first electrically controlled phase shifter based on ferrite materials was introduced in 1957 by Reggia and Spencer [48], which was followed by the first semiconductor diode phase shifters [49, 50] and many more.

Phase shifters can be divided into two classes: (1) discrete phase shifters, which can provide discrete steps of phase offset $\Delta\varphi_{step}$ and (2) continuous phase shifters, which can provide any value of phase offset $\Delta\varphi$. The easiest way to implement a phase shifter is by switching between transmission lines with different lengths. Figure 2.35 shows an example of a 3-bit phase shifter, which can realize $2^3 = 8$ different phase states (0°, 45°, 90°, 135°, 180°, 225°, 270°, 325°) by setting the three switches to the corresponding positions.

The resolution of the phase shifter is restricted by the number of bits. For an equally distance of phase shifts the resolution is

$$\Delta\varphi_{step} = \frac{360°}{2^n}. \tag{2.38}$$

These phase shifters make use of the propagation constant of a transmission line. They introduce the phase shift by varying the length of the transmission line.

$$\varphi = \mathfrak{J}(\gamma \cdot l) = \text{fct}(l). \tag{2.39}$$

The main drawback of this implementation is the achieved resolution of the phase shifters, which directly influences the application. For instance, the steering resolution

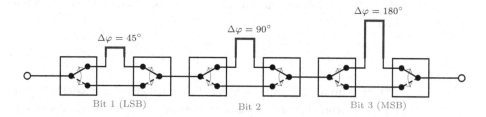

Figure 2.35 Example of a 3-bit discrete phase shifter, covering 360° of phase shift with the least significant bit (LSB) representing the resolution of the phase shifter.

of phased arrays is directly linked to the phase shifter resolution. Thus, if the application needs a precise and high resolved beam steering, the number of bits must be increased. For a 1° resolution, the phase shifter needs to have $n = \lceil \log_2 360 \rceil = 9$ segments. The occupied space of these phase shifter segments does not scale linearly, as the length of the transmission line differs for each stage. The total occupied space can be approximated by

$$A_\Sigma = n \cdot A_{\text{SPDT}} + \sum_n \frac{A_{\text{TL}, 180°}}{n}, \tag{2.40}$$

with the space requirements of the two single-pole double-throw (SPDT) switches per segment and of the 180° transmission line given by A_{SPDT} and $A_{\text{TL}, 180°}$, respectively. A second drawback is a nonconstant amplitude for different phase shifts, meaning that the insertion loss changes for different selected phase shifts. This is caused by the different lengths of the transmission lines switched together:

$$\alpha = \Re(\gamma \cdot l) = \text{fct}(l). \tag{2.41}$$

For low-loss transmission lines $\text{IL}_{\text{LSB}} \to 0$, this effect becomes negligible as the total range of insertion loss is in the range up to

$$\text{IL} = \text{IL}_0 + \frac{n(n+1)}{2} \text{IL}_{\text{LSB}} = \text{IL}_0 + \frac{n(n+1)}{2} \frac{\text{IL}_{180°}}{n}, \tag{2.42}$$

with the phase shift independent (constant) insertion loss IL_0.

Last, SPDT switches also introduce losses, thus lowering the FoM of such phase shifters. The FoM is an interesting metric to compare the performance of different passive phase shifter implementations. It is defined by the ratio of the maximum phase shift and the maximum insertion loss in all tuning states.

For the implementation of discrete phase shifters, the FoM is given by

$$\text{FoM} = \frac{360°}{\text{IL}_0 + \dfrac{\text{IL}_{180°} \cdot (n+1)}{2}} \tag{2.43}$$

and scales with the resolution.

Recalling Eq. (2.39), a phase shift can be realized in two different ways. The first implementation by changing the length of a transmission line has been discussed before. The second approach modifies the propagation constant γ of a transmission line:

$$\varphi = \Im(\gamma \cdot l) = \text{fct}(\gamma). \tag{2.44}$$

Starting from the transmission line theory and electrical model in Figure 2.36 the propagation constant γ and characteristic impedance Z_L of the transmission line can be calculated as

$$\gamma = \alpha + j\beta = \sqrt{(R' + j\omega L')(G' + j\omega C')} \tag{2.45}$$

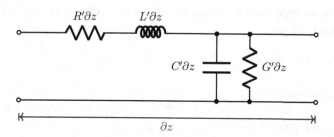

Figure 2.36 Equivalent circuit of a quasi-TEM transmission line unit cell.

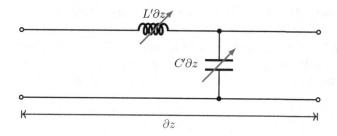

Figure 2.37 Equivalent circuit of a tunable ideal lossless quasi-TEM transmission line unit cell.

and

$$Z_L = \sqrt{\frac{R' + j\omega L'}{G' + j\omega C'}}, \tag{2.46}$$

with the primary transmission line properties loop resistance R', loop inductance L', insulator capacitance C', and insulator conductance G'. Simplifying to an ideal transmission line $R' \to 0$ and $G' \to 0$ results in a lossless transmission line $\alpha = 0$ with a phase constant of

$$\beta = \omega\sqrt{L'C'}. \tag{2.47}$$

Hence, the phase of a transmission line can be influenced by the loop inductance L' or by the insulator capacitance C' as shown in Figure 2.37.

With this approach, the transmission line's characteristic impedance can be kept constant, resulting in a phase constant of

$$\beta = \omega Z_L C'. \tag{2.48}$$

Keeping Z_L constant while changing C' implies the use of a tunable inductor. As will be seen later, the realization of tunable inductors is much more complicated than that of tunable capacitors. Hence, usually the inductance value is kept constant and only the insulator capacitance value is changed. The drawback of nonconstant transmission line characteristic impedance is accepted consciously. The phase shift of such a capacitively loaded line is

$$\Delta\varphi = \Im(\Delta\gamma \cdot l) = l \cdot \omega \cdot \left(\sqrt{LC_0} - \sqrt{LC_1}\right) = l \cdot \omega \cdot \sqrt{L \cdot C_0} \cdot \left(1 - \sqrt{1 - \tau_C}\right), \quad (2.49)$$

with the untuned capacitance C_0 and the relative tuning coefficient τ_C, which ideally ranges from 0 to 1, and l the physical length of the capacitively loaded line section. With this, the tuning efficiency with respect to a change in capacitance is

$$\frac{\partial(\Delta\varphi)}{\partial C} = \frac{l \cdot \omega \cdot \sqrt{L}}{2\sqrt{C}} = \frac{l \cdot \omega}{2} Z_L. \quad (2.50)$$

In the past few years artificial transmission lines, so-called metamaterials, have been studied in detail. These transmission lines have a capacitance in the loop of the transmission line and an inductance in the insulator branch, as shown in Figure 2.38.

If C_s' and L_p' are zero, we get the original transmission line, which follows the right-hand rule for the Poynting vector. If C_p' and L_s' are zero, the phase constant of the transmission line is

$$\gamma_{\text{LH}} = -j \frac{1}{\omega\sqrt{L_p' C_s'}}, \quad (2.51)$$

which results in a left-hand relation for the Poynting vector. Therefore, these transmission lines are also called left-handed (LH) transmission lines. By fully population of all four elements of Figure 2.38 the transmission lines show left- and right-handed behavior. Therefore, these transmission lines are called composite left-right-handed (CRLH) transmission lines. They show a phase constant and a characteristic impedance of

$$\gamma_{\text{CRLH}} = \pm j \sqrt{\left(\omega L_s' - \frac{1}{\omega C_s'}\right)\left(\omega C_p' - \frac{1}{\omega L_p'}\right)} \quad (2.52)$$

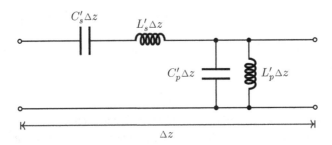

Figure 2.38 Equivalent circuit of a CRLH unit cell.

and

$$Z_{\text{CRLH}} = \pm \sqrt{\frac{\omega L'_s - \dfrac{1}{\omega C'_s}}{\omega C'_p - \dfrac{1}{\omega L'_p}}}.$$

(2.53)

In theory, these transmission lines can keep a constant characteristic impedance for different phase shift values, even if only the capacitances are tuned. But in practice the required capacitance ranges cannot be realized in an easy way. Moreover, the bandwidth of such implementations is very narrow.

In contrast to right-handed (conventional) transmission lines, which can be realized as longitudinally homogeneous transmission lines such as microstrip or coplanar waveguides, artificial transmission lines have to be realized with discrete lumped components. For this, a periodical unit cell of length Δz with discrete component values (instead of primary line properties) can be introduced with $L_s = L'_s \cdot \Delta z$, $L_p = L'_p / \Delta z$, $C_s = C'_s / \Delta z$, and $C_p = C'_p \cdot \Delta z$. For a periodically loaded transmission line, as shown in Figure 2.39, the propagation constant can be derived from

$$\gamma_{\text{PL}} = \pm \frac{1}{\Delta z} \cosh^{-1}\left(1 + \frac{ZY}{2}\right).$$

(2.54)

For the three transmission line's sections with periodic lumped components, it follows:

$$\gamma_{\text{PRH}} = \pm \frac{1}{\Delta z} \cosh^{-1}\left(1 - \frac{\omega^2 L_s C_p}{2}\right)$$

(2.55)

for the periodically loaded RH transmission line,

$$\gamma_{\text{PLH}} = \pm \frac{1}{\Delta z} \cosh^{-1}\left(1 - \frac{1}{2\omega^2 L_s C_p}\right)$$

(2.56)

for the periodically loaded LH transmission line, and

$$\gamma_{\text{PCRLH}} = \pm \frac{1}{\Delta z} \cosh^{-1}\left(1 - \frac{\omega^2}{2}\left(L_s - \frac{1}{\omega^2 C_s}\right)\left(C_p - \frac{1}{\omega^2 L_p}\right)\right)$$

(2.57)

for the simple composite right/left-handed (CRLH) transmission line.

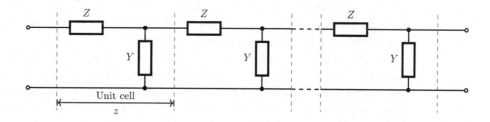

Figure 2.39 Transmission line with periodic structure.

To compare different realizations, the FoM must be evaluated. For a lossy right-handed transmission line, the propagation constant becomes

$$\gamma_{RH} = j\omega\sqrt{L'(1-j\tan\delta_{L'})\cdot C'(1-j\tan\delta_{C'})},\qquad(2.58)$$

with $\tan\delta_{L'}$ and $\tan\delta_{C'}$ the loss tangent of the inductance and the capacitance, respectively. In case of small losses $\tan\delta_{L'}\ll 1$ and $\tan\delta_{C'}\ll 1$, Eq. (2.58) can be approximated by

$$\gamma_{RH}\approx\underbrace{\omega\sqrt{L'C'}\,\frac{\tan\delta_{L'}+\tan\delta_{C'}}{2}}_{\alpha}+j\underbrace{\omega\sqrt{L'C'}}_{\beta},\qquad(2.59)$$

which further results in the phase shift of

$$\Delta\varphi=\frac{180°}{\pi}\left(\beta|_{C_2}\cdot\ell-\beta|_{C_1}\cdot\ell\right)=\frac{180°}{\pi}\omega\ell\sqrt{L'C'}\left(1-\sqrt{1-\tau_C}\right).\qquad(2.60)$$

For the insertion loss, the maximum value is used, which corresponds to the highest capacitance value:

$$IL_{max}=20\log_{10}e^{\alpha_{max}\ell}\ dB\approx 8.686\ dB\cdot\alpha_{max}\ell=8.686\ dB\cdot\omega\ell\sqrt{L'C'}\frac{\tan\delta_L+\tan\delta_C}{2}.\qquad(2.61)$$

Combining Eqs. (2.60) and (2.61) leads to the FoM:

$$FoM=\frac{\Delta\varphi}{IL_{max}}=13.193°/dB\cdot\frac{1-\sqrt{1-\tau_C}}{\tan\delta_L+\tan\delta_C}.\qquad(2.62)$$

The evaluation of a left-handed transmission line results in the identical Eq. (2.62). From the theoretical point of view, there is no difference in both realizations for a single-frequency operation. Figure 2.40 shows the achievable FoM for different tunability versus the quality factor of the transmission line elements.

If the FoM is the same for left- and right-handed phase shifters, there must be another difference in performance. Figure 2.41 shows simulation results for an arbitrarily chosen periodical phase shifter in a left- and a right-handed T-configuration.

From these plots, two differences can be derived: (1) for a fixed phase shift, the right-handed transmission line shows a much wider bandwidth compared to the left-handed one. (2) As the overall space occupation is the same for the two implementations and the frequency of operation is lower for the left-handed transmission line, a miniaturization is possible by using left-handed structures.

The realization of these homogenous phase shifters usually needs functional materials with tunable properties (see also Section 2.4) or lumped components. These concepts for phase shifters can be extended to reflection type phase shifters (RTPS). Here a coupler is combined with two identical tunable loads. Hence, the isolated port of the coupler exhibits a phase shifted replica of the input wave reflected by the loads. Details on this kind of phase shifters can be found in Section 3.4.2.4 of Chapter 3.

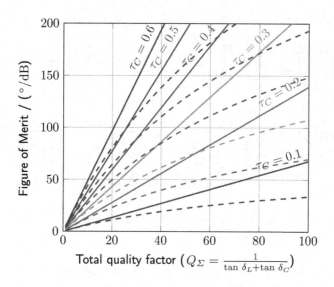

Figure 2.40 Figure of merit according to Eq. (2.62) for different tunability of the capacitor τ_C. Solid lines represent the combined quality factor, while dashed lines show the FoM for a fixed quality factor for the inductor of $Q_L = 100$.

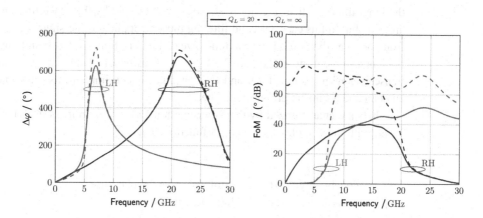

Figure 2.41 Simulation results for a left- and a right-handed periodic transmission line with ten T-shaped unit cells ($C_L = C_H = 0.3\ pF$, $Q_C = 20$, $L_L = L_H = 0.5\ nH$, $\tau_c = 50\%$) and different Q of the inductor.

2.3　Tunable Filter

The antenna is only the very first (or last) element in a transceiver chain. As important as the antenna are input filters (1) to block interference signals and (2) to limit the noise bandwidth of the system. The blocking of interferers can be further subdivided into two parts, (1a) blocking of unwanted signals such as image frequencies, which are converted down to the same frequency range as the wanted signal (out-of-band), and

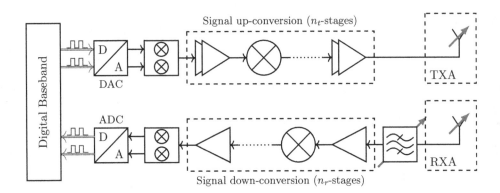

Figure 2.42 Basic tunable mm-wave transceiver front end consisting of tunable smart RX and TX antennas and a tunable preselection filter.

(1b) blocking of interfering in-band signals (in-band-blocking). A reconfigurable antenna already acts as a tunable space filtering element, removing many of the unwanted in-band interferers by simple space filtering. For a mm-wave system the out-of-band blocking must be considered, as the first element after the antenna is an amplifier followed by a mixing stage. Thus, the interferers have also to be blocked by the input filters. A first filtering is already accomplished by the space filtering of the antenna, but in this case a second filtering must be applied as the signals might come from the same direction. Hence a mm-wave communication system should implement tunability in the front end. Figure 2.42 shows a possible block diagram of a mm-wave transceiver with separated TX and RX antennas, which implements a tunable smart antenna and a tunable preselection filter.

Filters are characterized by their transmission function from input $X(j\omega)$ to output $Y(j\omega)$ with the complex transmission function $H(j\omega)$ by

$$Y(j\omega) = H(j\omega) \cdot X(j\omega). \tag{2.63}$$

The previously discussed phase shifters are a special case of filters with constant amplitude $|H(j\omega)|$ but variable phase variation:

$$H_{PS}(j\omega) \propto e^{j\omega\beta} = e^{j\omega \cdot \text{fct}(c)}, \tag{2.64}$$

which is dependent on the control signals c. A second special case of filters are matching networks, where also the port impedances Z_1 and Z_2 are considered in the network design:

$$H'(j\omega, Z_S, Z_L) = \frac{Y(j\omega)}{X(j\omega)}\bigg|_{Z_1=Z_S, Z_2=Z_L}. \tag{2.65}$$

For many applications, a fixed fractional bandwidth is standardized; hence, a filter independently tunable in center frequency f_c and bandwidth B is highly desired. Figure 2.43 sketches the transmission of three different classes of tunable filters for definition of the frequency tunability Δf and the bandwidth ΔB.

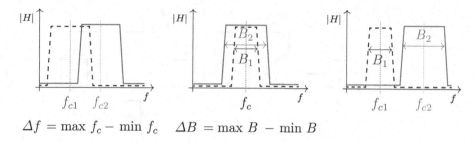

$$\Delta f = \max f_c - \min f_c \quad \Delta B = \max B - \min B$$

Figure 2.43 Three different classes of tunable filters: (left) center-frequency tuning (center) bandwidth tuning and (right) center-frequency and bandwidth tuning.

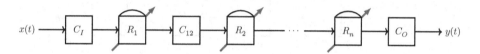

Figure 2.44 Structure of a linear coupled tunable filter with tunable resonators $R_1 \ldots R_n$, but fixed input- (C_I) and output- (C_O) as well as interresonator coupling ($C_{12} \ldots C_{(n-1)(n-2)}$).

General filter synthesis is well understood for nontunable filters with numerous publications. Tunable filters with external control of the transmission coefficient $H(j\omega)$ are usually discussed as side aspect only [51–53]. They bring unique properties but have large hurdles in designing them at the same time. First practical realizations of tunable filters were based mainly on cavity resonators with mechanically changeable dimensions. With this, the individual resonators of a filter can be tuned. Figure 2.44 shows an example of a simple linear coupled filter of $(2N - 1)$-th order with N resonators.

The main drawback of this kind of realization is the fixed coupling between the individual resonators. This results in a specific fractional bandwidth, which results in changing absolute bandwidth while tuning the center frequency. For many applications, a fixed fractional bandwidth is standardized. Hence, a filter independently tunable in center frequency f_0 and bandwidth B is highly desired. Such a realization requires the implementation of reconfigurable coupling coefficients. Moreover, for the introduction of additional nulls into the transmission coefficient, cross coupling between resonators is highly desired. Figure 2.45 shows the schematic of such a tunable filter with tunable cross coupling.

In a coupling matrix representation [54] this filter can be described by the coupling matrix **C**:

$$\mathbf{C} = \begin{bmatrix} R_1 & C_{12} & \cdots & C_{1N} \\ C_{12} & R_2 & & \\ \vdots & & \ddots & \\ C_{1N} & & & R_N \end{bmatrix}, \tag{2.66}$$

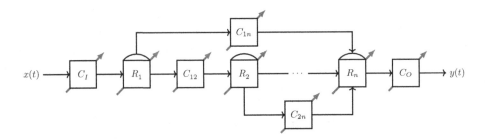

Figure 2.45 Structure of a linear coupled tunable filter with tunable resonators $R_1 \ldots R_n$, and tunable input (C_I) and output (C_O) as well as interresonator coupling (C_{mn}).

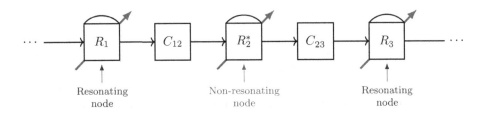

Figure 2.46 Realization example of a tunable coupling of two resonators R_1, R_3 by a nonresonating node R_2^* with fixed coupling C_{12}, C_{23}.

with the resonators on the main diagonal and the interresonator coupling on the first side diagonals. However, while the filter's synthesis is well described in the literature for fixed filters, synthesis methods for tunable filters are not well established, as underlined in [55].

Tunable filters have been known for a long time. However, the implementation of tunable filters is challenging, especially at microwave frequencies. The insertion loss of a filter is directly dependent on the resonators' quality factor. Hence for high-quality low-loss filters, cavity filters are mainly used [56–58]. Today, the miniaturization of microelectromechanical system (MEMS) devices allows an easy integration of tunable filters also in planar topologies [59, 60]. While the tuning of cavity resonators either by means of geometric changes or by tuning of the cavity's material is straightforward, the implementation of tunable coupling structures is more challenging. One example used for nonplanar filters is iris coupling with a variable iris [61, 62]. This implementation is efficient but increases system complexity. Also, system reliability suffers from mechanically moving parts. As alternative the variable couplings can be emulated by nonresonating nodes – resonators operated out of resonance – [57, 63, 64] as shown in Figure 2.46.

A similar concept is followed in planar structures where components of the J- and K-inverters of the coupling structures are absorbed in the resonator itself. Figure 2.47 shows an example of a simple third-order filter with the introduction of K-inverters and the absorption of its components into adjacent resonators. This method is often used, as the component values and the required tuning ranges of the components are too different for series and parallel resonators.

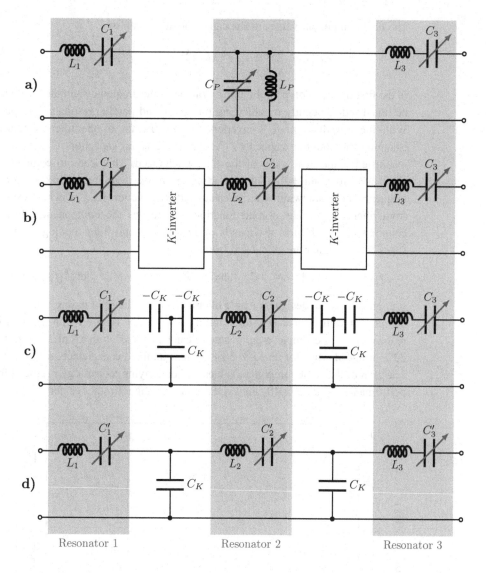

Figure 2.47 (a) Prototype of a common bandpass filter with (b) K-inverters avoiding parallel resonators. In (c) the realization of the K-inverters by capacitors is shown before the series capacitors are absorbed by adjacent resonators. (d) Final realization of the filter.

The components' values are getting closer, but this also leads to additional boundaries for the coupling elements as the negative capacitor value must be compensated by the series resonant circuit capacitors, resulting in

$$C_K < C_1, C_3 \quad \wedge \quad 2C_K < C_2 \quad \Rightarrow \quad C_K < \sqrt{\frac{C_1}{L_2 \omega_0^2}}. \tag{2.67}$$

The new component values of the capacitors are given by

$$C_1' = \frac{C_1 C_K}{C_K - C_1}, \quad C_2' = \frac{C_2 C_K}{C_K - 2C_2}, \quad C_3' = \frac{C_3 C_K}{C_K - C_3}. \tag{2.68}$$

In the first filter prototype in Figure 2.47a, the center frequency and the bandwidth can be tuned independently by changing the series and parallel resonators, respectively. With the introduction of K-inverters, the bandwidth of the filter is adjusted by changing the coupling capacitors C_K. Thus, the serial capacitors $C_1' \ldots C_3'$ must be readjusted if the bandwidth of the filter is changed. This leads to complex control algorithms to control the filter. Starting from the coupling matrix $\mathbf{C}$, the center frequency and bandwidth can be calculated by function fct_A, which is well known from filter theory. The second function fct_B relates the component values to the coupling matrix. Hence, it depends on technology and filter topology. The control states $\mathbf{S}$ of the filter are calculated from

$$f_c, B = \text{fct}_A\{\mathbf{C}\} \quad \wedge \quad \mathbf{C} = \text{fct}_B\{\mathbf{S}\} \quad \Rightarrow \quad \mathbf{S} = \text{fct}_B^{-1}\{\text{fct}_A^{-1}\{f_0, B\}\} \tag{2.69}$$

It might be easy to design a tunable filter and choose the right tuning. But the required component values for every single tuning state can be realized with (almost) infinite realizations. For example, a 2 pF varactor can be realized with a 8 pF varactor tuned to 25% or a 4 pF varactor tuned to 50% of their untuned capacitance, respectively. The selection of the right component values and underlying technology is crucial for filter performance. Figure 2.48 shows the tuning range in the center frequency and

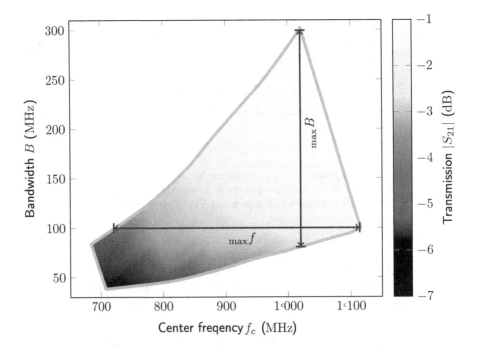

Figure 2.48 Bandwidth and center frequency tuning of a filter.

bandwidth of a tunable bandpass filter [65]. As the third parameter, the insertion loss of the filter is chosen as the cost function. Depending on the chosen tuning ranges, the average performance of the filter changes dramatically.

The work on optimizing the filter design and the respective control algorithms fct_B^{-1} is just at the beginning. First ideas include interesting approaches such as swarm optimization, machine learning, as well as online filter parameter extraction.

2.3.1   Impedance Matching

A special class of tunable filters are impedance matching networks. Here, the tuning of the filter elements is used in order to change the port's characteristic impedance. Common places for impedance matching circuits are in the vicinity of antennas and amplifiers. Here, tunability can be used, for example, to balance frequency-dependent behavior. The basic idea is to shift a narrow-band optimized circuit in frequency to allow broadband operation. This principle has been adopted at frequencies below 10 GHz in a variety of demonstrators such as tunable power amplifiers [66–74] and adaptive antenna matching [75–79] realized in different technologies.

2.3.2   Variable Attenuators

A special class of filters are variable attenuators. These can be used to attenuate a strong signal in order to fulfill input limits between different system blocks. In phased arrays, variable attenuators are required in purely passive systems beside phase shifters to realize complex weighting on the elementary signals. In hybrid or active arrays, attenuators can be replaced by variable gain amplifiers (VGA) when noise behavior and dynamic range are superior for amplifiers than for variable attenuators [80, 81]. Variable attenuators can be realized in different ways. Switchable attenuators comparable to discrete phase shifters (compare Figure 2.35) offer mechanically or electrically controllable attenuation states. By replacing the resistors in a fixed π attenuator by PIN diodes a continuously tunable attenuator can be realized [82]. Other approaches include hybrids with variable coupling or interference-based approaches [83].

2.4   Technologies for Tunable Components

Different technologies can be used for the realization of tunable devices. As described previously, most concepts are based on varying a capacitance. Thinking of a simplified parallel plate capacitor the capacitance can be calculated by

$$C = \varepsilon_0 \varepsilon_r \frac{A}{d}, \qquad (2.70)$$

with the vacuum permittivity ε_0, the material's relative permittivity ε_r, the size A, and distance d of the electrodes, respectively. As ε_0 is a constant, there exist two different possibilities in changing the capacitance value. The historic way was to use rotary capacitors to change the overlapping area of the electrodes. A similar mechanism is used today in MEMS devices, where the distance and overlapping area are used to control the capacitance value. Let's refer to this class of tunable capacitors, where the relation A/d is changed, as geometry-tuned devices. The realization of geometry-tuned devices offers many opportunities, because the dielectric material used can be of high quality. MEMS-based devices are described in Chapter 4. Using semiconductors, especially varactor diodes, have also to be considered in this class of geometry-tunable devices because of the modulation of the depletion zone by the applied bias voltage. This results in a change of the effective distance between the electrodes at the p–n junction. Another way consists of tuning a transistor, leading to n-type metal–oxide–semiconductor (NMOS) varactors. As semiconductors offer high volume production and rather easy design process, they have a high impact on commercial applications in modern communications, even if their quality factor is much lower as compared to MEMS. This technology is described in detail in Chapter 3.

As a second tuning mechanism, the change of a material's relative permittivity can be used to control the capacitance. The permittivity of the dielectric material results directly from the polarizability by

$$\varepsilon = \varepsilon_0 \varepsilon_r = (1 + \chi)\varepsilon_0, \tag{2.71}$$

with χ the electric susceptibility, which is related to the induced dielectric polarization density P by

$$P = \varepsilon_0 \chi E. \tag{2.72}$$

The mechanisms that result in material polarization are frequency dependent. Figure 2.49 shows a typical material behavior versus the operating frequency and names the different polarization mechanisms.

As can be seen, the polarization, and hence the permittivity, of the material drops with increasing frequency. The position of the frequency where a certain polarization mechanism stops working is material dependent. Most of the polarization mechanisms are used to realize tunable capacitors by utilizing different polarization methods:

Ionic polarization: In this class of materials, the ionic concentration of the material is changed to realize different permittivities. There are no materials known, which can utilize this functionality for electronic steering. Nevertheless, by changing the material, the ionic concentration can be adapted. There are some ideas around, which are based on liquid tuning by pumping dielectric or conducting materials in and out of cavities [84–86]. With this idea very high quality-factor tunable devices can be realized. But these devices are not within the focus of this book, as the tuning mechanism and tuning speed is more comparable to electromechanical tuning of capacitors.

Dipolar polarization: In this scenario the material shows an anisotropy of the permittivity tensor for different orientations of the molecules to the electric field. By aligning the tensor of anisotropy to the electric field, the polarization of the

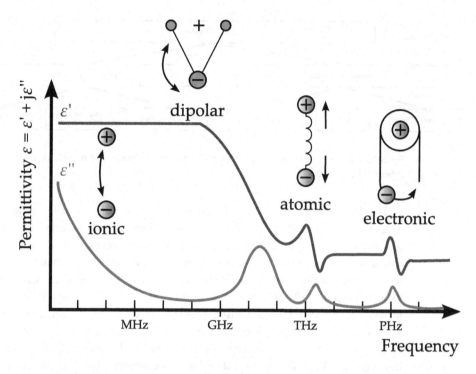

Figure 2.49 An illustration of the frequency response of various dielectric mechanisms in terms of the real and imaginary parts of the permittivity.
Based on work of Prof. Kenneth A. Mauritz

material changes and, hence, the capacitance value can be changed without changing the geometry. The liquid crystal technology is using this polarization mechanism. A detailed description can be found in Chapter 5.

Atomic polarization: By changing the position of (charged) atoms in a crystal lattice the polarization of the material can be changed as well. Ferroelectric materials from the Perovskite structure are the most prominent examples of this class of materials. Barium strontium titanate (BST) has a centric crystal lattice with a "free" titanium ion, which can be elongated by applying an electric field to the material. By changing the position of these ions in the lattice, the polarization of the material changes. This technology has been studied for some decades now with very great success in the lower frequency domain up to 20 or 30 GHz [73, 87–91]. There are some examples where BST has also been used at higher frequencies, such as 60 GHz [92, 93]. However, the demonstrated device performance suffers from high material loss at these frequencies. Therefore, these materials are usually not considered as candidates for commercial mm-wave applications.

As these capacitors use only a material's property, they are referred to as functional materials. Figure 2.50 gives an overview and classification of different tuning technologies, which can be used for electrically tuned capacitors.

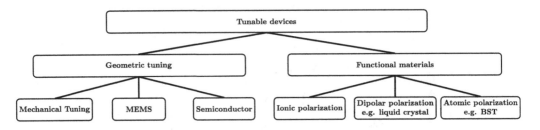

Figure 2.50 Overview of different technologies for the realization of tunable devices, especially capacitors.

The disadvantage of changing the capacitance value only is usually connected with a change of the matching, especially due to changing of the characteristic impedance of transmission line phase shifters, as already stated. In the past decade, research effort was and still is focused on the development of multifunctional materials, so-called multiferroics, which can adapt not only their permittivity but at the same time also their permeability [94–98]. With this technique, it would be possible to adapt both material parameters simultaneously, keeping the characteristic impedance constant while changing the propagation constant. As these materials are of great interest for commercial applications, huge effort has been put in the optimization of these materials such as $BiFeO_3$ or $YbMnO_3$ thin films. The performance of these research-grade materials is much lower than that of established materials such as semiconductors, BST, or liquid crystals.

2.4.1   Semiconductors

In general, semiconductor-based varactors are operated in a reverse-biased state, so no DC current flows through the device. The thickness of the depletion zone d at the p–n junction is controlled by the applied reverse bias voltage V_{bias} and therefore the varactor's junction capacitance C can be tuned. Generally, the depletion thickness is proportional to the square root of the applied voltage and inversely proportional to its capacitance. Thus, the capacitance is inversely proportional to the square root of the applied voltage:

$$d \propto \sqrt{V_{bias}} \quad \rightarrow \quad C \propto \frac{1}{d} \propto \frac{1}{\sqrt{V_{bias}}} \tag{2.73}$$

Figure 2.51 visualizes the operation principle of a simple varactor diode.

All diodes exhibit this variable junction capacitance, but varactors are manufactured to exploit the effect and increase the capacitance variation. Figure 2.51 shows a cross section of a varactor with the depletion layer formed of a p–n junction. This depletion layer can also be made of a metal–oxide–semiconductor (MOS) or a Schottky diode [99]. Varactor diodes are usually connected in an antiserial fashion in order to reduce nonlinear effects and to decouple bias voltage from the rest of the circuit.

By introducing an intrinsic semiconductor layer between the p- and n-region a PIN diode can be realized. This kind of diodes has negligible capacitance variation, but it can be used as switch even at high frequencies and especially at high power levels. In reverse or zero bias the intrinsic layer forms a small capacitance value between the p- and n-region and hence a large resistance for an applied RF field. By forward biasing the diode, a free flow of carriers is enabled resulting in a very low component resistance of about 1 Ω.

Details on the realization of varactor diodes and PIN diodes, their respective performance, as well as a detailed description of their fields of application can be found in Chapter 3.

2.4.2   Microelectromechanical Systems

MEMSs work with the same functional principle as semiconductor varactor diodes. In contrast to semiconductors, the change of capacitance is generated by mechanical movement of electrodes. By generating a galvanic contact, the MEMS can actually work as a switch. Several MEMS switches were developed by many different research laboratories and companies in the late 1990s and early 2000s. MEMS devices come with the main drawback of requiring hermetic packaging and of wear-out failures. The main pros are low insertion loss, high linearity, and high-power handling, which are far superior when compared to PIN- and metal–oxide–semiconductor field-effect transistor (MOSFET)-based switches. As costs are much higher for MEMS compared to classical semiconductor technologies, they are implemented only at places where cost can be justified by superior performance. Especially at high frequencies, MEMS become a relevant option. There are numerous implementations and different technologies on the market to realize MEMS. One very simple example of a cantilever-based ohmic MEMS switch is shown in Figure 2.52.

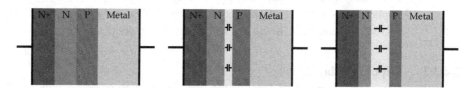

Figure 2.51 Operation principle of a semiconductor-based varactor: (left) doping profile (not to scale); (center) low-bias state with narrow depletion zone resulting in high capacitance; and (right) high-bias state with wide depletion zone resulting in low capacitance value.

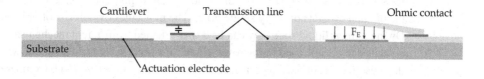

Figure 2.52 Operation principle of a cantilever MEMS switch.

Details on the realization of MEMSs, their respective performance, as well as a detailed description of their fields of application can be found in Chapter 4.

2.4.3   Functional Materials

For functional materials, the permittivity is a function of applied bias fields. For most applications, electric biasing is favorable because of ease of realization:

$$\varepsilon = \varepsilon_0 \varepsilon_r = \varepsilon_0 \cdot \text{fct}(E) \qquad (2.74)$$

In specific materials such as yttrium iron garnet (YIG) and ferrites, magnetic biasing is used. Owing to the weak coupling of magnetic to electric fields, the linearity of such materials is very high. The main drawback is the generation of variable magnetic bias fields, as electromagnets must be used to generate the fields. Compared to electrical biasing, this requires much more energy and leads to bulky solutions. Therefore, these kinds of materials are employed only where the energy consumption is not of critical importance or the device performance justifies an increased energy consumption.

2.4.3.1   Atomic Polarization

In the class of ferroelectric material, BST is one of the most investigated materials so far. A comprehensive overview was published in [87]. The crystal lattice of BST is composed in such a way that the $Ti4^+$ ion can freely be moved within the unit cell. By means of applying an external tuning field, the polarization and hence the permittivity of the material can be changed. As the material is still an insulator, no DC current flows and hence the material can be tuned in a passive way. The major drawback of BST is high loss, which hinders the realization of high-frequency devices. There are few examples for the implementation of BST at 60 GHz [93] but the main field of application is below 10 GHz. As this material plays only a very small role in mm-wave communication, it will not be discussed in more detail.

2.4.3.2   Dipolar Polarization

The main advantage of ferroelectric materials is high speed, as only small ions are elongated by the tuning bias field. But larger molecules can also be used for the realization of tunable microwave devices. As larger molecules will react more slowly on the bias field, the losses at mm-waves are also expected to be lower than for ferroelectrics. The most prominent material of this class, liquid crystal, has been in use for several decades, in particular in liquid crystal displays (LCDs), to switch between the on- and off-state of every single pixel. The molecules exhibit anisotropic behavior. Owing to their liquid state they can be easily oriented by different mechanisms. Liquid crystals can also be used in microwave and mm-wave systems in a similar way, when the anisotropy is utilized to generate tunable components [100, 101]. Compared to all other techniques, LCs have the unique property to be also applicable at higher frequencies with proven properties up to 8 THz [102]. Different possibilities of implementing tunable LC-based components and their respective performance can be found in Chapter 5.

2.4.3.3 Spintronics

YIG is a ferrite material that resonates at microwave frequencies when immersed in a DC magnetic field. This resonance is directly proportional to the strength of the applied magnetic field and has very linear tuning over multioctave microwave frequencies. By proper design, the coupling can be utilized for tunable filters. These filters are used for their high Q-factors, typically between 100 and 200, and can be tuned from 3 GHz up to 50 GHz. In such a way, they are often used as preselectors in spectrum analyzers.

References

1. International Telecommunication Union (ITU), "Attenuation by atmospheric gases," standard ITU-R P.676-11, 2016.
2. T. S. Rappaport, R. W. Heath, R. C. Daniels, and J. N. Murdock, *Millimeter Wave Wireless Communications.* Upper Saddle River, NJ: Prentice-Hall, 2014.
3. D. Dawn, P. Sen, S. Sarkar, B. Perumana, S. Pinel, and J. Laskar, "60-GHz Integrated transmitter development in 90-nm CMOS," *IEEE Transactions on Microwave Theory and Techniques,* vol. 57, pp. 2354–2367, 2009.
4. D. Zhao, S. Kulkarni, and P. Reynaert, "A 60-GHz outphasing transmitter in 40-nm CMOS," *IEEE Journal of Solid-State Circuits,* vol. 47, pp. 3172–3183, 2012.
5. K. Okada, R. Minami, Y. Tsukui, S. Kawai, Y. Seo, S. Sato, et al., "20.3 A 64-QAM 60GHz CMOS transceiver with 4-channel bonding," in *2014 IEEE International Solid-State Circuits Conference Digest of Technical Papers (ISSCC),* 2014, pp. 346–347.
6. T. LaRocca, Y. Wu, K. Thai, R. Snyder, N. Daftari, O. Fordham, et al., "A 64QAM 94GHz CMOS transmitter SoC with digitally-assisted power amplifiers and thru-silicon waveguide power combiners," in *2014 IEEE Radio Frequency Integrated Circuits Symposium,* 2014, pp. 295–298.
7. P. Wu, Y. Liu, B. Hanafi, H. Dabag, P. Asbeck, and J. Buckwalter, "A 45-GHz Si/SiGe 256-QAM transmitter with digital predistortion," in *2015 IEEE MTT-S International Microwave Symposium,* 2015, pp. 1–3.
8. D. Zhao and P. Reynaert, "A 40 nm CMOS E-band transmitter with compact and symmetrical layout floor-plans," *IEEE Journal of Solid-State Circuits,* vol. 50, pp. 2560–2571, 2015.
9. K. Khalaf, V. Vidojkovic, K. Vaesen, M. Libois, G. Mangraviti, V. Szortyka, et al., "Digitally modulated CMOS polar transmitters for highly-efficient mm-wave wireless communication," *IEEE Journal of Solid-State Circuits,* vol. 51, pp. 1579–1592, 2016.
10. U. Kodak and G. M. Rebeiz, "A 5G 28-GHz common-leg T/R front-end in 45-nm CMOS SOI with 3.7-dB NF and −30-dBc EVM with 64-QAM/500-MBaud modulation," *IEEE Transactions on Microwave Theory and Techniques,* vol. 67, pp. 318–331, 2019.
11. W. Tai, L. R. Carley, and D. S. Ricketts, "A 0.7W fully integrated 42GHz power amplifier with 10% PAE in 0.13µm SiGe BiCMOS," in *2013 IEEE International Solid-State Circuits Conference Digest of Technical Papers,* 2013, pp. 142–143.

12. A. Agah, J. A. Jayamon, P. M. Asbeck, L. E. Larson, and J. F. Buckwalter, "Multi-drive stacked-FET power amplifiers at 90 GHz in 45 nm SOI CMOS," *IEEE Journal of Solid-State Circuits,* vol. 49, pp. 1148–1157, 2014.
13. D. Zhao and P. Reynaert, "A 40-nm CMOS E-band 4-way power amplifier with neutralized bootstrapped cascode amplifier and optimum passive circuits," *IEEE Transactions on Microwave Theory and Techniques,* vol. 63, pp. 4083–4089, 2015.
14. M. Babaie, R. B. Staszewski, L. Galatro, and M. Spirito, "A wideband 60 GHz class-E/F$_2$ power amplifier in 40nm CMOS," in *2015 IEEE Radio Frequency Integrated Circuits Symposium (RFIC)*, 2015, pp. 215–218.
15. S. Shakib, H. Park, J. Dunworth, V. Aparin, and K. Entesari, "20.6 A 28GHz efficient linear power amplifier for 5G phased arrays in 28nm bulk CMOS," in *2016 IEEE International Solid-State Circuits Conference (ISSCC)*, 2016, pp. 352–353.
16. B. Welp, K. Noujeim, and N. Pohl, "A wideband 20 to 28 GHz signal generator MMIC with 30.8 dBm output power based on a power amplifier cell with 31% PAE in SiGe," *IEEE Journal of Solid-State Circuits,* vol. 51, pp. 1975–1984, 2016.
17. S. R. Helmi, J. Chen, and S. Mohammadi, "High-efficiency microwave and mm-wave stacked cell CMOS SOI power amplifiers," *IEEE Transactions on Microwave Theory and Techniques,* vol. 64, pp. 2025–2038, 2016.
18. J. A. Jayamon, J. F. Buckwalter, and P. M. Asbeck, "A PMOS mm-wave power amplifier at 77 GHz with 90 mW output power and 24% efficiency," in *2016 IEEE Radio Frequency Integrated Circuits Symposium (RFIC)*, 2016, pp. 262–265.
19. K. Datta and H. Hashemi, "75–105 GHz switching power amplifiers using high-breakdown, high-f$_{max}$multi-port stacked transistor topologies," in *2016 IEEE Radio Frequency Integrated Circuits Symposium (RFIC)*, 2016, pp. 306–309.
20. S. Y. Mortazavi and K. Koh, "A 43% PAE inverse Class-F power amplifier at 39–42 GHz with a λ/4-transformer based harmonic filter in 0.13-μm SiGe BiCMOS," in *2016 IEEE MTT-S International Microwave Symposium (IMS)*, 2016, pp. 1–4.
21. International Telecommunication Union (ITU), "Specific attenuation model for rain for use in prediction methods," standard ITU-R P.838-3, 2005.
22. International Telecommunication Union (ITU), "Attenuation due to clouds and fog," standard ITU-R P.840-7, 2017.
23. International Telecommunication Union (ITU), "Characteristics of precipitation for propagation modelling," standard ITU-R P.837-6, 2012.
24. L. J. Chu, "Physical limitations of omni-directional antennas," *Journal of Applied Physics,* vol. 19, pp. 1163–1175, 1948.
25. C. A. Balanis, *Antenna Theory Analysis and Design*, 4th ed. Hoboken, NJ: John Wiley & Sons, 2016.
26. L. Joseffsson and P. Persson, *Conformal Array Antenna Theory and Design*. Piscataway, NJ: IEEE Press, 2006.
27. R. J. Allard, D. H. Werner, and P. L. Werner, "Radiation pattern synthesis for arrays of conformal antennas mounted on arbitrarily-shaped three-dimensional platforms using genetic algorithms," *IEEE Transactions on Antennas and Propagation,* vol. 51, pp. 1054–1062, 2003.
28. J. Lei, J. Yang, X. Chen, Z. Zhang, G. Fu, and Y. Hao, "Experimental demonstration of conformal phased array antenna via transformation optics," *Scientific Reports,* vol. 8, p. 3807, 2018/02/28 2018.
29. S. Applebaum, "Adaptive arrays," *IEEE Transactions on Antennas and Propagation,* vol. 24, pp. 585–598, 1976.

30. B. Widrow and J. McCool, "A comparison of adaptive algorithms based on the methods of steepest descent and random search," *IEEE Transactions on Antennas and Propagation,* vol. 24, pp. 615–637, 1976.

31. B. Murmann.. *ADC Performance Survey.* Available at: http://web.stanford.edu/~mur mann/adcsurvey.html

32. M. Oppermann and R. Rieger, "Multifunctional MMICs – key enabler for future AESA panel arrays," in *2018 IMAPS Nordic Conference on Microelectronics Packaging (NordPac),* 2018, pp. 77–80.

33. P. N. Drackner and B. Engstrom, "An active antenna demonstrator for future AESA-systems," in *IEEE International Radar Conference, 2005,* 2005, pp. 226–231.

34. T. Chaloun, W. Menzel, F. Tabarani, T. Purtova, H. Schumacher, M. Kaynak, et al., "Wide-angle scanning active transmit/receive reflectarray," IET Microwaves, *Antennas & Propagation,* vol. 8, pp. 811–818, 2014.

35. D. F. Sievenpiper, J. H. Schaffner, H. J. Song, R. Y. Loo, and G. Tangonan, "Two-dimensional beam steering using an electrically tunable impedance surface," *IEEE Transactions on Antennas and Propagation,* vol. 51, pp. 2713–2722, 2003.

36. J. Huang and J. A. Encinar, *Reflectarray Antennas.* Piscataway, NJ: IEEE Press, 2007.

37. S. Bildik, S. Dieter, C. Fritzsch, M. Frei, C. Fischer, W. Menzel, et al., "Reconfigurable liquid crystal reflectarray with extended tunable phase range," in *2011 8th European Radar Conference,* 2011, pp. 404–407.

38. S. Bildik, S. Dieter, C. Fritzsch, W. Menzel, and R. Jakoby, "Reconfigurable folded reflectarray antenna based upon liquid crystal technology," *IEEE Transactions on Antennas and Propagation,* vol. 63, pp. 122–132, 2015.

39. S. V. Hum and J. Perruisseau-Carrier, "Reconfigurable reflectarrays and array lenses for dynamic antenna beam control: A review," *IEEE Transactions on Antennas and Propagation,* vol. 62, pp. 183–198, 2014.

40. J. J. Lee, "G/T and noise figure of active array antennas," *IEEE Transactions on Antennas and Propagation,* vol. 41, pp. 241–244, 1993.

41. D. M. Pozar, *Microwave Engineering,* 4th ed. Hoboken, NJ: John Wiley & Sons, 2011.

42. M. Nickel, O. H. Karabey, M. Maasch, R. Reese, M. Jost, C. Damm, et al., "Analysis of hybrid-passive-active phased array configurations based on an SNR approximation," in *European Conference on Antennas and Propagation (EUCAP),* 2017, pp. 852–856.

43. National Institute of Standards and Technology (NIST), "Guide for the use of the International System of Units (SI)," vol. NIST Special Publication 811, ed, 2008.

44. H. W. Schüßler and P. Steffen, "Halfband filters and Hilbert transformers," *Circuits, Systems and Signal Processing,* vol. 17, pp. 137–164, 1998/03/01 1998.

45. A. G. Fox, "An adjustable wave-guide phase changer," *Proceedings of the IRE,* vol. 35, pp. 1489–1498, 1947.

46. R. C. Hansen, *Microwave Scanning Antennas.* Westport, CT: Peninsula Publications, 1986.

47. L. Stark, "A helical line scanner for beam steering a linear array," *IRE Transactions on Antennas and Propagation,* vol. 5, pp. 211–216, 1957.

48. F. Reggia and E. G. Spencer, "A new technique in ferrite phase shifting for beam scanning of microwave antennas," *Proceedings of the IRE,* vol. 45, pp. 1510–1517, 1957.

49. J. F. White, "High power, p-i-n diode controlled, microwave transmission phase shifters," *IEEE Transactions on Microwave Theory and Techniques,* vol. 13, pp. 233–242, 1965.

50. J. F. White, "Review of semiconductor microwave phase shifters," *Proceedings of the IEEE,* vol. 56, pp. 1924–1931, 1968.

51. G. L. Matthaei, L. Young, and E. M. T. Jones, *Microwave Filters, Impedance-Matching Networks, and Coupling Structures.* New York: McGraw-Hill, 1964; Repr. 1980.

52. J. Uher, J. Bornemann, and U. Rosenberg, *Waveguide Components for Antenna Feed Systems Theory and CAD.* Norwood, MA: Artech House, 1993.

53. R. J. Cameron, "Advanced filter synthesis," *IEEE Microwave Magazine,* vol. 12, pp. 42–61, 2011.

54. R. J. Cameron, "Advanced coupling matrix synthesis techniques for microwave filters," *IEEE Transactions on Microwave Theory and Techniques,* vol. 51, pp. 1–10, 2003.

55. A. L. C. Serrano, F. S. Correra, T. Vuong, and P. Ferrari, "Synthesis methodology applied to a tunable patch filter with independent frequency and bandwidth control," *IEEE Transactions on Microwave Theory and Techniques,* vol. 60, pp. 484–493, 2012.

56. R. R. Mansour, "High-Q tunable dielectric resonator filters," *IEEE Microwave Magazine,* vol. 10, pp. 84–98, 2009.

57. C. Arnod, J. Parlebas, and T. Zwick, "Center frequency and bandwidth tunable waveguide bandpass filter with transmission zeros," in *2015 10th European Microwave Integrated Circuits Conference (EuMIC),* 2015, pp. 369–372.

58. B. Yassini, M. Yu, and B. Keats, "A Ka-band fully tunable cavity filter," *IEEE Transactions on Microwave Theory and Techniques,* vol. 60, pp. 4002–4012, 2012.

59. J. Brank, J. Yao, M. Eberly, A. Malczewski, K. Varian, and C. Goldsmith, "RF MEMS-based tunable filters," *International Journal of RF and Microwave Computer-Aided Engineering,* vol. 11, pp. 276–284, 2001.

60. K. Entesari and G. M. Rebeiz, "A 12-18-GHz three-pole RF MEMS tunable filter," *IEEE Transactions on Microwave Theory and Techniques,* vol. 53, pp. 2566–2571, 2005.

61. U. Rosenberg, R. Beyer, P. Krauß, T. Sieverding, A. Papanastasiou, M. Pueyo-Tolosa, et al., "Novel remote controlled dual mode filter providing flexible re-allocation of center frequency and bandwidth," in *2016 IEEE MTT-S International Microwave Symposium (IMS),* 2016, pp. 1–3.

62. U. Rosenberg, R. Beyer, P. Krauß, T. Sieverding, P. M. Iglesias, and C. Ernst, "Remote controlled high-Q cavity filters providing center frequency and bandwidth re-allocation," in *2017 IEEE MTT-S International Microwave Workshop Series on Advanced Materials and Processes for RF and THz Applications (IMWS-AMP),* 2017, pp. 1–3.

63. R. V. Synder, "Generalized cross-coupled filters using evanescent mode coupling elements," in *1997 IEEE MTT-S International Microwave Symposium Digest,* vol. 2, pp. 1095–1098, 1997.

64. S. Amari and G. Macchiarella, "Synthesis of inline filters with arbitrarily placed attenuation poles by using nonresonating nodes," *IEEE Transactions on Microwave Theory and Techniques,* vol. 53, pp. 3075–3081, 2005.

65. C. Schuster, A. Wiens, F. Schmidt, M. Nickel, M. Schüßler, R. Jakoby, et al., "Performance analysis of reconfigurable bandpass filters with continuously tunable center frequency and bandwidth," *IEEE Transactions on Microwave Theory and Techniques,* vol. 65, pp. 4572–4583, 2017.

66. K. Chen, X. Liu, and D. Peroulis, "Widely tunable high-efficiency power amplifier with ultra-narrow instantaneous bandwidth," *IEEE Transactions on Microwave Theory and Techniques,* vol. 60, pp. 3787–3797, 2012.

67. J. Fu and A. Mortazawi, "Improving power amplifier efficiency and linearity using a dynamically controlled tunable matching network," *IEEE Transactions on Microwave Theory and Techniques,* vol. 56, pp. 3239–3244, 2008.

68. Z. Haitao, G. Huai, and L. Guann-Pyng, "Broad-band power amplifier with a novel tunable output matching network," *IEEE Transactions on Microwave Theory and Techniques,* vol. 53, pp. 3606–3614, 2005.

69. H. M. Nemati, C. Fager, U. Gustavsson, R. Jos, and H. Zirath, "Design of varactor-based tunable matching networks for dynamic load modulation of high power amplifiers," *IEEE Transactions on Microwave Theory and Techniques,* vol. 57, pp. 1110–1118, 2009.

70. C. Sánchez-Pérez, M. Özen, C. M. Andersson, D. Kuylenstierna, N. Rorsman, and C. Fager, "Optimized design of a dual-band power amplifier with SiC varactor-based dynamic load modulation," *IEEE Transactions on Microwave Theory and Techniques,* vol. 63, pp. 2579–2588, 2015.

71. A. Tombak, "A ferroelectric-capacitor-based tunable matching network for quad-band cellular power amplifiers," *IEEE Transactions on Microwave Theory and Techniques,* vol. 55, pp. 370–375, 2007.

72. Y. Yoon, H. Kim, Y. Park, M. Ahn, C. Lee, and J. Laskar, "A high-power and highly linear CMOS switched capacitor," *IEEE Microwave and Wireless Components Letters,* vol. 20, pp. 619–621, 2010.

73. S. Preis, A. Wiens, H. Maune, W. Heinrich, R. Jakoby, and O. Bengtsson, "Reconfigurable package integrated 20 W RF power GaN HEMT with discrete thick-film MIM BST varactors," *Electronics Letters,* vol. 52, pp. 296–298, 2016.

74. S. Preis, N. Wolff, F. Lenze, A. Wiens, R. Jakoby, W. Heinrich, et al., "Load tuning assisted discrete-level supply modulation using BST and GaN devices for highly efficient power amplifiers," in *2018 IEEE/MTT-S International Microwave Symposium - IMS,* 2018, pp. 1230–1233.

75. Y. Chen, R. Martens, R. Valkonen, and D. Manteuffel, "Evaluation of adaptive impedance tuning for reducing the form factor of handset antennas," *IEEE Transactions on Antennas and Propagation,* vol. 63, pp. 703–710, 2015.

76. C. Dong-Hyuk and P. Seong-Ook, "A varactor-tuned active-integrated antenna using slot antenna," *IEEE Antennas and Wireless Propagation Letters,* vol. 4, pp. 191–193, 2005.

77. J. Park, Y. Tak, Y. Kim, Y. Kim, and S. Nam, "Investigation of adaptive matching methods for near-field wireless power transfer," *IEEE Transactions on Antennas and Propagation,* vol. 59, pp. 1769–1773, 2011.

78. N. J. Smith, C. Chen, and J. L. Volakis, "An improved topology for adaptive agile impedance tuners," *IEEE Antennas and Wireless Propagation Letters,* vol. 12, pp. 92–95, 2013.

79. I. Vasilev, V. Plicanic, and B. K. Lau, "Impact of antenna design on MIMO performance for compact terminals with adaptive impedance matching," *IEEE Transactions on Antennas and Propagation,* vol. 64, pp. 1454–1465, 2016.

80. B. Wang, H. Gao, R. v. Dommele, M. K. Matters-Kammerer, and P. G. M. Baltus, "A 60 GHz low noise variable gain amplifier with small noise figure and IIP3 variation in a 40-nm CMOS technology," in *2018 IEEE MTT-S International Wireless Symposium (IWS),* 2018, pp. 1–4.

81. D. Siao, J. Kao, and H. Wang, "A 60 GHz low phase variation variable gain amplifier in 65 nm CMOS," *IEEE Microwave and Wireless Components Letters,* vol. 24, pp. 457–459, 2014.

82. Avago Technologies, "A low-cost surface mount PIN diode π attenuator," Application Note 1048, 2010.

83. A. E. Prasetiadi, M. Jost, B. Schulz, M. Quibeldey, T. Rabe, R. Follmann, et al., "Liquid-crystal-based amplitude tuner fabricated in LTCC technology," in *2017 47th European Microwave Conference (EuMC)*, 2017, pp. 1085–1088.

84. M. Wang, C. Trlica, M. R. Khan, M. D. Dickey, and J. J. Adams, "A reconfigurable liquid metal antenna driven by electrochemically controlled capillarity," *Journal of Applied Physics*, vol. 117, p. 194901, 2015.

85. B. L. Cumby, D. B. Mast, C. E. Tabor, M. D. Dickey, and J. Heikenfeld, "Robust pressure-actuated liquid metal devices showing reconfigurable electromagnetic effects at GHz frequencies," *IEEE Transactions on Microwave Theory and Techniques*, vol. 63, pp. 3122–3130, 2015.

86. R. C. Gough, A. M. Morishita, J. H. Dang, W. Hu, W. A. Shiroma, and A. T. Ohta, "Continuous electrowetting of non-toxic liquid metal for RF applications," *IEEE Access*, vol. 2, pp. 874–882, 2014.

87. S. Gevorgian, *Ferroelectrics in Microwave Devices, Circuits and Systems Physics, Modeling, Fabrication and Measurements*. London: Springer, 2009..

88. H. Jiang, B. Lacroix, K. Choi, Y. Wang, A. T. Hunt, and J. Papapolymerou, "Ka- and Ku-Band tunable bandpass filters using ferroelectric capacitors," *IEEE Transactions on Microwave Theory and Techniques*, vol. 59, pp. 3068–3075, 2011.

89. G. Velu, K. Blary, L. Burgnies, A. Marteau, G. Houzet, D. Lippens, et al., "A 360°BST phase shifter with moderate bias voltage at 30 GHz," *IEEE Transactions on Microwave Theory and Techniques*, vol. 55, pp. 438–444, 2007.

90. Z. Zhao, X. Wang, K. Choi, C. Lugo, and A. T. Hunt, "Ferroelectric phase shifters at 20 and 30 GHz," *IEEE Transactions on Microwave Theory and Techniques*, vol. 55, pp. 430–437, 2007.

91. S. Gevorgian, "Agile microwave devices," *IEEE Microwave Magazine*, vol. 10, pp. 93–98, 2009.

92. D. Kuylenstierna, A. Vorobiev, and S. Gevorgian, "40 GHz lumped element tunable bandpass filters with transmission zeros based on thin $Ba_{0.25}Sr_{0.75}TiO_3$ (BST) film varactors," in *Digest of Papers. 2006 Topical Meeting on Silicon Monolithic Integrated Circuits in RF Systems*, 2006, pp. 342–345.

93. R. D. Paolis, F. Coccetti, S. Payan, M. Maglione, and G. Guegan, "Characterization of ferroelectric BST MIM capacitors up to 65 GHz for a compact phase shifter at 60 GHz," in *2014 44th European Microwave Conference*, 2014, pp. 492–495.

94. H. Bea, M. Bibes, G. Herranz, X. Zhu, S. Fusil, K. Bouzehouane, et al., "Integration of multiferroic $BiFeO_3$ thin films into heterostructures for spintronics," *IEEE Transactions on Magnetics*, vol. 44, pp. 1941–1945, 2008.

95. L. A. Makarova, V. V. Rodionova, Y. A. Alekhina, T. S. Rusakova, A. S. Omelyanchik, and N. S. Perov, "New multiferroic composite materials consisting of ferromagnetic, ferroelectric, and polymer components," *IEEE Transactions on Magnetics*, vol. 53, pp. 1–7, 2017.

96. L. Malkinski, M. McGehee, T. Gould, R. Eskandari, and A. Chalastaras, "Phase control of magnetic susceptibility of multiferroic composites," *IEEE Transactions on Magnetics*, vol. 51, pp. 1–3, 2015.

97. M. Vopsaroiu, M. Stewart, T. Fry, M. Cain, and G. Srinivasan, "Tuning the magneto-electric effect of multiferroic composites via crystallographic texture," *IEEE Transactions on Magnetics*, vol. 44, pp. 3017–3020, 2008.

98. X. Yang, Y. Gao, J. Wu, S. Beguhn, T. Nan, Z. Zhou, et al., "Dual H- and E-field tunable multiferroic bandpass filter at Ku-band using partially magnetized spinel ferrites," *IEEE Transactions on Magnetics,* vol. 49, pp. 5485–5488, 2013.

99. I. n. i. Gutiérrez, J. Meléndez, and E. Hernández. "Design and characterization of integrated varactors for RF applications." Available at: https://onlinelibrary.wiley.com/doi/book/10.1002/9780470035924

100. H. Maune, M. Jost, R. Reese, E. Polat, M. Nickel, and R. Jakoby, "Microwave liquid crystal technology," *Crystals,* vol. 8, 2018.

101. D. C. Zografopoulos, A. Ferraro, and R. Beccherelli, "Liquid-crystal high-frequency microwave technology: Materials and characterization," *Advanced Materials Technologies,* vol. 0, p. 1800447, 2018/12/13 2018.

102. C. Weickhmann, R. Jakoby, E. Constable, and R. A. Lewis, "Time-domain spectroscopy of novel nematic liquid crystals in the terahertz range," in *2013 38th International Conference on Infrared, Millimeter, and Terahertz Waves (IRMMW-THz),* 2013, pp. 1–2.

3   CMOS and BiCMOS Technologies

Philippe Ferrari, Sylvain Bourdel, Alfredo Bautista, and Thomas Quémerais

3.1   Introduction

The purpose of this chapter is to describe tunable circuits in complementary metal–oxide–semiconductor (CMOS) and bipolar CMOS (BiCMOS) technologies. First, a brief description of active and passive basic components available in these technologies is carried out. MOS and bipolar transistors are first described, followed by the description of passive components such as metal–oxide–metal (MOM) capacitors and transmission lines. Slow-wave transmission lines are described and compared to their microstrip lines counterparts, in terms of electrical performance and footprint. Next, tunable components are introduced, first varactors and switches, and then digital tunable capacitors, which could become an efficient approach in the future. Then, tunable transmission lines are described. Some examples of tunable inductances are also highlighted. Finally, two families of tunable circuits are addressed, phase shifter and the voltage-controlled oscillators, respectively. Both circuits are used in many systems, in beam-forming/steering concerning phase shifters, and in transceivers concerning voltage-controlled oscillators, respectively. For these two families of circuits, many design examples are given. Topologies based on the use of lumped components (varactors and inductances) are compared to those based on distributed components (tunable transmission lines). In the middle, hybrid approaches mixing lumped and distributed components can lead to very efficient solutions, both in terms of electrical performance and footprint. Many hybrid topologies are also described.

All the developments presented in this chapter can be used for the design of many other tunable circuits such as impedance tuners (briefly described in terms of perspectives), power dividers, and couplers.

3.2   Fundamentals of CMOS and BiCMOS Technologies for Millimeter Waves

3.2.1   Back-End-of-Line of Current CMOS/BiCMOS Technologies

3.2.1.1   Technologies with High Density of Integration

Technologies with high density of integration were originally optimized for digital applications. The increasing need for logical standard cells to design the digital functions has contributed to integrating several levels of metals to enable the transistor

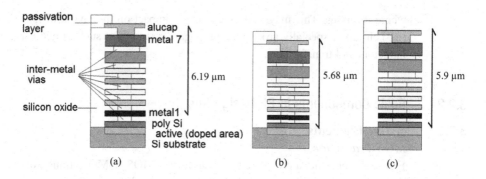

Figure 3.1 BEOL in technologies. (a) CMOS 65 nm. (b) CMOS 40 nm. (c) CMOS 28 nm.

interconnections. The Moore law (1965) has predicted such a density of integration. The increasing demand in terms of number of transistors inside the digital functions increases the demand in terms of integration. The increase of transistor performance and integration is linked to two effects: decrease of the transistor's gate length and decrease of the metal interconnection height (back end of line – [BEOL]). These two effects lead to a global reduction (shrink) on each dimension level. Unfortunately, this decrease of the BEOL height leads to a decrease the quality factors of the passive elements. For instance, for a given characteristic impedance, the width of the strip of a microstrip line decreases dramatically with the BEOL height, thus leading to an increase of attenuation constant. This behavior cannot be compensated by the increase of the level height of the higher metals, especially at mm-waves, since the skin-effect limits the current penetration to much less than 1 μm.

Figure 3.1 shows the technological evolution of the BEOL, from 65 nm down to 28 nm CMOS node.

The metal levels available in the CMOS technologies 65 nm, 40 nm, and 28 nm technologies, presented in Figure 3.1, are all constituted of five thin levels of copper (metal1 to metal5), two thick levels of copper (metal6 and metal7), and a top aluminum level called alucap. These metal levels are connected by copper vias. The intermediate dielectric layers are constituted by silicon oxide and by silicon nitride. The replacement of the aluminum by the copper in the interconnections leads to a lower resistivity and hence lower losses, but it requires the use of a damascene process [1]. This process allows depositing the copper in an area isolated by a barrier in an alloy of tantalum. This allows avoiding any contamination of the wafer by copper atoms. Indeed, their low atomic sections make them diffuse very easily in the silicon.

The consequence of using these technologies with a high density of integration is the respect of the density of metal. Indeed, for each of these levels, their density must be between 20% and 80% in windows of 50 μm by 50 μm in CMOS 65 nm and 20 μm by 20 μm in 28-nm CMOS fully depleted silicon on insulator (FDSOI), respectively. These rules allow avoiding any problem of erosion or cup shape due to the planarization done by the chemical mechanical polishing process (CMP). To respect these densities, it is thus necessary to make holes in the full levels of metal and to add metal where it misses. This may have a significant impact on the layout of radiofrequency (RF) circuits, especially around the propagation structures such as microstrip lines or inductors and transformers.

3.2.2   Basic Components in RF Design Kits

3.2.2.1   Active Elements

MOS Transistor

Figure 3.2 presents a cross section of an n-type MOS (NMOS) transistor. The drain-source current I_{ds} crossing the channel between the drain and the source is modulated by the vertical electric field applied to the transistor gate through the V_{gs} voltage, which manages the charge density in the channel. On the other hand, this current is modulated by the horizontal field applied between the drain and the source through the V_{ds} voltage, which manages the speed of the carriers mainly in the linear zone.

The drain current I_{ds} is a function of the physical and geometrical parameters of the transistor as well as the voltages V_{gs} and V_{ds}. In first approximation, this current can be expressed in the linear zone (low V_{ds}, i.e., $V_{ds} < V_{gs} - V_{th}$) by the relation

$$I_{ds} = \mu \cdot \frac{\varepsilon_{ox}}{t_{ox}} \cdot \frac{w}{l} \cdot \left((V_{gs} - V_{th}) \cdot V_{ds} - \frac{V_{ds}^2}{2} \right), \tag{3.1}$$

and in the saturation regime ($V_{ds} > V_{gs} - V_{th}$) by the relation

$$I_{ds} = \mu \cdot \frac{\varepsilon_{ox}}{t_{ox}} \cdot \frac{w}{l} \cdot (V_{gs} - V_{th})^2 (1 - \lambda V_{ds}), \tag{3.2}$$

where w is the width of the MOS channel, l its length, μ is the mobility of electrons, ε_{ox} the dielectric constant of the gate oxide, t_{ox} its thickness, V_{gs} and V_{ds} the gate-source and drain-source voltage, and V_{th} the transistor threshold voltage above which the transistor is in strong inversion regime, respectively.

These two expressions of the current allow to understand, according to the biasing of the transistor, what will be the main factors acting on the I_{ds} current value.

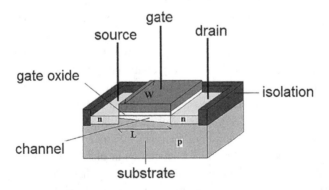

Figure 3.2 NMOS transistor cross section.

However, these expressions are quite simple and do not describe all the phenomena in the advanced CMOS technologies submicron transistors. Indeed, for the very short channel transistors complex physical phenomena appear, such as the modulation of the mobility by the vertical field, the saturation of the carrier velocity, the modulation of the channel, and the drain induce barrier lowering effect (DIBL). They must be considered to refine the modeling of the device, especially at high-frequency operation.

Hence intrinsic accurate models of MOS transistors are developed, which contain a very large number of parameters. They are optimized thanks to static and dynamic measurements of large sets of transistors with different dimensions. Today, two main models describing the operation of such devices, according to their physical and geometrical parameters and applied voltages, were developed: the surface-potential-based (PSP) [2] and the Berkeley Short-chmsnel IGFET Model (BSIM4) [3] models. The BSIM4 model was developed by the BSIM research group in the Department of Electrical Engineering and Computing Sciences at the University of California, Berkeley [4]. It is very close, in terms of performance, to the PSP model developed jointly by the University of Pennsylvania and Philips.

At low frequency, the intrinsic model of an MOS transistor can be modeled with a gate source capacitance C_{gs}, a gate drain capacitance C_{gd}, a drain source capacitance C_{ds}, and a drain source output resistor R_{ds} as shown in Figure 3.3. The main parameters are obtained as follows:

$$R_{ds} = \frac{\delta V_{ds}}{\delta I_{ds}}; \quad g_m = \frac{\delta I_{ds}}{\delta V_{gs}}; \quad C_{nm} = \frac{\delta Q_n}{\delta V_m}, \tag{3.3}$$

where C_{nm} is the capacitance seen between nodes n and m (drain, source, or gate), Q_n is the charge stored at node n, and V_m is the voltage at node m, respectively.

In the active region (saturation) the transconductance g_m can be approximated as follows:

$$g_m = \frac{2I_{ds}}{V_{gs} - V_t} = 2\sqrt{\mu \frac{\varepsilon_{rox}}{t_{ox}} \frac{w}{l} I_{ds}}. \tag{3.4}$$

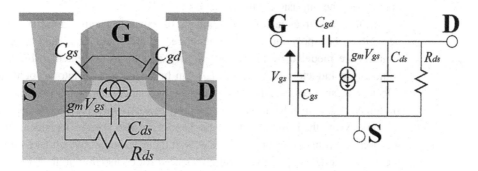

Figure 3.3 Intrinsic model of the MOS transistor at low frequency.

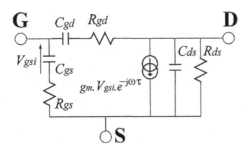

Figure 3.4 NQS model of the MOS transistor.

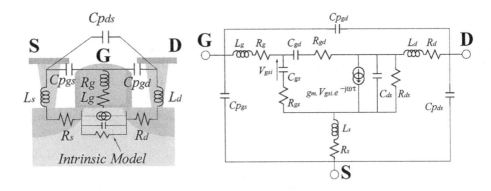

Figure 3.5 RF model of a MOS transistor with NQS effect.

At higher frequency such as in mm-waves, other effects must be introduced in the model. For instance, the response time of the channel charges to the control voltage of the gate cannot be considered as instantaneous anymore. Such effects are known as non-quasistatic (NQS) effects and can be modeled as described in Figure 3.4. R_{gd} and R_{gs} are added to introduce a delay in the charging of the capacitances C_{gd} and C_{gs}, respectively. Also, a delay τ is introduced in the transfer function between the drain current and the intrinsic control voltage v_{gsi}.

At mm-wave frequencies, the influence of extrinsic parameters is also no longer negligible. The polysilicon gate, the doped source, and drain and their metallic contacts are modeled by resistances R_g, R_s, and R_d as presented in Figure 3.5. In some cases, inductors (L_g, L_s, and L_d) can be added to take into account inductive effects in these accesses. Also, fringing capacitance between each port of the transistor should be added in an RF model ($Cp_{gs}, Cp_{gd}, Cp_{ds}$). All these parameters depend on the technology and the layout, but the capacitances are also voltage dependent. Mm-wave technologies propose accurate RF models in their design kit, which take into account the extrinsic effects up to the metal layers used for connection in the P-cell. All other connections added by the designer should be taken into account by the way of electromagnetic (EM) simulations.

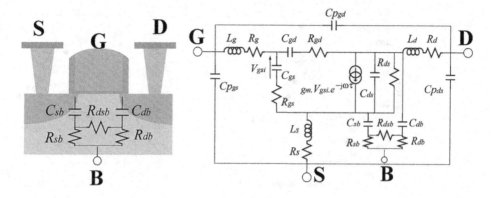

Figure 3.6 Bulk effects in the RF model of a MOS transistor.

The MOS transistor is physically a four-ports device and the substrate effects must be modeled when a bulk technology is used. As presented in Figure 3.6, these effects are modeled by the capacitances (C_{sb} and C_{sd}) of the N^{++}/P source-bulk and drain-bulk junctions. Resistors (R_{sb}, R_{db}, and R_{dsb}) are added to take into account the low resistivity of the substrate.

Three main RF MOS transistor figures of merit are defined, corresponding to the critical parameters of the MOS transistors dedicated to RF applications:

- The frequency of transition f_T
- The maximum oscillation frequency f_{max}
- The minimum noise factor NF_{min}

The expressions of these parameters are given below:

$$f_T = \frac{g_m}{2\pi C_{gs}\sqrt{1 + 2\dfrac{C_{gd}}{C_{gs}}}}, \tag{3.5}$$

$$f_{max} = \frac{g_m}{4\pi C_{gs}\sqrt{(R_g + R_s)\left(R_{ds} + g_m\dfrac{C_{gd}}{C_{gs}}\right)}}, \tag{3.6}$$

$$NF_{min} = 1 + K\frac{f}{f_T}\sqrt{g_m(R_g + R_s)}, \tag{3.7}$$

where g_m is the global transconductance of the transistor, C_{gs} and C_{gd} the gate-source and gate-drain capacitances, R_{ds} the dynamic drain-source resistance, R_s and R_g the source and gate resistances, K the Boltzmann constant, and f the working frequency, respectively.

The f_T and f_{max} frequencies are extrapolated from measurements. The frequency for which the value of the h_{21} current gain is equal to 1 defines the transition

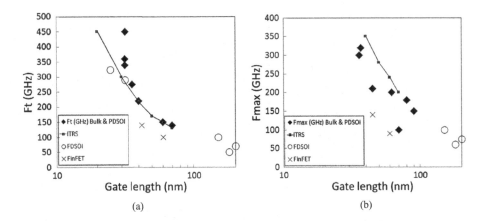

Figure 3.7 (a) f_T and (b) f_{max} of MOS transistors versus gate length.

frequency f_T. The maximal oscillation frequency f_{max} comes from an extrapolation of the unilateral Mason's gain [5]; it corresponds to a unitary gain.

The f_T and f_{max} of different transistors in several technologies versus the gate length are given in Figure 3.7. CMOS technologies can offer transistors with very high f_T and f_{max}. However, note that the MOS transistor with the maximum f_T value does not correspond to the one with the highest f_{max} value. MOS transistors can be optimized only to either get the highest f_T or highest f_{max}, other choices resulting from compromises.

The Bipolar Transistor

The bipolar transistor consists of two PN junctions giving three connecting terminals, the emitter (E), the base (B), and the collector (C). Bipolar transistors are current-regulated devices that control the amount of current flowing through them in proportion to the biasing voltage applied to their base terminal acting like a current-controlled switch. These devices are dedicated to high-frequency applications due to the very high carrier velocity in their base that leads to simultaneously high f_T and high f_{max}, as compared to the MOS transistors for which both figures of merit cannot be optimized simultaneously. Figure 3.8 shows a plot of f_{max} versus f_T for SiGe heterostructure bipolar transistors (HBTs) published in the literature. Bipolar transistors with $f_{max} = 400$ GHz and simultaneously $f_T = 300$ GHz are available.

Furthermore, bipolar transistors are biased on higher voltage than MOS transistors, which leads to a much higher compression point for an identical power gain at high frequency.

For example, in the STMicroelectronics BiCMOS 55-nm technology, SiGe HBTs feature a conventional double-polysilicon self-aligned (DPSA) architecture, for which emitter-base self-alignment is provided by the selective epitaxial growth (SEG) of the SiGe:C base. Three collector flavors, leading to different f_T/BV_{CE0} tradeoffs for the high-speed (HS), medium-voltage (MV), and high-voltage (HV) HBTs, are available.

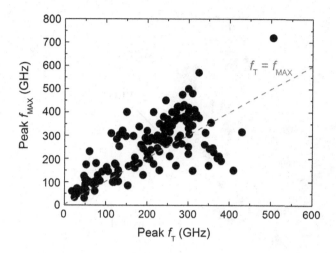

Figure 3.8 f_{max} versus f_T for SiGe HBT transistors.
Courtesy of Pascal Chevallier, STMicroelectronics, Crolles, France

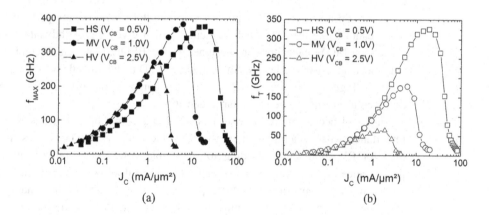

Figure 3.9 STMicroelectronics BiCMOS 55-nm bipolar transistor performance, f_{max} (a), and f_T (b).

The collector of the HS HBT is formed by a standard "buried layer + epitaxy + sinker/deep trenches/SIC" module. Emitters are scalable in width and length with a minimum area of 0.10×0.30 μm². The performance of the three SiGe HBT flavors is presented in Figure 3.9.

3.2.2.2 Passive Elements
MOM Capacitor
Integrated capacitors are critical elements of RF circuits. They should have a high capacitance density, a high quality factor (Q), and low parasitic series elements values. Three main integrated capacitors are currently used in RF: metal–oxide–metal (MOM)

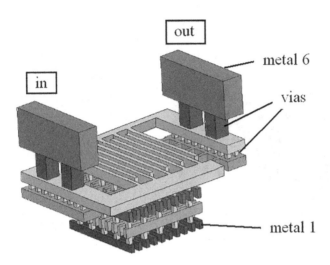

Figure 3.10 Cross section of a metal1/metal5 MOM capacitor.

capacitors, metal–insulator–metal (MIM) capacitors, and 3D trench capacitors. MOM capacitors do not need another process step like their MIM counterparts, or process modifications like the 3D trench capacitor, which gives a great advantage to these capacitors. MOM capacitors are designed in the BEOL, that is, interconnect levels with metal and insulator layers. An example of a 3D metal1/metal5 MOM capacitor is given in Figure 3.10. The metal fingers that compose the MOM capacitors were designed to maximize the capacitance density.

They still occupy more volume than MIM capacitors; however, this is of minor importance when dealing with mm-waves, since the capacitors' values are lower than for RF applications, leading to acceptable volumes. Moreover, even if the performance of these three types of capacitor are well known at low frequencies but rarely shown at mm-waves, MOM capacitors usually exhibit higher quality factors than for MIM capacitors at mm-waves, since the high-K oxides used for MIM capacitors have a higher loss tangent than the oxides used in the BEOL. Metallic losses are also greater for MIM capacitors, due to proximity effects, leading to high conductive losses.

The capacitance density performed with a metal1/metal5 MOM capacitor, as described in Figure 3.10, is 1 fF/μm^2. It seems to be preferable to use the whole stack of the BEOL to reach a higher quality factor. For instance, a 215 fF MOM capacitor performed with only metal5/metal6/metal7 exhibits a quality factor of 6.7 at 60 GHz, whereas a 1.15 pF MOM using all metal layers exhibits a quality factor higher than 10.2 at 60 GHz.

The electrical equivalent model of MOM capacitors is given in Figure 3.11. The substrate losses can be removed because of the very low capacitive coupling. This simple model can be easily used in computer-aided design software. It includes series elements R_0, L_0, and C_0, modeling the conductive and dielectric losses, the inductive part of the electrodes, and the intrinsic capacitance of the device, respectively.

Table 3.1 Extracted lumped elements of MOM capacitors with the metal stack, resonance frequency f_{res} and quality factor Q

Dens. (fF/μm²) // metals levels	C_0 (pF)	C_1 (fF)	R_0 (Ω)	L_0 (pH)	f_{res} (GHz)	Q @ 60 GHz
1 // M5/M7	0.129	1.97	0.4	3	160	3.2
1.1 // M5/M7	0.215	3.5	0.3	5	110	6.7
3.1 // M5/M6	0.629	37	0.45	1	80	9.5
3.2 // M1/M6	1.15	44	0.7	1.5	70	10.2
3.1 // M1/M6	5.6	195	1	7.4	25	×

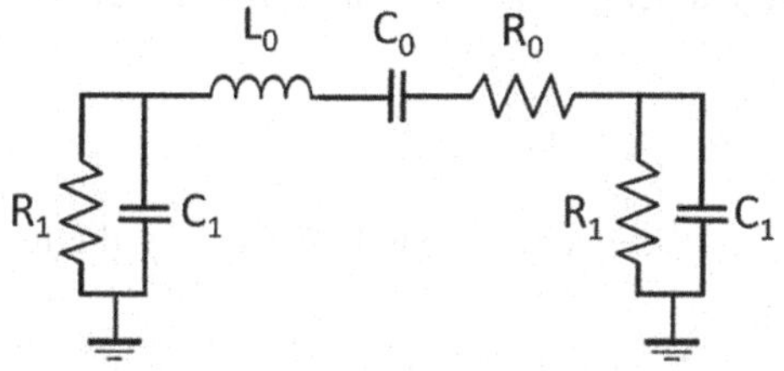

Figure 3.11 Equivalent electrical model of the MOM capacitor.

Two parallel capacitors C_1 are added to model the low parallel capacitive coupling. R_1 models the losses occurring in the bulk substrate.

The MOM quality factor Q can be extracted from measurement results using the relation

$$Q = \frac{1}{R_0 \cdot C_0 \cdot \omega},$$
(3.8)

where R_0 and C_0 are the series resistance and capacitance of the MOM capacitor, respectively.

Table 3.1 shows measured intrinsic MOM capacitors' lumped model elements, resonance frequency, and quality factor for different metal stack configurations and capacitance values. These MOM capacitors dedicated to mm-wave applications present high capacitance density from 1 to 3.2 fF/μm².

Transmission Lines

The propagation of the RF signal in an integrated circuit can be made via various structures: microstrip lines, coplanar waveguide (CPW), and their derivatives, transmission lines based on slow-wave concepts, such as the slow-wave CPW (S-CPW) or slow-wave coplanar stripline (S-CPS). To choose the optimal structure for a given application, many criteria can be considered. For interconnection purposes, the transmission line that offers the lowest attenuation constant, most of the time for a 50-Ω characteristic impedance, must be chosen. When dealing with transmission lines used to build circuits, such as matching networks, power dividers, couplers, or filters, one must consider the transmission line's quality factor, given by

$$Q = \frac{1}{2}\frac{\beta}{\alpha},$$ (3.9)

where α and β are the attenuation and propagation constants, in m^{-1}, respectively.

If the two terms of the expression of Q are multiplied by the transmission line's length l, it leads to

$$Q = \frac{1}{2}\frac{\beta \cdot l}{\alpha \cdot l} = \frac{1}{2}\frac{\theta}{\alpha \cdot l} = 4.34\frac{\theta}{IL},$$ (3.10)

where θ (in rad) is the electrical length and IL (in dB/m) is the insertion loss, expressed as

$$IL = 20 \cdot \log\left(e^{\alpha \cdot l}\right).$$ (3.11)

Hence, the quality factor Q gives the electrical length divided by the insertion loss, multiplied by a factor 4.34.

Circuits are built according to fixed electrical lengths, for instance, quarter-wavelength transmission lines to achieve classical power dividers or couplers, of half-wavelength transmission lines when dealing with coupled-lines couplers. Thus, to minimize the circuits' insertion loss, a designer may choose the transmission line that gives the highest quality factor Q, for a given characteristic impedance. As presented in the text that follows, in that case S-CPW are the most attractive, since they exhibit the highest quality factor, whatever the characteristic impedance. However, a second important parameter is the transmission line footprint, owing to the cost of the advanced technologies used for mm-wave applications. Finally, a tradeoff between electrical performance and footprint must be the right guide for circuit designers.

In the next two subsections, an overview of microstrip lines and S-CPWs is presented, and a brief comparison between these two types of transmission lines is carried out, to give some practical design rules. CPWs are not presented, since they cannot be used in standard bulk or FDSOI technologies, as they require high-resistivity substrates to avoid very high substrate losses.

Microstrip Lines
A classical microstrip line, with the definition of the main design parameters, is shown in Figure 3.12.

A very complete study was carried out in [6]. It compares the attenuation constant measurement of various microstrip lines, up to 40 GHz, with various ground planes.

As shown in Figure 3.13, the integrated microstrip line in a conventional CMOS/BiCMOS technology is composed of one alucap metal strip and a ground plane composed of metal1 and metal2 stacked layers. The use of the alucap layer permits to maximize the dielectric height. In that way, a given characteristic impedance is obtained for a larger strip width, thus decreasing the attenuation constant, even if the

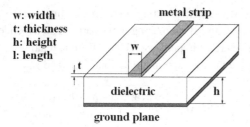

w: width
t: thickness
h: height
l: length

Figure 3.12 Microstrip line scheme showing geometrical parameters.

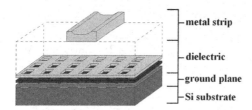

Figure 3.13 Scheme of an integrated alucap strip metal microstrip line with a metal1/metal2 ground plane.

alucap conductivity is lower as compared to copper metal layers. The ground plane includes holes that respect the metal density rules in such a way that the two metal layers shield completely the low-resistivity silicon substrate. The substrate between the metal strip and the ground plane is composed of silicon oxide.

The classical configuration shown in Figure 3.13 allows design of microstrip lines with a characteristic impedance range that depends on the BEOL height and metal density rules. For instance, characteristic impedances ranging from 15 Ω to 70 Ω can be achieved, in compliance with the CMOS 45 nm technology recommended rules. As already underlined, the limited height of the BEOL is the main limitation in terms of electrical performance. The higher the BEOL, the lower the attenuation constant and the higher the quality factor, since the strip width increases with the BEOL height, for a given characteristic impedance.

Three main analytical models are used to model the silicon-integrated microstrip lines. They were formulated by H. A. Wheeler [7], M. V. Schneider [8], and E. Hammerstad and Ø. Jensen [9], respectively. A study of these three models leads to choose the Hammerstad and Jensen one because it is valid for large w/h ratios and a broad range of dielectric constant values, ε_r. Furthermore, it fits well with the realistic conditions of integrated microstrip lines and provides the best accuracy.

The characteristic impedance $Z_{c0}(\varepsilon_r, w, h)$ and the effective relative dielectric constant $\varepsilon_{reff0}(\varepsilon_r, w, h)$, which depends only on the physical and geometrical parameters of the microstrip line, are established from [9]. Because the equations for Z_{c0} and ε_{reff0} assume the metal strip thickness as null ($t = 0$ µm), the strip width must be corrected. Indeed, growing the strip thickness is physically equivalent to enlarging the strip width so that w becomes $w + \Delta w$. Thus, modified formulas for the characteristic impedance and the effective relative constant applied to microstrip lines were established. Next, the dispersion was modeled in [8] by making these parameters dependent on the operation frequency. The characteristic impedance Z_c and effective relative constant ε_{reff} depending on the frequency f are then written in a function of Z_{c0} and ε_{reff0} as

$$Z_c(f,\varepsilon_r,w+\Delta w,h) = Z_{c0}(\varepsilon_r,w+\Delta wh) \cdot \sqrt{\frac{\varepsilon_{reff}(\varepsilon_r,w+\Delta w,h)}{\varepsilon_{reff}(f,\varepsilon_r,w+\Delta w,h)}} \cdot \frac{\varepsilon_{reff}(f,\varepsilon_r,w+\Delta w,h)-1}{\varepsilon_{reff0}(\varepsilon_r,w+\Delta w,h)-1},$$

(3.12)

and

$$\varepsilon_{reff}(f,\varepsilon_r,w+\Delta w,h) = \varepsilon_r - \varepsilon_{reff0}(\varepsilon_r,w+\Delta w,h)$$
$$\cdot \frac{1}{1+\dfrac{\pi^2 \cdot \mu_0 \cdot h \cdot f(\varepsilon_r-1)}{6Z_{c0}(\varepsilon_r,w+\Delta w,h) \cdot \varepsilon_{reff0}(\varepsilon_r,w+\Delta w,h)}} \cdot \frac{1}{\sqrt{\dfrac{2\pi \cdot Z_{c0}(\varepsilon_r,w+\Delta w,h)}{\sqrt{\dfrac{\mu_0}{\varepsilon_0}}}}} \quad (3.13)$$

In practice, the simple "thru-line" method [10] can be used for the extraction of the attenuation and propagation constants α and β of a transmission line. This method permits removal of pad parasitics and pad-line discontinuities. Also, from the de-embedded S-parameters, the characteristic impedance can also be extracted by using the method described in [11].

Measured and simulated characteristic parameters of microstrip lines are given in Table 3.2, at a 60-GHz working frequency, for different CMOS/BiCMOS technologies.

The good agreement between simulation and measurement results in Table 3.2 shows that the foregoing model is valid for mm-wave designs. In terms of accuracy, the model is comparable to commercial electromagnetic simulators.

Table 3.2 Measurement and simulation results for microstrip lines for different CMOS/BiCMOS technologies, at 60 GHz

Technology node	Metal strip	Width (μm)	Alpha (dB/mm) 60 GHz		$Z_c(\Omega)$ 60 GHz	
			Measurement	Simulation	Measurement	Simulation
CMOS 65 nm	AP*/M7	9	1.15	1.15	37	37
CMOS 65 nm	AP/M7	4.4	1.20	1.21	52	52
CMOS 65 nm	AP	9	0.97	0.99	47	47
CMOS 65 nm	AP	4.4	1.18	1.21	65	65
CMOS 45 nm	AP/M7	6	1.12	1.2	42	42
CMOS 45 nm	AP/M7	3	1.22	1.25	55	55
CMOS 45 nm	AP	9.2	1.04	1.05	43	43
CMOS 45 nm	AP	3	1.22	1.22	69	69
CMOS 32 nm	AP	9	1.1	1.11	44	44
CMOS 32 nm	AP	3	1.23	1.18	59	62
BiCMOS 55 nm	M8	7	0.6	0.6	50	50

AP, Alucap layer.

It can be noted from Table 3.2 that the lowest attenuation constant is obtained from BiCMOS 55 nm technology, which offers a thick metal layer (thicker than its CMOS counterparts).

Slow-Wave Transmission Lines: S-CPW and S-CPS

As explained earlier, except for interconnection purposes, the most critical point for a designer is not the physical length of a transmission line but rather its electrical length, that is, its phase shift, and its characteristic impedance.

If the phase velocity v_φ decreases, the guided wavelength λ_g decreases for a given operating frequency f, as illustrated by Eq. (3.14). Then, thanks to the slow-wave effect, the physical length l is shorter for the same electrical length θ, as shown by Eq. (3.15).

$$v_\varphi = \lambda_g \cdot f. \tag{3.14}$$

$$\theta = \frac{\omega}{v_\varphi} \cdot l. \tag{3.15}$$

From a circuit point-of-view, the slow-wave effect can also be expressed by the increase of the effective relative dielectric constant ε_{reff}, which is directly linked to the propagation constant β:

$$v_\varphi = \frac{C_0}{\sqrt{\varepsilon_{reff}}} = \frac{2\pi f}{\beta}, \tag{3.16}$$

with C_0 the light velocity in vacuum and f the operating frequency. So, the propagation constant β is increased with the slow-wave effect. If the attenuation constant is not deteriorated compared to classical transmission lines, the slow-wave propagation combines efficient miniaturization and high quality factor [12].

Finally, it can be interesting to define a figure of merit, called "slow-wave factor" (SWF), to evaluate this effect. Many different definitions can be found in the literature [13, 14]. The one given in Eq. (3.17) is used in this book; it is the ratio between the wave velocity in vacuum and the phase velocity of the considered transmission line:

$$SWF = \frac{c_0}{v_\varphi}. \tag{3.17}$$

S-CPWs were demonstrated in 2003 in CMOS technologies [15]. Since that date, many circuits were built with these transmission lines, since they offer higher quality factors than microstrip lines, whatever the characteristic impedance.

A classical S-CPW, with the definition of the main design parameters, is shown in Figure 3.14. The topology of the electrical and magnetic field lines is also shown.

As shown in Figure 3.14, an S-CPW is based on a classical CPW with central and ground strips under which perpendicular floating ribbons are placed. The principle to get slow-wave propagation consists in separating the electric and magnetic fields. This is the case in S-CPWs, since only the magnetic field can flow below the ribbons. In another way, it can be said that magnetic energy is stored below the floating ribbons,

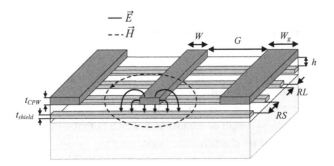

Figure 3.14 S-CPW scheme showing geometrical parameters.

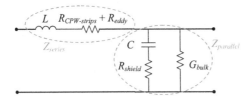

Figure 3.15 S-CPW electrical model topology.

decreasing the wave propagation velocity. From an electrical point of view, we can either consider that floating ribbons capture the electric field lines, leading to capacitively loaded CPW, or that the magnetic flux goes over the floating ribbons, increasing the inductance and leading to inductively loaded CPW.

An equivalent model was derived for S-CPWs, thus simplifying their use in circuits' design [16, 17]. The topology of the model is given in Figure 3.15 [16].

The classical *RLCG* electrical model does not fit well with the physical behavior of the S-CPW, especially when losses must be considered. Hence it was necessary to develop a new electrical model based on a fine study of the physical behavior of the S-CPW. However, the *RLCG* electrical model is still well suited for the phase velocity calculation, except at very high frequencies. As the magnetic field has the same spatial distribution in a S-CPW and in a classical CPW, the modeled inductance L is almost equal for both topologies having the same geometrical dimensions. Similarly, the capacitance C is almost the same for the S-CPW and the grounded CPW because the electric field is concentrated between the two metal layers in both transmission lines.

Nevertheless, the *RLCG* model is not adequate to evaluate the full behavior of S-CPWs when losses are considered. Losses do not come only from conductive losses in the CPW strips and substrate losses in the silicon. Conductive and eddy current losses also appear in the floating ribbons. The resistance $R_{\text{CPW-strips}} + R_{\text{eddy}}$ includes conductive losses in the CPW strips and eddy currents losses introduced by the floating ribbons. Due to the capacitance between the CPW strips and the floating ribbons, a current propagates in the floating ribbons, leading to conductive losses modeled as a resistance R_{shield}. In addition, the conductance G_{bulk} expresses the losses in the bulk silicon: both conductive and eddy current losses. In [18], an additional study pointing out the effect of the floating ribbons length on the loss distribution for both standard Bulk and HR-SOI substrates demonstrated that eddy current losses induced by the magnetic field into the substrate are negligible for these transmission lines for current gap width G (i.e., up to about 100 μm). Above 100–150 μm, eddy current losses occur in the low-resistivity silicon substrate. Hence, in practice, if the floating ribbons are efficient enough to ensure that no electrical field propagates through the substrate, G_{bulk} can be neglected.

Based on the model topology given in Figure 3.15, a simple method for calculus of each element of the model was proposed in [17].

The S-CPW topology is optimal for the design of medium to low characteristic impedance transmission lines (basically down to 15 Ω) with high slow-wave factors. It is more challenging to reach high characteristic impedance S-CPWs. Nevertheless, in many technologies, it is usually possible to reach 80 Ω, and sometimes 100 Ω, while maintaining a quality factor always higher than that of classical microstrip lines. High characteristic impedance is an issue when dealing with conventional microstrip lines. Values higher than 70–80 Ω are basically not possible due to process limitations (width resolution) and would result in dramatic increase of attenuation loss. Hence the use of S-CPWs can be a solution to skirt this problem a little bit.

Whatever the targeted characteristic impedance, the effective relative dielectric constant of a S-CPW, and hence its slow-wave factor, is always higher than the one of a classical microstrip line. As a rule of thumb, values around 20–25 can be achieved in quite all current technologies with high quality factors (greater than 30) and moderate footprint. When wide gaps are considered to get high slow-wave factor, it is important to check that the final surface area is still competitive compared to classical solutions. Moreover, a large total width may introduce connection issues.

Finally, high quality factors (roughly up to 40–45 at 60 GHz) can be reached regardless of the targeted characteristic impedance, from 35 Ω up to 70 Ω, as shown by the experimental results in Figure 3.16.

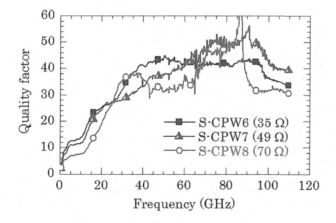

Figure 3.16 Quality factor for S-CPWs exhibiting characteristic impedances ranging from 35 to 70 Ω, in STMicroelectronics BiCMOS 130 nm technology.

Microstrip Lines and S-CPW Comparison

It is useful to compare microstrip lines and S-CPWs, in terms of electrical perform-ance and footprint, respectively, to derive design rules resulting from trade-off between performance and footprint.

For electrical performance comparison, the measured quality factor of many trans-mission lines is plotted versus the attenuation constant for a characteristic impedance around 50 Ω, and the benchmark is realized at 60 GHz (Figure 3.17). CPWs realized in high-resistivity Si substrates are also reported. This plot obviously exhibits a hyperbola for conventional topologies. Hence, the propagation constant β depends only on the intermetal relative dielectric constant and is roughly the same for all the considered CMOS technologies. The hyperbola directly comes from the quality factor definition. Through an increase in the propagation constant, the slow-wave phenom-enon frees the quality factor from the hyperbola, as shown in Figure 3.17. So, for a given attenuation constant, much higher quality factors may be obtained, which means a longitudinal miniaturization for a targeted electrical length.

At first sight, one thinks that the top left area in Figure 3.17 is the most interesting area, because it corresponds to a high quality factor with low attenuation constant. On further consideration, the top right area offers better performance, because the same quality factor is obtained with higher attenuation constant, meaning that a higher relative effective dielectric constant is obtained, hence leading to more compact transmission lines.

However, not only the length of the transmission line must be noted, but also its width, the footprint area being the parameter to consider, as stated earlier. In compari-son to conventional CPW or microstrip lines, the S-CPW topology gives more design degrees of freedom. So, several groups of geometrical dimensions give comparable results in terms of characteristic impedance or electrical length. At the same time, wide S-CPWs are required to reach strong slow-wave effects. So, depending on the application, a tradeoff must be defined between the targeted electrical performance and the maximum footprint. Figure 3.18 helps in understanding this point. It

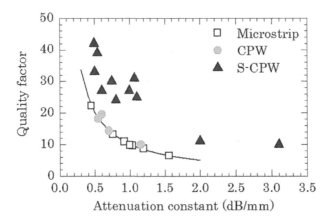

Figure 3.17 60-GHz experimental results: Quality factor versus attenuation constant for transmission lines in various technologies.

Table 3.3 Geometries and measurement results for microstrip lines, CPWs and S-CPWs, for different CMOS/BiCMOS technologies, at 60 GHz

| | | Geometrical dimensions (µm) | | | | | | Electric performance at 60 GHz | | | |
| | | Main strips | | | Floating ribbons | | | | | | |
	Technology	W	G	W_g	RL	RS	H	Z_c (Ω)	ε_{reff}	α (dB/mm)	Q
CPW_AMS	AMS	10	7	20				53	4.6	1.16	10
S-CPW1	0.35 µm	10	100	60	0.6	0.6	1	40	35	1.07	31
S-CPW2		18						31	47	1.17	33
S-CPW3			150					36	48	1.41	29
S-CPW4		7	50	10	0.7	0.7		53	24	0.99	27
S-CPW5					0.6	1		55	21.2	0.74	30
µstrip_B9MW	STM	12.2					8.3	51	3.7	0.5	20
S-CPW6	B9MW	5	50	10	0.16	0.64	0.4	35	25.6	0.66	43
S-CPW7	0.13 µm						1.15	49	13.3	0.49	42
S-CPW8			40		0.2	0.8	3.05	70	8.2	0.48	38
µstrip_65nm	STM	3.8					3.62	47	3.45	1.55	6.5
S-CPW9	65 nm	20	25	12	0.1	0.55	2.1	45	12.3	0.8	24

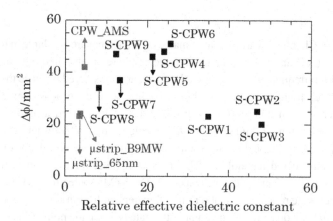

Figure 3.18 Electrical length per surface area versus the relative effective dielectric constant.

represents the electrical length per mm² corresponding to the microstrip lines, CPWs and S-CPWs, described in Table 3.3, versus the effective dielectric constant. The footprint (or surface area) was calculated as the physical length multiplied by the total width ($W + 2 \cdot G + 2 \cdot W\_g$; see Figure 3.14). The highest effective dielectric constant does not give the lowest footprint, because high effective dielectric constant is obtained thanks to wide gaps, leading to high footprint. CPWs and microstrip lines

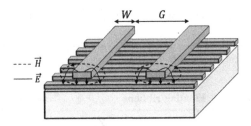

Figure 3.19 S-CPS.

lead to inferior performance, in terms of footprint and quality factor. Finally, the 50-Ω characteristic impedance S-CPW6 leads to the best electrical performance, with a quality factor greater than 40 along with the smallest footprint.

A variant of the S-CPW is the S-CPS, illustrated in Figure 3.19.

S-CPSs were not studied as much as S-CPWs. However, their differential behavior can be very interesting when dealing with differential circuits. They are used in standing wave oscillators described in Section 3.4.3.2. An electrical model was derived that simplifies their design [19].

3.3   Tunable Components

3.3.1   Varactors

As explained in Chapter 2, varactors are tunable elements that are available in RF design kits. Although there are different kinds of varactors, here is presented only the NMOS varactor. At high frequency, only N-doped devices are considered because the electron mobility is much higher than hole mobility, leading to a much better device quality factor. Furthermore, in this section we propose a description and a modeling of the varactor processed in the CMOS 28FDSOI technology from STMicroelectronics. The methodology used would be the same for all these kinds of devices and almost the same whatever the considered technology. The interest of these devices comes from their small gate length ($l = 32$ nm in the present case). The STMicroelectronics CMOS 28 nm FDSOI devices were processed on 300 mm SOI wafers with a buried oxide thickness of 25 nm. The final silicon film thickness under the gate is 7 nm, the nominal gate length is 30 nm, and the oxide thickness is 1.5 nm and 3.4 nm for thin and thick oxide transistors, respectively. The channel was left undoped, and Nwell and additional N-type ground plane (GP) implantations were coupled, as for the Pwell and P-type GP (GPP). Deep well implants are like the bulk ones. The threshold voltage V_{th} is adjusted only by placing the right Nwell/GPN or Pwell/GPP doping below the device. After well implantations and prior to gate patterning, the hybrid FDSOI/bulk areas are realized by removing the Si layer and BOX (Figure 3.20) using specific lithography steps. A raised source/drain (S/D) with specific implantations compared to bulk devices is made to reduce feeding resistance. This is necessary to increase the quality factor of RF devices.

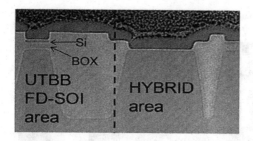

Figure 3.20 TEM figures of hybrid and SOI areas.

The developed accumulation MOS N+Poly/Nwell varactors have a 3.4 nm oxide thickness (thick oxide technology) and a minimum gate length of 43 nm. This subdesign-rule channel greatly improves the varactor Q-factor.

A top view and a cross section of the varactor are shown in Figure 3.21. The input and output metal accesses are integrated up to metal4 (Figure 3.21a). This device is processed on hybrid area (no SOI), which implies having a high doping density in the channel. Consequently, the channel resistance is lowered, and hence the Q-factor is increased.

Vector network analyzers (VNAs) are used to measure the varactor test structures S-parameters. Several calibration methods can be used. A common well-known calibration is the short-open-load-thru (SOLT), which was used for the measurement results presented in this book. After calibration, accesses must be de-embedded to extract S-parameters of the varactor from the test structures measurements. Here again, a well-known method as the pad-thru one can be used [20]. The extraction method of the varactor series resistance, capacitance, and inductance is based on the use of the admittance matrix [Y]. The substrate impedance is extracted by using the impedance matrix [Z]. Figures 3.22–3.24 present a comparison between measurement and simulation results of a 45-fF varactor (a) and a 140-fF varactor (b), respectively. Capacitance versus gate voltage (V_{Bias}) at 100 MHz is given in Figure 3.22. Resistance and Q-factor versus frequency, at a V_{Bias} of 1.8 V, are given in Figures 3.23 and 3.24, respectively. The Q-factor is given for C_{max}, that is, the maximum capacitance exhibited by the varactor corresponding to the accumulation mode. The Q-factor is greater at C_{min} corresponding to the depletion mode. The varactors' dimensions are $l = 43$ nm, $w = 0.9$ μm, $N = 30$ and $Mult = 5$ (a), and $l = 43$ nm, $w = 0.9$ μm, $N = 12$ and $Mult = 4$ (b), respectively.

These results show that a 46-fF varactor exhibits a series resistance of only 4.5 Ω, leading to a quality factor of 10 at 100 GHz. The quality factor of conventional RF and mm-wave accumulation MOS varactors in standard CMOS processes are reported in Table 3.4.

3.3.2   Switches

3.3.2.1   Basic Principle at RF Frequencies

At RF frequencies, antenna switches are unavoidable circuits in mobile communications applications. In the early days of mobile communications, they were only used

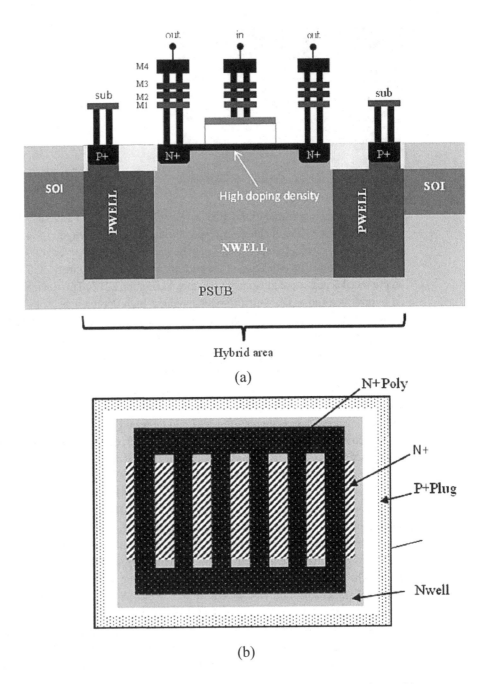

Figure 3.21 Cross section (a) and top view of the N+Poly/Nwell varactor layout (b).

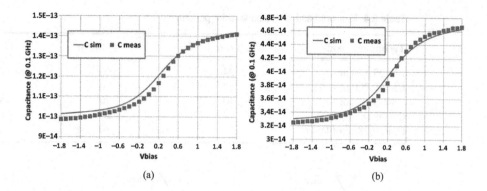

Figure 3.22 Measured and simulated capacitance versus bias voltage at 100 MHz.

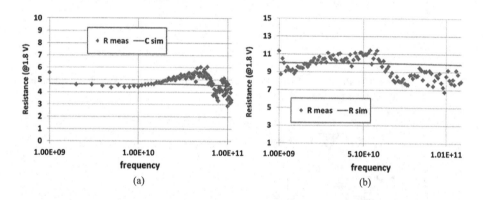

Figure 3.23 Measured and simulated resistance versus frequency at 1.8 V.

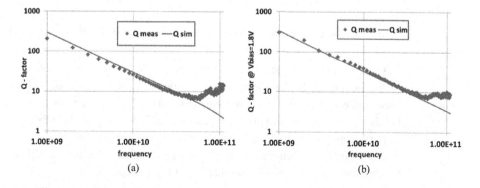

Figure 3.24 Measured and simulated Q-factor versus frequency at 1.8 V.

Table 3.4 Measured and simulated varactor performances

Technology node	Freq. (GHz)	Q-factor	C_{max}/C_{min}
CMOS 130 nm	24	100	<2
CMOS 200 nm	50	6	<2
CMOS 65 nm	10	20	4
CMOS 90 nm SOI	2	20	>8
CMOS 28 nm FDSOI	100	10	<2

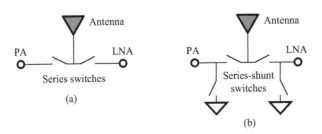

Figure 3.25 Principle of RF SPDT (single pole double throw) switch, based on (a) series switches, or (b) series-shunt switches.

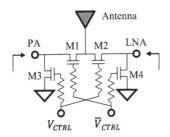

Figure 3.26 Schematic of a SPDT (single pole double throw) with series-shunt switches.

to switch between Rx and Tx transceiver paths (in general single pole double throw [SPDT] circuits). Today, in multistandard mobile applications, multiple antennas are used, and each antenna switch presents multiple paths, to cope with several frequency bands and different requirements, all for Rx and Tx paths (globally SPxT, with x > 2). Typical examples of lumped RF SP2T switch schematic is presented in Figure 3.25, with an SPDT switch that connects an antenna to a power amplifier (PA) for transmission or a low-noise amplifier (LNA) for reception.

Series switches configuration leads to minimum insertion loss but less isolation than series-shunt switches configuration. In practice, most designs result in a tradeoff between insertion loss and isolation; hence both configurations are used, depending on the design objectives.

The schematic of the SP2T switch in Figure 3.25b with four transistors is presented in Figure 3.26.

When the device operates in transmission, M3 and M2 transistors are in OFF state and M1 and M4 are in ON state, the RF signal crosses the M1 transistor from the PA to the antenna. When the device operates in reception, M1 and M4 transistors are in the OFF state, and M2 and M3 are in the ON state; hence the RF signal crosses the M2 transistor from the antenna to the LNA.

At mm-waves, the switch is an essential block in a transceiver front end, sharing Tx and Rx, but the design of high-performance mm-wave switches is still challenging. Two approaches can be considered, lumped and distributed. These two approaches are briefly described in the next sections.

3.3.2.2 Lumped Switches

The lumped approach is basically inspired by the RF SPDT described in Section 3.3.2.1. Specific topologies based on shunt switches and quarter-wavelength inverters are preferred at mm-waves to maintain high isolation while minimizing insertion loss. The basic principle is described in Figure 3.27. A T-junction composed of two quarter-wavelength transmission lines of 50-Ω characteristic impedance connects the input port 1 to the output ports 2 and 3, and two shunt switches, S1 and S2, are connected to the output ports, as shown in Figure 3.27a. If ideal switches are considered, modeled by a short circuit in the ON state and open circuit in the OFF state, the transmission between port 1 and port 2, port 3 being isolated, is achieved when S1 is turned off while S2 is turned on. In that case, the lower branch quarter-wavelength transmission line transforms the short circuit into an open circuit, leading to the simple scheme given in Figure 3.27b.

The mm-wave SPDT described in [21], fabricated in IBM's 90-nm 9HP technology, which features SiGe HBTs with peak f_T/f_{max} of 300/350 GHz, is based on the topology presented earlier in this section. It exhibits insertion loss equal to 1.05 dB and return loss equal to 22 dB at 94 GHz, respectively.

3.3.2.3 Traveling-Wave Switches

To meet the requirement of low insertion loss and broad bandwidth, specific topologies have been proposed, as traveling-wave switches [22–28]. A traveling-wave SPDT

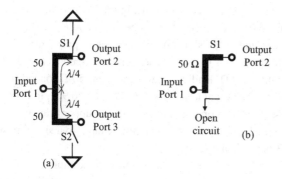

Figure 3.27 Principle of mm-wave SPDT switch.

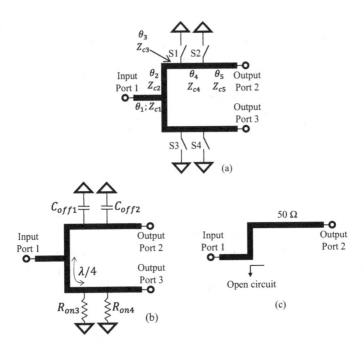

Figure 3.28 Principle of traveling-wave SPDT switch.

switch is composed of two distributed single-pole-single-throw (SPST) switches connected to a T-junction. Its principle is given in Figure 3.28. The input port is port 1, while ports 2 and 3 are output ports. The circuit is symmetric; hence only the transmission lines of the upper branch are defined, having an electrical length θ_i (with $i = 1$ to 5) and characteristic impedance Z_{ci}, as shown in Figure 3.28a. Two pairs of shunt switches are connected to the upper and lower branches, respectively. More switches can be used, depending on the tradeoff between insertion loss and return loss, and also circuit size. Let consider a transmission between ports 1 and 2, port 3 being isolated. In that case, switches S1 and S2 are turned on while S3 and S4 are turned off. In the OFF state, switches can be replaced, as a first approximation, by an OFF state capacitance, named C_{off1} and C_{off2}, as shown in Figure 3.28b, while in the ON state they can be replaced by an ON state resistance, named R_{on3} and R_{on4}. R_{on3} and R_{on4} are quite low and can be considered as shirt circuits as a first approximation. C_{off1} and C_{off2} are loading the transmission lines of characteristic impedance Z_{c2} to Z_{c5}. Z_{c3}, Z_{c4}, and Z_{c5} are higher than 50 Ω, leading to an equivalent 50-Ω loaded transmission line. On the other hand, the electrical length $\theta_2 + \theta_3$ is equal to 90°, thus transforming the short circuit ($R_{on3} \approx 0$) into an open circuit. Hence the upper branch is matched to 50 Ω, while the lower branch presents an open circuit, leading to a signal flowing from port 1 to port 2, port 3 being isolated.

The realization presented in Figure 3.29 constitutes an example of traveling-wave SPDT based on the principle described earlier. Its operating mode is detailed in the text that follows. SPDT input is port 1, while ports 2 and 3 are considered as outputs.

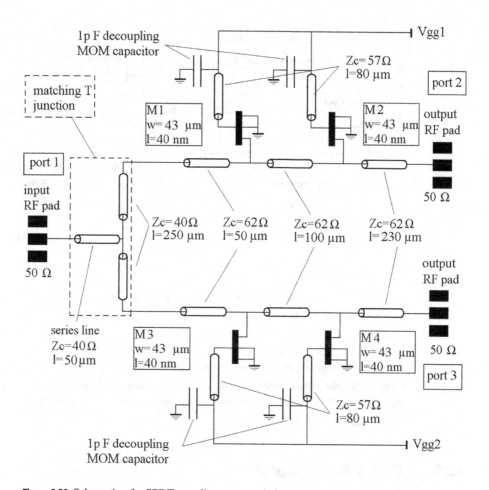

Figure 3.29 Schematic of a SPDT traveling-wave switch.

Let also consider a transmission from port 1 to port 2, while port 3 is isolated. In the transmit path, when the control voltage V_{gg1} applied to transistors M1 and M2 is set to 0 V, M1 and M2 are turned off and are equivalent to two capacitors of 50 fF. The integrated microstrip lines loaded with the OFF state capacitances of the parallel MOS form an artificial transmission line of 50-Ω characteristic impedance so that the signal flows from port 1 to port 2. In the isolated path, from port 1 to port 3, $V_{gg2} = 1$ V and transistors M3 and M4 are turned ON. They work in the linear region with low ON state resistance R_{on}. This shorts the signal to the ground. The T-junction branch of length 250 μm and microstrip line of length 50 μm are used as quarter-wavelength impedance inverter to transform port 3 into high impedance; thus ports 1 and 3 are isolated.

When the transistors are in the OFF state, small dimensions are needed to achieve low insertion loss due to capacitance coupling in the lossy substrate. When the transistors are in the ON state, large dimensions are needed to obtain a small ON state

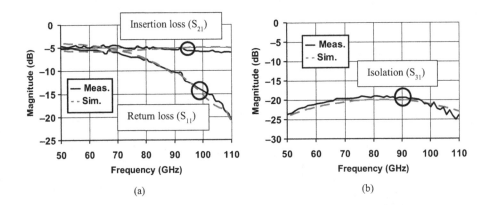

Figure 3.30 SPDT switch measured and simulated insertion loss (S_{21}) and return loss (S_{11}) (a) and isolation (b).

resistance R_{on}, and consequently a high isolation. Hence, the transistors' dimensions result from a tradeoff between (low) insertion loss and (high) isolation. Figure 3.30 shows the SPDT switch measured and simulated insertion loss (S_{21}) and return loss (S_{11}) (Figure 3.30a), and isolation (S_{31}) (Figure 3.30b), respectively.

From 90 GHz to 110 GHz, insertion loss is around 5–6 dB, with a return loss better than 10 dB and isolation better than 18 dB. Return loss is better above 100 GHz, since the 250-μm T-junction branch and 50-μm microstrip line are equal to a quarter-wavelength around 110–120 GHz.

3.3.3 Digitally Tunable Capacitance

The digitally tunable capacitor (DTC) could be considered as the "Holy Grail" of tunable components. Thanks to this concept, designers could have a simple varactor that is controllable with a digital signal. Digital control is interesting because it is more robust against noise, as compared to analog control. In this section, a concept of mm-wave DTC is described. Even if the electrical performance is not high, it gives an interesting component to mm-wave circuits' designers and can pave the way for future improvements in terms of performance, thanks to progress in CMOS/BiCMOS technologies.

Basically, a DTC is composed of switches and capacitors. In silicon technology, the transistor is the switch and the capacitor is an integrated MOM or MIM, preferably MOM for its higher quality factor, as explained in Section 3.2.2.2. The transistor is considered as a passive component, since its main function is to switch between two states, capacitive or resistive, by applying a control voltage on its input. The operation principle of the single-bit DTC is presented in Figure 3.31. A transistor, which is controlled by a control bit b_0, is connected in series with a capacitor (Figure 3.31a). When the control bit b_0 is equal to "1," meaning a positive gate bias, the transistor is in the ON state and can be basically replaced, at a first approximation, by a resistance R_{DS}. When the control bit b_0 is equal to "0," the transistor is in the OFF state and can be replaced by a capacitor C_{OFF}. Hence,

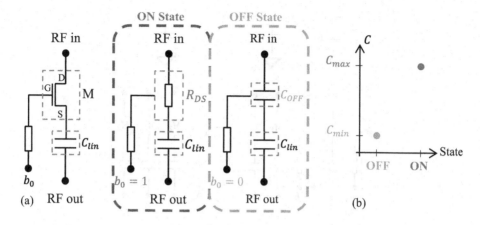

Figure 3.31 DTC operation principle.

two values of capacitors are realized, $C_{max} = C_{lin}$ in ON state and $C_{min} = \dfrac{C_{lin} \cdot C_{OFF}}{C_{lin} + C_{OFF}}$ in OFF state, respectively, as shown in Figure 3.31b.

Based on the simple single-bit DTC as described in Figure 3.31, the objectives of a multibits DTC are, from RF to mm-wave frequencies:

- To provide very important capacitance tuning range (TR)
- To synthesize several states (multibits DTC) with a linear response to the input control voltage

The 3-bit DTC architecture presented in Figure 3.32 results from the single-bit architecture, with several single-bit DTC being connected in parallel. Such an arrangement provides a configurable capacitance with linear steps, which is able to address the needs of much of tunable circuits. To vary the digital capacitance in a linear way in function of the input binary combinations, it is necessary to associate each bit to the correct capacitance value, knowing that the most significant bit corresponds to the largest capacitance. Each switch (transistor) must also be sized to present the right impedance at both OFF and ON states.

The architecture presented in Figure 3.32 has four branches. The transistors (M_i) operate in the ON state and OFF state to obtain an equivalent capacitance variation. A 2^N type power law is adopted to obtain a linear capacitance variation. The drain accesses of the transistors are connected to each other. They constitute the DTC input port. The digital control of the capacitance is achieved by the polarization of 1.2 V to 0 V of the transistor gate through a biasing resistor.

The conventional DTC topology presented in Figure 3.32 is limited to the RF range. This component achieves variable capacitance with high tuning ratio (TR > 4) for a range of 100 fF to several pF for frequencies below 30 GHz. To address mm-wave frequencies, traveling-wave DTCs (Figure 3.33) are recommended to keep a high tuning ratio up to at least 100 GHz. In that configuration, the shunt transistors periodically load a high characteristic impedance transmission line at an optimal distance to form a distributed-type switch.

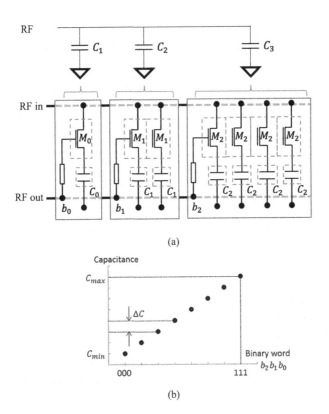

(a)

(b)

Figure 3.32 3-Bit DTC example. (a) Scheme. (b) Capacitance states.

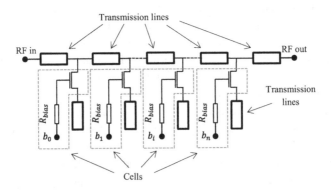

Figure 3.33 Multibits traveling wave DTC principle.

Experimental results of a 4-bit DTC using the architecture described in Figure 3.33 are given in [29]. The BiCMOS 130-nm from STMicroelectronics was used. The operating frequency extends up to 100 GHz. The least significant bit corresponds to the lowest capacitance $C_{LSB} = 3$ fF when all the transistors are in the OFF state ($V_g = 0$), while the most significant bit corresponds to the lowest capacitance $C_{MSB} = 41$ fF when all the transistors are in the ON state ($V_g = 1.2$), thus leading to a tuning ratio (TR) = 13.

The main drawback of such mm-wave DTC is the poor-quality factor of the tunable capacitance, which lies between 1 and 10. Future developments should address this issue before a wide use of such an efficient concept.

3.3.4   Tunable Transmission Lines

3.3.4.1   Needs and Principles

Needs

Using a lumped approach, based on inductors, capacitors, transformers, and varactors, is still possible at mm-waves to build tunable circuits. The quality factor of the inductors, expressed as $Q_L = \dfrac{L\omega}{R}$, comprises between 15 and 20 [30], which is comparable to the quality factor of microstrip lines described in Section 3.2.2.2. However, the quality factor of MOM capacitors, also described in Section 3.2.2.2, becomes very low, that is, lower than 10 above 100 GHz. Moreover, the modeling of inductors and transformers becomes tricky at mm-waves.

For these reasons, it is necessary to "think distributed" and use transmission lines instead of lumped components. Transmission line length decreases with frequency and becomes more and more acceptable at mm-waves, and their quality factor comprises between 15 (for microstrip lines) and 30 (for S-CPWs, Section 3.2.2.2). When dealing with circuits for which the electrical length of the transmission lines is determined by the design (contrary to interconnect lines), it is necessary to think in terms of quality factor that represents the insertion loss per electrical length and not attenuation constant, given in dB/mm. Hence, since the quality factor of a transmission line increases with frequency, at least from few GHz to more than 100 GHz, transmission lines' losses at mm-waves, for a given electrical length, decrease!

However, the major advantage in using transmission lines is perhaps linked to their easy modeling. First electrical models for microstrip lines were carried out at the beginning of the 1980s, whereas a first electrical model for S-CPWs, dedicated to CMOS/BiCMOS technologies, was carried out in 2015 [17]. Hence, from a designer point of view, it is much easier to design tunable mm-wave circuits by using distributed approaches instead of lumped ones. It is why microwave engineers and researchers have begun to work with their electrician colleagues since early 2000s to build mm-wave circuits and systems in silicon technologies.

A first illustration of the interest of distributed approaches was given in Figure 3.33, which describes a multibit traveling wave DTC principle. Many other examples of distributed mm-wave tunable circuits are given in Sections 3.4 and 3.5. But before we go further, it is necessary to give a clear definition of "what is a tunable transmission line?"

What Does Tunable Transmission Line Mean?

For the sake of simplicity, let consider a lossless transmission line defined by its characteristic impedance Z_c and electrical length θ. Z_c and θ can be expressed using the inductance L and capacitance C of the transmission line:

$$\theta = \omega\sqrt{LC} \cdot l, \tag{3.18}$$

and

$$Z_c = \sqrt{\frac{L}{C}}. \tag{3.19}$$

Tunable transmission line means that either Z_c and/or θ can be controlled, but most of the time only the variation of θ is considered. The control can be achieved by varying either L and/or C. If one wants to vary Z_c and θ independently, both L and C must be controlled independently. However, varying C is quite easy, thanks to the use of varactors, but varying L is a bigger challenge.

A solution based on fixed inductance in series with a varactor was proposed in [31] for printed circuit boards (PCBs) using surface-mounted components, but this solution would lead to cumbersome and low quality factor circuits when dealing with CMOS/BiCMOS technologies, with a quality factor always lower for the tunable inductor as compared to that of the series varactor. In [32], this solution was used to realize the LC tank of a mm-wave voltage-controlled oscillator (VCO). With a varactor quality factor equal to about 7, a tunable inductance with a quality factor equal to 6 was realized.

Another solution consists in varying the flux generated by a loop by using the concept of switched-surface loops, as illustrated in Figure 3.34. However, the concept could not be used to build an artificial tunable transmission line based on variable C and L, since the loops are too cumbersome and the switches' losses would dramatically decrease the transmission line quality factor.

The concept of switched lines could be implemented to tune the inductance of a transmission line, as illustrated in Figure 3.35. A coplanar stripline (CPS) is tuned by varying the distance between the strips, thanks to the use of switches at input/output of the transmission line. Even if a continuous tuning is not possible with this technique, it could help to reduce the characteristic impedance variation when varying the varactor capacitance of tunable loaded lines. Effort must be made to realize low-loss switches to maintain acceptable overall quality factor.

Hence, in practice only C is varied yet when talking about tunable transmission lines, and the variation of the electrical length θ leads to a variation of the characteristic impedance Z_c, as shown by Eq. (3.19). The "Grail" of having independent

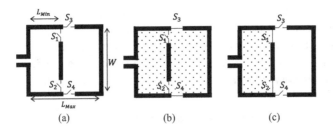

Figure 3.34 Tunable inductance based on the concept of switched-surface loop. (a) Switchable inductance. (b) Switches S_3 and S_4 ON (S_1 and S_2 OFF): maximum inductance with surface $L_{max} \cdot W$. (c) Switches S_1 and S_2 ON (S_3 and S_4 OFF): minimum inductance with surface $L_{min} \cdot W$.

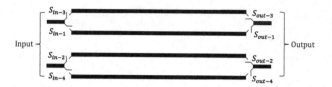

Figure 3.35 Switched-line tunable transmission line based on the switched-surface concept. Minimum inductance is obtained for switches S_{in-1} and S_{in-2} ON (S_{in-3} and S_{in-4} OFF); maximum inductance is obtained for switches S_{in-3} and S_{in-4} ON (S_{in-1} and S_{in-2} OFF).

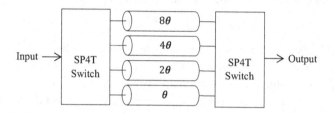

Figure 3.36 Example of a 2-bit tunable transmission line based on the switched-line concept. The signal passes from input to output, through a transmission line of electrical length θ, 2θ, 4θ, or 8θ, respectively.

continuous control of L and C of a transmission line is still a challenge. Later in this chapter, we consider the variation of θ when speaking about a tunable transmission line, without any particular control of the characteristic impedance Z_c.

Switched Lines and Loaded Lines

Two approaches can be envisaged to build tunable transmission lines, switched lines and loaded lines, respectively.

An example of switched lines is given in [33], where the concept was used to build a 2-bit phase shifter, with SP4T (single-pole four-throw) fabricated in MEMS technology. The concept is illustrated in Figure 3.36.

The concept as proposed in [33] gives satisfactory results in terms of electrical performance, with return loss better than 12 dB and low insertion loss, thanks to the use of a 500-µm quartz substrate. However, it suffers from large dimensions, even at mm-waves, and hence could not be used in CMOS/BiCMOS technologies, even if one could build efficient SP4Ts. Moreover, the complexity to achieve more than 2 bits would be very high, since SPkT switches are necessary to realize n bits, with $k = 2^n$. However, as will be discussed later in Section 3.3.4.3, the use of a mixed approach of switched lines and loaded lines could lead to efficient tunable transmission lines.

The concept of loaded lines, used by many authors in the literature, is based on a very simple idea. A transmission line is loaded by tunable components, that is, varactors acting as tunable capacitors, as illustrated in Figure 3.37. Its electrical length θ is given by Eq. (3.18).

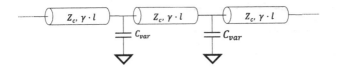

Figure 3.37 Periodically loaded-line concept. γ is the propagation constant of the transmission line, expressed as $\gamma = \alpha + j\beta$, where α is the attenuation constant, and β is the propagation constant, respectively.

If the transmission line is loaded by a lossless ideal varactor of capacitance C_{var}, the electrical length of the loaded transmission line is given by θ_l:

$$\theta_l = \omega\sqrt{L(C + C_{var}/l)} \cdot l. \qquad (3.20)$$

Variable C_{var} leads to variable θ_l, that is, tunable transmission line.

This concept leads to continuously variable phase shift, with simple design when lossless transmission lines are considered. The design of minimum-loss tunable transmission lines based on the loaded-line concept is trickier; this point is addressed in Section 3.3.4.2. However, the main drawback of loaded lines is linked to the variation of the characteristic impedance with the electrical length, the loaded characteristic impedance being given by

$$Z_{c-l} = \sqrt{\frac{L}{C + C_{var}/l}}. \qquad (3.21)$$

This will also be discussed in Section 3.3.4.2.

3.3.4.2 Loaded-Lines Design

As explained earlier, the loaded-line concept consists in periodically loading a transmission line with tunable components, that is, varactors, thus varying its electrical length θ. In this section, the Bragg frequency issue is first discussed, then the slow-wave effect in loaded line is briefly presented, and the expression of its quality factor is carried out and discussed. Design rules are established to optimize the electrical performance of loaded lines. Finally, examples of practical loaded lines in CMOS/BiCMOS technologies are given.

Bragg Frequency
Periodical structures face the cut-off Bragg frequency issue. A deep analysis of periodically loaded transmission lines was carried out in [34] and [35]. Let us consider an elementary cell consisting in a lossless transmission line of electrical length θ and characteristic impedance Z_c, loaded by an ideal capacitance C_{var}. One can make the equivalence with a transmission line having a characteristic impedance Z_{c-l} and an electrical length θ_l (Figure 3.38).

The four terms A_l, B_l, C_l, and D_l of the *ABCD* matrix of the loaded line shown in Figure 3.38 are given by

$$A_l = \cos(\theta) - Z_c \cdot C_{var} \cdot \pi \cdot f \cdot \sin(\theta), \qquad (3.22)$$

$$B_l = jZ_c(\cdot\sin(\theta) - Z_c \cdot C_{var} \cdot \pi \cdot f(1 - \cos(\theta))), \qquad (3.23)$$

$$C_l = j1/Z_c(\sin(\theta) + Z_c \cdot C_{var} \cdot \pi \cdot f(1 + \cos(\theta))), \tag{3.24}$$

$$D_l = A_l, \tag{3.25}$$

with f the operating frequency.

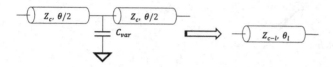

Figure 3.38 Loaded-line elementary ideal lossless cell.

Thanks to the equivalence of the two transmission lines in Figure 3.38, the following can be written

$$A_l = \cos(\theta_l). \tag{3.26}$$

Then two cases must be studied:

Case 1: The modulus $|A_l|$ of A_l is lower than 1. This case corresponds to an imaginary propagation constant $\gamma_l \cdot l = j\beta_l \cdot l = j\theta_l$. The wave propagates without losses through the loaded line. The condition $|A_l| \leq 1$ defines the permitted propagation frequency bands.

Case 2: $|A_l|$ is greater than 1. In that case propagation constant γ_l is real, $\gamma_l \cdot l = \alpha_l \cdot l$. The wave cannot propagate through the loaded line. Since a lossless ideal loaded line was considered, the power is totally reflected by the periodical structure.

To illustrate this mechanism, A_l was plotted in Figure 3.39 considering four cases, with Bragg frequencies (f_{Bragg}) equal to 100 GHz, 300 GHz, and 600 GHz, respectively. Case (d) corresponding to $f_{Bragg} = 600$ GHz is realistic when dealing with loaded lines working around 100 GHz. In that case, the Bragg effect can be neglected in the first design steps, since A_l is almost equal to 1 up to more than 100 GHz.

Figure 3.39 brings to the fore that the Bragg frequency increases when the electrical length θ and the loading capacitance C_{var} decrease. It is also obvious from Eq. (3.22) that the Bragg frequency also increases when the characteristic impedance Z_c decreases. Hence, to avoid Bragg effects, one must distribute small varactors over small pieces of transmission lines. The characteristic impedance Z_c must be as high as possible to get loaded-line characteristic impedance Z_{c-l} near 50 Ω ... if needed.

Next, the loaded-line characteristic impedance Z_{c-l} can be calculated by

$$Z_{c-l} = \sqrt{\frac{B_l}{C_l}} = \sqrt{\frac{Z_c \cdot \sin(\theta) - Z_c^2 \cdot C_{var} \cdot \pi \cdot f(1 - \cos(\theta))}{\frac{\sin(\theta)}{Z_c} + C_{var} \cdot \pi \cdot f(1 + \cos(\theta))}} = \tag{3.27}$$

$$Z_c \sqrt{\frac{1 - Z_c \cdot C_{var} \cdot \pi \cdot f \cdot \tan(\theta/2)}{1 + Z_c \cdot C_{var} \cdot \pi \cdot f \cdot \cotan(\theta/2)}}.$$

The plot of Z_{c-l} versus f/f_{Bragg} is given in Figure 3.40.

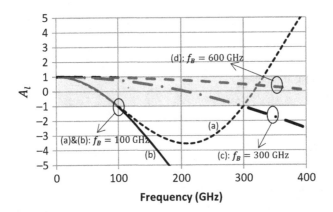

Figure 3.39 A_l coefficient for four cases, with $Z_c = 70\ \Omega$. (a) $f_{Bragg} = 100$ GHz with $\theta = 60°$ and $C_{var} = 79\ fF$; (b) $f_{Bragg} = 100$ GHz with $\theta = 30°$ and $C_{var} = 174\ fF$; (c) $f_{Bragg} = 300$ GHz with $\theta = 30°$ and $C_{var} = 58\ fF$; (d) $f_{Bragg} = 600$ GHz with $\theta = 30°$ and $C_{var} = 29\ fF$.

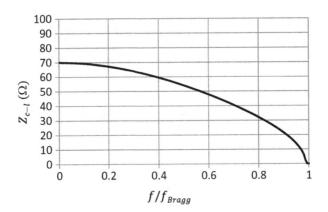

Figure 3.40 Loaded-line characteristic impedance Z_{c-l} versus normalized frequency f/f_{Bragg}.

Z_{c-l} decreases by 5% at 23% of the Bragg frequency. This kind of rule can be taken as a design rule for first-step design.

Slow-Wave Effect

Loading a transmission line with a capacitance decreases the phase velocity, given by

$$v_\varphi = \frac{1}{\sqrt{LC}}. \tag{3.28}$$

Hence, the electrical length of the loaded line (θ_l) is higher than that of the unloaded line (θ). The interesting consequence is the physical length reduction of the loaded line as compared to the unloaded line, for the same electrical length.

This is illustrated in Figure 3.41, where the ratio of θ over θ_l is plotted.

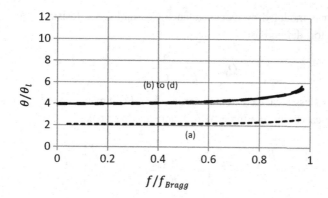

Figure 3.41 Ratio of the loaded-line electrical length θ_l and unloaded line electrical length θ, versus normalized frequency $f/f_{Bragg}.\Omega$. (a) $f_{Bragg} = 100$ GHz with $\theta = 60°$ and $C_{var} = 79\,fF$. (b) $f_{Bragg} = 100$ GHz with $\theta = 30°$ and $C_{var} = 174\,fF$. (c) $f_{Bragg} = 300$ GHz with $\theta = 30°$ and $C_{var} = 58\,fF$. (d) $f_{Bragg} = 600$ GHz with $\theta = 30°$ and $C_{var} = 29\,fF$.

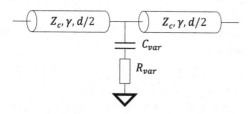

Figure 3.42 Loaded-line elementary lossy cell.

The ratio of θ over θ_l versus normalized frequency f/f_{Bragg} is directly linked to the choice of θ. The higher θ, the lower the slow-wave effect. This is easy to understand, since higher θ means weaker capacitive loading. As expected, curves for cases (b) to (d) in Figure 3.41 are superposed, since these three cases correspond to equal θ, and hence equal loading capacitor impedance, defined as

$$Z_{Cvar-eq} = \frac{1}{C_{var} \cdot f_{Bragg}}. \tag{3.29}$$

Loaded-Line Quality Factor

Lossy models must be used for the transmission line and varactor to derive an expression of the loaded-line quality factor. Let us consider the elementary cell given in Figure 3.42.

The transmission line is defined by its characteristic impedance Z_c, propagation constant γ, and physical length d. γ is defined as

$$\gamma = \alpha + j\beta, \tag{3.30}$$

with α the attenuation constant and β the propagation constant.

The varactor is modeled by its capacitance C_{var} and series resistance R_{var}, leading to a quality factor, defined as

$$Q_{var} = \frac{1}{R_{var} \cdot C_{var} \cdot \omega}. \tag{3.31}$$

As for the lossless ideal case treated earlier, the ABCD matrix of the loaded line can be obtained by cascading the three ABCD elementary matrixes:

$$\begin{bmatrix} A_l & B_l \\ C_l & D_l \end{bmatrix} = [T] \cdot \begin{bmatrix} 1 & 0 \\ \dfrac{jC_{var} \cdot \omega}{1 + jR_{var} \cdot C_{var} \cdot \omega} & 1 \end{bmatrix} \cdot [T], \tag{3.32}$$

with $[T]$ the ABCD matrix of the unloaded line, defined as

$$[T] = \begin{bmatrix} \cosh\left(\dfrac{\gamma \cdot d}{2}\right) & Z_c \cdot \sinh\left(\dfrac{\gamma \cdot d}{2}\right) \\ \dfrac{1}{Z_c} \sinh\left(\dfrac{\gamma \cdot d}{2}\right) & \cosh\left(\dfrac{\gamma \cdot d}{2}\right) \end{bmatrix}. \tag{3.33}$$

After calculus, the expression of A_l can be used to derive a relation linking the loaded-line propagation constant, defined as

$$\gamma_l = \alpha_l + j\beta_l, \tag{3.34}$$

and the parameters of the unloaded line and varactor, respectively. Equations (3.35) and (3.36) give the real and imaginary parts of A_l:

$$Real(A_l) = \cosh(\alpha_l \cdot d_l) \cdot \cos(\theta_l) = \cosh(a \cdot d) \cdot \cos(\theta) + \frac{1}{2} Y_{var} \cdot \tag{3.35}$$
$$Z_c[\sinh(a \cdot d) \cdot \cos(\theta) - Q_{var} \cdot \cosh(a \cdot d) \cdot \sin(\theta)].$$

$$Imag(A_l) = \sinh(\alpha_l \cdot d_l) \cdot \sin(\theta_l) = \sinh(a \cdot d) \cdot \sin(\theta) + \frac{1}{2} Y_{var} \cdot \tag{3.36}$$
$$Z_c[Q_{var} \cdot \sinh(a \cdot d) \cdot \cos(\theta) + \cosh(a \cdot d) \cdot \sin(\theta)],$$

with

$$Y_{var} = \frac{C_{var} \cdot \omega}{Q_{var} + 1/Q_{var}}, \tag{3.37}$$

the real part of the admittance of the varactor, and $\theta_l = \beta_l \cdot d$ the electrical length of the loaded line.

To express α_l and β_l, and hence the quality factor Q_l of the loaded line, expressed as

$$Q_l = \frac{1}{2} \frac{\beta_l}{\alpha_l}, \tag{3.38}$$

Eqs. (3.35) and (3.36) must be simplified. Low-loss transmission lines corresponding to practical cases are assumed, leading to

$$a \cdot d \ll 1, \tag{3.39}$$

and

$$\alpha_l \cdot d \ll 1. \tag{3.40}$$

Relations (3.39) and (3.40) are realistic in practice, since the physical lengths d and d_l are small as compared to the wavelength.

Then, first-order Taylor series development of Eqs. (3.35) and (3.36) leads to

$$Real(A_{ls}) = \cos(\theta_l) = \cos(\theta) + \frac{1}{2} Y_{var} \cdot Z_c[\alpha \cdot d \cdot \cos(\theta) - Q_{var} \cdot \sin(\theta)], \tag{3.41}$$

and

$$Imag(A_{ls}) = \alpha_l \cdot d \cdot \sin(\theta_l) = \alpha \cdot d \cdot \sin(\theta) + \frac{1}{2} Y_{var} \cdot Z_c[Q_{var} \cdot \alpha \cdot d \cdot \cos(\theta) + \sin(\theta)], \tag{3.42}$$

where A_{ls} corresponds to the simplified expression of A_l.

After some simple calculus, the expression of $\alpha_l \cdot d$ can be obtained:

$$\alpha_l \cdot d = \frac{\alpha \cdot d \cdot \sin(\theta) + 0,5 \cdot Y_{var} \cdot Z_c[Q_{var} \cdot \alpha \cdot d \cdot \cos(\theta) + \sin(\theta)]}{\sqrt{1 - (\cos(\theta) + 0,5 \cdot Y_{var} \cdot Z_c[\alpha \cdot d \cdot \cos(\theta) - Q_{var} \cdot \sin(\theta)])^2}}, \tag{3.43}$$

as the expression of $\theta_l = \beta_l \cdot d$:

$$\theta_l = \beta_l \cdot d = \arccos\{\cos(\theta) + 0,5 \cdot Y_{var} \cdot Z_c[\alpha \cdot d \cdot \cos(\theta) - Q_{var} \cdot \sin(\theta)]\}. \tag{3.44}$$

Then the quality factor of the loaded line can be derived:

$$Q_l = \frac{1}{2} \frac{\beta_l}{\alpha_l} =$$

$$\frac{1}{2} \frac{\arccos\{\cos(\theta) + 0,5 \cdot Y_{var} \cdot Z_c[\alpha \cdot d \cdot \cos(\theta) - Q_{var} \cdot \sin(\theta)]\}}{\alpha \cdot d \cdot \sin(\theta) + 0,5 \cdot Y_{var} \cdot Z_c[Q_{var} \cdot \alpha \cdot d \cdot \cos(\theta) + \sin(\theta)]} \cdot$$

$$\sqrt{1 - (\cos(\theta) + 0,5 \cdot Y_{var} \cdot Z_c[\alpha \cdot d \cdot \cos(\theta) - Q_{var} \cdot \sin(\theta)])^2}. \tag{3.45}$$

Q_l depends on

- The characteristic impedance of the unloaded line, Z_c
- The length of the unloaded line, d
- The attenuation constant of the unloaded line, α,
- The real part of the varactor admittance, Y_{var}
- The quality factor of the varactor, Q_{var}.

Next, to calculate Q_l, the parameters of the unloaded line and varactor must be fixed, that is,

- The characteristic impedance of the unloaded line, Z_c
- The length of the unloaded line, d
- The min/max values of the varactor capacitance, that is, $C_{var-min}$ and $C_{var-max}$

Then the value of the loaded-line electrical length θ_l, and its variation $\Delta\theta_l$ can be calculated, along with the loaded-line characteristic impedance Z_{c-l}.

Note that the higher the average capacitance defined by

$$C_{var-avg} = \sqrt{C_{var-max} \cdot C_{var-min}}, \tag{3.46}$$

and the capacitance ratio defined by

$$\eta = \frac{C_{var-max}}{C_{var-min}}, \tag{3.47}$$

the shorter the loaded line, for a given electrical length variation $\Delta\theta_l$.

Similarly, the higher the unloaded line characteristic impedance Z_c, the shorter the loaded line, for a given electrical length variation $\Delta\theta_l$, since higher varactor capacitance values can be used.

The calculus of $\Delta\theta_l$ and Z_{c-l} are carried out by neglecting the losses, since losses do not modify $\Delta\theta_l$ in a low-loss context. After some simple calculus, the following can be obtained:

$$\Delta\theta_l = \arccos\{\cos(\theta) - 0.5 \cdot Z_c \cdot C_{var-min} \cdot \omega \cdot \sin(\theta)\} \\ - \arccos\{\cos(\theta) - 0.5 \cdot Z_c \cdot C_{var-max} \cdot \omega \cdot \sin(\theta)\}, \tag{3.48}$$

and

$$Z_{c-l} = \sqrt{\frac{B_l}{C_l}} = Z_c \sqrt{\frac{1 - 0.5 \cdot Z_c \cdot C_{var} \cdot \omega \cdot \tan\left(\dfrac{\theta}{2}\right)}{1 + 0.5 \cdot Z_c \cdot C_{var} \cdot \operatorname{cotan}\left(\dfrac{\theta}{2}\right)}}. \tag{3.49}$$

From Eq. (3.49), the min/max and mean values of Z_{c-l}, that is, $Z_{c-l-min}$, $Z_{c-l-max}$, and $Z_{c-l-avg}$, respectively, can be derived:

$$Z_{c-l-min} = Z_c \sqrt{\frac{1 - 0.5 \cdot Z_c \cdot C_{var-max} \cdot \omega \cdot \tan\left(\dfrac{\theta}{2}\right)}{1 + 0.5 \cdot Z_c \cdot C_{var-max} \cdot \omega \cdot \operatorname{cotan}\left(\dfrac{\theta}{2}\right)}}, \tag{3.50}$$

$$Z_{c-l-max} = Z_c \sqrt{\frac{1 - 0.5 \cdot Z_c \cdot C_{var-min} \cdot \omega \cdot \tan\left(\dfrac{\theta}{2}\right)}{1 + 0.5 \cdot Z_c \cdot C_{var-min} \cdot \omega \cdot \operatorname{cotan}\left(\dfrac{\theta}{2}\right)}}, \tag{3.51}$$

$$Z_{c-l-avg} = Z_c \sqrt{\frac{1 - 0.5 \cdot Z_c \cdot C_{var-avg} \cdot \omega \cdot \tan\left(\dfrac{\theta}{2}\right)}{1 + 0.5 \cdot Z_c \cdot C_{var-avg} \cdot \omega \cdot \operatorname{cotan}\left(\dfrac{\theta}{2}\right)}}. \tag{3.52}$$

Then the varactor capacitance average $C_{var-avg}$ can be calculated from Eq. (3.52):

$$C_{var-avg} = \frac{Z_c^2 - Z_{c-l-avg}^2}{0.5 \cdot Z_c \cdot \omega \cdot \left(Z_c^2 \cdot \tan\left(\frac{\theta}{2}\right) + Z_{c-l-avg}^2 \cdot \cotan\left(\frac{\theta}{2}\right) \right)}. \tag{3.53}$$

$Z_{c-l-avg}$ can also be defined as

$$Z_{c-l-avg} = \sqrt{Z_{c-l-min} \cdot Z_{c-l-max}}. \tag{3.54}$$

Both definitions of $Z_{c-l-avg}$ in Eqs. (3.52) and (3.54) give almost the same result for small variations of C_{var} ($\eta < 5$ in practice), which leads to small variations of Z_{c-l}.

From the capacitance ratio η, it is now possible to calculate the varactor's min/max capacitance values:

$$C_{var-max} = C_{var-avg}\sqrt{\eta} \quad \text{and} \quad C_{var-min} = \frac{C_{s-avg}}{\sqrt{\eta}}, \tag{3.55}$$

and the unloaded line electrical length:

$$\tan\left(\frac{\theta}{2}\right) = \frac{\dfrac{Z_c^2 - Z_{c-l-avg}^2}{0.5Z_c \cdot \omega \cdot C_{var-avg}} \pm \sqrt{\left(\dfrac{Z_c^2 - Z_{c-l-avg}^2}{0.5Z_c \cdot \omega \cdot C_{var-avg}}\right)^2 - 4Z_c^2 \cdot Z_{c-l-avg}^2}}{2 \cdot Z_c^2}. \tag{3.56}$$

From Eq. (3.54), a last equation that makes the link between Z_{c-l} and Z_c can be derived:

$$\frac{Z_{c-l}^4}{Z_c^4}\left\{ 1 + 0.25Z_c^2 \cdot C_{var-max} \cdot C_{var-min} \cdot \omega^2 \cdot \cotan^2\left(\frac{\theta}{2}\right) + 0.5Z_c \cdot \right.$$

$$\left. (C_{var-min} + C_{var-max}) \cdot \omega \cdot \cotan\left(\frac{\theta}{2}\right) \right\} = 1 + 0.25Z_c^2 \cdot C_{var-max} \cdot$$

$$C_{var-min} \cdot \omega^2 \cdot \tan^2\left(\frac{\theta}{2}\right) - 0.5Z_c \cdot (C_{var-min} + C_{var-max}) \cdot \omega \cdot$$

$$\tan\left(\frac{\theta}{2}\right). \tag{3.57}$$

Now, let's consider an example with

- An operating frequency $f = 60$ GHz
- A targeted phase shift equal to $180°$
- A varactor from a BiCMOS 55 nm technology is chosen, with $C_{var-min} = 20\,fF$ ($Q_{var-max} = 22$) and $C_{var-max} = 36\,fF$ ($Q_{var-min} = 13$), leading to $\eta = 1.8$
- $Z_c = 80\,\Omega$ (realistic practical value when considering S-CPWs, more optimistic for microstrip lines)

Solving Eq. (3.57), the following is obtained:

- $\theta = 38.1°$
- $\Delta\theta_l = \theta_{l-max} - \theta_{l-min} = 63° - 53° = 10°$. Hence 18 sections are necessary to achieve a $180°$ phase shifter.

These values are confirmed by circuit simulations, carried out with real transmission lines and varactor.

From Eqs. (3.50) and (3.51), the min/max loaded-line characteristic impedance can be derived:

- $Z_{c-l-max} = 55.3\ \Omega$
- $Z_{c-l-min} = 45\ \Omega$

Next, the loaded-line quality factor Q_l can be derived. A S-CPW was considered for the realization of the unloaded line, with

- $Q = 40$
- An effective dielectric constant ε_{reff} equal to 9

We obtain

- $\alpha = \dfrac{\beta}{2 \cdot Q} = \dfrac{\omega}{80 \cdot C_0}\sqrt{\varepsilon_{reff}} = 15\pi = 47.1\ \text{m}^{-1}$, or $\alpha\ (\text{dB/mm}) = 0.41\ \text{dB/mm}$
- $d = \dfrac{\theta}{\beta} = 176.4\ \mu\text{m}$

From the expression of Q_l given by Eq. (3.45), the following are calculated:

- $Q_{l-min} = 12.4$ for $C_{var-max} = 36\ fF$
- $Q_{l-max} = 19.9$ for $C_{var-min} = 20\ fF$

Finally, the insertion loss can be easily calculated from the expression of the loaded-line quality factor:

$$Q_l = \frac{1}{2}\frac{\beta_l}{\alpha_l} = \frac{1}{2}\frac{\beta_l \cdot d_l}{\alpha_l \cdot d_l} = \frac{1}{2}\frac{\theta_l}{IL_l} 8.68 \tag{3.58}$$

with IL_l the insertion loss of the loaded line:

$$IL_l = \frac{1}{2}\frac{\theta_l}{Q_l} 8.68. \tag{3.59}$$

Hence:

- $IL_{l-max} = \dfrac{\frac{1}{2}\theta_{l-max}}{Q_{l-min}} 8.68 = 6.05\ \text{dB},$

- $IL_{l-min} = \dfrac{\frac{1}{2}\theta_{l-min}}{Q_{l-max}} 8.68 = 3.05\ \text{dB}.$

The calculus carried out above does not take into account the return loss due to impedance mismatch. Hence added insertion loss due to return loss was added to the maximum insertion loss IL_{l-max} for the calculus of the figure of merit (FoM), by considering added insertion loss as

$$Added\ Insertion\ Loss = 10 \cdot \log\left(1 - RL^2\right) \tag{3.60}$$

The return loss (RL) was calculated as follows, by considering the worst case:

$$Maximum\ Return\ Loss = RL = 10 \cdot \log\left(\frac{\Gamma}{1 + \Gamma^2}\right), \tag{3.61}$$

Table 3.5 Insertion loss and FoM of tunable transmission lines, for two loaded-line mean characteristic impedances, i.e., 50 Ω and 35 Ω, respectively. $Z_0 = 80\ \Omega$, $C_{var-max} = 36\ fF$, $C_{var-min} = 20\ fF$, $\eta = C_{var-max}/C_{var-min} = 1.8$, $Q_{var-min} = 13$, $Q_{var-max} = 22$. $\Delta\theta_l = 180°$

$Z_{c-l-avg} = 50\ \Omega$ Total physical length: 3.19 mm (18 sections of 176.4 μm) Maximum electrical length: 1139° Maximum return loss 9.9 dB						$Z_{c-l-avg} = 35\ \Omega$ Total physical length: 1.56 mm (29 sections of 53.7 μm) Maximum electrical length: 854° Maximum return loss 9.2 dB					
Q = 40			**Q = 20**			**Q = 40**			**Q = 20**		
IL_{min} (dB)	IL_{max} (dB)	FoM (°/dB)	IL_{min} (dB)	IL_{max} (dB)	FoM (°/dB)	IL_{min} (dB)	IL_{max} (dB)	FoM (°/dB)	IL_{min} (dB)	IL_{max} (dB)	FoM (°/dB)
3.05	6.05	29.8	4.5	7.6	23.8	2.6	5.2	34.4	3.4	6.2	29.1
+0.47 (RL)	27.6 (RL)		+0.47 (RL)	22.3 (RL)		+0.56 (RL)	31.2 (RL)		+0.56 (RL)	26.6 (RL)	

with Γ the reflection coefficient:

$$\Gamma = \frac{Z_{c-l-max} - Z_{c-l-avg}}{Z_{c-l-max} + Z_{c-l-avg}}. \tag{3.62}$$

The effect of the return loss is not negligible. For $Z_{c-l-avg} = 50\ \Omega$ and $Q = 40$, the FoM, defined as

$$FoM = \frac{Max\ (electrical\ length)}{Max\ (Insertion\ loss)}, \tag{3.63}$$

decreases from 29.8° to 27.6°/dB.

Table 3.5 gives the insertion loss of loaded lines considering different unloaded line and varactor quality factors, for $\eta = 1.8$ and $\Delta\theta_l = 180°$.

The equations derived in this section can help designers choose the several parameters constituting a tunable transmission line.

However, one must be aware of the limitations that are evidenced from results given in Table 3.5.

First, the length of the tunable transmission line is large when a mean 50-Ω characteristic impedance ($Z_{c-l-avg}$) is targeted, since the maximum characteristic impedance of the unloaded line that can be achieved is limited to 70–80 Ω (even less for advanced nodes) in the case microstrip lines and around 100 Ω when considering S-CPWs. Figure 3.43 gives the value of $Z_{c-l-avg}$ versus elementary section electrical length θ, for two values of B, that is a function of $C_{var-avg}$.

When θ increases, the mean characteristic impedance ($Z_{c-l-avg}$) increases as well, since the loading effect decreases. Hence, longer electrical length is necessary to achieve higher mean characteristic impedance.

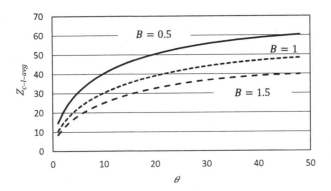

Figure 3.43 Mean loaded-line characteristic impedance $Z_{c-l-avg}$ versus elementary section electrical length θ. $Z_{c-l-avg} = Z_c \sqrt{\frac{1-0.5 \cdot B \cdot \tan{(\theta/2)}}{1+0.5 \cdot B \cdot \cot an(\theta/2)}}$, with $B = Z_c \cdot C_{var-avg} \cdot \omega$.

Similarly, when B (and hence $C_{var-avg}$ for given Z_c and operation frequency f) increases, the mean characteristic impedance decreases, and here again longer electrical length is necessary to achieve higher mean characteristic impedance.

Second, the FoM is limited to less than 30 in practice for a mean characteristic impedance equal to 50 Ω, even by considering very high quality factor S-CPWs and varactors. This leads to a more than 12 dB insertion loss for 360° phase shifters.

Third, the insertion loss imbalance between lowest and highest insertion loss is high, that is, 3 dB in the preceding example (for $Z_{c-l-avg} = 50$ Ω and $Q = 40$, and a phase shift of only 180°). This is a critical issue when dealing with phase shifters used in phased arrays.

In conclusion, loaded-line phase shifters can be interesting when small phase shifts are targeted, for example, 45°. In that case, the simplicity of the approach in terms of design is interesting, and insertion loss and insertion loss imbalance become acceptable for many applications. Their use in mixed switched- and loaded-lines phase shifters, as discussed in Section 3.3.4.3, gives them some interest.

Implementation in Silicon Technologies
The concept of loaded line is illustrated in Figure 3.37. A transmission line defined by its characteristic impedance Z_c and propagation constant γ is periodically loaded by a varactor of capacitance C_{var}. Implementation in silicon technologies will be discussed in the next sections dealing with phase shifters (Section 3.4.2) and VCOs (Section 3.4.3), where several implementations will be described.

3.3.4.3   Mixed Switched and Loaded Lines
The study of the quality factor of loaded lines, carried out in Section 3.3.4.2, clearly shows that the insertion loss is on the order of 6–7 dB when high phase shift variation is ambitioned, that is, 180°. In that case, a mixed approach combining switched fixed transmission lines and tunable varactor-loaded transmission lines could have some interest. This concept is illustrated in Figure 3.44. An ON/OFF 0/90° switched line is associated with a 90° continuously tunable transmission line. In the 0° state, the phase

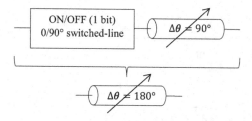

Figure 3.44 Mixed switched- and loaded-lines approach.

shift achieved by the tunable transmission line covers 0°–90°, whereas the range 90°–180° is covered in the 90° state, thus leading to a 180° phase shifter.

If the insertion loss associated to the 0/90° switched line is lower than half of the insertion loss of a loaded-line 180° phase shifter, the mixed approach is interesting. Such an approach is discussed in Section 3.4.2.2, which is dedicated to passive phase shifters.

3.4   Applications

3.4.1   Design Flow

At mm-wave frequencies, several effects are difficult to take into account during the design. First, the propagating effects in the interconnections between different circuits are not taken into account in classical design flow. Post layout extractors only give R, RC, or RCC lumped models. The impact of interconnections of a complex design can have dramatic effects if they are not taken into account. On the other hand, if interconnections are known, they can be used in the design to achieve better performance. Moreover, electromagnetic coupling effects between circuits are not taken into account in standard post layout extractors. For these reasons, custom design flows are used.

In mm-wave integrated circuits using CMOS or BiCMOS technologies, there are two critical issues for the designer. The first one is to know if the propagation effect in the interconnection can be neglected or not. The second is to know how to simulate interconnections with a sufficient precision without spending a long time under a 3D EM simulator. Because these interconnections highly impact the performance, they must be considered as soon as possible in the design flow.

To address the first issue, the following rule of thumb, illustrated in Figure 3.45, can be applied. If the length of the interconnection is higher than $\lambda/10$ or $\lambda/20$, it should be considered with a distributed model for better accuracy. If the length of the interconnection is lower than $\lambda/100$ it should be considered with a lumped model. In-between ($\lambda/100 < d < \lambda/20$), only the experience of the designer can help for the choice of the model type. In that case, the distributed model is often preferred at the cost of time-consuming simulations.

Once the interconnections to be considered are known, they must be modeled and simulated. Several cases must be considered. If the interconnection is on the top metal

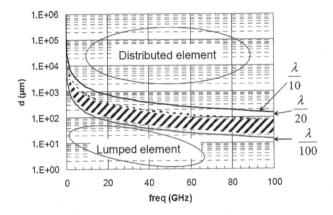

Figure 3.45 Model type as a function of the wavelength. Courtesy of A. Siligaris (IDESA seminars, CMOS front-end design at mm-wave frequencies – Alexandre Siligaris (CEA LETI), http://cordis.europa.eu/docs/projects/cnect/6/246906/080/deliverables/001-download.pdf)

layer, it is then possible to use a transmission line given by the design kit (DK) or a transmission line that was developed with a 3D EM simulator. For other kind of interconnections and especially when vias are used, it is necessary to build a model. A solution consists in simulating the interconnection with a 3D EM simulator. To save time, a layout can be directly simulated with MOMENTUM™ 3D planar simulator from VIRTUOSO™ thanks to GOLDENGATE™ software. In that case, the designer must be aware of the 3D planar limitations especially when the vias are longer than the total length of the interconnection. For better accuracy a full 3D EM simulator should be used. Unfortunately, there is no link between VIRTUOSO™ and such simulators and custom bridges must be developed. From the EM simulator, it is possible to generate a S-parameters file, which can be used in the electric simulation of the circuit.

When simulating an interconnection to a device that is available in a DK (a MOS transistor for example), it is important to know which effects are described in the model of the device. For example, the first metal layers are often modeled in accurate (or RF) models of the devices and the designer must be aware of that by reading the DRM (Design Rule Manual). When a custom model from EM simulator is used, post layout extraction of the interconnection must be avoided.

In any case, this part of the design in mm-wave-IC using CMOS or BiCMOS technologies is the most critical point and is highly dependent on the know-how of the designers.

3.4.2   Phase Shifters

Tunable phase shifters are key building blocks in phased arrays, which constitute mandatory systems in many mm-wave systems to focus a signal in various directions, due to the difficulty to generate high power at mm-waves, and also simply to save

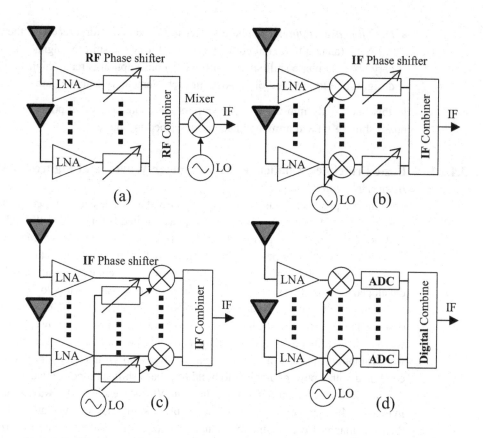

Figure 3.46 Phased array feeding network architectures, with different electronic phase shifting architectures.

energy in any cases! Four main architectures can be considered to build a phased array feeding network using phase shifters, as described in [36]; see Figure 3.46:

- *RF phase shifting*: The phase shifter is placed near the antenna, in the RF/mm-wave path.
- *IF phase shifting*: The phase shifter is placed after (for a receiver) the mixer. The main drawback of this architecture is that a mixer is necessary for each elementary antenna path, thus leading to high complexity and power consumption when dealing with large phased arrays. The main advantage is that the FoM of phase shifters is higher, since they are working at lower frequencies.
- *LO phase shifting*: The phase shifter is placed in the LO path. Main advantage and drawback are the same as for IF phase shifting, but specific advantages are also those given in [37]: "Phase-shifting in the LO-path is considered advantageous since the circuits in the LO-path operate in saturation and therefore it is relatively simple to ensure that the gain of each element does not vary with the phase shift setting. Additionally, the requirements on phase-shifter linearity, noise figure and bandwidth are substantially reduced when LO-path phase shifting is adopted."

- *Digital phase shifting*: The phase shifter is placed after (for a receiver) the mixer. An ADC (analog to digital converter) is used to phase shift the signals, leading to very accurate phase shift steps, but the DC consumption is really high in practical designs, due to the need of an ADC in each elementary antenna path.

In this book, only mm-wave phase shifters are addressed, corresponding to the RF phase shifting phased array architecture described in Figure 3.46.

3.4.2.1   Basic Principles, Challenges at Millimeter-Waves, and Figures of Merit
Basic Principles

The aim of a phase shifter is to modify the phase of a signal. The basic scheme is given in Figure 3.47. A signal entering port 1 is phase shifted before exiting through port 2.

Two main families of phase shifters can be identified, the passive and the active ones. In this book, only passive phase shifters will be described in detail.

A great deal of literature exists concerning **mm-wave active phase shifters** that were developed in the past 10 years. Among all the active phase shifters topologies, the vector modulation is the most popular one, because of its simplicity. It is based on the scheme described in Figure 3.48. The input signal is split in two ways with a power splitter. One way is phase shifted by 90°, thus producing I and Q ways that are in quadrature. Each way is amplified thanks to a Variable gain amplifier (VGA), and finally the two ways are combined with a power combiner to build the phase-shifted output signal.

The phase shift is adjusted by the gain of the VGAs in each I and Q way, as described in Figure 3.49. In this example, two different phase shifts, $\Delta\varphi_1$ and $\Delta\varphi_2$, are realized with the same magnitude, since the magnitude of I_1 (I_2) is the same as that of Q_2 (Q_1). By modifying the magnitude of Q_1 and I_1 by the same manner, that is, keeping the ratio Q_1/I_1 constant, the magnitude of the output signal can be adjusted for a constant phase shift. This constitutes an interesting characteristic, since the independent variation of phase shift and signal magnitude leads to very efficient beam-steering/forming

Figure 3.47 Basic phase shifter scheme.

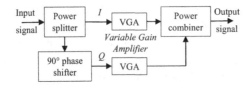

Figure 3.48 Active vector modulation phase shifter scheme.

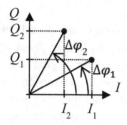

Figure 3.49 Vector phase modulation principle.

capabilities. For instance, secondary lobes can be greatly reduced by optimizing the magnitude of the signal feeding each elementary antenna of a phased array.

Passive phase shifters are based on the use of varactors and switches, as those described in Sections 3.3.1 and 3.3.2, respectively. The use of switches leads to digital phase shifters or mixed analog/digital phase shifters. The main topologies of passive phase shifters are described in detail in Section 3.4.2.2.

Figure of Merit and Challenges at Millimeter-Waves
As for any integrated circuits, the quality of a phase shifter depends on its electrical performance, footprint, and power consumption. However, contrary to VCOs that are described in Section 3.4.3, only one FoM is used for phase shifters, which depends only on specific electrical performance. It is expressed as the ratio of the maximum phase shift and the maximum insertion loss, as defined in Eq. (3.63).

The other characteristics giving the performance of a phase shifter, which correspond to so many challenges, are the following:

- *Return loss*: A return loss equal to 10 dB is acceptable in many applications.
- *Amplitude imbalance*: Amplitude imbalance measures the variation of the magnitude of the phase shifter outgoing signal, versus the phase shift. This is an important feature when dealing with phased arrays, since a variation in magnitude for the signal feeding the antenna elements changes the radiation pattern. In particular, it generates secondary lobes.
- *Bandwidth*: This is totally linked to the application.
- *Footprint*: It is very important when dealing with integrated circuits used in phased arrays. Phase arrays with more than 1 000 antenna elements can be used at mm-waves. If a classical architecture employing one phase shifter per antenna element is used, this means 1 000 phase shifters, rendering their size very critical.
- *Power consumption*: This is an important measure for active phase shifters, for example, those realized with vector modulation schemes, as described in [37]. However, only passive phase shifters are considered in this book, for which power consumption is negligible.
- *Linearity*: Even if power is much lower at mm-waves as compared to RF, linearity will be more and more considered in the future, with the advent of more and more mm-wave applications.

3.4.2.2   Passive Phase Shifters

Passive phase shifters can be classified in three categories, with digital (switched) or continuous phase shifting, and mixed digital and continuous phase shifting, as described in Figure 3.50. In this section, the use of the term "digital" does not mean a phase shift control achieved after an analog to digital conversion, but rather a switched scheme carried out in the RF path. In theory, all the topologies referenced in Figure 3.50 could be used either for continuous or digital phase shifting. However, in practical realizations, only the RTPS reflection type phase shifter (RTPS) is really used in both phase shifting schemes, that is, digital and continuous.

It is quite difficult to say what is the best choice among digital or continuous phase shifting, since it is really application dependent. For instance, for communication purposes, between a transmitter and a receiver, with antenna aperture of more than 10°, digital phase shifting could be considered as the best choice, since it is not necessary to have a very fine tuning to get a good communication. In practice, 5- or 6-bit phase shifters over a 360° phase shift variation are sufficient, leading to 11.25° (360/25) and 5.625° of phase shift resolution, respectively. On the contrary, when dealing with very small antenna aperture of 1° or less, in the case of big phased arrays, it is then preferable to use continuous phase shifting, since the number of bytes that would be required for digital tuning would be too high.

Continuous and mixed digital/continuous tuning phase shifters are discussed in detail in this book. Digital tuning phase shifters are discussed less deeply, but they are addressed in Section 3.4.2.7. Continuous tuning is interesting when very precise phase shift variation is necessary. When used with built-in-self-test circuits, it offers the possibility to modify the control of the tuning elements (i.e., varactors) according to temperature variation and technology dispersion.

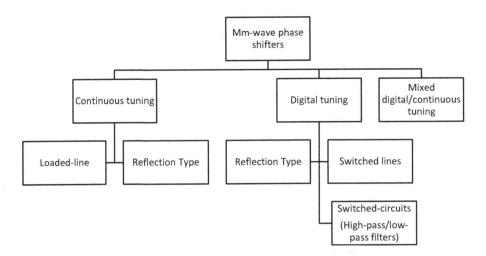

Figure 3.50 Millimeter-wave phase shifters classification.

3.4.2.3   Continuous Tuning Loaded-Line Phase Shifter

The loaded-line phase shifter is probably the most intuitive way to realize phase shifters. They are simply loaded lines, as described in Section 3.3.4.2. A transmission line is periodically loaded by varactors, leading to a hybrid tunable transmission line. Since the phase shift of the tunable transmission line depends on the overall capacitance, it acts as a tunable phase shifter. However, two main drawbacks make this approach not suitable, at least when dealing with mm-wave integrated circuits:

- When varying the loading varactor capacitance, not only the phase shift is varied, but also the loaded-line characteristic impedance, as explained in Section 3.3.4.2. Hence a compromise exists between the maximum phase shift and the return loss, for a given physical length of loaded line.
- The FoM of loaded-line phase shifters is poor, for two reasons:
  - The quality factor of the varactors is poor. However, this is an issue not only for loaded-line phase shifters, but also for all phase shifters based on varactors, that is, all continuous phase shifters.
  - There exists a compromise between maximum phase shift and return loss, which inevitably leads to high physical length of practical loaded-line phase shifters. The result is twofold: both area and insertion losses are high.

Recently, a new topology of loaded-line phase shifter with constant characteristic impedance was proposed in [38] and [39]. In both approaches, the authors proposed switching simultaneously the ground, to modify the inductance of the transmission line, and a floating strip, to modify the capacitance of the transmission line. In this way, the phase velocity can be changed, by increasing (or decreasing) simultaneously the capacitance and inductance, without any change in characteristic impedance. The principle proposed in [39] is described in Figure 3.51.

A CPW configuration is used. Two possibilities of ground strips permit modification of the inductance L. A switchable intermediate MIM capacitor permits modification of the capacitance C.

The maximum phase shift is obtained in the L_{max}, C_{max} configuration (Figure 3.51, top scheme). Switch C is ON, whereas switches S and L are OFF. The inductance is maximum, since the distance between the signal strip and ground strips is maximum, thus maximizing the magnetic flux. The MIM capacitor, connected to the signal strip by a via, is connected to the ground strips through perpendicular ribbons connected with the C switch.

The minimum phase shift is obtained in the L_{min}, C_{min} configuration (Figure 3.51, bottom scheme). Switches S and L are ON, whereas switch C is OFF. The inductance is minimum, since the distance between the signal strip and ground strips is minimum. The MIM capacitor is floating in that configuration, and the capacitance is given by the capacitance of a classical grounded CPW.

In Figure 3.51, perpendicular ribbons must be designed by respecting some rules that are described in [16]: (1) distance between ribbons must be lower than the height between CPW strips and ribbons, to avoid electrical field to go through the bulk

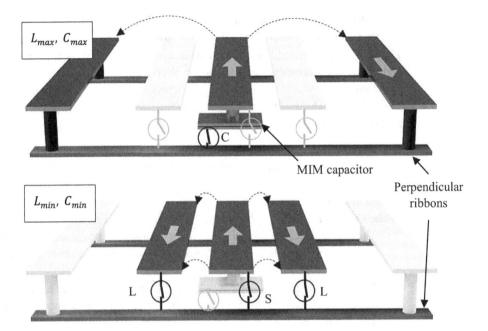

Figure 3.51 Constant characteristic impedance loaded-line phase shifter principle.

silicon, which would lead to conductive losses, and (2) width of ribbons must be minimized (0.5 μm is a good deal) to reduce eddy current losses.

By maintaining $Z_c = \sqrt{L_{max}/C_{max}} = \sqrt{L_{min}/C_{min}}$, the characteristic impedance remains constant, leading to a good return loss. Moreover, as the ratios L_{max}/L_{min} and C_{max}/C_{min} can be high, the length of the phase shifter can be reduced, as compared to classical loaded-line phase shifter.

3.4.2.4   Continuous Tuning Reflection-Type Phase Shifter
Principle

RTPS are based on the scheme presented in Figure 3.52. It consists in a 3-dB coupler loaded by two identical tunable reactive reflective loads.

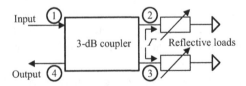

Figure 3.52 RTPS scheme, Γ, is the reflection coefficient looking to the reflective loads from the coupler output ports ② and ③.

An ideal 3-dB coupler is considered, defined by its S-matrix linking input (a_i) and output (b_i) waves:

$$\begin{bmatrix} b_1 \\ b_2 \\ b_3 \\ b_4 \end{bmatrix} = -\frac{1}{\sqrt{2}} \begin{bmatrix} 0 & j & 1 & 0 \\ j & 0 & 0 & 1 \\ 1 & 0 & 0 & j \\ 0 & 1 & j & 0 \end{bmatrix} \cdot \begin{bmatrix} a_1 \\ a_2 \\ a_3 \\ a_4 \end{bmatrix}. \tag{3.64}$$

From the S-matrix, the following equations can be derived:

$$b_1 = -\frac{1}{\sqrt{2}}(ja_2 + a_3); \quad b_2 = -\frac{1}{\sqrt{2}}(ja_1 + a_4)$$
$$b_3 = -\frac{1}{\sqrt{2}}(a_1 + ja_4); \quad b_4 = -\frac{1}{\sqrt{2}}(a_2 + ja_3). \tag{3.65}$$

The reflection coefficient Γ is defined by

$$\Gamma = \frac{a_2}{b_2} = \frac{a_3}{b_3}. \tag{3.66}$$

By considering input signal at port ① and output at port ④, and matching at each coupler port, the RTPS transmission parameter can be derived as

$$S_{41} = j\Gamma. \tag{3.67}$$

The transmission magnitude is equal to the modulus of Γ whereas the transmission phase variation is equal to the phase variation of Γ. Hence, it is necessary to get a reflection coefficient as high as possible with a maximum phase variation to achieve high FoM, as explained in the next sections.

3-dB Coupler

Two kinds of couplers are used in practice, branch-line couplers and coupled-line couplers, as shown in Figure 3.53. The Lange coupler [40] can be considered as an extension of coupled-line couplers.

When port ① is considered as the input port, port ② is the through port, port ③ is the coupled port, and port ④ is the isolated port. Signal exiting ports ② and ③ are in quadrature. These ports are loaded by the reflective loads, as shown in Figure 3.52.

Branch-line and coupled-line couplers exhibit the same behavior at the working frequency corresponding to quarter-wavelength lengths (Figure 3.53). However, their

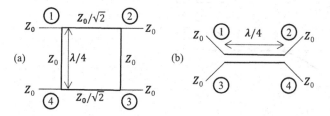

Figure 3.53 (a) Branch-line coupler. (b) Coupled-line coupler.

behavior differs when one moves away from this working frequency. Figures of merit that can be used to compare ideal couplers are ports matching, and phase and amplitude imbalance between through and coupled ports. In that context, coupled-line couplers exhibit great advantages in terms of bandwidth due to wider matching and lower phase imbalance and amplitude imbalance. However, asymmetric branch-line couplers can lead to larger phase variation for the RTPS. All these points are addressed in the sections that follow by considering ideal 3-dB couplers working at 60 GHz.

Return Loss:
The return loss of coupled-line couplers is infinite for ideal couplers, whatever the frequency. The return loss of branch-line couplers is given in Figure 3.54. It is degraded when one moves away from the center frequency. The 10-dB bandwidth extends from 50 to 70 GHz, that is, ± 20%.

Phase imbalance:
For coupled-line couplers, signals exiting ports ② and ③ are in quadrature whatever the operating frequency, whereas for branch-line couplers the phase imbalance equals 5° at 20% of the working frequency, that is, at 50 and 70 GHz, respectively, and degrades very quickly as shown in Figure 3.55.

Amplitude imbalance:
Amplitude imbalance is given in Figure 3.56. Comparison between coupled-line and branch-line couplers depends on the maximum amplitude imbalance that can be accepted. In practice, amplitude imbalance larger than 1 dB is penalizing; hence the comparison is

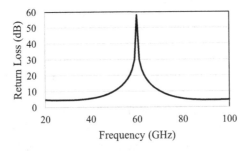

Figure 3.54 Return loss for branch-line coupler.

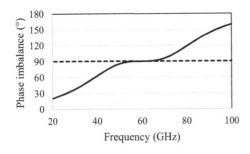

Figure 3.55 Phase imbalance for branch-line (plain line) and coupled-line (dotted line) couplers.

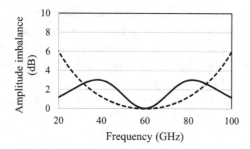

Figure 3.56 Amplitude imbalance for branch-line (plain line) and coupled-line (dotted line) couplers.

carried out for 1 dB of amplitude imbalance. In that context, coupled-line couplers exhibit better performance, since the 1-dB amplitude imbalance bandwidth ranges from 42 to 78 GHz, as compared to 52–68 GHz for the branch-line coupler.

From the foregoing study, it can be concluded that coupled-line couplers should be preferred to branch-line couplers when large bandwidths (roughly $> \pm 20\%$) are targeted. Moreover, the size of coupled-line couplers is smaller, thus leading to lower fabrication costs.

However, many reasons explain why branch-line couplers are still used by many authors.

First, narrow-band systems do not need bandwidths larger than 10%.

Second, the implementation of 3-dB coupled-line couplers is not so simple. When considering microstrip lines, the design rules of current technologies dramatically limit the minimum gap between two parallel strips. In advanced technologies that can be used at mm-waves; that is, from CMOS 65 nm or BiCMOS 0.18-μm nodes, this limits couplings above 5 to 10 dB. To circumvent this issue, broadside couplers are designed [41]. However, their design is a bit tricky because metals thicknesses are different between layers, and phase velocity is not the same for the two microstrip lines being fabricated in different layers, since the dielectric environment is different. Coupled slow-wave CPWs were also developed [42]. They offer a simple design, higher electrical performance, and better compactness as compared to microstrip lines. They were implemented in [43] to achieve a 45–50 GHz RTPS in BiCMOS 55-nm technology.

Third, branch-line couplers can be made asymmetric, that is, with different matching impedance at ports ①– ④ and ②– ③. This was exploited in PCB technology at RF frequency (2.45 GHz) to increase the phase shift variation, for given reflective loads [44]. The same technique could be easily used at mm-waves in silicon technologies.

Reflective Load

The reflective load is based on varactor diodes. One can consider two figures of merit for the varactors: the ratio between maximum and minimum capacitance, $\eta = \dfrac{C_{max}}{C_{min}}$, and the quality factor, Q.

There exists a compromise between the ratio η and the quality factor Q. However, at mm-waves, due to low-Q varactors, the quality factor is always considered first. This leads to ratio η around 2 in practice. This low ratio leads to small phase shift variation $\Delta\Phi$. To improve $\Delta\Phi$, an inductance can be placed in series with the varactor to increase the ratio η, as shown in the text that follows.

Let us consider a varactor capacitance in series with an inductance, Figure 3.57.

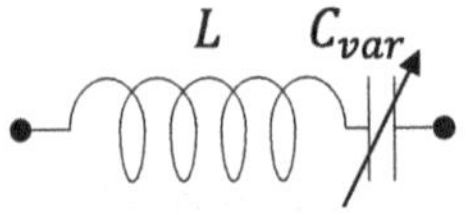

Figure 3.57 Ideal capacitance in series with an ideal inductance.

The equivalent capacitance C_{eq} is expressed as

$$C_{eq} = \frac{C_{var}}{1 - LC_{var}\omega^2}. \tag{3.68}$$

The relative variation of C_{eq} can be expressed as

$$\frac{\Delta C_{eq}}{C_{eq}} = \frac{\Delta C_{var}}{C_{var}} \frac{1}{1 - LC_{var}\omega^2}. \tag{3.69}$$

Since the second part of the right term in Eq. (3.69) is greater than 1, $\dfrac{\Delta C_{eq}}{C_{eq}} > \dfrac{\Delta C_{var}}{C_{var}}$.

Thus, the ratio η is improved. However, the price to pay is an increase in insertion loss, since the inductance is not ideal. Hence, it must be determined when this technique is really interesting by considering lossy varactor and inductance. Let consider the simple models given in Figure 3.58.

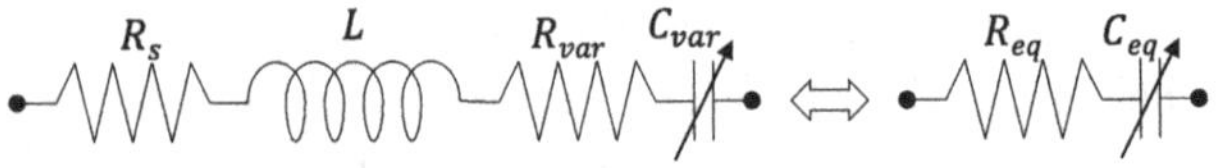

Figure 3.58 Lossy capacitance in series with a lossy inductance.

The complex impedance of R_{eq} in series with C_{eq} can be expressed as

$$Z_{eq} = R_{eq}\left(1 - jQ_{eq}\right), \tag{3.70}$$

with Q_{eq} the equivalent quality-factor:

$$Q_{eq} = \frac{1}{R_{eq}C_{eq}\omega}. \tag{3.71}$$

The reflection coefficient defined in Eq. (3.66) is expressed as

$$\Gamma = \frac{Z_{eq} - Z_0}{Z_{eq} + Z_0}, \tag{3.72}$$

with Z_0 the coupler ports impedance.

Using Eqs. (3.70) and (3.72) gives

$$\Gamma = \frac{R - j\left(\dfrac{Z_0 - R_{eq}}{Q_{eq}}\right)}{R + j\left(\dfrac{Z_0 + R_{eq}}{Q_{eq}}\right)}. \tag{3.73}$$

From Eq. (3.73), the phase and magnitude of the transmission parameter S_{41} (3.67) can be calculated. If $Z_0 \gg R_{eq}$ holds, the expression of the phase of S_{41} is simply

$$\varphi(S_{41}) = +90° - 2\tan^{-1}(Z_0 C_{eq}\omega). \tag{3.74}$$

From circuit simulations, carried out by considering an ideal 3-dB coupler, the interest in using an inductance in series with the varactor can be verified, without tedious calculus. For this purpose, a working frequency of 60 GHz was considered, without any lack of generality, and varactors corresponding to the BiCMOS 55-nm from STMicroelectronics were considered.

Without added series inductance, a varactor with the following characteristics was considered:

- $C_{vmax-wo} = 72$ fF; $Q_{min} = 13$, leading to $R_{var} = 2.85\ \Omega$
- $C_{vmin-wo} = 40$ fF; $Q_{max} = 22$, leading to $R_{var} = 3\ \Omega$

For simplification, R_{var} was considered equal to 3 Ω in the simulations, whatever the capacitance value.

With added series inductance, a varactor exhibiting a capacitance value divided by 2 was considered:

- $C_{vmax-w} = 36$ fF; $Q_{min} = 13$, leading to $R_{var} = 5.7\ \Omega$
- $C_{vmin-w} = 20$ fF; $Q_{max} = 22$, leading to $R_{var} = 6\ \Omega$

For simplification, R_{var} was considered equal to 5.85 Ω in the simulations, whatever the capacitance value.

The added inductance value was calculated so that the mean value of C_{eq} (called C_{eq-avg}) is equal to the mean value of the varactor selected for the RTPS without series inductance (called C_{v-avg}). Hence:

$$C_{eq-avg} = C_{v-avg} = \sqrt{C_{vmax-wo} \cdot C_{vmin-wo}} = 53.7\ \text{fF}. \tag{3.75}$$

By using Eq. (3.68) a value of 131 pH was found for the added inductance. Next, by considering an inductance quality-factor equal to 20, resulting in $R_s = 2.5\ \Omega$, the equivalent series resistance is equal to $R_{eq} = R_{var} + R_s = 8.35\ \Omega$.

Simulation results are summarized in Table 3.6.

As expected, the phase shift variation is greatly improved thanks to the use of the series inductance. It is almost multiplied by 2, from 33.2° to 66°.

When considering an ideal coupler, the FoM without inductance is much better, that is, 48.8°/dB instead of 27.7°/dB. However, in practice insertion loss between 1 and 2 dB can be expected, depending on the CMOS/BiCMOS technology and

Table 3.6 Phase shift variation, insertion loss, and FoM of two RTPSs carried out with or without series inductance

	Insertion loss (dB) (Ideal coupler)	Phase shift variation (°)	FoM (°/dB) Ideal coupler	FoM (°/dB) (1-dB insertion loss coupler)
Without inductance	0.68	33.2	48.8	19.8
With inductance	2.38	66	27.7	19.5

The working frequency is 60 GHz.

coupler topology, as shown in Section 3.4.2.7. By considering 1-dB insertion loss for the coupler, the FoM with or without inductance become almost equal, that is, 19°/dB for both approaches. In that case adding a series inductance can be an interesting way if the FoM/area is considered … after calculating the surface area of a 131-pF inductance. When a 2-dB insertion loss coupler is considered, the FoM drops to 15.1°/dB for the load with inductance and 12.4°/dB without inductance, respectively. Hence adding the inductance becomes clearly interesting.

Note that these values of FoM (from 12°/dB to 19°/dB) are in good agreement with those published in literature, as shown in Section 3.4.2.7.

3.4.2.5 Digital Tuning (Switched) Phase Shifters

Digital tuning (or switched) phase shifters can be achieved in many ways, as described in Figure 3.50.

First, any analog phase shifter carried out with varactors for continuous tuning can be transformed to a digital phase shifter by replacing the varactors by switches or bank of switched capacitors as described in Section 3.3.3, or by simply controlling the varactors as switched capacitances with two values corresponding to the maximum and minimum capacitance (C_{max}, C_{min}) of the varactor. The interest in controlling varactors as switched capacitances is to avoid capacitance variation due to control voltage variation, since the capacitance curve versus voltage is flat for C_{max} and C_{min}. An example of digitally controlled RTPS is given in [45]. In this article, the reflective load was realized by N varactors (corresponding to a N-bit RTPS) placed in parallel and individually controlled.

Second, switched networks and switched transmission lines can be used. Switched transmission lines were already described in Section 3.3.4. Only switched networks will be briefly described in the text that follows.

Switched networks are based on filter switching. They are more compact as compared to switched transmission lines, but they can lead to higher insertion loss when the switched delay (phase shift) is high, since the order of the filters increases with the delay, thus increasing the insertion loss. Two main configurations can be used:

- Switched low-pass filter [45–47]: a low-pass filter composed of series inductances and parallel capacitances is placed in series and can be bypassed by a switch placed in parallel, as shown in Figure 3.59.

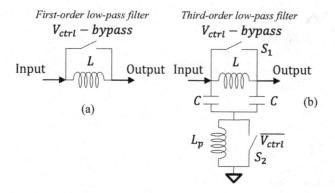

Figure 3.59 (a) First-order switched low-pass filter (series inductance). (b) Third-order switched low-pass filter (T-type network).

A first-order low-pass filter is described in Figure 3.59a. It is based on a single-series inductance. When the transistor is controlled by the voltage control, V_{ctrl} is "ON," the switch is closed, and the inductance is bypassed, the delay (phase shift) between input and output ports is minimum. When the switch is open, the series inductance creates a delay, as it acts as a first-order low-pass filter. Hence this simple structure acts as a switched delay.

Based on the same principle, a third-order low-pass filter is described in Figure 3.59b. In the bypass state, switch S_1 is closed and S_2 is open. Input and output ports are connected through S_1 and the delay is minimum. The parallel inductance L_p is necessary to cancel the effect of the OFF state parasitic capacitance C_{OFF} coming from the transistor used to realize the switch S_2. L_p must be calculated to resonate with C_{OFF} at the working frequency. In the delay state, the parallel inductance L_p is bypassed, and the network acts as a third-order low-pass filter, leading to phase delay. Note that pi-type network could be used. However, the pi-type network involves the use of two series inductances, which in general tend to increase the insertion loss of the network, since the quality factor of inductances is lower as compared to capacitances [48].

First-order low-pass filters can be used to realize small delays, whereas third-order low-pass filters must be preferred for larger delays. For instance, if a 5-bit 360° switched-network phase shifter was considered, least significant bits corresponding to 11.25° and 22.5° could be carried out with first-order switched low-pass filters, whereas 45° and 90° phase shifts could be realized with third-order switched low-pass filters. The realization of the 180° phase shift can be achieved by the use of two cascaded third-order switched low-pass filters.

- Switched high-pass low-pass filter ([47]): This switched network is interesting for the realization of large phase shifts, for instance for the realization of the 180° switched delay, instead of using two cascaded switched low-pass filters as described earlier. This concept was proposed in [47]; it is described by the scheme in Figure 3.60.

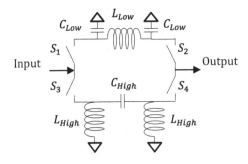

Figure 3.60 Switched high-pass low-pass filter.

Two third-order filters are used in Figure 3.60. When switches S_1 and S_2 are closed (S_3 and S_4 are open), the third-order low-pass filter formed by C_{Low} and L_{Low} is connected between input and output ports. On the contrary, when switches S_3 and S_4 are closed (S_1 and S_2 are open), the third-order high-pass filter formed by C_{High} and L_{High} is connected between input and output ports. The phase difference between low- and high-pass filters permits achieving an overall phase shift of 180° with only one switching network. Note that a T-network can be used to realize the high-pass filter.

3.4.2.6   Mixed Digital/Continuous Tuning Phase Shifters

As already mentioned, continuous tuning phase shifters are interesting because they allow very precise phase shift variation. However, the FoM of loaded-line or RTPS phase shifters is limited to less than 30°/dB, as explained in previous sections. This means insertion loss higher than 12 dB for a 0°–360° phase shift variation, which is unacceptable in practical applications. Mixing continuous and digital phase shifters is a solution to build continuous phase shifters with higher FoM. The digital cells can be achieved by switched-lines, switched filters, or switched high-pass/low-pass filters already discussed earlier. An example of 0°–360° mixed digital/continuous tuning phase shifter is given in Figure 3.61.

It is composed of three switched circuits (transmission lines or filters) that achieve 3 bits with a phase shift resolution of 45°, and a continuous phase shifter achieving 0°–45° phase shift. The command of the switched phase shifters to achieve 0°–360° continuous phase shifter is given in Table 3.7.

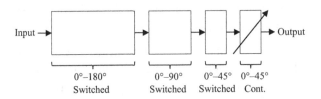

Figure 3.61 0°–360° mixed digital/continuous tuning phase shifter principle.

Table 3.7 Command for 0°–360° continuous phase shifter

Phase shift variation	0°–180° Switched	0°–90° Switched	0°–45° Switched
0° → 45°	0	0	0
45° → 90°	0	0	45°
90° → 135°	0	90°	0
135° → 180°	0	90°	45°
180° → 225°	180°	0	0
225° → 270°	180°	0	45°
270° → 315°	180°	90°	0
315° → 360°	180°	90°	45°

Several solutions are possible to build a phase shifter based on the scheme proposed in Figure 3.61. As an example, a solution based on the use of switched transmission lines to achieve the 0°–180° switched phase shifter, and switched filters to achieve the 0°–90° and 0°–45° switched phase shifters [51], is described in the text that follows. The continuous 0°–45° phase shifter is based on an RTPS such as that described in [43]. The 0°–180° switched phase shifter is built on the scheme described in Figure 3.62. The working frequency in the considered example is equal to 80 GHz.

The SPDT described in [21], fabricated in IBM's 90-nm 9HP technology, which features SiGe HBTs with peak f_T/f_{max} of 300/350 GHz, can be used for this application. It exhibits an insertion loss equal to 1.2 dB in E-band.

The expected FoM of such a topology is estimated in the following.

Switched transmission lines: The realization of switched transmission lines in CMOS faces two challenges:

1. The size
2. The amplitude imbalance, since longer transmission lines mean higher insertion loss.

A solution to address these challenges consists in using both S-CPWs and microstrip lines. As shown in [5, 17], 50-Ω characteristic impedance S-CPWs having a quality-factor higher than 30 and dielectric constant of about 20 can be achieved in standard CMOS/BiCMOS technologies BEOL. Their width is about 120 μm. 50-Ω characteristic impedance microstrip lines exhibit quality factors lower than 15 and dielectric constant of about 3.5. Their width is around 10 μm. By mixing these two types of transmission lines, straight lines can be used, thus avoiding meandering that reduces the design accuracy. The physical length l of the transmission lines is calculated by considering a phase shift difference equal to 180°:

$$\Delta\varphi = 180° = \Delta\beta \cdot l = \frac{\omega}{C_0}\sqrt{\varepsilon_{reff-S-CPW} - \varepsilon_{reff-\mu strip}} \cdot l, \qquad (3.76)$$

with C_0 the light velocity in vacuum, $\varepsilon_{reff-S-CPW}$ the S-CPW dielectric constant, and $\varepsilon_{reff-\mu strip}$ the microstrip line dielectric constant. Considering a working frequency

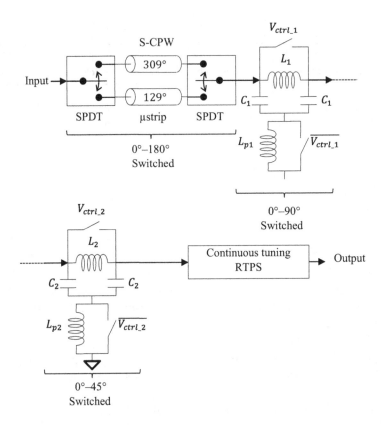

Figure 3.62 $0°$–$360°$ mixed digital/continuous tuning phase shifter example.

equal to 80 GHz, it leads to $l = 720$ μm. The total phase shift of the S-CPW and microstrip sections are:

$$\varphi_{\text{S-CPW}} = \frac{\omega}{C_0} \sqrt{\varepsilon_{\text{reff-S-CPW}}} \cdot l = 309°$$

$$\varphi_{\text{μstrip}} = \frac{\omega}{C_0} \sqrt{\varepsilon_{\text{reff-μstrip}}} \cdot l = 129°. \tag{3.77}$$

The insertion loss can be calculated from the quality factor:

$$Q = \frac{1}{2}\frac{\beta}{\alpha} = \frac{1}{2}\frac{\beta \cdot l}{\alpha \cdot l} = \frac{1}{2}\frac{\theta}{IL}8.68 \rightarrow IL = \frac{1}{2}\frac{\theta}{Q}8.68. \tag{3.78}$$

It leads to $IL_{\text{S-CPW}} = 0.78$ dB and $IL_{\text{μstrip}} = 0.65$ dB, that is, an amplitude imbalance equal to only 0.13 dB. If only microstrip lines were used, the amplitude imbalance would correspond to an electrical length difference equal to 180°, leading to an amplitude imbalance equal to 0.9 dB.

Finally, the overall width of the two transmission lines can be estimated to 120 μm (S-CPW) + 70 μm (microstrip line of width 10 μm with 30 μm free on each side), leading to a surface equal to $0.190 \times 0.75 = 0.14$ mm².

The insertion loss of the two switches (1.2 dB) must be added to the S-CPW insertion loss to calculate the overall insertion loss of the 0–180° stage, leading to a total insertion equal to about 3.2 dB (2×1.2 dB + 0.78 dB).

Switched filters:
The maximum insertion loss of the switched filters occurs when the series switches are in their ON state and is then equal to 1.2 dB. From [51], the size of the switched filters can be estimated to less than 0.02 mm².

RTPS:
The RTPS published in [43] exhibits maximum insertion loss equal to 4.2 dB at 40 GHz, for a phase shift equal to 60°. Its size is equal to $0.63 \times 0.280 = 0.18$ mm². For a phase shift variation equal to 45° at 80 GHz, the insertion loss would be almost the same, since the increase of insertion loss due to the lower quality factor of the varactors is compensated by the use of lower C_{max}/C_{min} ratio, which leads to an increase of the varactor's quality factor. Moreover, the insertion loss of the S-CPW coupler would be reduced at 80 GHz, since the quality factor of S-CPWs is greater at 80 GHz. The size of the RTPS would be reduced by a factor 2, since the main area is given by the coupler, leading to an area equal to $0.32 \times 0.28 = 0.09$ mm².

0°–360° phase shifter performance:
From the foregoing calculus, the overall electrical performance and size of the 0°–360° continuous phase shifter is given in Table 3.8.

Table 3.8 Overall electrical performance and size of the 0°–360° continuous phase shifter

Phase shift variation	Max. IL (dB)	FoM (°/dB)	Area (mm²)
0° → 360° continuous	$3.2 + 2 \times 1.2 + 4.2$ $= 9.8$ dB	36.7	$0.14 + 2 \times 0.02 + 0.09$ $= 0.27$

The FoM and area obtained in Table 3.8 correspond to the state-of-the-art of digital mm-wave phase shifters with 4 to 5 bits. Hence, a high FoM with moderate size could be obtained by using mixed analog/digital approaches. For that, designers must select the best option for each phase shifter stage.

3.4.2.7 State-of-the-Art

It is not the first goal of this book to make a complete state-of-the-art concerning the circuits that are discussed, including phase shifters. A state-of-the-art is just a photo of what was achieved at a given time. However, it gives (or not!) some trends that can help the readers to choose a certain technology and/or a certain circuit topology.

Hence, the current state-of-the-art is given in Table 3.9. It is organized in three main topologies:

1. Loaded-lines phase shifters
2. Reflection-type phase shifters
3. Switched networks phase shifters

Table 3.9 Millimeter-wave passive phase shifters state-of-the-art

Freq. (GHz)	FoM (°/dB)	Phase Shift (°)	Analog / digital Bits (resolution °)	IL variation (dB)	Area (mm²)	Area (mm²) for 360°	Technology	Topology	Reference
60	7	20	Analog		0.02	0.36	CMOS 90 nm	Loaded-line (S-CPW)	[51]
60	12.5	156	4 (22.5)	7	0.2	0.46	CMOS 65 nm	Loaded-line (Differential grounded CPW)	[52]
28	20	180	6 (4.75)	0.5	0.18	0.36	BiCMOS 130 nm	Loaded-line (Constant Zc / Switched)	[39]
45	24	79			0.072	0.33	CMOS SOI 45 nm	Loaded-line (Constant Zc / Switched)	[38]
60	24	180	Analog	3.3	0.15	0.30	BiCMOS 130 nm	RTPS (Lange Coupler)	[53]
60	11.2	90	Analog	2	0.08	0.32	CMOS 90 nm	RTPS (Broadside coupler)	[54]
60	24.2	150	Analog	2.2	0.33	0.79	BiCMOS 130 nm	RTPS (Slow-wave coupled lines)	[55]
60	10	72	Analog		0.5		BiCMOS 130 nm	RTPS (Branch-line coupler)	[56]
60	22	180	Analog	3.3	0.031	0.06	CMOS 65 nm	RTPS (Lumped Branch-line coupler)	[57]
60	27.8	147	Analog	2.2	0.048	0.12	CMOS 65 nm	RTPS (Lumped Branch-line coupler)	[57]
62	36	367	Analog	6.6	0.16	0.16	BiCMOS 130 nm	RTPS (Lumped Branch-line coupler)	[58]
80	18	360	5 (11)				BiCMOS 130 nm	Switched networks (Low-pass)	[46]
28	47	360	4 (22.5)		0.23	0.23	CMOS 65 nm	Switched networks (High-pass/Low-pass)	[48]
60	22	360	5 (11)		0.094	0.09	CMOS 65 nm	Switched networks (Low-pass)	[45]
60	20	360	5 (11)	6	0.34	0.34	CMOS 90 nm	Switched networks (Low-pass)	[48]

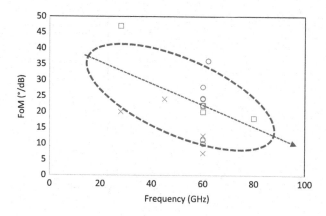

Figure 3.63 FoM versus frequency. Crosses: loaded-lines, squares: switched networks, circles: RTPS.

Working frequency, FoM, achieved phase shift, type (analog or digital), insertion loss variation, area, and area for 360° of phase shift, technology, and topology are given.

The area for 360° of phase shift is simply calculated by multiplying the actual area by the ratio between 360° and the achieved phase shift.

From the state-of-the-art given in Table 3.9, it seems very difficult and unrealistic to derive what would be the best technology or topology to achieve high FoM or small area.

From Table 3.9, the FoM is plotted versus working frequency in Figure 3.63. The trend (dot arrow and ellipse in Figure 3.63) shows that the FoM decreases with frequency. This is mainly due to the decreased performance of varactors or switches versus frequency, as discussed in Section 3.3. However, as can be observed for the phase shifters carried out around 60 GHz, the FoM can vary from 7 up to 36°/dB. Hence, it is not sufficient to deal with general topologies, but the design approaches carried out by different authors can make all the difference.

In that context, let consider the work carried out in [58]. In this article, the RTPS was realized with a lumped branch-line coupler, which leads to a high FoM equal to 36°/dB at 62 GHz, which is the best FoM obtained so far with passive phase shifters around 60 GHz. Hence, a lumped approach for the realization of the RTPS coupler seems to be an effective way at this frequency. Nevertheless, this conclusion would be perhaps different at higher frequency, with design complexity that increases for lumped components, due to parasitic effects.

The FoM versus area (normalized for 360° phase shift) is given in Figure 3.64. FoM higher than 20°/dB can be achieved for areas lower than 0.2 mm². Thanks to the use of a lumped approach, here again, the work carried out in [58] exhibits very good tradeoff between performance (FoM) and area, with a FoM equal to 36 for an area of 0.16 mm².

Many other criteria could be derived for the choice of the right topology and technology when dealing with passive phase shifters, as

- Digital versus analog approach: In many topologies, both approaches can be used, except for switched networks; the varactors can be replaced by switches, leading to digital phase shifters instead of analog ones.

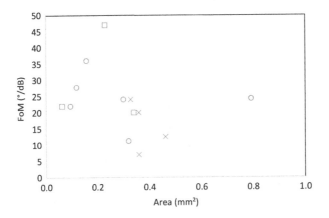

Figure 3.64 FoM versus area. Crosses: loaded-lines, squares: switched networks, circles: RTPS.

- Return loss
- Insertion loss variation, which is a critical point when dealing with antenna arrays
- Technology cost

3.4.3   Voltage-Controlled Oscillators

Voltage-controlled oscillators (VCOs) are key building blocks in receivers, emitters, and transceivers. They must be able to deliver a sinusoidal wave featuring low phase noise and high power efficiency, over a wide tuning range. Modern communication systems require both very sensitive receivers able to capture very small signals, thus reducing the power required for the emitter, or increasing the communication distance, and high efficiency to improve mobility. In this section, basic principles of VCOs are first reviewed, then several topologies used at mm-waves are described.

3.4.3.1   **Basic Principles, Challenges at Millimeter-Waves, and Figures of Merit**

Principle

In mm-wave IC, most of the oscillators (*LO* for "local oscillator") are based on resonators. Some topologies based on inverters exist but are dedicated to very specific applications and will not be discussed here. Ideally, a lossless resonator would oscillate indefinitely. Unfortunately, its losses avoid such a behavior. Compensating the losses of a resonator is a solution to turn this resonator into an oscillator. For a given frequency, an *LC* tank can be modeled as an inductor L_p in parallel with a capacitor C_p as represented in Figure 3.65. The losses in the resonator can be represented by a parallel resistor R_p.

To compensate for the losses (R_p), a negative resistance (R^-) must be added in parallel, such as the total resistor ($R_p//R^-$) is lower than zero. In practice, the R^- is a two-port active device having a current i_x flowing from the negative voltage to the positive voltage as represented in Figure 3.66. The cross coupled pair (CCP), represented in Figure 3.66, is commonly used in integrated circuits.

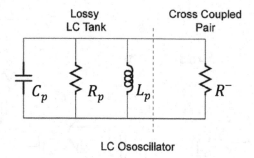

Figure 3.65 *LC* oscillator principle.

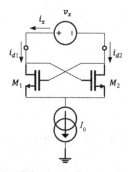

Figure 3.66 Cross coupled pair (CCP)-based negative resistor.

Because the current i_x and the voltage v_x are defined as follows:

$$i_x = i_{d1} - i_{d2} \quad \text{and} \quad v_x = v_{gs2} - v_{gs1}, \tag{3.79}$$

the equivalent resistor seen from the CCP port is

$$R^- = \frac{v_x}{i_x} = \frac{-2}{g_m}. \tag{3.80}$$

To oscillate, $R_p // R^-$ must be equal or lower than zero. This condition leads to

$$\frac{R^- \cdot R_p}{R_p + R^-} \leq 0 \Rightarrow R_p + R^- \geq 0 \Rightarrow R_p \geq \frac{2}{g_m} \Rightarrow g_m \geq \frac{2}{R_p}. \tag{3.81}$$

This latter condition is the "oscillation condition." In practice a loss compensation factor is used to ensure the starting of the LO. The g_m of the *CCP* depends mainly on the size of the transistors used (W/L) and on the value of the DC current (I_0):

$$g_m = 2\sqrt{2\mu_n C_{ox} \frac{W}{L} I_0}, \tag{3.82}$$

with μ_n the electron mobility and C_{ox} the oxide capacitance.

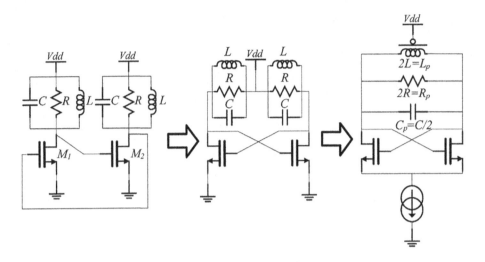

Figure 3.67 Looped representation of an *LC* oscillator.

Because the Q-factor of a resonator decreases when the losses increase (i.e., when R_p decreases), the value of the g_m needed to compensate for the losses grows with the losses. Therefore, low Q-factor means large CCP (i.e., large W) and/or large power consumption (i.e., large I_0).

It is worth mentioning that a CCP can be seen as two looped amplifiers loaded by a part of the resonator, as represented in Figure 3.67. Such representation explains how the oscillation occurs. When the LO is turned on, the drain current noise is amplified and filtered. Because the loopback is positive, the magnitude of the noise component, which is centered at the center frequency of the resonator, grows and the oscillations start. This is the reason why the settling time of such LO is dominated mainly by the transconductance of the CCP and R, both of which set the gain of the amplifiers.

With the representation in Figure 3.67, it is easy to apply the Barkausen criterion. The Barkausen criterion states that an oscillation occurs in a positive feedback system if the open loop gain phase is null and its module is greater than zero. In the system described in Figure 3.67, at the resonant frequency $F_0 = 1/(2\pi\sqrt{LC})$, the loads of the common source amplifier (M_1 and M_2) are real and equal to $R = R_p/2$. Each amplifier introduces 180° phase shift (which gives a positive feedback with a phase equal to zero) and the system can oscillate if the open loop gain module is greater than 1. Expressing the open loop gain gives the following oscillation condition:

$$\|g_m \cdot Z(j\omega_0)\|^2 = (g_m \cdot R_p/2)^2 \geq 1 \rightarrow g_m \geq 2/R_p. \tag{3.83}$$

To turn an LO into a VCO, it is necessary to vary the center frequency of the resonator. Depending on the nature of the resonator, several solutions can be used. However, in integrated circuits, the most convenient device is the varactor. A varactor is a reverse-biased diode, as described in Section 3.3.1. The capacitance of a reverse-biased diode varies with the bias voltage of the diode. However, varactors achieve a poor quality factor, as shown in Figure 3.24, especially at mm-waves. This is discussed in the next section.

Issues in Millimeter-Wave Design

A first issue concerns the frequency tuning range (FTR), which represents the ability of the VCO to generate different frequencies around its center frequency. The CCP adds parasitic capacitors (C_{par}) to the resonator as represented in Figure 3.68. The total capacitance of the network is then the sum of the capacitance of the varactor (C_{var}) and C_{par}. These parasitic capacitors reduce the working frequency F_o, defined as

$$F_o = \frac{1}{2\pi\sqrt{L\left(C_{var\_avg} + C_{par}\right)}}, \tag{3.84}$$

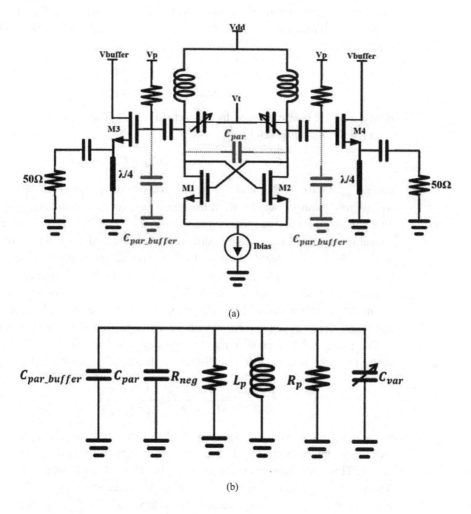

(a)

(b)

Figure 3.68 *LC*-VCO with parasitic capacitors (a) and its equivalent circuit (b).

where C_{var_avg} is the average value of C_{var}. However F_o can be tuned with C_{var}, and it is not an issue.

But these parasitic capacitors finally also reduce the FTR defined by

$$FTR = \frac{F_{max} - F_{min}}{F_o} = \frac{\Delta F_o}{F_o}. \tag{3.85}$$

Let us call C_{total} the total capacitance and C_{total_avg} its average value:

$$C_{total\_avg} = C_{par\_avg} + C_{var}. \tag{3.86}$$

The relative variation of C_{total} is simply written as

$$\frac{\Delta C_{total}}{C_{total\_avg}} = \frac{\Delta C_{var}}{\left(C_{par} + C_{var\_avg}\right)} = \frac{\Delta C_{var}/C_{var\_avg}}{1 + C_{par}/C_{var\_avg}}. \tag{3.87}$$

Hence the total capacitance variation is reduced by the factor $1 + C_{par}/C_{var\_avg}$.

Next, by replacing $C_{var\_avg} + C_{par}$ by C_{total_avg} in Eq. (3.84), the FTR can be expressed as

$$FTR = \frac{1}{2}\frac{\Delta C_{total}}{C_{total\_avg}} = \frac{1}{2}\frac{\Delta C_{var}/C_{var\_avg}}{1 + C_{par}/C_{var\_avg}}. \tag{3.88}$$

Hence the FTR is also reduced by the factor $1 + C_{par}/C_{var\_avg}$.

In a CCP, C_{par} is given by

$$C_{par} = 2C_{gd} + \left(C_{gs} + C_{gb} + C_{db} + C_{ds}\right), \tag{3.89}$$

where C_{gd}, C_{gs}, C_{gb}, C_{db}, and C_{ds} are the extrinsic capacitors of an MOS transistor, which grows with the width (W) of the transistor. Low Q factors lead to high g_m and finally large parasitic capacitors, which reduces the FTR. This issue is by far the most critical in mm-wave-IC, since it is difficult to achieve a high Q factor. For the same reasons, the Q factor highly impacts the power consumption through the g_m, since a high current is necessary to produce high g_m. Moreover, buffers are needed either for measurement purposes or to isolate the VCO from the other part of the system driven by the VCO. These buffers also add parasitic capacitance that reduce the FTR. An equivalent circuit taking into account all the above parasitics is given in Figure 3.68b.

The phase noise is another issue in the design of mm-wave-IC VCO. Leesson showed that the noise transfer function depends on the Q factor as follows [59]:

$$\left\|\frac{n_0(f_{OL} + f_n)}{n_i(f_n)}\right\|^2 \approx \frac{1}{4Q^2}\left(\frac{f_{OL}}{f_n}\right)^2, \tag{3.90}$$

where $n_o(f)$ and $n_i(f)$ are the spectrum of the output noise and the input noise of the looped system, respectively. The input noise comes from the CCP and the power supply. This latter formula describes simply the fact that the baseband noise from the devices is transposed around the oscillation frequency (f_{LO}). As described in Figure 3.69, the $1/f$ noise generates $1/f^3$ slope close to f_{LO} and the white noise generates a $1/f^2$ slope at $f_{LO} + f_c$, where f_c is the $1/f$ corner frequency.

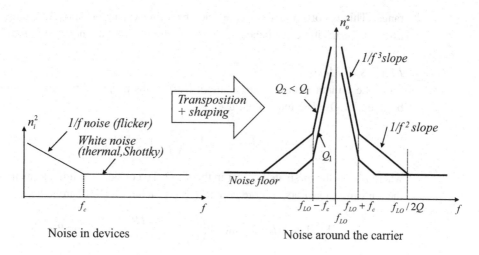

Figure 3.69 Output noise of *LC* oscillators.

A more complete formula based on experimental results was proposed by Leesson to consider additional aspect such as the output power:

$$L(f_n) = 10\log\left[\left(\left(\frac{f_{LO}}{2Q \cdot f_n}\right) + 1\right)\left(\frac{f_c}{f_n} + 1\right)\left(\frac{FkT}{P_{out}}\right)\right], \qquad (3.91)$$

where $L(f_n)$ is the phase noise relative to the carrier in dBc/Hz, P_{out} is the output power, and F is a fitting parameter related to the noise factor of the amplifiers used in the oscillator.

Whereas the design of LNA or mixers is matter of tradeoff between different characteristics (i.e., linearity and noise), the design of a VCO is highly dominated by the performance of the resonator. Improving the Q factor improves consumption, FTR, and noise. From this point of view, and because the performance of the active devices in terms of f_t is not a limiting aspect in modern technology, the major issue in the design of a mm-wave-IC VCO is the performance of the resonator in terms of losses. It is even more an issue since tunable resonators are needed to tune the oscillation frequency. This is the reason why the major effort in research is led by the exploration of high-performance tunable passive devices.

The quality factor of the inductance of the *LC* tank is the main limiting factor at RF frequencies, up to few GHz. On the other hand, at mm-waves the Q-factor of varactors becomes the main limiting factor, with values lower than about 6–10 at frequencies greater than 50 GHz, for C_{max}/C_{min} ratios lower than 2 (see Table 3.4 in Section 3.3.1). In that context of low quality factor, the size of the *CCP* needs to be increased to guarantee the oscillations, which in turn increases the transistor's parasitic capacitance C_{par}. This dramatically limits the frequency tuning range, since C_{par} becomes significant as compared to the capacitance required for the *LC* tank. Hence there is clearly a critical tradeoff between the maximum oscillation frequency and the tuning range. This is solved at RF frequencies by using capacitor-bank structures, but the associated parasitic capacitance of these structures prohibits their use at mm-waves.

Figures of Merit

From these main issues, figure of merit (FoM) was proposed to make a comparison between VCOs. The most often used is

$$\text{FoM} = L(f_n) - 20\log\left(\frac{f_{LO}}{f_n}\right) + 10\log(P_{diss}), \tag{3.92}$$

where P_{diss} is the dissipated power in the VCO. To consider the limitations induced by the parasitic capacitors, a more complex FoM including FTR can be used:

$$\text{FoM} = L(f_n) - 20\log\left[\left(\frac{f_{LO}}{f_n}\right)\cdot\left(\frac{FTR}{10}\right)\right] + 10\log(P_{diss}). \tag{3.93}$$

3.4.3.2 Architectures for Millimeter-Waves

Some solutions have been proposed by different authors to address the issue mentioned earlier, that is, permit to increase the maximum oscillation frequency without sacrificing the tuning range. They will be presented in this section.

Millimeter-wave VCOs can be classified into two main categories: "lumped" and "distributed." Lumped means that VCOs are based on an *LC* tank, whereas distributed VCOs are based on the use of transmission lines. Figure 3.70 illustrates this simple classification, along with subclasses. Among *LC*-tank VCOs, those based on fundamental frequency and those based on the use of harmonics, such as push–push

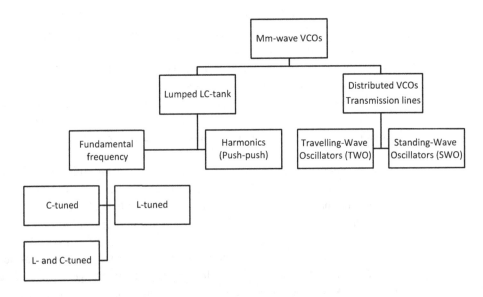

Figure 3.70 Millimeter-wave VCO classification.

topology, constitute two distinct subclasses. By the same manner, fundamental-frequency VCOs can be separated into three categories, depending on the element that is tuned. Next, two subclasses can be distinguished among distributed VCOs, that is, VCO-based traveling-wave topology and those based on standing-wave topology.

Most of the mm-wave VCOs reported in the literature so far concern the 60-GHz band (57–65 GHz), since this specific band addresses many consumer applications, that is, wireless HDMI cable replacement, uncompressed video and audio transmission, and short-distance, high-speed uncompressed data transfer, and probably 5G in the future. Hence, most of the concepts based on published papers and described in the next subsections concern 60-GHz band applications.

Lumped LC-Tank VCOs

The classical schematic used for the design of mm-wave *LC*-tank VCOs is given in Figure 3.71. It is composed of an *LC* tank that can be tuned either by the capacitance (V_{tune}), the inductance, or both simultaneously, and a cross-coupled pair (M1 and M2) that serves to compensate the *LC*-tank losses.

Due to the aforementioned issues concerning the tradeoff between tuning range and operation frequency, only few VCOs based on classical *LC* tanks were reported so far. In [60], a classical *LC*-tank approach was used in a 130 nm CMOS technology. Thanks to a careful design of the MOS varactors, a tuning range of 5.8 GHz was obtained between 53.6 GHz and 59.4 GHz. But the tuning range is reduced to 200 MHz at 105.2 GHz for the same architecture. A Colpitts scheme was used in [61] as an alternative to cross-coupled pairs, in a 32-nm CMOS technology. An oscillation frequency up to 240 GHz was achieved, with a tuning range of 5.6%, that is, more than 13 GHz.

To increase the VCOs' operation frequency, push–push topologies have been proposed by many authors. For example, a 60-GHz VCO was carried out in [62]. It exhibits a tuning range of 13.9 GHz, that is, 26.5%. However, the drawback of

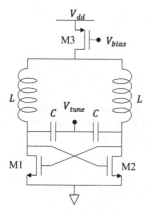

Figure 3.71 Classical schematic used for the design of mm-wave *LC*-tank VCOs.

push–push topologies is linked to the output power, which is generally limited, since frequency conversion efficiency is low.

Hybrid approaches were used to increase the tuning range with operation frequencies near the 60-GHz band, to cover the whole 57–65 GHz spectrum. In [63] and [64], a digitally controlled varactor bank was used in parallel with continuous tuning analog control varactors, as shown in Figure 3.72. A tuning range of 17% was achieved. By the same manner, switched MOM capacitors and continuous tuning was used in [65].

In parallel with the development of hybrid approaches described earlier, designers tried to develop tunable inductances. The idea is to replace the varactors that exhibit a low quality factor by tunable inductances with higher quality factor. In [32], a varactor was placed in series with an inductance in order to realize a tunable inductance, as proposed in [31] and illustrated in Figure 3.73. However, the overall quality factor is

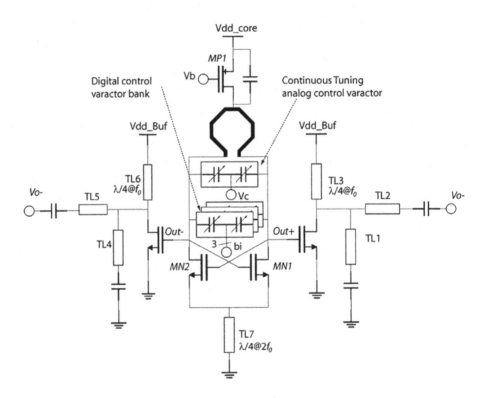

Figure 3.72 Hybrid digital/analog approach: digitally controlled varactor bank is used in parallel with continuous tuning analog control varactors in order to increase the tuning range. Figure courtesy of Jose-Luis Gonzalez Jimenez, CEA-Leti, Grenoble, France.

Figure 3.73 Tunable inductance realized by an inductance in series with a varactor.

lower than that of the poorer component, that is, the varactor. Moreover, the relative variation of the equivalent tunable inductance is lower than that of the varactor, thus limiting the overall tuning range, as explained in the text that follows. Let us first consider lossless components, an inductance L in series with a varactor of capacitance C_{var}. The overall impedance is equal to

$$Z = jL\omega + \frac{1}{jC_{var}\omega} = j\frac{\left(\frac{\omega}{\omega_r}\right)^2 - 1}{C_{var}\omega} = jL_{var}\omega, \tag{3.94}$$

with

$$\omega_r = \frac{1}{\sqrt{L \cdot C_{var}}} \quad \text{and} \quad L_{var} = L\frac{L \cdot C_{var}\omega^2 - 1}{L \cdot C_{var}\omega^2}. \tag{3.95}$$

Hence L_{var} is lower than L.

Next, the variation of L_{var} $\left(\frac{\delta L_{var}}{L_{var}}\right)$ versus the variation of C_{var} $\left(\frac{\delta C_{var}}{C_{var}}\right)$ is given by

$$\frac{\delta L_{var}}{L_{var}} = \frac{1}{L \cdot C_{var}\omega^2 - 1}\frac{\delta C_{var}}{C_{var}}. \tag{3.96}$$

Since $LC_{var}\omega^2 \gg 1$ for proper operation

$$\frac{\delta L_{var}}{L_{var}} =< \frac{\delta C_{var}}{C_{var}}. \tag{3.97}$$

This indicates that the proposed technique fails in increasing the VCO tuning range.

In [66], a transformer-based tunable inductance was proposed to realize a multi-band VCO. The concept is based on the use of a transformer working as an impedance inverter, as shown in Figure 3.74. A transistor M_v loading a transformer (L_1, L_2, k) works as a tunable resistance R_v in parallel with a parasitic capacitance C_v. From the primary coil, the equivalent circuit is a tunable inductance L_{eq} in parallel with a tunable resistance R_{eq}. The main interest of such a concept is that the tunable

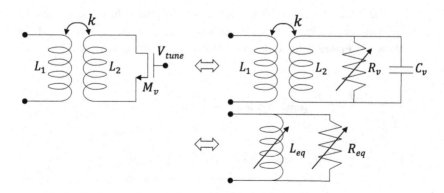

Figure 3.74 Transformer-based tunable inductance [66].

inductance does not suffer from any parasitic, which is the case for a classical LC-tank where the varactor is in parallel with the cross-coupled pair parasitic capacitance that dramatically limits the tuning range, as explained earlier.

The tunable inductance quality factor is between 3.6 and 6.7 in [66]. No additional capacitance was added to realize the LC tank, the LC-tank capacitance being realized with the cross-coupled pair transistors' parasitic capacitance.

Based on this principle, by using many transistors in parallel at the secondary coil, each controlled by an independent voltage, as shown in Figure 3.75, multiband operation can be achieved. Fine tuning can be achieved by controlling the first transistor stage in a continuous way.

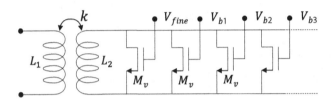

Figure 3.75 Independent voltage-controlled parallel transistors at the secondary coil for multiband operation [66]. Fine tuning is realized with the first stage.

The complete VCO scheme is described in [66].

Thanks to the concept of tunable inductance, very high performance was achieved in [66], with a tuning range of 14% from 53.1 to 61.3 GHz, and a low phase noise of -118.75 dBc/Hz at 10 MHz of the center frequency.

The main limitation of the tuning range improvement is the limited coupling coefficient of the transformer, since the theoretical limit of tuning range is given by $k^2/2$, with k the coupling factor. An innovative way could be to replace the transformer by a coupled-line coupler. As demonstrated in [42], coupling coefficient higher than 3 dB can be achieved by using a S-CPW coupled-line coupler. Hence, replacing the lumped transformer by a distributed coupler could lead to a tuning range as high as about 25%.

The same concept of transformer-based tunable inductance was used in [67], the main difference being the introduction of the concept of distributed resistance, as shown in Figure 3.76, to develop a more linear tuning scheme.

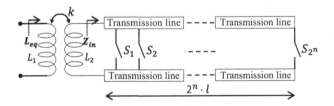

Figure 3.76 N-bit linear variable inductor using the concept of distributed resistance.

As presented in [67], the real part of the input impedance Z_{in} of the distributed resistance is equal to

$$Re(Z_{in}) \approx 0.5Z_0\left(1 + \frac{\pi}{2} - 2\beta l\right),\tag{3.98}$$

with Z_0 the characteristic impedance of the transmission line, l the length of a transmission line section, and β its propagation constant.

Equation (3.98) shows that the equivalent resistance is almost proportional to l, thus leading to a linear operation, Z_{in} being controlled by positioning the switches to reconfigure the effective length of the transmission line, as explained in [67].

Another way for the realization of tunable inductor is the manipulation of the magnetic flux. This can be done in two different manners, as explained in the text that follows.

First, since the inductance of a loop is proportional to its area, changing the area modifies the inductance. The modification of the loop area needs the use of at least two switches in series with the inductance strip, as illustrated in Figure 3.77. When switches S_1 and S_2 are in their ON state (S_3 and S_4 OFF), the area of the loop is maximum, equal to the sum of the areas A_1 and A_2. When switches S_3 and S_4 are in their ON state (S_1 and S_2 off), the area of the loop is minimum, equal to the single area A_1. It is thus a very simple way to achieve tunable inductance. Unfortunately, in practice the losses added by the switches dramatically decrease the overall quality factor, thus increasing the phase noise.

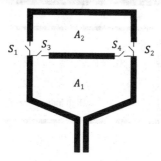

Figure 3.77 Modification of the loop area by using switches.

The second way to achieve tunable inductance by modifying the flux is more efficient. Adding a satellite loop around the main loop modifies the inductance, since the current induced in the satellite loop creates a flux that is in opposite direction of the main flux, due to the Lenz law, thus decreasing the inductance of the main loop, as illustrated in Figure 3.78. The main current i creates a flux considered as positive through the loop, whereas the induced current i_{ind} creates an opposite flux, hence considered as negative. From an electrical point of view, it can also be said that the coupling coefficient between the loops is negative.

The modification of the satellite loop area leads to the modification of the inductance, thus creating the opportunity to achieve a tunable inductance. Here again, two switches are necessary to modify the satellite loop area. However, since the main

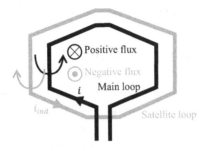

Figure 3.78 Satellite loop modifying the inductance of a main coil.

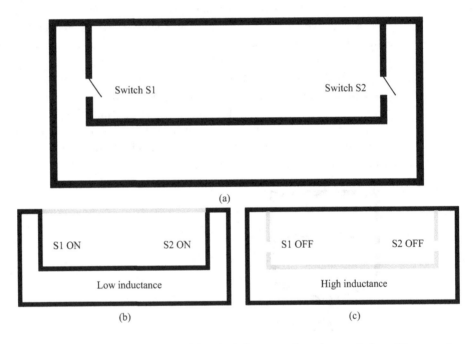

Figure 3.79 Single satellite loop modifying the inductance of a unique main loop. The ground ring is modified thanks to switches S1 and S2 (a), providing two types of guard ring structures, namely, (b) and (c), with low and high inductance area, respectively.

current does not flow in the satellite loop, the impact of the switch losses on the loop quality factor is lower.

The concept based on the modification of the satellite loop area was proposed in [68] and [69]. A variable guard ring was considered in [68], as illustrated in Figure 3.79, whereas a more general scheme was proposed in [69]. However, the concepts are based on the same flux manipulation.

In [68], the main inductance was realized by an S-CPS (slow-wave coplanar stripline) that leads to a high quality factor greater than 40 in CMOS technologies, as proposed in [70].

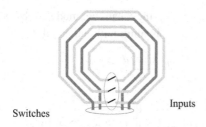

Switches   Inputs

Figure 3.80 Tunable inductance proposed in [68].

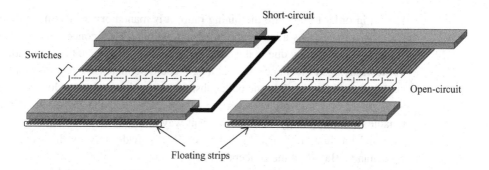

Figure 3.81 Tunable S-CPS carried out to achieve tunable components. (Left) Tunable inductance realized with a short-circuited tunable S-CPS, and (right) tunable capacitance realized with an open-circuited tunable S-CPS.

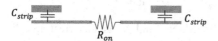

Figure 3.82 Tunable S-CPS maximum capacitance.

A more general scheme of the tunable inductance was proposed in [69], with two main coupled loops surrounded by three satellite loops, as illustrated in Figure 3.80.

Referring to Figure 3.70, the last family of LC-tank VCOs is based on both tunable inductance and capacitance. The interest in using both tunable components is the increased tuning range, since the tuning range offered by the tunable inductance roughly adds to the one offered by the tunable capacitance. A realization based on this concept was proposed in [71] and [72]. In this example, both tunable elements were achieved thanks to the use of tunable S-CPSs, as shown in Figure 3.81.

Floating strips are cut in the middle and switches link both sides of the strips. When the switches are in their ON state, that is, they can be replaced by a series resistance R_{on}, the capacitance of the S-CPS is maximum C_{max}, as illustrated in Figure 3.82.

C_{max} is expressed as

$$C_{max} = \frac{C_{strip}}{2}. \tag{3.99}$$

When the switches are in their OFF state, they can be modeled by a parasitic capacitance C_{off}, and the S-CPS capacitance is minimum and equal to

$$C_{min} = \frac{C_{strip} \cdot C_{off}/d}{C_{strip} + 2C_{off}/d}. \tag{3.100}$$

where d is the length of a section of floating strips.

The important parameter is the capacitance tuning ratio, given by

$$TR = \frac{C_{max}}{C_{min}} = 1 + \frac{1}{2}\frac{C_{strip}}{C_{off}/d}. \tag{3.101}$$

Hence, in order to increase the tuning ratio, it is mandatory to maximize the ratio of C_{strip} over C_{off}/d. As already presented in Section 3.4.2.2 concerning passive phase shifters, C_{off} is mainly due to the coupling capacitance between the cut floating strips. To decrease this capacitance, the cut length could be increased, but this would lead to electrical field flowing into the bulk substrate, hence increasing the loss of the S-CPS. Thus, there is a tradeoff between the tuning range and the losses for this kind of tunable S-CPS topology. A more efficient way is to put the switches between a CPS strip and the floating strips. In that case, C_{off} is made only by the switch parasitic capacitance. Based on the concept of ON/OFF switched floating strips of the S-CPS, a tunable inductance is made when the tunable S-CPS is terminated by a short circuit acting as a shorted stub, whereas a tunable capacitance is made when the tunable S-CPS is terminated by an open circuit acting as an open stub. Indeed, a short-ended stub presents an input impedance equal to

$$Z_{in,open} = -jZ_c/\tan(\theta). \tag{3.102}$$

with Z_c the stub characteristic impedance and θ its electrical length. For small electrical length θ, $Z_{in,open}$ can be simplified as

$$Z_{in,open} \approx \frac{Z_c}{j\theta} = \frac{1}{jC_{eq}\omega}. \tag{3.103}$$

where $C_{eq} = \dfrac{\theta}{\omega Z_c}$ is the stub's equivalent capacitance.

By the same manner, a short open-ended stub presents input impedance equal to

$$Z_{in,short} \approx jZ_c \cdot \theta = L_{eq}\omega. \tag{3.104}$$

where $L_{eq} = \dfrac{Z_c \cdot \theta}{\omega}$ is the stub's equivalent inductance.

Distributed VCOs

As presented in Figure 3.70, distributed VCOs are based on transmission lines. Two families of distributed VCOs can be identified, based either on traveling-wave oscillators (TWOs) or standing-wave oscillators (SWOs). Among these two architectures of distributed VCOs, SWOs exhibit better performance in terms of efficiency; their design is also much easier. They were studied by many groups in the last few years. Hence only this distributed architecture will be presented in detail in this book.

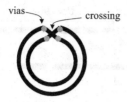

Figure 3.83 Crossing in rotary TWOs.

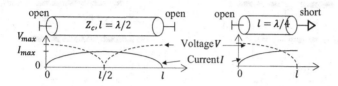

Figure 3.84 Resonators for the realization of standing-wave oscillators. (Left) Half-wavelength. (Right) Quarter-wavelength.

TWOs need to be terminated by termination resistors, where some portion of the energy may be absorbed, explaining their lower efficiency. To improve the efficiency, rotary TWOs were proposed, as in [73], that do not need termination resistors. However, they need to design crossing transmission lines, as illustrated in Figure 3.83. The crossing is an issue at mm-waves, since the parasitic coupling capacitor must be modeled accurately, as the vias that are necessary to realize the crossing, thus leading to more complex designs as compared to SWOs that are described in the text that follows.

SWO architectures are based either on half- or quarter-wavelength resonators, as illustrated in Figure 3.84.

The advantage of a half-wavelength resonator is the absence of short-circuit, which causes losses (and can be tricky to design). Thus half-wavelength resonators exhibit a higher quality factor. However, there length can be prohibitive, even at mm-waves, since half-wavelength is equal to half an mm at 100 GHz by considering a slow-wave transmission line (S-CPS) with effective dielectric constant equal to 9. It is near 1 mm when considering microstrip lines. In practice, the overall quality factor of an SWO is imposed mainly by the quality factor of the varactors, thus lowering the impact of the short circuit. That explains why a great majority of authors have preferred the use of quarter-wavelength resonators.

Many architectures are possible to achieve SWOs, depending on the number of cross-coupled pairs (CCPs) and varactors that are used:

- Only one CCP and one varactor
- Only one CCP and periodically distributed varactors
- Periodically distributed CCP and one varactor
- Fully distributed: periodically distributed CCP and varactors

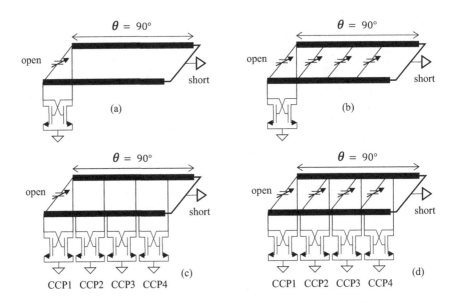

Figure 3.85 Different architectures of SWOs. (a) Lumped CCP and varactor. (b) Lumped CCP and periodically distributed varactors. (c) Periodically distributed CCP and lumped varactor. (d) Fully distributed: periodically distributed CCP and varactors.

These architectures are described in Figure 3.85. Note that the electrical length of the SWO is equal to 90°, whereas the physical length of the resonator is lower than $\lambda/4$, since loading the resonator by the CCPs and the varactors reduces its physical length.

In Figure 3.85a, corresponding to lumped CCP and varactor, the CCP and the varactor are both placed at the open-ended side of the resonator, where the voltage is maximum. In Figure 3.85b, the varactors are distributed along the resonator. Obviously, no varactor is placed at the short. The simplest way consists in periodically distributing the varactors along the resonator, by using T or π networks. However, a careful study of the value and position of each varactor could probably lead to optimum resonator quality factor. Figure 3.85c scheme is the dual of Figure 3.85b, that is, the CCPs are distributed along the resonator, whereas only one varactor is used. Finally, the scheme in Figure 3.85d represents the fully distributed SWO, with both distributed CCP and varactors. As in case (b), the position and value of varactors and CCP must be selected smartly to get the maximum resonator quality factor.

The use of differential transmission lines, that is, two coupled strips with or without ground plane, is the simplest way to realize the resonator. The best performance is obtained when using S-CPS, described in Figure 3.19.

Quarter-wavelength SWOs were proposed in [70, 74]. In both papers, S-CPSs were used.

In [70], the shape of the S-CPS was tapered to maximize the overall resonator quality factor, as illustrated in Figure 3.86.

The optimization of the tapering shape was done to minimize the dissipated power along the S-CPS. The tapering concept is based on the analysis of current and voltage

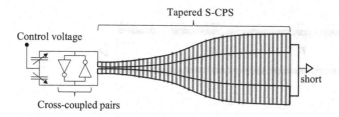

Figure 3.86 Tapered S-CPS proposed in [70] to maximize the overall quality factor of the quarter-wavelength SWO.

magnitude along the resonator, in parallel with the analysis of the S-CPS equivalent circuit model. An *RLCG* model was considered in [70]. As explained in [19], *RLCG* is not a consistent model for S-CPS, but the analysis carried out in [70] is still valid over a small frequency band. The current is maximum at the short end; thus losses are mainly due to the series branch of the *RLCG* model, that is, *R*. To minimize the losses, it is therefore opportune to increase the width of the CPS strips near the short-ended side of the resonator. By the same manner, the voltage is maximum at the cross-coupled pair side of the resonator; thus losses are mainly due to the parallel branch of the *RLCG* model, that is, *G*. In that case, the losses are mainly due to the current flowing in the floating strips, as explained in [16]. To reduce these losses, it is therefore opportune to decrease the gap *S* between the CPS strips. By the same time, as explained in [70], it is necessary to maintain constant characteristic impedance along the resonator in order to prevent local reflections, thus leading to the typical shape given in Figure 3.86.

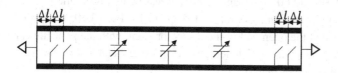

Figure 3.87 Wide band tunable resonator using multiband concept [75].

By controlling the resonator length by switches, as illustrated in Figure 3.87, multiband or very wide band operation can be achieved. This concept was proposed in [75], achieving an FTR equal to 20% around 40 GHz. The authors used the concept of fully distributed SWO illustrated in Figure 3.85d, with a resonator built with coupled microstrip lines.

A half-wavelength resonator was used in Figure 3.87, but the concept could be used with quarter-wavelength resonators, even if less degree of tuning would be available, since only one side of the resonator would be controlled.

Dual-band SWOs can also be achieved. Such a concept was proposed in [76]. It is based on the multimodes operation of a quarter-wavelength resonator, as illustrated in Figure 3.88. The resonance frequencies of quarter-wavelength resonator are given by

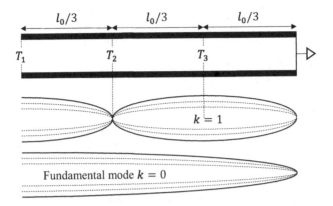

Figure 3.88 Dual-band SWO concept [76].

$$f_{2k+1} = (2k+1)f_0. \tag{3.105}$$

where f_0 is the fundamental frequency.

The two first resonance modes are described in Figure 3.88, the fundamental mode at f_0 ($k = 0$), and the first harmonic at $3f_0$ ($k = 1$).

In [76], the authors have shown that placing a tunable capacitance at the open end of the resonator (position T_3 in Figure 3.88) and cross-coupled pairs at positions T_1 and T_3, it is possible to excite either the fundamental frequency f_0 or the first harmonic frequency $3f_0$, without any instable switching issue between these two modes. However, this technique is limited to frequencies near those given by Eq. (3.105). In [76], 24-GHz and 60-GHz bands were addressed, that is, a ratio of 2.5 between the two bands, different from the ratio of 3 given by Eq. (3.105). This is due to the capacitive load at the open end of the resonator along with the parasitic capacitance of the cross-coupled pairs, which led to a dispersive behavior of the resonator, thus leading to ratios lower than 3.

3.4.3.3   State-of-the-Art and Trends

As already discussed concerning phase shifters, a state-of-the-art is the photo of the performance of a technology, a circuit, a system, at a given time. However, it is interesting to make a state-of-the-art in a book, since some trends can be derived. This is the modest objective of this section. The performance of several VCOs is reported in Table 3.10. The first part of Table 3.10 concerns SWOs, whereas LC-tank VCOs are reported in the second part of Table 3.10.

$$\text{FoM} = PN(\Delta f) + 20\log\left(\frac{f_0}{\Delta f}\right) - 10\log\left(\frac{P_{diss}}{1\ \text{mW}}\right)$$

$$\text{FoM}_T = PN(\Delta f) + 20\log\left(\frac{f_0}{\Delta f}\cdot\frac{FTR}{10\%}\right) - 10\log\left(\frac{P_{diss}}{1\ \text{mW}}\right)$$

Table 3.10 State-of-the-art VCOs

Ref.	Process	Topology	f_0 (GHz)	FTR (%)	V_{DD} (V)	P_{diss} (mW)	PN (dBc/Hz) @Δf	FOM	FOM_T	Die area (mm²)
[70]	0.18 μm CMOS	SWO	14.2	—	—	—	−110@1MHz	—	—	0.8
[74]	90 nm CMOS	SWO	60	0.2	1	1.9	−100@1MHz	−193	—	0.015
[75]	0.18 μm CMOS	SWO	40	20	1.5	27	−100.2@1MHz	−177	−183.9	0.62
[76]	0.13 μm CMOS	SWO	24	10.8	0.8	11	−120@10MHz	−177	—	0.05
[36]		Dual-band	60	7.2	1.2	24	−114@10MHz	−176	—	—
[36]	0.13 μm CMOS	LC-tank	59	9.8	1.5	9.8	−89@1MHz			—
			98.5	2.5	1.5	15	−102.7@10MHz			—
			105.2	0.2	1.2	7.2	−97.5@10MHz			—
[32]	0.18 μm CMOS	LC-tank tunable L	49	2.2	1.8	45	−101@1MHz			0.72[a]
[66]	90 nm CMOS	LC-tank tunable L	57.2	14.3	0.7	8.7	−118.8@10MHz	−184.3	−187.4	0.1
[63]	65 nm CMOS	LC-tank tunable L	56	17	1.2	15	−99.4@1MHz	−182	−187	0.98
[65]	0.13 μm CMOS	LC-tank	45			4	−103@3MHz		−173	
[71]	90 nm CMOS	LC-tank tunable L and C	59	10	1.2	12	−93@1MHz		−177.9	0.16*
[68]	90 nm CMOS	LC-tank tunable L	55.7	17	0.9	10.2	−119@10MHz	−183.4	−188	0.07
[67]	90 nm CMOS	LC-tank tunable L	56.7	16	0.7	8.7	−118.8@10MHz	−184.3	−187.4	0.1
[69]	90 nm CMOS	LC-tank tunable L	73.8	41		8 to 10.8	−104.6 to −112.2@10MHz	−172 to −180	−184.2 to −192.2	0.03
[61] [72]	32 nm CMOS SOI Same as [71]	LC-tank	240	5.6		13				0.055

[a] Including pads.

191

3.5   Challenges and Perspectives

The development of high-performance tunable transmission lines could open the way for the development of many tunable circuits. Phase shifters and VCOs, discussed in Sections 3.4.2 and 3.4.3, respectively, clearly need tunability, with already many realizations at the time of writing the book. But many other circuits could be envisaged, to build future flexible mm-wave systems, thus decreasing their overall cost. Tunable power dividers could lead to optimization of power distribution to control the spurious lobes in phased arrays. Tunable coupled-line couplers could permit to modify the bandwidth of tunable bandpass filters. More generally, tunable filters could be envisaged, in particular notch filters that could be useful in the future when the mm-wave spectrum will be overfilled. The main challenge to open this way of developing tunable circuits and systems is the development of high-performance tunable transmission lines, and hence the improvement of the varactors' quality factor, since it is much lower as compared to that of high-performance transmission lines such as S-CPW or S-CPS.

However, for some particular applications, the realization of on-wafer tunable circuits permits improvement of the performance of the overall system. This is, for example, the case for on-wafer impedance tuners that will be discussed in Section 3.5.1, since external tuners lead to very poor performance.

Finally, a way to achieve a multiband distributed standing-wave oscillators is briefly proposed in Section 3.5.2.

3.5.1   Impedance Tuner

Thanks to the advanced micro and nano silicon technologies, the operating frequency of transistors is increasing with maximum frequency (f_{max}) greater than 300 GHz for CMOS technology (Figure 3.7) and greater than 450 GHz for BiCMOS technology (Figure 3.9), respectively. This progress allows the design of integrated circuits in the mm-wave band beyond 110 GHz. In a transceiver, the wide frequency band generation is one of the critical blocks in transmit and receive chains. Increasing system operating frequency allows higher bandwidth and therefore higher data rate or better image resolution in the context of imaging and radar system. Consequently, accurate models of active components such as CMOS or bipolar transistors including their noise parameters (Γ_{opt}, NF_{min} and R_n) and their large signal performance (compression point, power-added efficiency, and power gain) are required at mm-waves.

Beyond 75 GHz, noise and power measurements remain a huge challenge. On-wafer noise and power characterization setups with commercial mechanical impedance tuners are available. But these benches revealed important losses between the tuner and the device under test (DUT) that significantly limit the maximum reflection coefficient Γ to be synthetized, and also their bandwidth. Consequently, the use of external tuners makes very difficult the optimum matching of transistors, needed to reach minimum noise or maximum power, due to the required optimum impedances (Z_{opt}) that are far from 50 Ω.

These constraints motivate the development of integrated tuners in the W-band (75–110 GHz). An electrical schematic of a typical impedance tuner is presented in

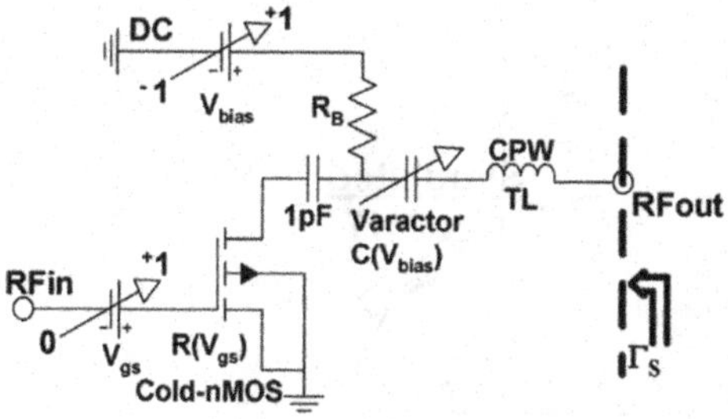

Figure 3.89 Electrical representation of the on-wafer tuner in 65-nm HR SOI.

Figure 3.89 [77]. It was implemented in CMOS 65 nm HR SOI technology. This tuner was developed in the frame of the common lab between STMicroelectronics and IEMN laboratory, France. The architecture of the tuner uses lumped elements, cold-nMOS (as variable resistance) in series with a varactor (as variable capacitance) and a transmission line (CPW). Only passive components were used.

The CPW was realized on HR SOI substrate using six metallization levels for analog applications. The characteristic impedance Z_c of the CPW is close to 50 Ω. It was obtained for a gap $S = 70$ μm, and strip width $W = 26$ μm. The varactor is based on N+Poly/N well structure. It was made of five poly fingers in parallel. Each finger has a length of 0.35 μm and a width of 3 μm. Finally, variable resistance was realized thanks to a cold-FET (Field Effect Transistor) obtained by using an MOS transistor with floating drain bias. It is constituted of 80 poly fingers in parallel, contacted on both sides: each finger has a length of 0.12 μm and a width of 0.5 μm, respectively.

The results of the synthesized admittance Y_s with the tuner are presented in Figure 3.90. Three zones can be identified corresponding to three different transmission line lengths. The lengths were chosen to move on the Smith Chart, avoiding the areas of instability and moving across different noise figure circles to carry out subsequent characterizations of the nMOS available in the technology. Inside each zone obtained for a given transmission line length, the different admittances are obtained through the controlled biases of $R(V_{gs})$ for the cold-FET and of $C(V_{bias})$ for the varactor.

As illustrated with squares in Figure 3.90, the length $L = 350$ μm covers the whole interesting area on the Smith Chart, allowing the extraction of the four noise parameters of the transistor.

In D band or higher frequencies, tuner traveling wave topologies must be used to enlarge the Smith Chart coverage, and DTC must be used to enlarge the variable capacitance tuning range. In Figure 3.91, a traveling wave tuner schematic is given.

The central microstrip line (between RFin and RFout) is periodically loaded by degenerated MOS transistors, which can be turned in an ON or OFF state. The transmission line strip width and length, and MOS dimension, were determined to reach the targeted coverage of the Smith Chart.

In Figure 3.92, the tuner along with the tuner synthetized impedance and insertion loss versus the frequency are shown, from 130 GHz to 170 GHz. Such a tuner exhibits very high losses and must be coupled with an LNA for noise characterization, but can be directly used for load-pull characterization.

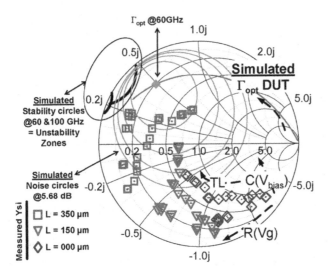

Figure 3.90 Impedance synthesis with on-wafer tuner, and 65-nm HR SOI 0.06×40 μm² nMOS simulated noise characteristics for $V_{gs} = 0.72$ V and $V_{ds} = 1.2$ V.

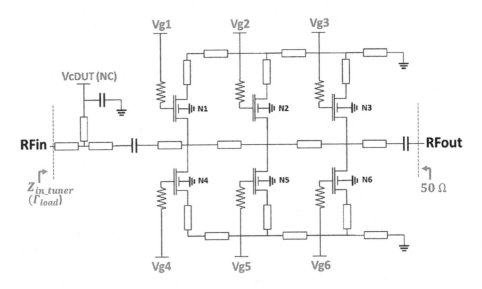

Figure 3.91 Traveling wave tuner schematic.

3.5.2   Multiband Distributed and Tapered Standing-Wave Oscillators

By mixing the concept of distributed SWO as proposed in Figure 3.85d and the concept of tapered resonator illustrated in Figure 3.86, a fully distributed and tapered SWO could be achieved. This could lead to low-loss, thanks to the tapered resonator; wide tuning range, thanks to the use of periodically loaded resonator absorbing the parasitic capacitances of the cross-coupled pairs; and low consumption, since the cross-coupled pair8s dimension would be adapted along the resonator.

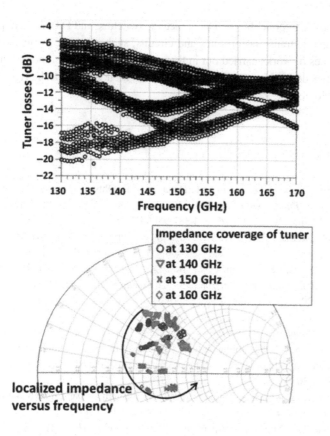

Figure 3.92 Tuner suppress measurement results.

By using switched resonators as described in Figure 3.87, multiband SWOs could be achieved.

3.6   Conclusion

In this chapter, the main active and passive devices and some tunable elements integrated in CMOS and BiCMOS technologies were presented. Today, transistor cutoff frequencies f_T and f_{max} are well above 300 GHz, enabling the design of very high frequency applications. Furthermore, passive devices are key elements for mm-wave design especially by considering their losses and quality factor.

First of all, the integrated transmission lines were presented with their associated model and their characterization on several CMOS BEOL. The MOM capacitors were also described with their associated model to be used as coupling and decoupling capacitor for mm-wave applications. The last passive device presented was the MOS varactor, especially the one integrated in the STMicroelectronics CMOS 28 nm FDSOI technology. Its associated model with measured performance shows a high quality factor up to 110 GHz to be used inside local oscillators, which is a key bloc of RF and mm-wave transceivers.

Next, tunable components, such as varactors, switches, and DTCs, were presented, along with tunable transmission lines and circuits, such as phase shifters and VCOs. Many references of works carried out in the recent literature were reported, and trends were sometimes derived to help future designers in their choice for the right technologies and topologies.

References

1. Y. Shacham-Diamand, T. Osaka, M. Datta, and T. Ohba, *Advanced Nanoscale ULSI Interconnects: Fundamentals and Applications*, 2009, New York: Springer Science +Business Media. DOI 10.1007/978-0-387-95868-2.

2. G. Gildenblat, X. Li, W. Wu, et al., "PSP: An advanced surface-potential-based MOSFET model for circuit simulation," *IEEE Transactions on Electron Devices*, vol. 53, no. 9, Sept. 2009.

3. K. M. Cao, W. -C. Lee, W. Liu, et al., "BSIM4 gate leakage model including source-drain partition," in *International Electron Devices Meeting, 2000. IEDM '00*, pp. 815–818, Dec. 10–13, 2000, San Francisco, CA, USA.

4. www-device.eecs.berkeley.edu/bsim/

5. B. C. Kuo, *Automatic Control System*, 1982. Englewood Cliffs, NJ: Prentice-Hall.

6. S. Pruvost, "Etude de faisabilité de circuits pour systèmes de communication en bande millimétrique, en technologie BiCMOS SiGeC 0,13 μm," PhD thesis, University of Lille 1, France, Nov. 2005.

7. H. A. Wheeler, "Transmission-line properties of parallel strips separated by a dielectric sheet," *IEEE Transactions on Microwave Theory Techniques*, vol. 13, no. 2, pp. 172–185, Mar. 1965.

8. M. V. Schneider, "Microstrip dispersion," *Proceeding of the IEEE, Letters*, vol. 60, no. 1, pp.144–146, Jan. 1972.

9. E. Hammerstad and Ø. Jensen, "Accurate models for microstrip computer-aided design," in *1980 IEEE MTT-S International Microwave Symposium*, pp. 407–409, May 28–30, 1980, Washington, DC..

10. P. Ferrari, B. Fléchet, and G. Angénieux, "Time domain characterization of lossy arbitrary impedance transmission lines," *IEEE Microwave and Guided Wave Letters*, vol. 4, no. 6, pp. 177–179, June 1994.

11. A. M. Mangan, S. P. Voinigescu, Y. Ming-Ta, and M. Tazlauanu, "De-embedding transmission line measurements for accurate modeling of IC designs," *IEEE Transactions on Electron Devices*, vol. 53, no. 2, pp. 235–241, Feb. 2006.

12. D. Kaddour, H. Issa, A.-L. Franc, et al., "High-Q slow-wave coplanar transmission lines on 0.35-μm CMOS Process," *IEEE Microwave Wireless Component Letters*, vol. 19, no. 9, pp. 542–544, Sep. 2009.

13. H. Hasegawa and H. Okizaki, "M.I.S. and Schottky slow-wave coplanar striplines on GaAs substrates," *Electron. Letters*, vol. 13, pp. 663–664, Oct. 1977.

14. A.-L. Franc, E. Pistono, N. Corrao, D. Gloria, and P. Ferrari, "Compact high-Q, low-loss mmW transmission lines and power splitters in RF CMOS technology," in *2011 IEEE MTT-S International Microwave Symposium*, June 5–10, 2011, Baltimore, MD.

15. T. S. D. Cheung, J. R. Long, K. Vaed, et al., "On-chip interconnect for mm-wave applications using an all-copper technology and wavelength reduction," in *Proceedings of the IEEE International Solid-State Circuits Conference*, pp. 396–397, Feb. 2003, San Francisco, CA.

16. A.-L. Franc, E. Pistono, G. Meunier, D. Gloria, and P. Ferrari, "A lossy circuit model based on physical interpretation for integrated shielded slow-wave CMOS coplanar waveguide structures," *IEEE Transactions on Microwave Theory and Techniques*, vol. 61, no. 2, pp. 754–763, Feb. 2013.

17. A. Bautista, A.-L. Franc, and P. Ferrari, "Accurate parametric electrical model for slow-wave CPW and application to circuits design," *IEEE Transactions on Microwave Theory and Techniques.*, vol. 63, no. 12, pp. 4225–4235, Dec. 2015.

18. X.-L. Tang, A.-L. Franc, E. Pistono, et al., "Performance improvement versus CPW and loss distribution analysis of slow-wave CPW in 65 nm HR-SOI CMOS technology," *IEEE Transactions Electron Devices*, vol. 59, no. 5, pp. 1279–1285, May 2012.

19. A. Bautista, A.-L. Franc, and P. Ferrari, "A predictive model for slow-wave coplanar striplines in integrated technologies," in *IEEE MTT-S International Microwave Symposium*, May 17–22, 2016, San Francisco, CA.

20. J. Dang, A. Noculak, F. Korndörfer, C. Jungemann, and B. Meinerzhagen, "A semi-distributed method for inductor de-embedding," in *2014 International Conference on Microelectronic Test Structures (ICMTS)*, March 24–27, 2014, Udine, Italy.

21. R. L. Schmid, P. Song, C. T. Coen, A. C. Ulusoy, and J. D. Cressler, "On the analysis and design of low-loss single-pole double-throw W-band switches utilizing saturated SiGe HBTs," *IEEE Transactions on Microwave Theory and Techniques*, vol. 62, no. 11, pp. 2755–2767, Nov. 2014.

22. S. F. Chao, H. Wang, C. Y. Su, and J. G. J. Chern, "A 50 to 94-GHz CMOS SPDT switch using traveling-wave concept," *IEEE Microwave and Wireless Components Letters*, vol. 17, no. 2, pp. 130–132, Feb. 2007.

23. J. Kim, W. Ko, S.-H. Kim, J. Jeong, and Y. Kwon, "A high-performance 40–85 GHz MMIC SPDT switch using FET-integrated transmission line structure," *IEEE Microwave and Wireless Components Letters*, vol. 13, no. 12, pp. 505–507, Dec. 2003.

24. K.-Y. Lin, W.-H. Tu, P.-Y. Chen, H.-Y. Chang, H. Wang, and R.-B. Wu, "Millimeter-wave MMIC passive HEMT switches using traveling-wave concept," *IEEE Transactions on Microwave Theory and Techniques*, vol. 52, no. 8, pp. 1798–1808, Aug. 2004.

25. C. M. Ta, E. Skafidas, and R. J. Evans, "A 60-GHz CMOS transmit/receive switch," in *2007 IEEE Radio Frequency Integrated Circuits Symposium*, pp. 725–728, June 3–5, 2007, Honolulu, HI.

26. M. C. Yeh, Z. M. Tsai, R. C. Liu, K. Y. Lin, Y. T. Chang, and H. Wang, "A millimeter-wave wideband SPDT switch with traveling-wave concept using 0.13-μm CMOS process," in *2005 IEEE MTT-S International Microwave Symposium Digest*, pp. 53–56, June 12–17, 2005, Long Beach, CA.

27. X.-L. Tang, E. Pistono, P. Ferrari, and J. M. Fournier, "A travelling wave CMOS SPDT using slow-wave transmission lines for millimeter wave application," *IEEE Electron Devices Letters*, vol. 34, no. 9, pp. 1094–1096, Sept. 2013.

28. Z.-M. Tsai, M.-C. Yeh, H.-Y. Chang, et al., "FET-integrated CPW and the application in filter synthesis design method on traveling-wave switch above 100 GHz," *IEEE Transactions on Microwave Theory and Techniques*, vol. 54, no. 5, pp. 2090–2097, May 2006.

29. R. Debroucke, A. Pottrain, D. Titz, et al., "CMOS digital tunable capacitance with tuning ratio up to 13 and 10dBm linearity for RF and millimeterwave design," in *2011 IEEE Radio Frequency Integrated Circuits Symposium (RFIC)*, June 5–7, 2011, Baltimore, MD.

30. R.-J. Chan, and J.-C. Guo, "Analytical modeling of proximity and skin effects for millimeter-wave inductors simulation and design in nano Si CMOS, in *2014 IEEE MTT-S International Microwave Symposium*, June 1–6, 2014, Tampa Bay, FL.

31. A. Jrad, A.-L. Perrier, R. Bourtoutian, J.-M. Duchamp and P. Ferrari, "Design of an ultra compact electronically tunable microwave impedance transformer," *Electronics Letters*, vol. 41, no. 12, pp. 123–125, June 9, 2005.

32. H. Hsieh, Y.-H. Chen, and L.-H. Lu, "A millimeter-wave CMOS LC-tank VCO with an admittance-transforming technique," *IEEE Transactions on Microwave Theory and Techniques*, vol. 55, no. 9, pp. 1854–1860, June 2007.

33. S. Gong, H. Shen, and N. S. Barker, "A 60-GHz 2-bit switched-line phase shifter using SP4T RF-MEMS switches," *IEEE Transactions on Microwave Theory and Techniques*, vol. 59, no. 4, pp. 894–900, April 2011.

34. D. Kaddour, E. Pistono, J.-M. Duchamp, J.-D. Arnould, P. Ferrari, and R. G. Harrison, "A compact and selective low-pass filter with reduced spurious responses, based on CPW tapered periodic structures," *IEEE Transactions on Microwave Theory and Techniques*, vol. 54, no. 6, pp. 2367–2375, June 2006.

35. D. Kaddour," Conception et réalisation de filtres RF passe-bas à structures périodiques et filtres Ultra Large Bande, semi localisés en technologie planaire." PhD thesis, Grenoble, France, July 11, 2007.

36. Z. Cao, Q. Ma, A. Bernardus Smolders, et al., "Advanced integration techniques on broadband millimeter-wave beam steering for 5G wireless networks and beyond," *IEEE Journal of Quantum Electronics*, vol. 52, no. 1, pp. Jan. 2016.

37. A. Natarajan, A. Komijani, X. Guan, A. Babakhani, and A. Hajimiri, "A 77-GHz phased-array transceiver with on-chip antennas in silicon: Transmitter and local LO-path phase shifting," *IEEE Journal of Solid-State Circuits*, vol. 41, no. 12, pp. 2807–2819, Dec. 2006.

38. P. Song, and H. Hashemi, "Wideband mm-wave phase shifters based on constant-impedance tunable transmission lines," in *2016 IEEE MTT-S International Microwave Symposium*, May 22–27, 2016, San Francisco, CA.

39. Y. Tousi, and A. Valdes-Garcia, "A Ka-band digitally-controlled phase shifter with sub-degree phase precision," in *2016 IEEE Radio Frequency Integrated Circuits Symposium*, pp. 356–359, May 22–27, 2016, San Francisco, CA.

40. J. Lange, "3.5 interdigitated stripline quadrature hybrid (Correspondence)," *IEEE Transactions on Microwave Theory and Techniques*, Vol. 17, No. 12, pp. 1150–1151, Dec. 1969.

41. Leijun Xu, Henrik Sjöland, Markus Törmänen, Tobias Tired, Tianhong Pan, and Xue Bai, "A miniaturized Marchand balun in CMOS with improved balance for millimeter-wave applications," *IEEE Microwave and Wireless Components Letters*, vol. 24, no. 1, pp. 53–55, Jan. 2003.

42. J. Lugo-Alvarez, A. Bautista, F. Podevin, and P. Ferrari, "High-directivity compact slow-wave coPlanar waveguide couplers for millimeter-wave applications," in *44th European Microwave Conference*, EuMC'14, Oct. 6–9, 2014, Rome, Italy.

43. Z. Iskandar, J. Lugo-Alvarez, A. Bautista, E. Pistono, F. Podevin, V. Puyal, A. Siligaris, and P. Ferrari, "A mm-Wave Ultra-Wideband Reflection-Type Phase Shifter in BiCMOS 55 nm Technology," in *Proceedings of the 46th European Microwave Conference*, October 3–7, 2016, London.

44. F. Burdin, Z. Iskandar, F. Podevin, and P. Ferrari, "Design of simple reflection type phase shifters with high figure-of-merit until 360°," *IEEE Transactions on Microwave Theory and Techniques*, vol. 63, no. 6, pp. 1883–1893, June 2015.

45. F. Meng, K. Ma, K. S. Yeo, and S. Xu, "A 57-to-64-GHz 0.094-mm² 5-bit passive phase shifter in 65-nm CMOS," *IEEE Transactions on Very Large Scale Integration (VLSI) Systems*, vol. 24, no. 5, pp. 1917–1925, May 2016.

46. S. Y. Kim, and G. M. Rebeiz, "A low-power BiCMOS 4-element phased array receiver for 76–84 GHz radars and communication systems," *IEEE Journal of Solid-State Circuits*, vol. 47, no. 2, pp. 359–367, Feb. 2012.

47. W.-T. Li, Y.-C. Chiang, J.-H. Tsai, H.-Y. Yang, J.-H. Cheng, and T.-W. Huang, "60-GHz 5-bit Phase Shifter With Integrated VGA Phase-Error Compensation," *IEEE Trans. Microwave Theory and Tech.*, vol. 61, no. 2, pp. 1224–1235, Mar. 2013.

48. G.-S. Shin, J.-S. Kim, H.-M. Oh, S. Choi, C. W. Byeon, J. H. Son, J. H. Lee, and C.-Y. Kim, "Low Insertion Loss, Compact 4-bit Phase Shifter in 65 nm CMOS for 5G Applications," *IEEE Microw. Wireless Compon. Lett.*, vol. 26, no. 1, pp. 37–39, Jan. 2016.

49. C. Hoarau, P.-E. Bailly, J.-D. Arnould, P. Ferrari, and P. Xavier, "A RF Tunable Impedance Matching Network with a Complete Design and Measurement Methodology," *37th European Microwave Conference*, EuMC'07, Oct. 9–11, 2007, München, Germany.

50. A.-L. Franc, Ph. D. Thesis, "Lignes de propagation intégrées à fort facteur de qualité en technologie CMOS - Application à la synthèse de circuits passifs millimétriques," July 6, 2011, Grenoble Alpes University (in French).

51. T. LaRocca, S.-W. Tam, D. Huang, et al., "Millimeter-wave CMOS digital controlled artificial dielectric differential mode transmission lines for reconfigurable ICs," in *2008 IEEE MTT-S International Microwave Symposium*, pp. 181–184, June 15–20, 2008, Atlanta, GA.

52. Y. Yu, P. Baltus, A. van Roermund, et al., "A 60GHz digitally controlled RF-beamforming receiver front-end in 65nm CMOS," in *34th European Solid-State Circuits Conference (ESSCIRC)*, Sept. 15–19, 2008, Edinburgh, UK.

53. M.-D. Tsai, and A. Natarajan, "60GHz passive and active RF-path phase shifters in silicon," in *2009 IEEE Radio Frequency Integrated Circuits Symposium (RFIC)*, June 7–9, 2009, Boston, MA.

54. B. Biglarbegian, M. R. Nezhad-Ahmadi, M. Fakharzadeh, and S. Safavi-Naeini, "Millimeter-wave reflective-type phase shifter in CMOS technology," *IEEE Microwave and Wireless Components Letters*, vol. 19, no. 9, pp. 560–562, Sep. 2009.

55. H. Krishnaswamy, A. Valdes-Garcia, and J.-W. Lai, "A silicon-based, all-passive, 60 GHz, 4-element, phased-array beamformer featuring a differential, reflection-type phase shifter," in *2010 IEEE International Symposium on Phased Array Systems and Technology*, Oct. 12–15, 2015, Waltham, MA.

56. D. Titz, F. Ferrero, C. Luxey, et al., "Reflection-type phase shifter integrated on advanced BiCMOS technology in the 60 GHz band," in *2011 IEEE 9th International New Circuits and systems conference (NEWCAS)*, June 26–29, 2011, Bordeaux, France.

57. CM. Tabesh, A. Arbabian, and A. Niknejad, "60GHz low-loss compact phase shifters using a transformer-based hybrid in 65nm CMOS," in *2011 IEEE Custom Integrated Circuits Conference (CICC)*, Sept. 19–21, 2011, San Jose, CA.

58. T.-W. Li, and H. Wang, "A millimeter-wave fully differential transformer-based passive reflective-type phase shifter," in *2015 IEEE Custom Integrated Circuits Conference (CICC)*, Sept. 28–30, 2015, San Jose, CA.

59. T. H. Hee and A. Hajimiri, "Oscillator phase noise: A Tutorial," *IEEE J. Solid-State Circuits*, vol. 35, no. 3, pp. 326–336, Mar. 2000.

60. C. Cao, and K. K. O, "Millimeter-wave voltage-controlled oscillators in 0.13-μm CMOS technology," *IEEE Journal of Solid-State Circuits*, vol. 41, no. 6, pp. 1297–1304, June 2006.

61. N. Landsberg, and E. Socher, "240 GHz and 272 GHz fundamental VCOs using 32 nm CMOS technology," *IEEE Transactions on Microwave Theory and Techniques*, vol. 61, no. 12, pp. 4461–4471, Dec. 2013.

62. T. Nakamura, T. Masuda, K. Washio, and H. Kondoh, "A 59GHz push-push VCO with 13.9GHz tuning range using loop-ground transmission line for a full-band 60GHz transceiver," in *Proceedings of the IEEE International Solid-State Circuits Conference*, Feb. 8–12, 2009, pp. 496–497, San Francisco, CA.

63. J. L. Gonzalez Jimenez, F. Badets, B. Martineau, and D. Belot, "A 56GHz LC-tank VCO with 17% tuning range in 65nm bulk CMOS for wireless HDMI applications," in *2009 IEEE Radio Frequency Integrated Circuits Symposium (RFIC)*, pp. 481–484, June 7–9, 2009, Boston, MA.

64. J. L. Gonzalez, F. Badets, B. Martineau, and D. Belot, "A 56-GHz LC-Tank VCO with 17% tuning range in 65-nm bulk CMOS for wireless HDMI," *IEEE Transactions on Microwave Theory and Techniques*, vol. 58, no. 5, pp. 1359–1366, May 2010.

65. G. Huang, S.-K. Kim, Z. Gao, S. Kim, V. Fusco, and B.-Sung Kim, "A 45 GHz CMOS VCO adopting digitally switchable metal-oxide-metal capacitors," *IEEE Microwave and Wireless Components Letters*, vol. 21, no. 5, pp. 270–272, May 2011.

66. C.-Y. Yu, W.-Z. Chen, C.-Y. Wu, and T.-Y. Lu, "A 60-GHz, 14% tuning range, multi-band VCO with a single variable inductor," in *IEEE Asian Solid-State Circuits Conference, 2008. A-SSCC '08*, Nov. 3–5, 2008, Fukuoka, Japan.

67. T.-Y. Lu, C.-Y. Yu, W.-Z. Chen, and Chung-Yu Wu, "Wide Tuning range 60 GHz VCO and 40 GHz DCO using single variable inductor," *IEEE Trans. Circuits and Systems*, vol. 60, no. 2, pp. 257–267, Feb. 2013.

68. P.-L. You, and T.-H. Huang, "A switched inductor topology using a switchable artificial grounded metal guard ring for wide-FTR MMW VCO applications," *IEEE Transactions on Electron Devices*, vol. 60, no. 2, pp. 759–766, Feb. 2013.

69. J. Yin, and H. C. Luong, "A 57.5–90.1-GHz magnetically tuned multimode CMOS VCO," *IEEE Journal of Solid-State Circuits*, vol. 48, no. 8, pp. 1851–1861, Aug. 2013.

70. W. F. Andress, and D. Ham, "Standing wave oscillators utilizing wave-adaptive tapered transmission lines," *IEEE Journal of Solid-State Circuits*, vol. 40, no. 3, pp. 638–650, March 2005.

71. W. Wu, J. R. Long, R. B. Staszewski, and J. J. Pekarik, "High-resolution 60-GHz DCOs with reconfigurable distributed metal capacitors in passive resonators," in *2012 IEEE Radio Frequency Integrated Circuits Symposium*, pp. 91–94, May 17–19, 2012, Montréal, Canada.

72. W. Wu, J. R. Long, R. B. Staszewski, "High-resolution millimeter-wave digitally controlled oscillators with reconfigurable passive resonators," *IEEE Journal of Solid-State Circuits*, vol. 48, no. 11, pp. 2785–2794, Nov. 2013.

73. J. Wood, T. C. Edwards, and S. Lipa, "Rotary travelling-wave oscillator arrays: A new clock technology," *IEEE Journal of Solid State Circuits*, vol. 36, no. 11, pp. 1654–1665, Nov. 2001.

74. D. Huang, W. Hant, N.-Y. Wang, et al., "A 60GHz CMOS VCO using on-chip resonator with embedded artificial dielectric for size, loss and noise reduction," in *Proceedings of IEEE International Solid-State Circuits Conference*, Feb. 6–9, 2006, San Francisco, CA.

75. J.-C. Chien, and L.-H. Lu, "Design of wide-tuning-range millimeter-wave CMOS VCO with a standing-wave architecture," *IEEE Journal of Solid-State Circuits*, vol. 42, no. 9, pp. 1942–1952, Sept. 2007.

76. L. Wu, A. W. L. Ng, L. L. K. Leung, and H. C. Luong, "A 24-GHz and 60-GHz dual-band standing-wave VCO in 0.13µm CMOS process," in *2010 IEEE Radio Frequency Integrated Circuits Symposium*, pp. 145–148, May 23–25, 2010, Anaheim, CA, USA.

77. Y. Tagro, D. Gloria, S. Boret, G. Dambrine, "MMW Lab In-Situ to Extract Noise Parameters of 65nm CMOS Aiming 70~90GHz Applications," in *2009 IEEE Radio Frequency Integrated Circuits Symposium*, pp. 397–400, June 7–9, 2009, Boston, MA.

4 RF MEMS Technology

Gustavo P. Rehder, Ariana L. C. Serrano, Marcelo N. P. Carreño, Mehmet Kaynak, Selin Tolunay Wipf, Matthias Wietstruck, and Alexander Göritz

4.1 Introduction

Microelectromechanical systems (MEMS) are now part of our daily lives. They were responsible for a small revolution in several applications, where accelerometers, gyroscopes, digital light processors (DLPs), pressure sensors, microphones, and others are key components. The advent of MEMS for the realization of these components using conventional microelectronic processing with batch fabrication significantly reduced the cost and allowed their integration with control and processing electronics on the same chip, achieving major miniaturization. With low cost and small footprint, these components are everywhere and are responsible for a multi-billion-dollar market.

In the radiofrequency (RF) market, the development of MEMS-based devices was motivated mostly by the performance benefits of the RF MEMS switch. The main goal was to replicate the performance of the bulky electromechanical switches (coaxial or waveguide) using planar structures that are miniaturized and easy to integrate with other planar components, focusing mainly on low-cost devices for consumer applications.

Several RF MEMS switches were developed by many different companies and research laboratories in the late 1990s and early 2000s. In general, they showed low insertion loss (0.1–0.2 dB), high linearity with low intermodulation (IIP$_3$ greater than +66 dBm), and high power handling (up to 10 W), which are far superior when compared to other planar switches such as the semiconductor-based PIN and field-effect transistor (FET) switches. The RF MEMS switches consume virtually no power because of their electromechanical actuation, but their mechanical constitution limits their switching speed to a few microseconds. Their use is still appealing, though, for mode switching, antenna tuning, and antenna beam steering with phase shifters and others.

In the late 2000s, commercial switches were fabricated by TeraVicta, Radant, and Omron. Several issues related to fabrication, encapsulation, and especially reliability had to be resolved for these commercial switches. Then, the industry was focused on producing components with one single switch (SPST to SP6T) either on a die encapsulation (Radant) or in a ball grid array package (TeraVicta and Omron). The encapsulation, in general, increased the insertion loss considerably (0.5–1 dB), and increased the cost prohibitively for consumer applications. For this reason, no major product used these commercial switches and their fabrication was discontinued.

In the early 2010s, driven by the need to increase cellular phone performance due to the recent degradation of the connection quality, RF MEMS came into play using a different approach that was proposed by Wispry and Cavendish Kinetics. The goal was to use MEMS-based switched capacitor arrays to improve antenna matching. Instead of having a single switch in one encapsulation, this new approach consists in combining in a single packaged chip a network of MEMS capacitors, high-voltage charge pump, driver for MEMS actuation, logic circuits, and serial interface for control. To integrate all these components in a single die, both companies based their devices on complementary metal–oxide–semiconductor (CMOS) technology. In 2011, Samsung's Focus Flash smartphone was released using Wispry's Tunable Digital Capacitor Array, and in 2015, Cavendish Kinetics announced that ZTE's Nubia Z9 smartphone was using their SmartTune antenna. The advent of these components in cellular phones showed that RF MEMS technology was mature for consumer applications. Hence its benefits in the microwave frequency range are evident.

Analog Devices released at the end of 2016 a new package MEMS (ADGM1304 and ADGM1004) switch operating from DC to 14 GHz. The Lead Frame Chip Scale Package includes a die with an SP4T MEMS switch and separate die with a charge pump and driver. This switch has a different approach, with a focus on testing and high-performance switching applications. It has shown a high reliability and was qualified in several different standards [1].

In spite of the new advances of the RF MEMS, switching and tuning at microwave frequencies is mainly done by FET transistors, varactor diodes, or MOS varactors, since their performance is acceptable and encapsulation costs are reduced. However, as the frequency increases into the millimeter-wave (mm-wave) range, their quality factor is considerably reduced and MEMS switches and varactors become a relevant option.

4.2   Fundamentals of RF MEMS for Millimeter-Wave Frequencies

In general, depending on the application requirements, MEMS actuators can be divided in four groups: magnetic, piezoelectric, thermal, and electrostatic. For RF MEMS, however, the vast majority of devices are based on electrostatic actuation for several reasons. Unlike magnetic and thermal, electrostatic and piezoelectric actuation consume virtually no power, which is especially important in mobile applications. Second, electrostatic actuation requires only materials that are readily available in any fabrication process, that is, metals and dielectric films, which allows an important compatibility with conventional CMOS processes. Piezoelectric actuation requires special materials that are normally deposited/sintered at high temperatures and incompatible with CMOS technology. Further, most RF MEMS are parallel-plate-based structures that can yield the highest operating frequency when compared to other actuation methods [2]. A drawback of the electrostatic actuation is the high voltages involved. Charge pumps can be used to increase the voltage from a few volts to several tens of volts, and they can be easily integrated in CMOS [3].

Attempts have been made to reduce the high voltage involved in the electrostatic actuation. For example, a combination of thermal and electrostatic actuation was demonstrated, which also reduced the power dissipated in the thermal actuation [4]. In this approach, thermal heating is responsible for the commutation of an RF MEMS switch, while the electrostatic force holds the switch at the commutated state. While reducing the voltage level to 10 V, the power dissipated during commutation is still high (50 mW), when compared with the electrostatic actuation, but depending on the application this can also be a viable solution.

Another RF MEMS switch reduced power consumption of thermal actuation by using a latching mechanism [5]. Despite the good RF performance up to 40 GHz, the power consumption was still elevated (0.4 W). The footprint of this switch was also considerably large, which would prevent its use in several applications such as in phased arrays, where many switches are necessary. Further, the switching time of a thermal actuated switch is long, on the order of milliseconds.

Besides the aforementioned examples and several other ones in the literature, electrostatic actuation is the primary mechanism when considering RF MEMS for most applications. All the commercial devices are based on this type of actuation. Therefore, the basics of the electrostatic actuation will be discussed in the next section. The focus is on the aspects required to understand the electrostatic actuation principle and how they relate to the performance and limitations of RF MEMS in the mm-wave frequency range. The electromechanical behavior of RF MEMS switches and varactors has been covered extensively in the literature [6–9], and it should be consulted if detailed information is required.

4.2.1   Electromechanical Behavior of RF MEMS Switches

An RF MEMS switch can be used in two configurations, as illustrated in Figure 4.1: series and shunt. In series, the closed switch (down state) connects the input to the output, ideally without insertion or return loss. The open switch (up state) reflects the signal at the input and no signal reaches the output. In a shunt switch, the open switch allows the signal to reach the output, while the closed switch directs the signal to the ground.

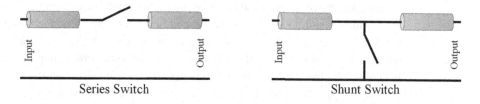

Series Switch   Shunt Switch

Figure 4.1 RF switch configuration: series and shunt.

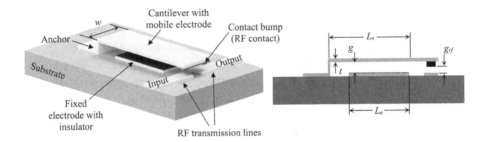

Figure 4.2 Illustration of a cantilever MEMS switch.

In general, RF MEMS switches are planar devices fabricated on a silicon or glass substrate using mostly standard microelectronic processes, as will be discussed in the next section. A series switch is normally open and it is formed by two unconnected sections of transmission lines that are closely spaced. A small section of transmission line on a miniaturized mechanical structure, such as a microcantilever or a micro-bridge, is placed above the unconnected input and output lines, separated by a small gap g. To close the switch, the microstructure is forced onto the unconnected lines, connecting the input to the output. The shunt switch operates in a similar manner; however, it is normally closed to the ground. An example of a cantilever-based RF MEMS switch is illustrated in Figure 4.2.

Both series and shunt switches can have a resistive or a capacitive RF contact. In the resistive case, there is a metal-to-metal contact between the input and the micro-structure and the output and the microstructure. This type of switch can be used from DC to several tens of gigahertz. To reduce the contact resistance, which reduces the transmission loss, contact bumps are used to increase the contact force applied by the microstructure on the transmission lines, thus reducing the contact resistance. The metals used in the resistive contact are essential to reduce contact degradation, which is a common failure mechanism of this type of switch. In the capacitive contact case, a thin dielectric layer is placed between the microstructure and the input and output transmission lines. For this reason, this switch works only above a certain frequency, depending on the contact capacitance.

In most RF MEMS switches, as discussed earlier, the microstructure is actuated electrostatically. This actuation is realized by placing one electrode on the micro-structure (mobile electrode) and the other one on the substrate (fixed electrode). In a cantilever-based structure, the RF contact is normally placed on the free end of the cantilever and the electrode on the middle section. For a bridge-based switch, the RF contact is placed on the middle and two electrodes on each side. The microstructure is mechanically anchored to the substrate. For a simplified mechan-ical analysis, the anchor can be considered as a ridged structure. If sufficient voltage is applied to the electrodes, the electrostatic force becomes stronger than the structure's spring force and it will collapse onto the fixed electrode and RF contact. To avoid a short-circuit between electrodes, a dielectric layer is deposited over one of them. The voltage required for this actuation is known as pull-in voltage, V_p. It is given by

$$V_p = \sqrt{\frac{8 \cdot k \cdot g^3}{27 \cdot \varepsilon_0 \cdot w \cdot L_e}}, \qquad (4.1)$$

where k is the spring constant, g the gap, ε_0 the free space permittivity, w the width of the microstructure, and L_e the length of the electrode. The dimensions are illustrated in Figure 4.2. Equation (4.1) is also valid for bridge or cantilever-based structures, using the corresponding spring constant.

The spring constant for beams, such as cantilevers (beam with fixed-free supports) and bridges (beam with fixed-fixed supports), can be obtained using Euler–Bernoulli beam theory [10]. By solving the beam equation, it is possible to obtain the displacement of the beam for a given load and support conditions. Several different cases were solved and tabulated in the literature and a good example is [11]. A detailed discussion on the possible configurations of electrostatic actuation of RF MEMS switches and their corresponding spring constant is presented in [6] and [9].

The spring constant is important to determine the pull-in voltage, but it is also responsible for restoring the structure to its initial position, when the applied voltage is removed. If it is too low, surface adhesion forces might prevent the switch to open. Even though V_p is inversely proportional to w, as shown in (4.1), k is also proportional to w; therefore the width of the electrode does not influence the pull-in voltage, but it contributes to increase the restoration force.

The maximum pull-in voltage is determined, initially, by the breakdown voltage of the dielectric on the actuation electrode. This breakdown voltage can range from 300 V/μm to 1000 V/μm for materials such as SiO_2 and Si_3N_4, usually used in RF MEMS switches. Considering a dielectric thickness ranging from 100 nm to 300 nm, the maximum pull-in voltage will range from 30 V to 300 V. For an integrated switch with charge pump, the maximum voltage of the transistors and capacitors used in the charge pump must be considered. An integrated charge pump with output voltage of 70 V was demonstrated using a 0.13 μm bipolar CMOS (BiCMOS) technology [3], and this voltage can be considered the maximum desired pull-in voltage.

To get low actuation voltages (<70 V), the separation between electrodes ($g + t_d$), from an actuation point of view, should be as small as possible (<5 μm). From an RF point of view, this separation will correspond to a capacitance that will limit the switch operation at mm-waves, as will be seen in the next section, and it should be as large as possible.

Owing to the mechanical aspect of the RF MEMS switch, its switching time is considerably longer when compared to the semiconductor-based switches, such as FET transistors and PIN diodes. In general, the switching time of a MEMS switch varies from 3 to 40 μs, whereas that of the semiconductor switches varies from 1 to 100 ns. However, some fast RF MEMS switches have been demonstrated with switching times between 150 and 400 ns [12]. This was achieved using a small gap of 0.3 μm, which in turn will reduce the isolation at higher frequencies and hence limit the use of the switch for RF frequencies.

To design an RF MEMS switch, all these compromises have to be taken into account. Special attention should be paid to the constraints imposed by the chosen technology.

4.2.2   Millimeter-Wave MEMS Technologies

From the technology point of view, several approaches have been proposed to fabricate RF MEMS, each of them with specific advantages. The key factor behind them, and which will ultimately determine its adoption in mm-wave systems, is the compromise between cost and performance. Aspects that include the size of devices at mm-waves (particularly for passive components, such as antennas and phase shifters), the utilized materials (because Si is not always the best solution for very high frequencies), the integration with other circuits, and the size of the overall system (a main issue in cell phone applications, for example) must be considered. In other words, the adopted technology to fabricate an RF MEMS device is determined not just by its intrinsic characteristics, but also by the form in which it will be integrated in the final mm-wave system.

On the other hand, is not easy to obtain a monolithic integrated high-performance system, because for that, devices and components fabricated with different materials and technologies are usually required. For instance, one can utilize MEMS technology to fabricate efficient active and passive RF components and standard Si CMOS technology for the control and drive electronic circuits. However, both technologies (MEMS and CMOS) are just partially compatible and a full integration is still a challenge.

Highly integrated complete transceivers operating at 120 GHz have been demonstrated by Silicon Radar on SiGe BiCMOS technology [13]. In fact, Si technologies allow the fabrication of high performance active circuits and systems at mm-waves, based on transistors with very high f_t (505 GHz) and f_{max} (720 GHz) [14], as discussed in Chapter 3. However, for passive components such as antennas and phase shifters in phased array systems, for example, the area required is considerably large and a compromise between cost and performance must be found. This compromise is evident in those 120 GHz transceivers from Silicon Radar (models TRA_129_002 and TRX_120_01 [13]). In the first one, the transmitting and receiving antennas are integrated on chip. On-chip antennas in general have a low gain; however, the entire system fits in a 5 mm × 5 mm package. To increase the antenna gain, the second transceiver uses two antenna arrays for receiving and transmitting, each with four patch antennas. The size of each array is at least four times the size of the active components on the transceiver; thus they were realized on a different substrate to reduce costs and increase performance. Therefore, as well for the choice of the mm-wave MEMS technology, the best implementation depends on the application requirements and this is not a straightforward comparison.

Another possibility is to integrate on the same substrate a number of different RF components and circuits fabricated with the best suited technologies and materials. To reduce the required area and power loss in the interconnections between these circuits,

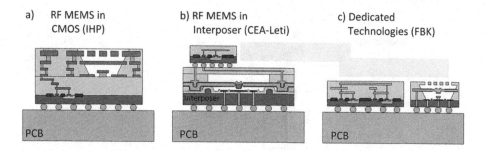

a)   RF MEMS in   b) RF MEMS in   c) Dedicated
      CMOS (IHP)   Interposer (CEA-Leti)   Technologies (FBK)

Figure 4.3 RF MEMS technologies and integration.

this has evolved into a 3D heterogeneous integration using an interposer. An interposer is an interlayer placed between different technologies, mostly between the master board (a printed circuit board [PCB]) and on-chip circuits (realized in various technologies: silicon, GaAs, InP, photonics, MEMS, etc.), that can interconnect their signals.

The same dilemma involves the choice of technology for RF and mm-wave MEMS and their integration with the rest of the system. There is not a unique approach to implement this integration. Figure 4.3 illustrates some possibilities. MEMS can be integrated with active circuits in the same chip utilizing the same Si technology, as demonstrated by IHP Microelectronics and illustrated in Figure 4.3a. In this approach, mm-wave MEMS switches are used to realize several reconfigurable circuits, such as oscillators and low-noise amplifiers [15]. Other possibility is to fabricate MEMS on the interposer, as conceptually illustrated in Figure 4.3b and demonstrated by CEA-Leti [16, 17]. Here, larger circuits such as phase shifters (see Section 4.5) for phased array applications can be fabricated. Also, the availability of through-silicon vias (TSVs) allows 3D integration with active circuits fabricated on other technologies. Another possibility is the separated fabrication of MEMS on RF dedicated technologies, which can be reported on the same PCB, as demonstrated by FBK MicroSystems Technology [18] and illustrated in Figure 4.3c.

4.2.2.1   CMOS-Based Technologies: IHP

CMOS-based technologies correspond to top-down approaches. IHP has available two SiGe BiCMOS technologies that allow the fabrication of RF MEMS switches: the 0.25 μm and 0.13 μm. Both technologies are well suited for mm-wave applications, due to their heterojunction bipolar transistors (HBTs) with f_t/f_{max} up to 505/720 GHz [14]. Their through-silicon via and backside etch capabilities are other interesting features for mm-wave SoC development. The Silicon Radar 120 GHz transceiver, mentioned earlier, was developed using IHP technology. Figure 4.4 shows the top view of the schematics of the coplanar waveguide (CPW)-based mm-wave MEMS switch (the encapsulation of the switch is hidden).

The cross section of the IHP's 0.13 μm Front-End-Of-Line (FEOL) with the detailed BEOL module is shown in Figure 4.5. In comparison to the five metallization

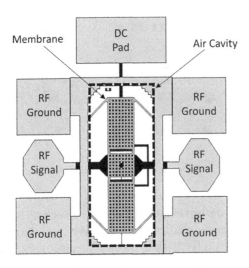

Figure 4.4 Top view of the IHP mm-wave MEMS switch.

layers of the IHP's 0.25 µm Back-end-of-line (BEOL), the 0.13 µm BEOL consists of seven metallization layers with different metal thicknesses and distances between them [19, 20]. The thin metal layers are named from bottom to top as Metal1 (M1), Metal2 (M2), Metal3 (M3), Metal4 (M4), and Metal5 (M5). The two thick metal layers are TopMetal1 (TM1) and TopMetal2 (TM2). Additionally, a metal–insulator–metal (MIM) capacitor is available between M5 and TM1 in order to achieve a high capacitance density of 1.5 $fF/\mu m^2$. The passivation on top includes a thickness of 1.5 µm SiO_2 and 0.4 µm Si_3N_4.

In the 0.25 µm technology, the RF MEMS switches are realized by using M1 as the high-voltage (HV) electrodes, M2 as the RF signal line, and M3 as the suspended movable membrane [21]. With IHP's 0.13 µm technology, the capacitive RF MEMS switch was developed between M4 and TM2. To minimize the substrate couplings, the RF signal line is shifted up to M5 instead of the M2. Moreover, the movable membrane is thicker in 0.13 µm technology and it significantly changes the electro-mechanical behavior, the RF performance, and the fabrication process.

Figure 4.5 shows the process flow of the embedded RF MEMS switch in IHP's back-end-of-line (BEOL) of the 0.13 µm SiGe BiCMOS technology. The developed RF MEMS switch consists of two M4 HV electrodes, an M5 RF signal line, a TM1 movable membrane, and a TM2 plate with release holes. The TM2 plate is placed on top of the switch for the wafer-level encapsulation. The integration of the RF MEMS switch includes supplementary process developments additional to the standard BEOL flow of the passives, which mainly include lateral and vertical etch stops and the planarized oxide on the TM2 plate. Initially, TM2 layers of the RF and DC pads are reached with the passivation opening (Figure 4.5a). Afterward, with an additional mask, the RF MEMS switches are released through the TM2 plate holes until the M4 HV electrodes by hydrofluoric acid vapor phase etching (HFVPE) (Figure 4.5b).

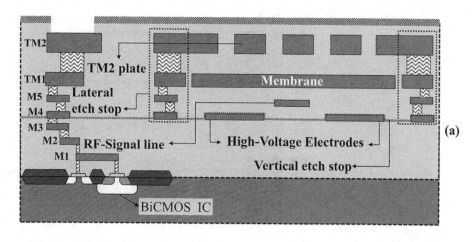

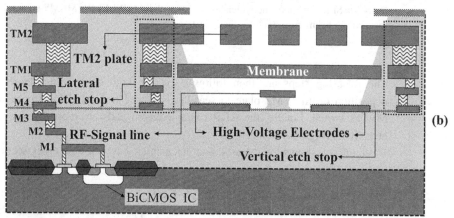

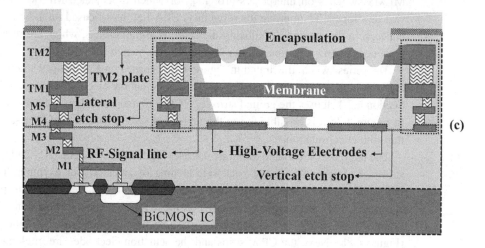

Figure 4.5 The process flow of the RF MEMS switch
(IHP 0.13 μm technology).

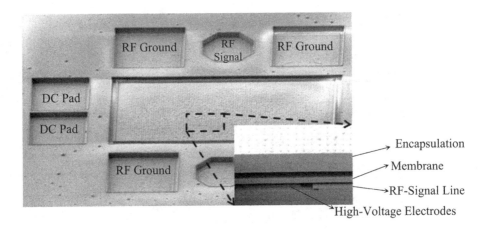

Figure 4.6 SEM image of the wafer-level encapsulation (WLE) RF MEMS switch including a focused ion beam cross section image [19].

The releasing steps are then followed by the high deposition rate (HDR) oxide deposition on the TM2 plate to seal the releasing holes and have encapsulated RF MEMS switch. Finally, the HDR oxide–covered pads are reopened by reactive ion etch (RIE) (Figure 4.5c).

Figure 4.6 shows the released and encapsulated switch with cross-section cut.

4.2.2.2   MEMS-Based Technologies: CEA-Leti

CEA-Leti, in France, has utilized a bottom-up approach to develop high-reliability RF MEMS switches on interposers [16, 17], for which a dedicated MEMS technology was developed. To improve the reliability, CEA-Leti has focused on a dielectric-less concept, in the attempt of reducing dielectric charging effects, which are responsible for a drift on the actuation voltage, leading to the switch failure.

The series switch, illustrated in Figure 4.7a, is based on a suspended silicon nitride (SiN) bridge, with metallic RF contact and actuation electrodes, as described in Section 4.3.1. It uses the ground strips of a CPW transmission line as fixed electrodes. In Figure 4.7b it is possible to see the fabricated switch (courtesy of CEA-Leti) with thin film encapsulation.

The interposer illustrated in Figure 4.8 uses high-resistive silicon ($\rho > 2 \text{ k}\Omega \cdot \text{cm}$) as substrate to allow TSVs and back side etching of antennas, for example. The first fabrication steps include a series of substrate oxidation and wet etching to promote substrate isolation, a recessed cavity for the CPW line, and production of small dimples that work as stoppers for the electrostatic actuation of the switch (Figure 4.8a). Next, the CPW strips and the actuation electrodes are implemented in gold (Au). To prevent stiction and mechanical damage of the RF contact and to avoid hillocks that could lead to a short-circuit in the actuation region, a thin layer of ruthenium (Ru) is used (see Figure 4.8b).

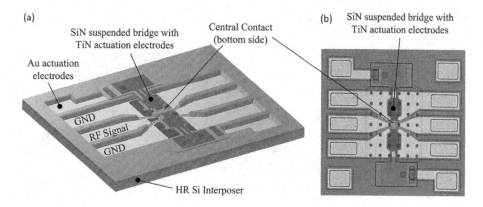

Figure 4.7 CEA-Leti RF MEMS switch.

Before the realization of the silicon nitride bridge, the recessed cavity is filled with a Si sacrificial layer. To obtain a flat bridge the Si layer is planarized; then, the TiN actuation electrode and the RF metallic contact are deposited and patterned (Figure 4.8c). Next, the SiN bridge itself and an upper TiN are defined (Figure 4.8d). This upper TiN layer is used as a stress compensation to increase thermal stability of the switch. Finally, the sacrificial Si is removed, releasing the SiN bridge. Figure 4.8e shows a cross section of the final device.

4.2.2.3   Dedicated Technologies: FBK

The FBK process was established to attain the flexibility required to allow the fabrication of a variety of RF MEMS components from capacitive switches, varactors, and inductors, to more complex circuits, like phase shifters and reconfigurable antennas, which can operate up to 100 GHz.

The FBK process is based on a surface micromachining modular approach, which adds steps to a basic sequence of processing to attend the specific requirements of different devices. This process allows, for example, the fabrication of capacitive and resistive RF MEMS switches with wafer-level encapsulation (chip and thin film capping). A quartz substrate can also be used to reduce substrate losses for applications up to 120 GHz. The series switch has an electrostatic actuated suspended membrane that contacts the interrupted signal strip as illustrated in Figure 4.9.

The basic sequence utilizes eight photolithographic masks. The switches are fabricated on a thermally oxidized Si substrate, over which a poly-Si layer is deposited and masked to obtain the polarization strips and electrodes for electrostatic actuation as shown in Figure 4.10a. This poly-Si must be doped to obtain the required resistance for different devices/applications. For the RF MEMS switches, it is interesting to have high-resistance strips to reduce losses. Small dimples are also defined on poly-Si to increase the contact force and guarantee a low contact resistance.

The poly-Si is isolated with a LPCVD TEOS SiO_2 layer and the accesses to the underlying poly-Si electrodes are opened in a second photolithography step. Next, a

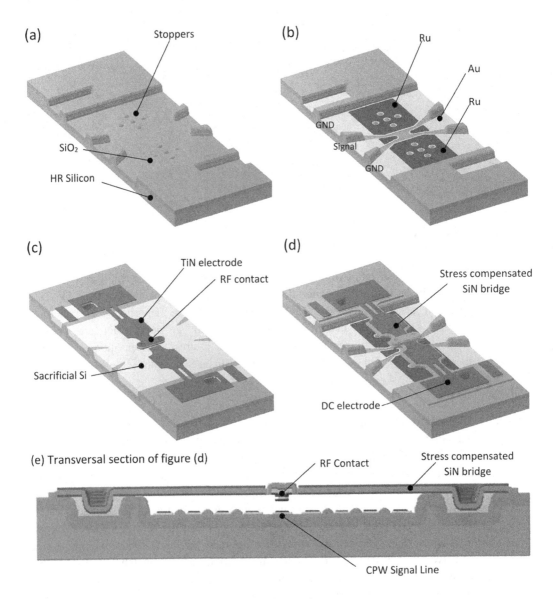

Figure 4.8 CEA-Leti RF MEMS switch fabrication process.

metal multilayer of Ti/TiN/Al/Ti/TiN is deposited and patterned to connect the contact region (with the dimples) to the signal strip of the CPW transmission line. A third photolithography step is utilized to define its geometry (Figure 4.10b). This metal multilayer is covered by a thin (100 nm) low-temperature SiO_2 film that is used as an insulator in the capacitive contact switch. This SiO_2 films is patterned by a fourth lithography step.

The next process step is to deposit an Au layer to work as metallic floating electrode (Flomet) over the small dimples created with the poly-Si. At this stage, a

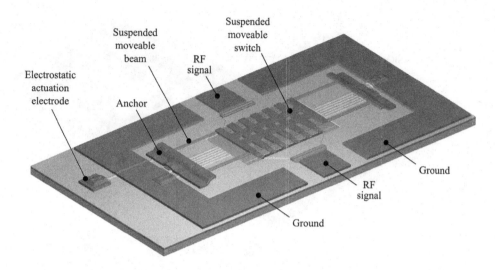

Figure 4.9 FBK RF MEMS switch.

metal–insulator–metal (multilayer metal–low-temperature SiO_2–Au) capacitor is formed over the poly-Si dimples. This Flomet is used in capacitive switches to create a reproducible contact capacitance, since the metallic membrane contacts the Au layer and eliminates capacitance variations due to contact deformation. In resistive switches, the thin low-temperature SiO_2 layer is removed and the Flomet contacts the multilayer metal directly. The Flomet geometry is defined by the fifth lithography step (Figure 4.10c).

The electric mobile contact of the switch is a suspended Au membrane. To obtain the gap between the Au membrane and the electrodes underneath, a thick photoresist layer that works as sacrificial material (geometry defined at the sixth photolithography step) is utilized. The Au membrane itself is obtained by electroplating technique in two steps, each of them with different geometries. The first electroplating step produces a 2-μm-thick Au layer. The second one produces a thicker Au layer of 3.5 μm. The electroplated gold layers are also used to create the CPW transmission line. The last step is the removal photoresist "spacer" to release the suspended Au membrane (Figure 4.10d).

Note that, since the geometry of the Au electroplating steps is defined by two separate photolithography steps (seventh and eighth), the final membrane has regions with different thicknesses (2 or 5.5 μm) and, consequently, with different electrical and mechanical properties. This increases the flexibility of the design, allowing an extra degree of freedom. Also, the contact region can be made flexible to increase the contact force on the poly-Si dimples. The CPW strips are made by superimposing the thin and thick Au layers (Figure 4.9).

4.3 Millimeter-Wave MEMS Switches

In RF MEMS, switches are the most basic components. They were studied, modeled, extensively tested, packaged, and successfully integrated. Several different

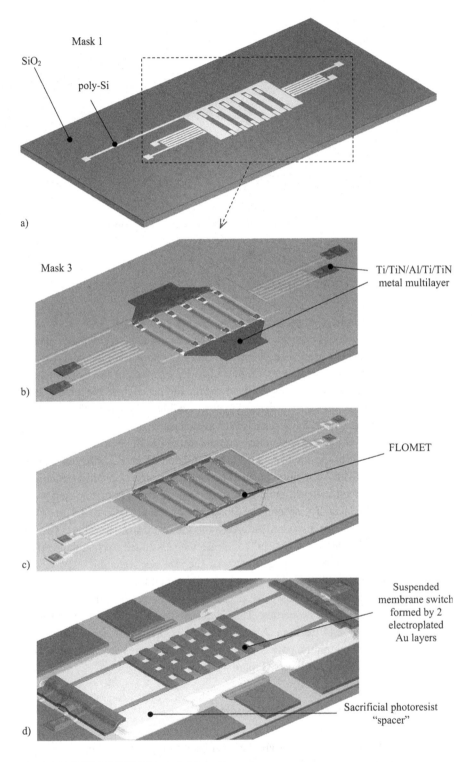

Figure 4.10 FBK RF MEMS switch fabrication process.

applications at microwave frequencies were demonstrated and their benefits and drawbacks contrasted in several publications. We will now analyze their utilization at mm-wave frequencies.

MEMS structures integrated in waveguides have been demonstrated with very promising results at mm-wave and terahertz frequencies [22, 23] with operation up to 700 GHz [24]. However, this type of MEMS will not be the focus of the coming sections, since it constitutes a different type of switch with a limited set of applications, owing to its considerable size and cost.

At lower frequencies, the RF MEMS switches are electrically small. For example, on a fused silica substrate ($\varepsilon_r = 3.9$) the guided wavelength (λ_g) of a 50-Ω CPW transmission line at 1 GHz is approximately 160 mm compared to the switches that are generally not longer than a few hundreds of microns.

At mm-wave frequencies, the dimensions of the switch need to be carefully taken into account. Considering the same fused silica substrate, at 60 GHz and 200 GHz, λ_g is equal to 3.3 mm and 0.8 mm, respectively. This means that to avoid unwanted reflection of the signal, either the switch has to be smaller than about $\lambda_g/10$ or its impedance must be carefully matched to the input and output transmission lines through a matching network. This can be somewhat challenging if microstrip lines are used. A 50-Ω microstrip line on a 500-μm fused silica substrate is 1.3 mm wide. In this case, either the switch has the same approximate width, which could be mechanically disadvantageous, or the line should be tapered to the width of the switch, which would increase drastically the length of the device if the switch is longer than 330 μm (i.e., $\lambda_g/10$) at 60 GHz and 80 μm at 200 GHz. A possibility to use microstrip lines for RF MEMS switches is to reduce the thickness of the substrate, which will reduce the width of the microstrip (~210 μm for a 100-μm-thick substrate) and relax the constraints. However, the thinning of the substrate increases fabrication complexity and cost. Further, shunt switches would require a through-substrate via, which would also increase costs, or large capacitors to ground, or radial stubs, which would increase the surface area of the switch. A possibility is to use the back-end-of-line of a CMOS process for microstrip-based switches. In this case, the thickness of the substrate is the distance between the first and last metallic layers, which ranges from 5 μm to 10 μm for current technologies. In a CMOS process, vias are available with no cost increase, but the thin substrate (height of the BEOL; see Chapter 3) reduces the quality factor of microstrip lines.

For the reasons discussed earlier, most of the mm-wave MEMS switches presented in the literature are based on CPW transmission lines. The characteristic impedance of CPWs is practically independent of the substrate thickness and the ground contact is readily available.

Because of their constitution, RF MEMS switches are essentially broadband devices; however, for mm-wave applications, the influence of the parasitics, such as the up-state capacitance and the down-state resistance (for resistive contact switches) or capacitance (for capacitive contact switches), represented in Figure 4.11 as the impedance Z and admittance Y, can limit their operation.

Equations (4.2) and (4.3) express the S-parameters of the series and shunt switches related to their reference impedance Z_0 and admittance Y_0 and to the switch impedance Z:

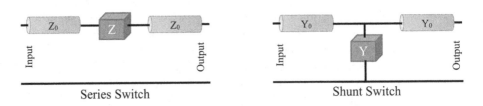

Figure 4.11 RF switch configuration: series and shunt.

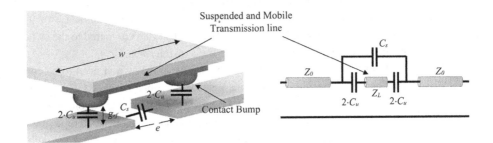

Figure 4.12 Series resistive switch and its up-state model.

$$S_{series} = \begin{bmatrix} \dfrac{Z}{Z + 2Z_0} & \dfrac{2Z_0}{Z + 2Z_0} \\[2ex] \dfrac{2Z_0}{Z + 2Z_0} & \dfrac{Z}{Z + 2Z_0} \end{bmatrix}$$ (4.2)

$$S_{shunt} = \begin{bmatrix} \dfrac{-Y}{Y + 2Y_0} & \dfrac{2Y_0}{Y + 2Y_0} \\[2ex] \dfrac{2Y_0}{Y + 2Y_0} & \dfrac{-Y}{Y + 2Y_0} \end{bmatrix}.$$ (4.3)

In the next sections, the contribution of the parasitics for the degradation of the performance of the RF MEMS switch will be discussed.

4.3.1   Up-State Capacitance

4.3.1.1   Series Switch

In the up state, a series switch can be modeled by a combination of capacitors (C_s and C_u) and a small section of suspended transmission line with characteristic impedance Z_L, as illustrated by the simplified structure in Figure 4.12.

The separation e between the input and output transmission lines in a series switch can seriously degrade its isolation (S_{21} parameter in the up state) at mm-waves, as shown in Figure 4.13a. For an isolation of 30 dB at 60 GHz, 120 GHz, and 200 GHz, the separation should be at least 70 μm, 100 μm, and 120 μm, which represents a series capacitance C_s of 0.8 fF, 0.4 fF, and 0.2 fF, respectively, or an impedance of 3.3 kΩ.

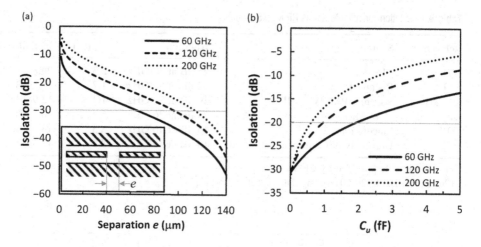

Figure 4.13 (a) Influence of the separation e between the input and output on the isolation of a series switch. (b) Isolation degradation as a function of C_u with $e = 70$ μm for 60 GHz and $e = 120$ μm for 200 GHz.

At 60 GHz, the separation is not critical but it should be accounted for, but at 200 GHz, it imposes a width w for the switch that is not electrically short. This indicates that the mobile part of the switch should be matched the input and output impedances.

Even if this separation can be important, the controlling aspect of the isolation in mm-wave MEMS switches is the C_u capacitance. This capacitance is the result of the overlap region A_c between input/output and the suspended transmission line in the up state. C_u can be further increased by the contact bumps in resistive switches. Figure 4.13b shows the degradation of the isolation as a function of C_u. The C_u capacitance was simulated considering a separation from 70 μm, 100 μm, and 120 μm, at 60 GHz, 120 GHz, and 200 GHz, respectively. It is possible to see that a capacitance of 0.6 fF reduces the isolation to 20 dB at 200 GHz because the impedance becomes lower (~1.3 kΩ). At 60 GHz, $C_u = 2$ fF produces the same effect.

The magnitude of the isolation can be expressed by Eq. (4.4), by neglecting the impedance of the suspended transmission line Z_L, which is often the case for most switches up to 50 GHz. The effect of the suspended transmission line will be discussed in a following section.

$$\text{Isolation}_{series} = |S_{21}| = \left| \frac{2Z_0}{Z + 2Z_0} \right| = \left| \frac{2Z_0}{Z_{C_s} \| Z_{C_u} + 2Z_0} \right| = \frac{2Z_0}{\sqrt{4Z_0^2 + \dfrac{1}{\omega^2 \cdot (C_s + C_u)^2}}}, \quad (4.4)$$

where ω is the angular frequency and Z_0 the characteristic impedance of the input and output ports. If $4Z_0^2 \ll \dfrac{1}{\omega^2 \cdot (C_s + C_u)^2}$ and $C_s \ll C_u$, which is normally the case for switches in the up state with good isolation, Eq. (4.4) can be reduced to

$$\text{Isolation}_{series} = 4 \cdot \pi \cdot f \cdot Z_0 \cdot C_u. \quad (4.5)$$

Table 4.1 Isolation for reliable series RF MEMS switches

Reference	Switch	Isolation	C_u (fF)	Isolation (Eq. 4.5)
[17]	CEA-Leti	25 dB at 40 GHz	2.3[a]	24.8 dB
[25]	Rockwell	25 dB at 50 GHz	1.75	25.2 dB
[26, 27]	Radant – RMSW200HP	14 dB at 20 GHz	16[b]	13.9 dB
[28]	U. of Michigan	23 dB at 26 GHz	6–8	20.2 dB
[29]	Lincoln Labs	15 dB at 40 GHz	8.5	13.4 dB
[30]	University of California San Diego	15 dB at 40 GHz	8.5	13.4 dB

[a] Fitted from data provided by CEA-Leti.
[b] Fitted from DC to 20 GHz using data presented in [26].

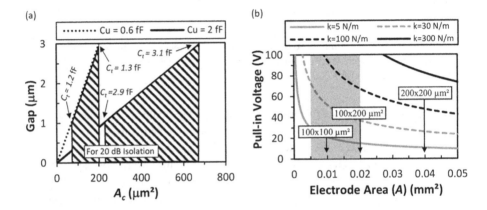

Figure 4.14 (a) Input/output and suspended transmission line overlap area A_c (C_u area) variation for 0.6 fF and 2 fF, which yields a 20 dB isolation at 200 GHz and 60 GHz, respectively. (b) Influence of the spring constant k and electrode area A on the pull-in voltage V_p for $g = 3$ μm. Hatched region indicates normally used area in reliable switches.

This relation can be verified with Table 4.1, which shows the isolation and C_u of other reliable and well-studied series switches. In these switches, the capacitance C_u was extracted by the authors, in most cases, from their measurements using a series capacitance model. The isolation was also calculated using Eq. (4.5), agreeing well with the provided data.

Considering the input/output and the suspended transmission line as a parallel plate capacitor, as a first approximation (neglecting the fringing field), it is possible to estimate a range of areas A_c for C_u that depends on the gap g_{rf}. For most of the reliable switches, g_{rf} is between 1 and 3 μm, as shown in Table 4.2, which provides the hatched areas in Figure 4.14a. For a 20 dB isolation ($C_u = 0.6$ fF at 200 GHz or $C_u = 2$ fF at 60 GHz) considering this range of g_{rf}, the overlap area A_c should be between 70 and 200 μm^2 at 200 GHz and between 230 and 680 μm^2 at 60 GHz.

Taking into account the fringing field capacitance, the total capacitance C_t was numerically simulated (Maxwell 3D – Ansys) for different overlap areas and gaps. From the simulated values, it is possible to verify that the fringing field capacitance

Table 4.2 Electromechanical characteristics of reliable series RF MEMS switches

Ref.	Switch	g_{rf} (µm)	k (N/m)	A (µm x µm)	V_p (V)
[32]	CEA-Leti	0.9	182	$2 \times 40 \times 115$	22
[25]	Rockwell	2.5	12–15	75×75	50–60
[26, 27]	Radant – RMSW200HP	1.0	60–100	20×35	60–80
[28]	U. of Michigan	1.5	79–119	100×70	45–55
[30]	U. of Cal. San Diego	0.85	6020	150×150	67–74
[33]	HRL	1.5	23[a]	100×100	30
[34]	Raytheon (Memtronics)	2.5	6–30	120×80	22
[35]	Omron	3.0	350–620[a]	1.4 mm^2	15–20

[a] Estimated using Eq. (4.1).

can be of the same order or greater than C_u, especially for small areas and large gaps. C_t of 1.2–1.3 pF was found for the range at 200 GHz and 2.9–3.1 pF for the range at 60 GHz, shown in Figure 4.14a.

To achieve a total capacitance C_t of 0.6 fF to keep a 20 dB isolation, the area A_c should be reduced to 25 µm^2 for a 1-µm-gap and to 50 µm^2 for a 3-µm-gap (not shown in Figure 4.14a). However, this can depend heavily on the geometry of the switch and existence of contact bumps and/or dielectric layer and it might be complicated to achieve such a low C_t. Some creative switch designs can help reduce this capacitance. One example is the fork tip design from XCOM [31], which allowed a considerable reduction of C_t from 10 to 3.8 fF.

As seen, the influence of the gap in the isolation is quite important. However, it also has a strong influence on the pull-in voltage V_p, as seen in Figure 4.13b and Eq. (4.1). It is desired to have the lowest possible pull-in voltage with the highest gap possible to increase isolation. A low V_p can be obtained with a small spring constant k, but k needs to be large enough to prevent stiction of the mobile electrode in the down state. Further, increasing the electrode area A ($A = L_e \cdot w$, Figure 4.2) can reduce the pull-in voltage, but it increases the overall area of the device.

The spring constant of reliable switches (Table 4.2) can vary drastically, from a few up to thousands of Newton-meters, depending on the design. The variation of the electrode area A, on the other hand, is not great (excluding the Omron and Radant switch), ranging from 0.007 to 0.0225 mm^2. The Omron switch, however, needs a very large A to obtain a low V_p, because of the combination of large gap and high k. All of the switches listed in Table 4.2 show a pull-in voltage V_p smaller than 80 V, which is important to able integrated CMOS-based charge pumps.

Figure 4.14b estimates the required V_p for different k and A, for a gap of 3 µm. A larger gap will relax some of the design constrains in order to achieve low C_u. For high k designs, a larger electrode area will be required.

4.3.1.2 Shunt Switch

In the up state, a shunt switch can be modeled by a combination of a capacitor (C_u) and two sections of suspended transmission lines with characteristic impedances $2 \cdot Z_L$

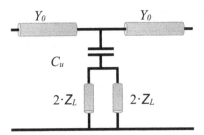

Figure 4.15 Shunt switch and its up-state model.

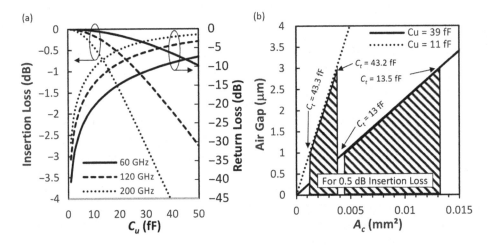

Figure 4.16 (a) Influence of C_u on the insertion loss of a shunt switch. (b) C_u area variation for 11 fF and 39 fF, which yields a 0.5 dB insertion loss at 200 GHz and 60 GHz, respectively.

connected to ground, as illustrated in the model in Figure 4.15. In shunt switches, C_u is the main parasitic responsible for the insertion loss (IL_{shunt}) that can be expressed by Eq. (4.6), neglecting the effect of the electrical length of the two sections of suspended transmission lines:

$$IL_{shunt(C_u)} = |S_{21}| = \left|\frac{2Y_0}{Y + 2Y_0}\right| = \left|\frac{2Y_0}{Y_{C_u} + 2Y_0}\right| = \frac{2Y_0}{\sqrt{4Y_0^2 + \omega^2 \cdot C_u^2}}. \tag{4.6}$$

These will be analyzed in Section 4.3.4.

Figure 4.16a shows the insertion loss as a function of C_u for 60, 120, and 200 GHz without considering resistive loss, which is discussed in the following sections. For an insertion loss of 0.5 dB, C_u should be 39 fF at 60 GHz and 11 fF at 200 GHz. In shunt switches, the overlap area constrains are considerably more relaxed as shown in Figure 4.16b. Figure 4.16 also shows the simulated (ANSYS Maxwell) total capacitance C_t for different gaps and areas, where it is possible to see that the fringing field capacitance is little significant in these larger capacitances. Table 4.3 corroborates this information with data from reliable shunt switches. It is possible to see that C_u for

Table 4.3 Insertion Loss for reliable shunt RF MEMS switches

Reference	Switch	Insertion loss	C_u (fF)
[34]	Raytheon (Memtronics)	0.06 dB at 35 GHz	30–40
[36]	University of Michigan	0.4 dB at 40 GHz	70
[32]	CEA-Leti	0.35 dB at 40 GHz	40

these switches ranges from 30 to 70 fF. Further, A_c is much larger when compared to the series switch, ranging from 0.00125 mm² (1250 μm²) to 0.0035 mm² (3500 μm²) and from 0.0045 mm² (4500 μm²) to 0.0135 mm² (13 500 μm²) for an insertion loss of 0.5 dB at 200 GHz and 60 GHz, respectively.

4.3.2   Down-State Resistance

4.3.2.1   Series Switch

In the following discussion the resistance/metallic losses associated with the access lines are not considered.

The down-state resistance R_d is responsible for most of the power loss in RF MEMS switches. The contribution of R_d to the insertion loss can be expressed by Eq. (4.7). R_d has two main components: (1) the contact resistance R_c associated with the physical contact between the suspended transmission line and the access lines; and (2) the series resistance R_s associated with the metallic losses in the suspended transmission line. This equation can be used to extract R_d from measured insertion loss.

$$ IL_{series(R_d)} = |S_{21}| = \left| \frac{2Z_0}{Z + 2Z_0} \right| = \frac{2Z_0}{R_d + 2Z_0}. \tag{4.7} $$

R_c is inversely proportional to the contact force F_c and also depends on the material and surface roughness. Most demonstrated reliable RF MEMS switches use gold or gold alloys as contact material, mostly because gold (1) is a soft metal, which is important for low-contact force designs, yielding small R_c; (2) has a high melting point, allowing higher currents through the switch; (3) has a high electrical conductivity, which reduces losses; and (4) does not oxidize, being also less prone to contamination.

For gold-based switches, a stable electrical contact is obtained with contact forces higher than 100 μN, which yields contact resistances from 0.15 Ω to 0.4 Ω for reliable switches [37]. The contact force can be calculated analytically as shown in [38], but it depends on several technological and design aspects. In general, to achieve 100 μN, the electrode area ranges from 0.01 mm² to 0.016 mm².

Despite the good qualities of gold as contact material, it requires high restoration forces (necessary force to separate contacts) because of its high contact adherence. Adhesion forces for gold-gold contacts can reach 2.7 mN [39]. To separate the contacts, the restoration force should be greater than the adhesion force. The

Table 4.4 Down-state resistance for reliable series resistive RF MEMS switches

Ref.	Switch	Contact Force (μN)	Insertion Loss at DC	R_d (Ω)[a]	$R_{d\ Calc}$ (Ω)[b]
[25]	Rockwell	100–150	0.08	1.0	0.9
[26, 27]	Radant - RMSW200HP	200	0.10	1.0	1.1
[30]	U. of California San Diego	1200	0.2	1.6	2.3
[33]	HRL	300	0.12	1.6	1.4
[35]	Omron	4500	0.05	0.5	0.6

[a] Extracted by the authors.
[b] Calculated using Eq. (4.7).

restoration force equals the spring constant times the displacement of the mobile electrode. Assuming that the displacement is equal to the gap g, the restoration force for a switch with $k = 100$ N/m and $g = 3$ μm is only 300 μN. For this reason, several switches use gold alloys such as AuNi5 that has much lower contact adhesion forces, reaching a maximum of 300 μN. Harder materials such as rhodium yield even lower, below 100 μN, contact adhesion forces [39]. However, these other materials and alloys require higher contact forces, 300 μN for AuNi5 and 600 μN for Rh, to achieve a stable electric contact, and their contact resistance is also higher. Table 4.4 shows the contact force for some reliable switches.

At DC, the series resistance R_s (metallic losses in the suspended transmission line) is represented by

$$R_{s\ DC} = \frac{1}{\sigma} \cdot \frac{e}{t \cdot w},\qquad (4.8)$$

where σ is the conductivity of the metal, e the length between the input and output, t is the thickness, and w the width of the suspended transmission line. As it can be seen in Figure 4.17a, $R_{s\ DC}$ is quite small in most RF MEMS switches. For cross section areas $(t \cdot w)$ as small as 12 μm^2 ($t = 0.5$ μm and $w = 24$ μm) that are easily achieved, $R_{s\ DC} = 0.2$ Ω and $|S_{21}| = 0.02$ dB, which is negligible, following a near liner relationship, reaching 0.5 dB for 6 Ω. Therefore, at DC, the contact resistance is the dominant contributor to the insertion loss. Further, as seen in Figure 4.17a, metals with different conductivity ($\sigma_{au} = 4.1 \times 10^7$ S/m and $\sigma_{al} = 3.5 \times 10^7$ S/m) show almost no influence on the resistance.

As the frequency goes up, the skin effect reduces the effective cross section of the transmission line that carries current, and R_s increases as shown by Eq. (4.9) and illustrated in Figure 4.17b for a rectangular conductor:

$$R_s = \frac{1}{\sigma} \cdot \frac{L_s}{10\delta(t + w - 10\delta)},\qquad (4.9)$$

where L_s is the length of the suspended transmission line and δ is the skin depth given by

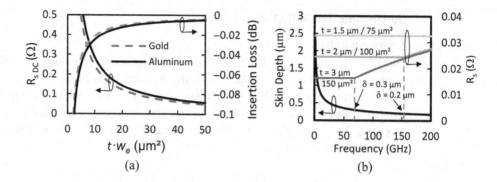

Figure 4.17 (a) DC resistance $R_{s\,DC}$ and respective insertion loss at DC of a 100-μm-long metallic strip as a function of the cross-sectional area $t \cdot w$. (b) Skin depth for a gold conductor and respective resistance R_s for a transmission line with w of 50 μm and L_s of 100 μm, with different thicknesses t.

$$\delta = \sqrt{\frac{1}{\pi \cdot f \cdot \sigma \cdot \mu}}, \tag{4.10}$$

where f is the frequency, and μ is the magnetic permeability of the medium. It is assumed that approximately 99% of the current in a conductor is confined within a shell with thickness of $5 \cdot \delta$. For this reason, the skin effect increases the resistance when the thickness t of the conductor is smaller than $5 \cdot \delta$. In general, the width w is determined by the impedance of the transmission line and is much larger than its thickness.

The proximity effect (signal and ground) can further reduce the effective area, confining even more the metallic area where current flows in the switch. This effect can be better analyzed with numerical simulations, but it results in an increase in R_s.

In any case, the metallic losses associated with the transmission line are small compared to the losses due to R_c and in most cases can be neglected. Table 4.4 compares the down-state resistance for several reliable switches. It is possible to see that the R_d extracted by the authors of the different switches agree quite well with the R_d resistance calculated using Eq. (4.5).

4.3.2.2   Shunt Switch

The down-state resistance degrades the isolation in resistive shunt switches as expressed in Eq. (4.11). An R_d of 0.5 Ω, 1 Ω, and 1.5 Ω results in a constant isolation across the frequency spectrum of 34 dB, 28 dB, and 25 dB, respectively. These values would correspond to the isolation of a shunt switch if R_d was the only parasitic element. In reality, the small suspended transmission line and any other parasitic should be considered in the isolation model and will be discussed later in this section.

$$\text{Isolation}_{shunt(R_d)} = |S_{21}| = \left| \frac{2Y_0}{Y + 2Y_0} \right| = \frac{2Y_0}{1/R_d + 2Y_0}. \tag{4.11}$$

4.3.3 Down-State Capacitance

4.3.3.1 Series Switch

For capacitive contact switches, the area A_c between the input/output and the suspended transmission lines (same area of the isolation capacitance C_u) is important. Besides controlling the isolation (in the up state), it will also determine the low cutoff frequency and dictate the insertion loss in this type of switch. Equation (4.12) expresses the contribution of the down-state capacitance C_d on the insertion loss of the series capacitive switch:

$$IL_{series(C_d)} = \frac{2Z_0}{\sqrt{4Z_0^2 + \dfrac{1}{\omega^2 \cdot C_d^2}}}. \tag{4.12}$$

Figure 4.18 illustrates the relation between A_c and the insertion loss (0.1 dB and 0.5 dB) at different frequencies, for two different insulator thicknesses ($t_i = 100$ nm and 200 nm, respectively, with $\varepsilon_r = 7$), considering a parallel plate model. In the down state, the total capacitance considering the fringing field (simulation with ANSYS Maxwell) is only 5% higher than the parallel plate approximation; thus the fringing field can be neglected for the areas and dielectric thicknesses involved.

For a 20 dB isolation in the up state, the overlap area should be between 70 μm^2 and 200 μm^2 at 200 GHz and between 230 μm^2 and 680 μm^2 at 60 GHz (not considering the fringing field), as shown in Figure 4.14a. Based on Figure 4.18, Table 4.5 summarizes the information for the A_c range mentioned earlier. As we can see, there is a tradeoff between the area, dielectric thickness, and the insertion loss. As the capacitance increases (A_c increases and t_i reduces) the insertion loss decreases, as expected, since the impedance of the series capacitance decreases with frequency. It is

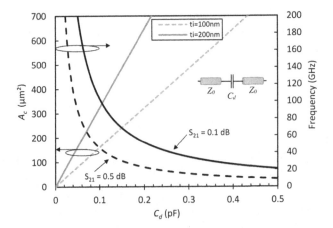

Figure 4.18 Correlation between the down-state capacitor area A_c, the capacitance C_d itself, and the frequency for insertion loss of 0.1 dB and 0.5 dB, respectively.

Table 4.5 Relation between insertion loss (IL), capacitance area A_c, and dielectric thickness t_i

Isolation 20 dB	A_c (μm²)	t_i = 200 nm		t_i = 100 nm	
		IL 0.5 dB	IL 0.1 dB	IL 0.5 dB	IL 0.1 dB
At	230	60 GHz	150 GHz	30 GHz	75 GHz
60 GHz	680	20 GHz	50 GHz	10 GHz	25 GHz
At	70	220 GHz	480 GHz	110 GHz	240 GHz
200 GHz	200	70 GHz	160 GHz	35 GHz	80 GHz

possible to see that the dielectric thickness has a strong impact on the insertion loss. For the A_c required to obtain 20 dB isolation in the up state at 60 GHz and 200 GHz, it is possible to see that an insertion loss of 0.5 dB in the down state is not difficult to obtain.

However, the actual C_d is highly dependent on the surface quality of the dielectric and the metal layers. The surface roughness can seriously degrade the capacitance and should be taken into account.

An obvious approach to guarantee a higher capacitance, thus lower insertion loss, would be to reduce even further t_i, thinner than 100 nm, but that would increase the pinhole density and reduce the dielectric breakdown voltage in traditionally used plasma enhanced chemical vapor deposition (PECVD) silicon nitride (SiN$_x$) films. SiN$_x$ is often used as dielectric material because it is easily obtained and CMOS compatible. However, unconventional materials with high relative dielectric constant, such as atrontium titanate (STO; ε_r ~120) and barium strontium titanate (BST; ε_r > 200), have been used successfully to increase C_d [40, 41]. Nonconventional deposition techniques such as high-density inductively coupled plasma chemical vapor deposition (HDICP CVD) are also used to obtain thin (25 nm) low-roughness SiN$_x$ films with high-breakdown electric fields [42].

4.3.3.2   Shunt Switch

The down-state capacitance degrades the isolation according to

$$\text{Isolation}_{shunt(C_d)} = \frac{2Y_0}{\sqrt{4Y_0^2 + \omega^2 \cdot C_d^2}}. \tag{4.13}$$

Figure 4.19 illustrates the relation between A_c and frequency for an isolation range of 15 dB–25 dB, for two different insulator thicknesses (t_i = 100 nm and 200 nm with ε_r = 7), considering a parallel plate model. From Figure 4.19, it is possible to see that an isolation of 20 dB at 200 GHz can be obtained with A_c of approximately 530 μm² (for t_i = 100 nm) and 1000 μm² (for t_i = 200 nm), which is much smaller than the required A_c to obtain an insertion loss of 0.5 dB at the up state. This is also true for the isolation at 60 GHz. The same considerations made for the series capacitive switch related to the surface roughness should be considered for the shunt switch.

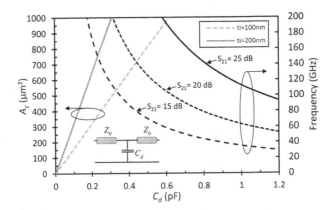

Figure 4.19 Correlation between the down-state capacitor area A_c; the capacitance C_d itself; and the frequency for an isolation of 15 dB, 20 dB, and 25 dB.

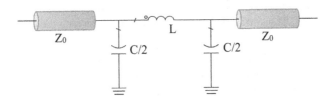

Figure 4.20 Lumped model for an electrically short ($L_s < \lambda_g/10$) suspended transmission line.

4.3.4   Suspended Transmission Line Section

4.3.4.1   Series Switch

At low frequencies, where the length of the suspended region is shorter than $\lambda_g/10$, the suspended line can be modeled as a series inductor and shunt capacitor in a Π configuration, as illustrated in Figure 4.20. The inductance L and capacitance C in this model are related in Eq. (4.14) to the characteristic impedance Z_L, the effective dielectric constant ε_{reff}, the speed of light c, and the length L_s of the suspended transmission line. For a length L_s of $\lambda_g/10$, the insertion loss error of the lumped model reaches a maximum of approximately 0.2 dB, as can be seen in Figure 4.21, where the Π lumped model is compared to an ideal transmission line model (TLine) with $Z_L = 95\ \Omega$, $\varepsilon_{reff} = 5.5$ and $L_s = 100\ \mu m$. It is possible to see that the Π model agrees quite well in insertion and return loss and phase below $\lambda_g/10$. A single series inductor (L-Series) with value from Eq. (4.14) can be used as a rough estimation at low (microwave) frequencies and high characteristic impedances, but as can be seen in Figure 4.21, at mm-waves the series inductance model deviates quite fast from the transmission line and Π models.

$$L(H) = \frac{Z_L \cdot \sqrt{\varepsilon_{reff}} \cdot L_s}{c} \quad \text{and} \quad C(F) = \frac{L}{Z_L^2}. \tag{4.14}$$

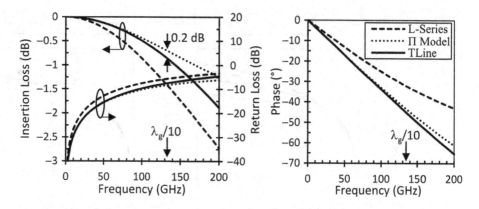

Figure 4.21 Comparison between the L-series, Π model, and transmission line model with $Z_L = 95\ \Omega$, $\varepsilon_{reff} = 5.5$, and $L_s = 100\ \mu m$.

In the switches, the signal strip of the transmission line is suspended and surrounded by air; therefore its effective dielectric constant ε_{reff} is reduced compared to a nonsuspended signal strip. Over a quartz substrate, the ε_{reff} of a CPW (ground strips on the substrate and signal strip floating above it) varies from 2.1 to 1.3 for an air gap of 0.5 μm and 5 μm, considering a signal strip width between 20 μm and 60 μm. Thus, for this case, $\lambda_g/10$ at 60 GHz, 100 GHz, and 200 GHz can be as small as 345 μm, 210 μm, and 100 μm, respectively. So, if the length of the suspended transmission line is kept below $\lambda_g/10$ for the desired frequency, it can be modeled accurately using the Π model.

On the other hand, for high dielectric constant substrates, such as silicon ($\varepsilon_r = 11.9$), ε_{reff} varies from 5.5 to 3.2 with the same gap variation and $\lambda_g/10$ at 60 GHz, 100 GHz, and 200 GHz becomes as small as 213 μm, 128 μm, and 64 μm, respectively. Depending on the substrate, geometry of the switch, and operating frequency, the suspended transmission line should be modeled as a transmission line in order to accurately predict its behavior. Figure 4.22 shows the simulation (Keysight's ADS) of the isolation of a series switch using a transmission line model for the suspended transmission line. The isolation is modeled with a capacitor $C_u = 3$ fF in series with a transmission line ($\varepsilon_{reff} = 5.5$, $L_s = 150$ μm, and 450 μm). Note that the resistance of the suspended transmission line has no effect in the isolation of a series switch and it is much smaller that the contact resistance R_c, as discussed previously.

In Figure 4.22, the effect of the characteristic impedance on the isolation is evident, particularly at mm-waves. It is possible to see that low characteristic impedance transmission lines can increase the isolation. However, low characteristic impedance transmission lines require a narrow separation between the signal strip and the ground strips in CPWs, and a wide strip in microstrip lines. The required dimensions for low characteristic impedance transmission lines can be difficult to implement due to fabrication process or mechanical constraints and will result in high insertion loss due to the impedance mismatch in the actuated state of the switch.

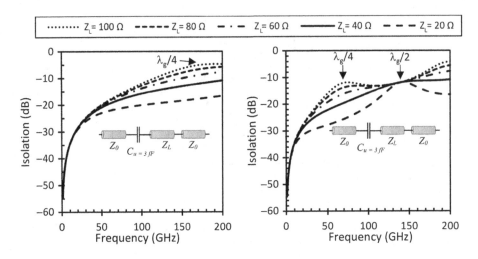

Figure 4.22 Isolation of a series resistive switch with ε_{reff} = 5.5 and L_s = 150 μm (left) and L_s = 450 μm (right) for different characteristic impedances of the suspended line Z_L.

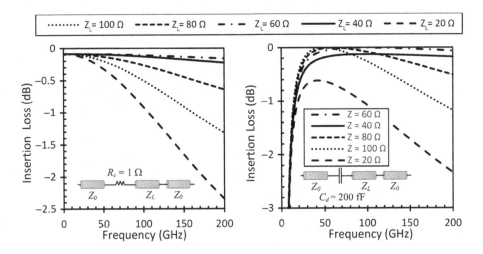

Figure 4.23 Insertion loss of a resistive (left) and capacitive (right) series switch with ε_{reff} = 5.5 and L_s = 150 μm for different characteristic impedances of the suspended line Z_L.

The length of the suspended transmission line plays an important role in the isolation. At $\lambda_g/4$, where the impedance mismatch between the input and output reaches a maximum value, the transmission line strongly affects the isolation, while at $\lambda_g/2$, where the input and output are matched disregarding the characteristic impedance, it has no influence, as seen in Figure 4.22.

The impedance mismatch of the suspended transmission line can also result in high insertion loss, as illustrated in Figure 4.23 for resistive and capacitive switches. At mm-waves, the suspended transmission line characteristic impedance should be as close to 50 Ω as possible (for $Z_0 = 50\ \Omega$). For the resistive switch illustrated in Figure 4.23, an insertion loss of 0.2 dB due to the impedance mismatch (not considering the influence of R_c) is obtained at 200 GHz for a 60-Ω transmission line, while the same insertion loss occurs at 65 GHz for an 80-Ω line.

For capacitive switches, characteristic impedances lower than 50 Ω results in higher insertion loss, owing to their capacitive nature, and should be avoided. For characteristic impedances higher than 50 Ω, the impedance mismatch can result in important insertion loss depending on the electrical length of the device. For the capacitive switch illustrate in Figure 4.23, an insertion loss of 0.1 dB is obtained at 160 GHz for a 60-Ω transmission line, while the same insertion loss occurs at 70 GHz for an 80-Ω line.

When $Z_L = 50\ \Omega$, for the resistive switch, the insertion loss is constant regardless the frequency; for the capacitive switch, the insertion loss is zero for the frequencies above the capacitor short-circuit behavior frequency.

The maximum insertion loss related to the impedance mismatch that occurs at $\lambda_g/4$ can be calculated using

$$IL = \sqrt{1 - \left(\frac{Z_L^2 - Z_0^2}{Z_L^2 + Z_0^2}\right)^2}. \tag{4.15}$$

4.3.4.2 Shunt Switch

In shunt switches the suspended region of the switch can be modeled as a shunt inductor. In resistive switches, as the frequency increases, the impedance of the inductor increases as well, thus reducing the isolation. This behavior is illustrated in Figure 4.24, comparing simulation and model results. Shunt switches were simulated in ANSYS HFSS using a CPW configuration with 50-Ω characteristic impedance.

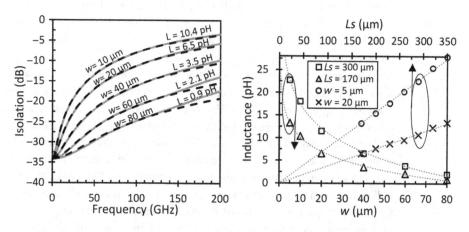

Figure 4.24 (Left) Isolation as a function of the switch width, w. Numerical simulation (Ansys HFSS), solid line; LR model, dashed line. (Right) geometric influence on the shunt inductance.

The width of the switch, w, was varied in order to change the equivalent inductance L of the switch. The results of the simulated switches were fitted in Ansys Circuit Designer to a shunt LR model. It is possible to see that the model agrees quite well with the simulation for most of the inductances. For low inductances (2.1 and 0.9 pH) for frequencies above 150 GHz, the model starts to deviate due to second-order phenomena that is not modeled by the simple LR circuit. The value of the inductance is not straightforward to calculate and it is better extracted from simulated or measured switches; however, it is interesting and important to analyze the general tendency of the inductance as a function of the switch geometry. Figure 4.24 also shows several simulated switches as a function of the switch length L_s and width w. The inductance shows a logarithmic behavior as a function of w, and a linear behavior as a function of L_s. The variation of the thickness showed little influence on the value of the inductance. Changing t from 0.5 μm to 3 μm resulted in a variation of only 0.5 pH.

The inductance values presented as a result of numerical simulation are related to a specific switch and should not be taken as universal values. The value of the extracted inductance depends on the technology of the fabricated switch, other parasitic elements, air gap height and so on; however, the important aspect to keep in mind while designing RF switches is their behavior as a function of their geometry. Numerical tools should be used in order to model the switch correctly and account for the effects of the shunt inductance.

In capacitive shunt switches the combination of the equivalent inductance and the contact capacitance C_d results in a transmission zero at the resonance frequency f_0, which is given by

$$f_0 = \frac{1}{2\pi\sqrt{L \cdot C_d}}. \tag{4.16}$$

This association can be used to increase the isolation of the switch at f_0. Figure 4.25 illustrates how the L–C_d combination can be used to tune the isolation of capacitive shunt switches. This effect can also be used in resistive shunt switches by deliberately including a shunt capacitor in series with the switch. Normally, this is done by including a thin dielectric layer between the switch and the ground, while maintaining the resistive contact between the switch and the signal strip.

Considering the inductance variation obtained in the numerical simulation presented in Figure 4.24 and a C_d of 0.65 pF, it is possible to change the transmission zero from approximately 50 GHz to above 200 GHz, without considering other parasitic elements that are related to switch technology and geometry.

The inductance of the shunt switch also degrades the insertion loss, as shown in Figure 4.26. It can be seen that the effect becomes evident after 100 GHz. However, the degradation of the insertion loss is limited. At 200 GHz, by increasing the inductance from 1 pH to 15 pH, the insertion loss increases only 0.26 dB, considering an up-state capacitance C_t of 10 fF. As the capacitance increases, the insertion loss also increases as shown in Figure 4.16, and the effect of the switch inductance becomes more pronounced also at lower frequencies. With C_t of 16 fF, the same inductance variation results in a 0.6 dB variation at 150 GHz, not shown in the figure.

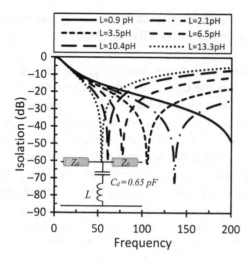

Figure 4.25 Simulated effect of the switch inductance on the isolation of capacitive shunt switches. Circuit simulated in Ansys Circuit Designer.

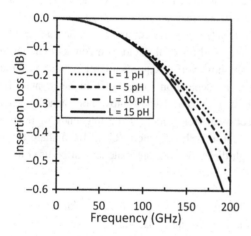

Figure 4.26 Simulated effect of the switch inductance on the insertion loss of shunt switches.

4.3.5   Guidelines for Millimeter-Wave Switches Design

By analyzing the behavior of the parasitic elements discussed in the previous sections, it is possible to draw some general guidelines for the design of mm-wave switches. Specific dimensions and characteristics of the switch are deeply dependent on the available technology and on the application of interest.

4.3.5.1   Series Switch

In general, series switches tend to be simpler to implement and show broadband response with low insertion loss, but their isolation is somewhat degraded at mm-waves, which can limit the range of applications that one can use them. They are easy to be implemented in both microstrip or CPW topologies.

Isolation It is controlled by the up-state capacitance C_u, although the capacitance related to the separation between the input and output transmission lines (C_s) can further degrade the isolation. Since the impedance of these capacitances decreases with frequency, special attention should be paid at mm-waves. The switch air gap should be increased as much as possible, while the overlap area A_c should be kept small to reduce C_u. However, the gap increase results in an increase of the pull-in voltage V_p that should be kept below 80 V (integrated charge pump limit). The increase of the electrode area A_e and the reduction of the spring constant k of the switch reduces the V_p. The increase of A_e increases the switch footprint, which can prevent its use in some applications, such as in switched-line phase shifters (see Section 4.5.1), and reducing k, can lead to stiction. The characteristic impedance of the suspended transmission line Z_L should be close to 50 Ω to avoid further degradation of the isolation.

Insertion Loss If the characteristic impedance Z_L is close to 50 Ω, then the insertion loss is mainly due to the down state resistance R_d or capacitance C_d, in resistive or capacitive switches, respectively. R_d is mainly composed of the contact resistance R_c and depends on the contact force and material. Gold as a contact material results in a small R_c with low contact force; however, it requires a large separation force (high k). Rhodium requires a much smaller separation force, but a higher contact force and a higher R_c. A good compromise is a gold–Nickel alloy. For capacitive switches, C_d should be large enough to act as a short circuit at the frequency of interest, without compromising the isolation because both capacitors (C_d and C_u) share the same area. To relax some constraints, high dielectric constant material can be used to increase C_d without increasing C_u.

Shunt Switch This type of switch allows a high isolation for a narrow band using the switch inductance and the down-state capacitance to achieve resonance. CPW-based switches are cumbersome to implement.

Insertion Loss C degrades the insertion loss with little influence of the switch inductance at frequencies below 100 GHz. Large air gaps and small overlap area A result in small insertion loss. Same mechanical aspects as the ones described for the series switch apply.

Isolation It can be degraded by the contact resistance R and switch inductance L, for resistive switches. Contact materials also play an important role in the mechanics of the switch, as discussed for the series switch. The down-state capacitance combined with switch inductance can provide a good isolation by adding a transmission zero at the frequency provided in Eq. (4.16).

4.3.6 Broadband RF MEMS Switch

Figure 4.27 shows a broadband series resistive switch developed by CEA-Leti, based on the technology described in Section 4.2.2. This switch shows an excellent perform-ance from DC to 40 GHz, with a low insertion loss of 0.3 dB and a high isolation of 25 dB at 40 GHz. This result illustrates the general behavior of the performance in this type of broadband switches. The isolation of the switch fits quite well with the series capacitor model, resulting in a $C_u = 2.3$ fF. The series LR model for the insertion loss, on the other hand, does not agree with the measurements, and a detailed analysis of the switch structure, considering its geometry, acess lines, tapering, and technology (material characteristics), should be carried out in order to correctly predict the insertion loss.

Table 4.6 summarizes the state-of-the-art of the broadband mm-wave switches. As it can be seen, the maximum isolation achieved is around 25 dB at 50 GHz.

Since the up-state capacitance controls the isolation, the latter will continuously decrease with frequency. This effect limits its utilization in the mm-wave range as a switching circuit, such as a duplexer. However, circuits such as distributed MEMS phase shifter can use this type of switch with interesting results, as shown in Section 4.5.2.

In [32] 22 dB isolation was achieved at 100 GHz by decreasing C_u using a double air gap configuration. However, this switch shows high insertion loss at mm-waves. The insertion loss in [31] is close to 0.5 dB at 100 GHz with an isolation of 15 dB, showing a compromise between isolation and insertion loss.

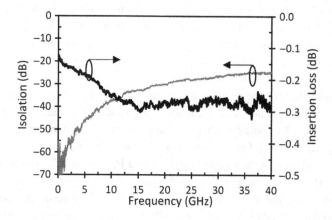

Figure 4.27 Measured isolation and insertion loss of the CEA-Leti series resistive switch from [17]. Measured data are courtesy of CEA-Leti.

Table 4.6 State-of-the-art for mm-wave MEMS switches

Ref.	Switch	Topology	Type	Frequency (GHz)	Isolation (dB)	Insertion loss (dB)
[17] [a]	CEA-Leti	Series	Resistive	40	25	0.3
[25]	Rockwell	Series	Resistive	50	25	0.35
[26]	Radant	Series	Resistive	40	13	1.0
[29]	Lincoln Labs	Series	Capacitive	40	15	0.7
[30]	U. of California San Diego	Series	Resistive	40	15	0.6
[34]	Raytheon	Shunt	Capacitive	35	15	0.06
[43]	Motorola	Series	Resistive	30	25	0.3
[32]	CEA-Leti	Shunt	Resistive	75–100	22	1.4
[31]	XCOM	Series	Resistive	100	15	0.5
[44]	U. of Michigan	Series/ Shunt	Res./Cap.	40	35	0.5-2
[45]	Seoul National University	Series	Resistive	50/100	15.5/10	0.34/0.5
[46]	FBK	Series	Resistive	50	15	0.3

[a] Data courtesy of CEA-Leti.

The isolation of the broadband switch can be increased by using a series/shunt combination, as demonstrated in [44]. In this case, the measured isolation was 35 dB at 40 GHz. However, this solution increases the overall size of the switch and the insertion loss that can reach 2 dB.

4.3.7 Narrowband RF MEMS Switch

The narrowband switch is a shunt switch with a transmission zero to achieve high isolation. This transmission zero is created by the resonance of the series combination of the parasitic down-state capacitor and inductor of this type of switch, as illustrated in the model in Figure 4.25. In resistive shunt switches, capacitor can be added intentionally in order to obtain the same behavior.

The concept is to tune the resonance frequency f_0 that creates a transmission zero, by changing this inductance and/or capacitance so that the resonance [Eq. (4.16)] occurs at the desired frequency. This is clearly illustrated by the mm-wave switch developed by IHP [47, 48]. Figure 4.28 shows four switches fabricated with IHP 0.25 μm technology, similar to the technology described in Section 4.2.2. Each anchor of the four supports of the membrane is loaded with an inductor fabricated in the BEOL, which combined with the down-state capacitance of 0.22 pF, resulting in transmission zeroes at 30 GHz, 50 GHz, 80 GHz, and 100 GHz. The inductors on this type of switch are created outside the suspended region of the membrane and do not influence the mechanical behavior of the switch, which is not the case if the membrane or the membrane's legs are changed. By changing the membrane, the electrode area ($w \cdot L_e$) changes and so does the pull-in voltage. If the membrane's supports are changed to

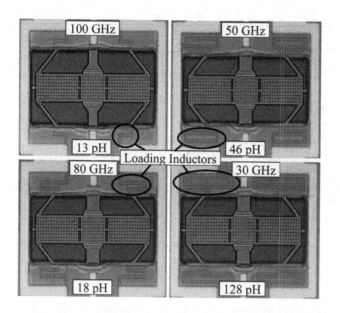

Figure 4.28 IHP narrowband mm-wave switches loaded with equivalent inductances of 128 pF, 46 pF, 18 pF, and 13 pF for transmission zeroes at 30 GHz, 50 GHz, 80 GHz, and 100 GHz, respectively [47, 48].

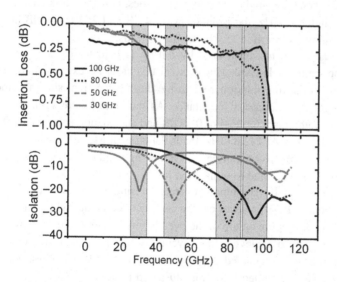

Figure 4.29 Insertion loss and isolation for the narrowband mm-wave switches from IHP [47, 48].

increase or decrease the switch inductance, this changes the spring constant of the switch, changing not only the pull-in voltage, but also the restoration force.

The isolation and insertion loss for the IHP mm-wave switches are shown in Figure 4.29. By loading inductively the switch, it was possible to obtain isolations of 20 dB and 24 dB at 30 GHz and 50 GHz, respectively. This is not much of an

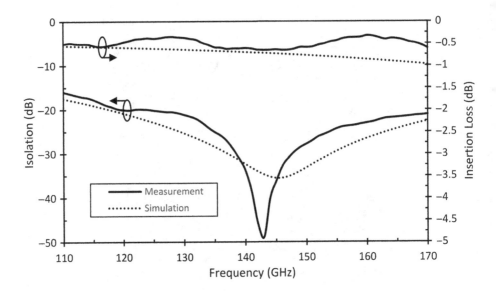

Figure 4.30 Comparison of the measured and simulated S-parameters of the WLE D-band RF-MEMS switch [19].

improvement when compared to broadband switches. However, at higher frequencies, i.e., 80 GHz and 100 GHz, the isolation was 34 dB and 25 dB (32 dB at 94 GHz), while maintaining an insertion loss better than 0.3 dB.

The same approach has been used by IHP in [19] to realize a capacitive shunt switch for operation at D-band. However, in this case, no external inductance was needed to obtain the transmission zero in the desired location, since the membrane supports provided the necessary inductance. As shown in Figure 4.30, the fabricated switch has minimum insertion loss of 0.65 dB and an isolation better than 30 dB from 138 GHz to 148 GHz. At 142.8 GHz, the isolation is of 51.6 dB. This shows that the narrow band switch approach can be used at the high end of the mm-wave spectrum, providing good isolation and low insertion loss.

Two switches can be combined to create a single-pole–double-throw (SPDT) switch. In general, SPDT switches are the component coming right after the antenna; thus adding its noise figure on top of the overall noise figure of the system. Therefore, low loss is required to achieve a high overall system performance. Nowadays, this type of performance is achieved only by using RF MEMS switches. The MEMS-based SPTD switch presented in [49] is illustrated in Figure 4.31.

This switch is based on the 0.13 μm SiGe BiCMOS IHP technology and uses a tee junction with λ/4 microstrip lines on each side to feed two RF MEMS switches (Switch #1 and Switch #2). Figure 4.32 shows its performance for the measured and the EM simulated cases, where the noncontacted third port was also left as an open circuit, as in the case of the measurements. To show the negligible effect of leaving the third port open, the same setup was simulated in the case of all three ports were terminated with 50 Ω. The performed three-port EM simulations showed that the

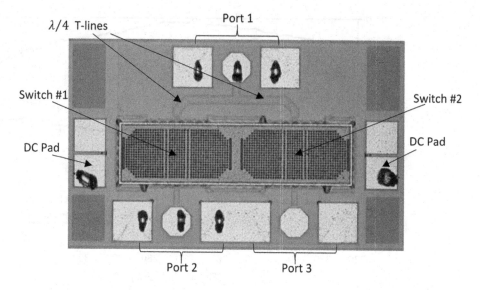

Figure 4.31 Micrograph of the D-band RF-MEMS SPDT switch in 0.13 μm SiGe BiCMOS technology [49].

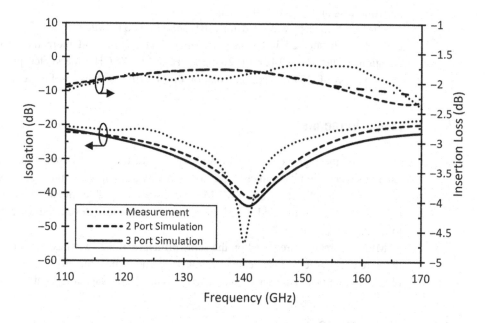

Figure 4.32 Comparison of the measured and simulated S-parameters of the D-band RF-MEMS SPDT switch in 0.13 μm SiGe BiCMOS technology: insertion loss (Switch #1 off-state, Switch #2 ON-state) and isolation (Switch #1 ON-state, Switch #2 OFF-state) [49].

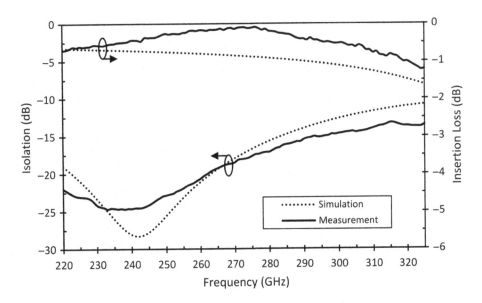

Figure 4.33 Comparison of the measured and simulated S-parameters of the J-band RF-MEMS switch [50].

insertion loss and isolation did not change significantly in the frequency band of interest, as compared to the two-port simulations. This switch showed 1.92 dB of insertion loss and 54.5 dB of isolation at 140 GHz.

The same concept was used for a RF-MEMS SPST switch in the THz frequency range [50]. It showed insertion loss better than 1.2 dB and more than 13 dB of isolation in the whole J-band (220–325 GHz). At 240 GHz, the switch provides an insertion loss of 0.44 dB and an isolation of 24.6 dB, as shown in Figure 4.33.

4.4   MEMS Varactors

Because of the high quality factor associated with the MEMS varactors, they were used in several circuits at microwave and mm-waves, mainly for tunability. Examples of mm-wave applications using MEMS varactors are tunable filters [51] and impedance tuners [52]. A detailed description and comparison of developed MEMS varactors can be found in [53].

MEMS varactors are based on the same principle of the switches. Most commonly they are actuated electrostatically, with a fixed and a movable electrode in a parallel plate configuration. There are mainly two types of MEMS varactors: analog and digital.

4.4.1   Analog MEMS Varactor

When applying a voltage between the parallel plates of the MEMS varactor, as in the MEMS switches, the electrostatic force developed moves the free electrode, thus

Table 4.7 State-of-the-Art for analog MEMS varactors

Ref.	Varactor	Ratio	Actuation voltage (V)	Quality factor	SRF (GHz)
[55]	U. of Michigan	1:1.46	18–25	95–100 @ 34 GHz	83 GHz
[56]	U. of Calgary	1:6.2	0–75	50 at 30 GHz	–
[57]	Purdue University	1:3.15	0–55	>80 at 40 GHz[a]	103 GHz[b]

[a] From [58].
[b] Calculated using Eq. (4.16) and model values from the authors.

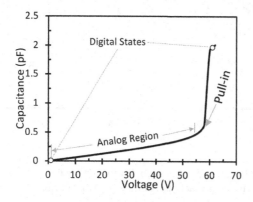

Figure 4.34 Operating region of the analog and digital MEMS varactors.

changing the capacitance between the parallel plates. As the applied voltage increases, the distance between the plates reduces and the capacitance increases, until pull-in. This behavior is illustrated in Figure 4.34. In this figure, parasitic capacitances that reduce the actual capacitance variation in the analog region were not taken into account. For this reason, the off-state capacitance is considerably small (~0.2 fF in this example). Despite the continuous change, the capacitance variation for conventional parallel plate varactor is quite small. A capacitance ratio of 1:1.5 (50%) is the limit for this type of device. Several authors worked on extending this ratio achieving quite large variations, up to 1:7 [54], by developing curled structures, structures with more than two electrodes or structures with no pull-in. Table 4.7 shows the relevant mm-wave state-of-the-art for the analog MEMS varactor.

The quality factor of a MEMS varactor can be calculated using Eq. (4.17). It is limited by the losses related to the varactor series resistance, which is the same as R_s for the MEMS switch. It is given by Eq. (4.9). SRF is the self-resonant frequency and it is related to the varactor's maximum capacitance (C_{max}) and associated inductance, such as in the MEMS switch, given by

$$Q = \frac{1}{2\pi f \cdot R_s \cdot C_{max}}. \tag{4.17}$$

Analog varactors might be difficult to control because the capacitance variation is not linear with the applied voltage and displays hysteresis and self-biasing with RF

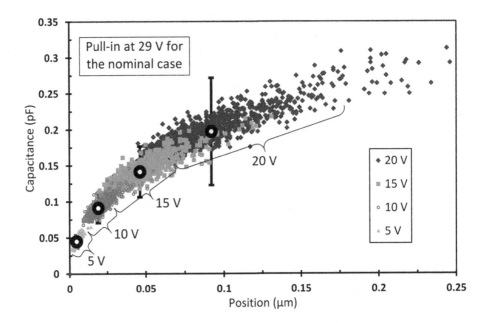

Figure 4.35 Monte Carlo simulation using 5% Gaussian distribution for the thickness of the mobile electrode and the gap between parallel plates of the MEMS varactor.

power. However, the limiting aspect for the wide spread of the analog varactor in commercial applications is the capacitance reproducibility and predictability due to in-wafer and wafer-to-wafer film thickness variation. Considering a 5% thickness variation for the mobile electrode and for the sacrificial material (gap distance), which is not uncommon in commercial technologies, a considerable variation of the behavior of the analog varactor is obtained, as shown in Figure 4.35, which illustrates a Monte Carlo simulation of a parallel-plate varactor's capacitance as a function of the applied voltage, and the effect on the physical position of the mobile electrode. A thousand simulations using a behavioral model with lumped elements were realized for each voltage using a Gaussian distribution [59]. The black circles indicate the average capacitance for each voltage and the error bars indicate two standard deviations ($2\sigma = 95\%$ of all variations). At 20 V, 1σ and 2σ represent a capacitance dispersion from 0.16 fF to 0.23 fF and 0.12 fF to 0.27 fF, respectively. Despite the high quality factor and the promising results presented in the literature, this dispersion is prohibitive for practical applications.

4.4.2   Digital Varactor

On the other hand, digital MEMS varactors use the well stablished MEMS switch technology to implement a switched capacitor bank. In this approach, the ON-state capacitance of capacitive MEMS switches are modified to have the desired capacitance and associated in parallel to obtain several capacitance states in a step of C_{LSB}

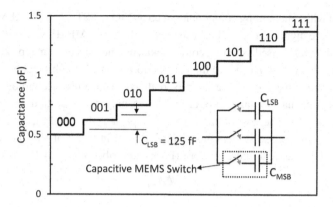

Figure 4.36 3-Bit digital varactor example.

(for least significant byte) and a maximum of C_{MSB} (for most significant byte), as illustrated in the example of a 3-bit digital varactor in Figure 4.36.

This type of varactor is highly scalable and has been implemented in commercial devices with 4 and 5 bits of precision by Wispry (WS1040 [60]) and Cavendish Kinetics (32CK417R [61]), respectively. These commercial products focus on improving mobile reception by actively matching the antenna. Despite the low frequency application (up to 3 GHz), the high quality factor of the MEMS switches can be used for mm-wave applications. In [55] the authors suggest that the quality factor of their digital varactor is similar to the one of their analog varactor, which is 95 at 34 GHz. Their utilization in commercial products suggests that this type of varactor is promising as a tunable component at mm-waves, even if there are not many tunable circuits using MEMS varactors in the literature.

4.5   Phase Shifters

To achieve longer communication range with mobile terminals or sensors, many mm-wave developments require phased arrays with beam-steering/forming capabilities. When dealing with low-power-consumption and high-performance systems, these phased arrays must be based on passive phase shifters, which constitute a major challenge that is still to be overcome. Even if they offer good electrical performance, active phase shifters lead to high power consumption and will not be discussed here.

Passive phase shifters have been developed in several technologies based on different tuning components, such as MEMS switches, MOS varactors, barium strontium titanate (BST) and liquid crystal (LC). A comparison between phase shifters is presented later in Section 4.5.5.

At mm-waves, there are three main types of phase shifters: switched line, distributed MEMS transmission line (DMTL), and reflected type, each with its advantages and disadvantages, as will be discussed in the next sections.

In contrast with other tuning components such as MOS, BST, and LC, MEMS-based phase shifters are digital in nature, because they are based on MEMS varactors. Analog MEMS varactor can be used as tuning components, but they suffer from reproducibility, as explained in Section 4.4.1; hence they are seldom found in the literature.

A careful comparison of phase shifters realized in all these technologies is not simple, since one has to take into account not only the size and the electrical performance, but also the cost, reliability, and, especially for MEMS, the packaging challenge. Here, the phase shifters are compared only in terms of size and electrical performance. The classical figure of merit (FoM), described in Eq. (4.18), is first used.

$$\text{FoM} = \frac{\Delta\phi_{max}}{IL_{max}},\tag{4.18}$$

where $\Delta\phi_{max}$ is the maximum phase shift and IL_{max} is the maximum insertion loss. A second FoM giving an idea of the tradeoff between size and electrical performance is defined in

$$\frac{\text{FoM}}{Area} = \frac{\Delta\phi_{max}}{IL_{max}} \cdot \frac{\lambda_0^2}{Area}.\tag{4.19}$$

In this case, the area of the phase shifter is normalized by the square of the free space wavelength at the device operating frequency. An idea of low and high FoM and FoM/Area is given in Section 4.5.5.

4.5.1   Switched-Line Phase Shifters

This type of phase shifter is the simplest one because it is based on switching between transmission lines of different electrical lengths (delay lines), as illustrated in Figure 4.37. This figure represents a 2-bit device with four phase states: 0° (00), 90° (01), 180° (10), and 270° (11). In this type of phase shifter, RF MEMS switches at the input and output are used to select the desired phase shift. The reference path with length L_1 represents a 0° of phase shift; it is made as small as possible to reduce losses. All the other paths have incremental length ΔL that corresponds to the phase shift, $\Delta\phi$, of that specific path at a given frequency f, as expressed by

$$\Delta\phi = 2\pi f \cdot \frac{\Delta L}{c},\tag{4.20}$$

where c is the speed of light.

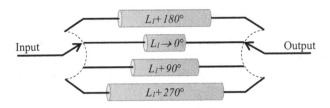

Figure 4.37 Switched line phase shifter concept.

SPMT switches have been developed and used in switched-line phase shifters [62–65]. These multithrow switches use multiple individual SPST switches connected to a common input through a carefully designed power divider (T junction). Figure 4.38 illustrates a SP4T series switch with the Input connected to Output 1, while the other switches are not activated.

If series switches are used, the unconnected branches are terminated in an open circuit, which limits the utilization of switched-line phase shifter at mm-waves, since they act as open-ended stubs and create a transmission zero at $\lambda_g/4$. For this reason, the length of the unconnected branches should be shorter than $\lambda_g/10$ at the operation frequency. For CPWs on quartz ($\varepsilon_{reff} = 2.3$), the length of the open-ended stubs should be shorter than 345 μm, 207 μm, and 103 μm at 60 GHz, 100 GHz, and 200 GHz, respectively. If those stubs were to be fabricated on silicon ($\varepsilon_{reff} = 6.3$), their lengths would have to be almost half the length on quartz. Instead of series switches, shunt switches are used, then the unconnected branches should have $\lambda_g/4$ at the frequency of interest, as illustrated in Figure 4.38.

To increase the operating frequency of the SPMT, miniaturized switches that allow the reduction of the length of the unconnected sections should be used, as proposed in [65]. Figure 4.38 illustrates the concept of a miniaturized SP4T presented in [65].

The impedance of the switched-line phase shifter depends only on the characteristic impedance of the transmission lines for every phase state. The insertion loss, on the other hand, varies with phase state, as can be seen in Table 4.8. This table presents the

Table 4.8 Variation of the insertion loss as function of the phase shift in switched-line phase shifters

	60 GHz		200 GHz	
Phase shift	Differential length (μm)	Differential insertion loss (dB)	Differential length (μm)	Differential insertion loss (dB)
45°	400	0.09	122.5	0.06
90°	800	0.18	245	0.11
135°	1200	0.26	367.5	0.15
180°	1600	0.33	490	0.19
225°	2000	0.42	612.5	0.24
270°	2400	0.51	735	0.29
315°	2800	0.59	857.5	0.33

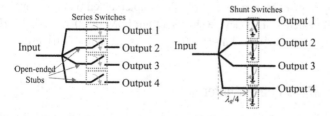

Figure 4.38 Schematic of a SP4T with Output 1 connected to the Input.

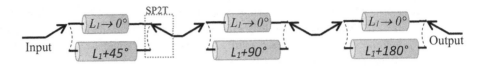

Figure 4.39 3-bit switched-line phase shifter with cascaded bits.

physical length difference required to obtain the specified phase shift for a 3-bit phase shifter, as well as the insertion loss related to these transmission lines. These transmission lines were simulated considering a straight CPW, quartz substrate, and gold metal.

The length difference from the least and most significant bits, 45° and 315°, is important. This poses a design challenge, since all these transmission lines should be connected to the same input and output. For this reason, the longer transmission lines should be meandered, which tends to increase even more the insertion loss of the longer transmission lines. In general, the insertion loss variation of a switched-line phase shifter is not more than 1 dB and the total insertion loss at 60 GHz is around 2–3 dB.

Further, the footprint of this type of phase shifter is quite large and increases with the resolution, as does the complexity to implement higher resolution phase shifters. For example, a 2-bit device that has four phase states (0°, 90°, 180°, and 270°) requires four transmission lines with different lengths and two SP4T switches. However, 3-bit devices with eight phase states, that is, eight transmission lines and two SP8T switches, are much more complex to implement using the concept presented in Figure 4.37. For this reason, 3- and 4-bit switched-line phase shifters use cascaded bits, as illustrated in Figure 4.39 and presented in [66]. The desired phase shift in the illustrated 3-bit phase shifter is obtained by selecting the right path. The signal either goes through the reference path or the delayed one and all eight phase states are possible. In this topology, only SP2Ts are required; however, the same unconnected branch (open-ended stub in shunt SPMT) restriction applies. In general, the footprint of the cascaded topology is larger than the parallel approach.

A combination of both topologies is presented in [67] to obtain a more compact 4-bit phase shifter. In this approach, two sections of parallel lines, a "coarse" bit and a "fine" bit, are used. To achieve 4-bit resolution, four SP4Ts are used to select between transmission lines with phase shift of 90°, 180°, and 270° in the coarse bit, and between 22.5°, 45°, and 67.5° in the fine bit.

The 2-bit phase shifter presented in [68] is implemented on quartz substrate using SP4T MEMS switches. This device shows a high FoM of 90°/dB at 60 GHz, but its footprint is large (4 mm^2), leading to a low FoM/Area ratio of 561°/dB.

4.5.2   Distributed MEMS Phase Shifter

In this type of phase shifter, a transmission line is periodically loaded with MEMS varactors, normally digital ones, in a shunt configuration. The same approach is used

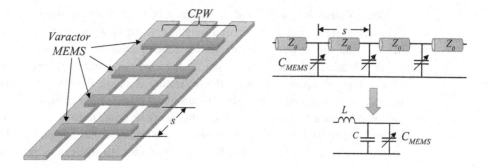

Figure 4.40 DMTL and its equivalent model.

in phase shifters with MOS and BST varactors as loads and FET switches to commute between loads. Figure 4.40 illustrates the distributed MEMS phase shifter and its equivalent model. As suggested by the model, the loading varactor changes the equivalent distributed capacitance, thus changing the phase velocity and the phase of the signal. Equation (4.21) gives the phase shift as a function of the difference in the characteristic impedance of the transmission line with the MEMS varactors in the up and down states, respectively:

$$\Delta\phi = s\cdot\omega\cdot Z_0\cdot\frac{\sqrt{\varepsilon_{reff}}}{c}\left(\frac{1}{Z_d}-\frac{1}{Z_u}\right),\qquad(4.21)$$

where Z_0 is the characteristic impedance of the unloaded line, which is normally around 80–100 Ω; ε_{reff} is its relative effective dielectric constant; s is the period of the loaded line (defined in Figure 4.40); and Z_u and Z_d [defined in Eq. (4.22)] are the characteristic impedance of the loaded transmission line with MEMS varactor in the up and down states, respectively. To avoid high return loss, a general rule is to keep Z_u and Z_d between 70 Ω and 35 Ω, so that the average characteristic impedance is close to 50 Ω.

$$Z_d = \sqrt{\frac{L}{C+\dfrac{C_d}{s}}}$$

$$Z_u = \sqrt{\frac{L}{C+\dfrac{C_u}{s}}}.\qquad(4.22)$$

The Bragg effect must be considered when designing this type of phase shifter, since it is a periodic high–low characteristic impedance structure. The distance s between MEMS varactors determines the Bragg frequency (f_B), given by

$$f_B = \frac{1}{\pi\cdot s\sqrt{L\left(C+\dfrac{C_d}{s}\right)}}.\qquad(4.23)$$

This equation is a simplified version that considers only L and C, and C_d, the inductance and capacitance per unit length, and the down-state capacitance of the MEMS varactor, respectively, neglecting other parasitics of the MEMS varactor. A more detailed approach considering the varactor inductance is proposed in [69].

This type of topology leads to the realization of phase shifters with higher resolution, as demonstrated by the 4-bit MEMS-based phase shifter presented in [70]. The high-performance MEMS switches and the quartz substrate used in this device lead to a high FoM of 91°/dB at 65 GHz; however, the occupied area is equal to 11.9 mm^2, leading to a considerably low FoM/Area ratio of 164°/dB that can be prohibitive for many applications needing large phased arrays.

A better FoM of 102°/dB at 78 GHz was obtained in [69] with a smaller footprint (9.7 mm^2), with similar FoM/Area ratio.

4.5.3   Slow-Wave MEMS Phase Shifters

As it will be discussed in Section 4.5.5, the distributed MEMS and switched-line phase shifters present the best FoM (dB/°) of all phase shifters, but their footprint is significantly large. Taking advantage of the high quality factor and the miniaturization due to the slow-wave effect of the shielded-CPW (S-CPW), described in Chapter 3, the slow-wave MEMS phase shifter proposes a good compromise between surface area and performance. The concept of this family of phase shifter is illustrated in Figure 4.41. Each floating ribbon of the shielding layer of the S-CPW can be considered as a bridge-type capacitive MEMS switch in shunt configuration.

The floating ribbons of the S-CPW (placed orthogonally to the direction of propagation) serve to create the slow-wave effect. They are normally placed below the CPW strips in CMOS/BiCMOS technologies in order to shield the CPW from the silicon substrate and because the CPW strips are fabricated on the uppermost metallic layer of the technology, which is normally the thickest one. These ribbons capacitively load the transmission line in a distributed manner, while the magnetic field is practically unperturbed. This leads to a slow-wave effect, which is particularly

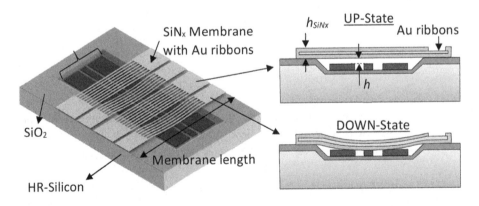

Figure 4.41 S-CPW MEMS phase shifter.

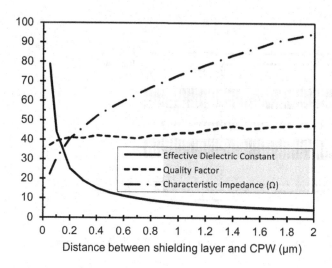

Figure 4.42 Variation of the gap between the CPW strips and the ribbons of the shielding layer.

interesting for the development of phase shifters, because it can reduce their size (length mostly) and insertion loss per phase shift compared to distributed MEMS phase shifters based on classical transmission lines.

In S-CPWs, the electric field is confined between the shielding ribbons and the CPW strips; hence their distance essentially controls the capacitance per unit length. By releasing the shielding ribbons, as proposed in [71], it is possible to move them with the application of a DC voltage between the CPW strips and themselves. The developed electrostatic force pushes the shielding ribbons closer to the CPW strips, thus reducing the phase velocity and increasing the phase shift.

The result of a continuous variation of the distance between the shielding ribbons and the CPW strips is illustrated in Figure 4.42. As the gap between the CPW strips and the ribbons is reduced, the effective dielectric constant (ε_{reff}) increases greatly. As a consequence, the characteristic impedance of the phase shifter is reduced. The quality factor ($\beta/2\alpha$, where β is phase constant and α is the attenuation constant), on the other hand, is mostly unchanged, which means that the insertion loss does not vary considerably with the phase change.

In order to get an optimal average return loss for all phase states, the minimum and maximum characteristic impedances Z_{min} and Z_{max}, corresponding to the actuated and rest positions of the shielding ribbons, respectively, must be fixed as expressed in

$$Z_0/Z_{min} = Z_{max}/Z_0. \qquad (4.24)$$

With $Z_0 = 50°\Omega$, Z_{max} could be around 70 Ω, and Z_{min}, around 35 Ω, for example.

As in MEMS switches, the ribbons can be displaced continuously from rest up to two-thirds of the gap by controlling the applied DC voltage, as discussed in Section 4.4.1 for the analog varactors. Even though it is possible to control the displacement of the shielding ribbons by controlling the applied DC voltage, this is not an adequate approach

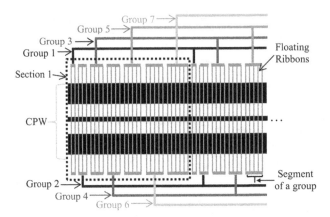

Figure 4.43 Illustration of the principle of a 3-bit S-CPW MEMS phase shifter.

because their position would be highly sensitive to the fabrication process variations and the phase variation would be small, as indicted by the small variation of ε_{reff}.

The pull-in voltage might change due to process variation, but the position of the shielding ribbons can be predictable. For this reason, the digital varactor approach, where the shielding ribbons are actuated only between rest and pull-in is used in the slow-wave MEMS phase shifter. This would result in a one-bit device, as presented in [71].

To obtain higher resolution, the shielding ribbons can be divided in groups that can be actuated independently. Ideally, to obtain n bits of resolution, the shielding ribbons would have to be divided in n groups. However, the phase shift is not linear with respect to the length of each group, because of the characteristic impedance steps between actuated and unactuated sections, leading to standing waves formation, and hence nonlinear phase variation. Therefore, to obtain n bits of resolution, there should be $2^n - 1$ groups.

The Bragg effect must be considered when designing the slow-wave MEMS phase shifter, since a periodic-like high–low characteristic impedance structure may appear for certain states. To minimize the Bragg effect, the groups need to be subdivided in segments and intercalated forming sections, as exemplified in Figure 4.43. In this case, the intercalated sections are equal in length.

For example, for a 3-bit phase shifter with segments of the same length, linear phase variation, and precise phase states, seven groups are necessary due the non-linearity of the phase states.

The design of the S-CPW can be carried out using the S-CPW electrical model presented in [72]. An elementary section corresponding to only one shielding ribbon is calculated with the model, for both actuated and rest positions. Then, the whole phase shifter can be optimized with Keysight's ADS. The length of each of the seven groups was defined to obtain 3 bits of resolution with a maximum phase shift of 315° (i.e., 360°–360°/8). The length and number of ribbons for each group is given in Table 4.9. Each group provides an incremental phase shift of 45° and they were designed to be

Table 4.9 Length of groups and segments of a 3-bit S-CPW MEMS phase shifter

	Phase shift (°)	Group length (μm)	Segment length (μm)	Number of ribbons in each segment
Group 1	45	121	16	5
Group 2	90	152	19	6
Group 3	135	176	22	7
Group 4	180	200	25	8
Group 5	225	224	28	9
Group 6	270	248	31	10
Group 7	315	248	31	10

Table 4.10 Linearized phase shifter with three commands for 3 bits

Phase state	Phase shift (°)	Phase error (°)
001	45	<0.1
010	90	<0.1
011	135	<0.1
100	180	<0.1
101	225	0.3
110	270	0.4
111	315	0.3

activated in combination with the previous groups, for example, to obtain a 270° phase shift, groups 1 through 6 should be activated. The principle of the designed phase shifter is given in Figure 4.43.

Considering a floating ribbon width of 1 μm and spacing between ribbons of 2 μm, the simulated length for each group and section, and the number of ribbons in each segment for the phase shift, are listed in Table 4.9. To achieve a maximum phase shift of 315° a total length of 1.375 mm was required, based on the CEA-Leti technology for this type of phase shifter, described in the text that follows.

If a higher resolution phase shifter is envisioned, it is necessary to increase the number of commands according to $2^n - 1$, where n is the number of bits. This can lead to a very complicated biasing scheme. A linearization method can be implemented to reduce the number of commands by varying the length of each group in each segment, thus minimizing the phase error for each state.

Using a quasi-Newton optimization algorithm in ADS (Keysight), it was possible to achieve the linearized phase states with only three commands, that is, three groups, for three bits. Table 4.10 shows phase error for each phase state using this linearization approach.

The length of each segment for the linearized approach is presented in Table 4.11. In the optimization, to minimize the Bragg effect, the length of each group was limited to 200 μm. A minimum length of 10 μm was assumed. To achieve a small phase error in state 111, a fourth group was added to the device, but this group is never actuated. Although a

Table 4.11 Length of each segment S# (in μm) for the linearized phase shifter

Group	S1	S2	S3	S4	S5	S6	S7	S8	S9	S10	S11	S12	Total
1	74	184	174	12	24	144	85	19	180	153	145	153	1347
2	65	17	117	39	11	25	154	142	13	16	89	18	706
3	11	11	25	156	28	10	11	12	12	10	11	18	315
4*	106	10	55	26	44	10	10	96	10	41	24	111	543
						Total							2911

simplified command is achievable, the total length of the device was increased considerably to 2.91 mm. As a consequence, the insertion loss was also increased.

The first results obtained for the slow-wave MEMS phase shifter in [73] using the low-cost CMOS technology 0.35 μm from AMS and post-process showed the possibility of implementing this type of phase shifter on commercial CMOS technology. The fabricated 2-bit MEMS phase shifter resulted in a 36°/dB FoM and 0.58 mm^2 of surface area, leading to a high FoM/Area ratio of 1552°/dB. However, the lack of control on the technology resulted in limited phase shift, due to incomplete release of the shielding layer. Still, the small footprint and small insertion loss for all phase states are, nonetheless, promising for this type of device.

To overcome the MEMS release issues encountered in [73], the slow-wave MEMS phase shifter was fabricated using a dedicated RF MEMS process based on the MEMS switch process of CEA-LETI, described in Section 4.2.2. In this technology, the classic S-CPW was inverted, that is, the CPW strips were placed below the shielding ribbons. Further, a stress compensated 1-μm-thick SiN$_x$ membrane also used in the CEA-Leti switch process was used to hold the thin 200-nm-thick ribbons that were separated from the CPW by a 1.2-μm airgap. The fabricated phase shifter with seven commands is shown in Figure 4.44.

Figure 4.45 shows the measurement results for the slow-wave MEMS phase shifter presented in Figure 4.44, from DC to 67 GHz. A maximum insertion loss of 3 dB was obtained at 60 GHz for a maximum phase shift of 152°, leading to a FoM of 51°/dB and a FoM/Area ratio of 2660°/dB. The insertion loss has a maximum variation of 1.3 dB at this frequency. The return loss is better than 13 dB for all phase states. Due to fabrication imprecisions, the characteristic impedance for the up and down states was not the one used for the phase shifter design. The up-state characteristic impedance was approximately 75 Ω, instead of 70 Ω, and the down-state characteristic impedance was approximately 50 Ω, instead of 35 Ω, respectively. In that case, the average characteristic impedance is equal to $Z_{avg} = \sqrt{75 \cdot 50} = 61\ \Omega°$. For this reason, the measured S-parameters were normalized to 61 Ω.

4.5.4   Reflection Type Phase Shifter

The reflection type phase shifter (RTPS) uses a 3-dB coupler that is loaded with two variable loads that can be considered as phase shifters, as illustrated in Figure 4.46. The signal is injected through Port 1 of the RTPS, divides equally in Ports 2 and 3,

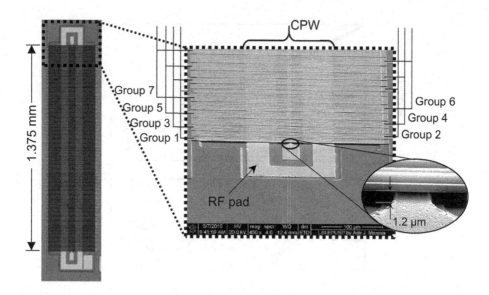

Figure 4.44 Optical image of the fabricated slow-wave MEMS phase shifter with seven commands and detailed view using a scanning electron microscope.

where it encounters the phase shifters. Because the phase shifters are terminated in a short or open circuit, the signal is reflected back through the phase shifters, entering the coupler (Ports 2 and 3). The once divided signal is then recombined in Port 4, the output of the RTPS.

In RTPSs the input and output are naturally matched to 50 Ω (or other reference impedance) for an ideal coupler, the condition being to have same loads at Ports 2 and 3. This is not the case for distributed MEMS and slow-wave MEMS phase shifters, for which the input impedance varies with the phase shift.

The loading phase shifters in the RTPS can be realized in several different ways. Transmission lines with different lengths can be switched using MEMS switches, as described in Section 4.5.1. MEMS varactors can also be used in distributed MEMS type loads. Series MEMS switches can be used to increase progressively the electrical length of the load.

A typical response of a RTPS can be viewed in Figure 4.47, which is the result of an ideal circuit simulation of a 2-bit phase shifter. This device uses a load composed of three series switches that increase the electrical length of the load, as depicted in the detail in Figure 4.47. This phase shifter was designed to operate at 60 GHz. Due to the frequency response of the 3-dB coupler, the phase variation of the RTPS is not linear with frequency, leading to a narrow band device. Considering a relative phase error between phase states of 5°, the bandwidth of this ideal RTPS is 1.4 GHz (59.3–60.7 GHz).

A RTPS phase shifter using a CMOS technology with MEMS was presented in [74]. The idea was to avoid the use of MOS varactors in order to reach a higher FoM. Indeed, the FoM was considerably improved, reaching 45°/dB. However, even if this device uses a small-footprint coupler, the large comb-like MEMS varactors result in a

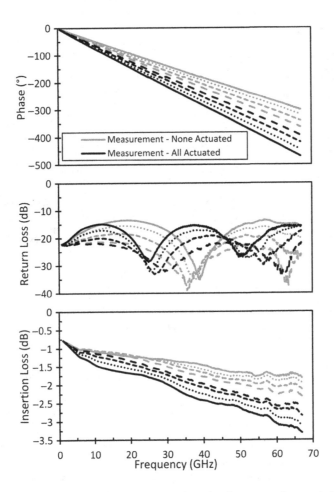

Figure 4.45 Measured and simulated insertion loss, return loss, and phase shift of the distributed-MEMS phase shifter.

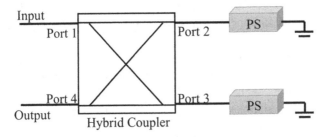

Figure 4.46 Reflection type phase shifter.

considerable relative increase in surface (1.04 mm²) when compared to the MOS varactor-based RTPS phase shifters. Moreover, only three phase states were possible in the phase shifter. Apart from the cost issue due to the device size along with the phase states issue, the latter example clearly illustrates the tradeoff between size and

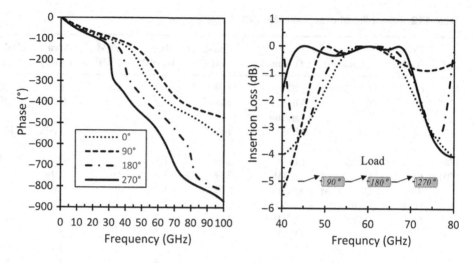

Figure 4.47 Circuit simulation (Keysight's ADS) of a 60-GHz ideal 2-bit RTPS using a 3-dB branch-line coupler and phase shifting load with three ideal switches.

electrical performance. When compared to the other topologies, the RTPS with MEMS shows a good compromise between FoM and surface area, showing a moderate FoM/Area ratio equal to 922°/dB.

4.5.5   Millimeter-Wave Phase Shifters State-of-the-Art

Several planar technologies have been used to develop passive phase shifters at mm-waves, including CMOS/BiCMOS, BST, liquid crystal, and MEMS (discussed in the previous sections). Table 4.12 shows some of the relevant published results.

Contrary to MEMS phase shifters discussed earlier, their CMOS/BiCMOS counterparts offer a much smaller footprint, but their electrical performance is poor, since MOS varactors used as tuning elements exhibit quality factors limited to about 10–15 at mm-waves [81].

In [76], an RTPS was realized in a 0.13-nm SIGe BiCMOS technology. Its area is only 0.33 mm^2, and its FoM is among the highest for MOS varactor-based phase shifters (25°/dB). The reduced area yields a high FoM/Area ratio of 1894°/dB; however, the 6.2 dB insertion loss need an additional amplification, thus leading to considerable increase of the power consumption. An equivalent FoM (26°/dB) was obtained in [77] with a much smaller footprint of only 0.048 mm^2. This leads to much higher FoM/Area ratio of 13 432°/dB; however, here again with a considerably high insertion loss of 5.7 dB.

In general, BST-based devices show good electrical performance at RF frequencies. A high FoM equal to 85°/dB was obtained at 30 GHz in [82]. However, the BST loss tangent dramatically increases with frequency, thus limiting the phase shifter's electrical performance at mm-waves. For instance, at 60 GHz a FoM of 23°/dB with a footprint of 1.2 mm^2 was reported in [75], leading to a poor FoM/Area ratio of 521°/dB.

LC-based phase shifters can be promising at higher frequencies, since LC loss tangent decreases with frequency [83]. A first result presented in [78] demonstrated a

Table 4.12 State-of-the-art of passive phase shifters

Ref.	Freq. (GHz)	Topology	Technology	Phase shift (°)	Max. IL (dB)	FoM (°/dB)	Area (mm²)	FoM/Area ratio (°/dB)
[75]	60	Distributed	BST	150	5.9	25	1.2	521
[76]	60	RTPS	MOS varactor	156	6.2	24	0.33	1894
[77]	60	RTPS	MOS varactor	147	5.7	26	0.048	13432
[78]	61	RTPS	LC	246	11.7	20.8	33	15
[70]	65	Distributed	MEMS	329	3.6	91	11.9	164
[69]	78	Distributed	MEMS	316	3.2	99	9.67	151
[68]	60	Switched	MEMS	269	3	90	4	561
[74]	65	RTPS	MEMS	144	3.2	45	1.04	922
[73]	60	Slow wave	MEMS	25	0.7	36	0.58	1552
–[a]	60	Slow wave	MEMS	152	3	51	0.47	2695
[79]	76	Distributed	MEMS and LC	92	2.2	42	0.65	1007
[80]	45	Slow Wave	MEMS and LC	275	5.35	51	0.38	5965

IL, insertion loss; LC, liquid crystal; MEMS, microelectromechanical systems; MOS, metal–oxide–semiconductor; RTPS, reflection type phase shifter.
[a] Unpublished results presented in Section 4.5.3.

high insertion loss (14 dB) at 61 GHz and large area (33 mm²), resulting in a low FoM of 20.8°/dB and even lower FoM/Area ratio of 15°/dB. However, a much better FoM of 42°/dB at 76 GHz was reported in [79] with a combination of LC and MEMS. The area was 0.65 mm², leading to an FoM/Area ratio of 1007°/dB. The switching time is also quite slow, that is, a few millisecond, thus limiting the application field. To solve these issues, in [80] an S-CPW was combined with MEMS and LC to take advantage of their high FoM, while reducing the size and response time of the phase shifter. A FoM of 51°/dB at 45 GHz with a 0.38 mm² footprint was obtained, thus resulting in a high FoM/Area ratio of 5965°/dB.

Figure 4.48 shows a comparison of the FoM of mm-wave planar phase shifters using different technologies. This figure summarizes the foregoing description and clearly illustrates the compromise between performance and surface area for different technologies and topologies. It can be seen that slow-wave-based phase shifters present a considerable reduction in size with a moderate FoM. If these devices can be optimized in terms of performance a considerable advance in the state-of-the-art could be achieved.

4.6   Switched Circuits

4.6.1   60–77 GHz Switchable Low-Noise Amplifier

The RF MEMS switch from IHP was used to realize a switchable low-noise amplifier (LNA) that is described in detail in [84] and illustrated in Figure 4.49. The LNA works

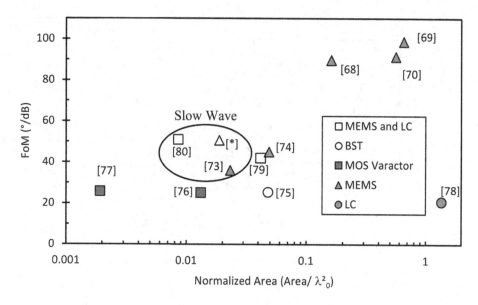

Figure 4.48 State-of-the-art of passive mm-wave phase shifters.

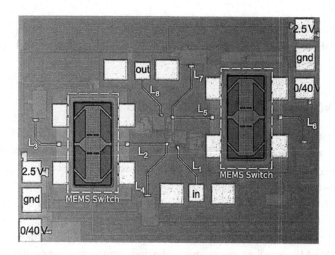

Figure 4.49 Optical image of the 60–77 GHz switchable LNA from IHP [84].

in the 60 GHz and 77 GHz frequency bands. It is based on the SG25 IHP MEMS technology.

Figure 4.50 shows a simplified schematic of the LNA. It is based on a two-stage cascode amplifier. The RF MEMS switch is used to modify the load of each stage. The capacitance of the tuned load is varied from 22 fF in the OFF state of the switch to 220 fF in the ON state. This change in capacitance changes the 3 dB bandwidth from 51–60 GHz to 66–78 GHz. The inductors L_1 to L_8 are realized using microstrip lines with an equivalent inductance of 120 pH for L_4 and L_7 and 80 pH for the others.

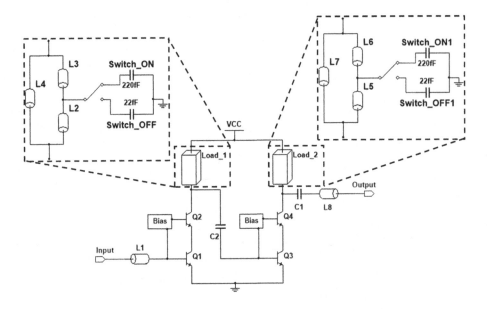

Figure 4.50 Schematic of the dual-stage cascade-based LNA with RF MEMS switches.

The switchable LNA presented a noise figure of 6.8–7.3 dB in the 60 GHz band and 7.6–8.4 dB in the 77 GHz band, respectively, which compares well with other fixed band LNAs in the literature.

4.6.2   Dual-Band Voltage-Controlled Oscillator

A dual-band voltage controlled oscillator (VCO) operating from 48 to 52 GHz and from 64 to 72 GHz was fabricated using IHP's BiCMOS technology [85] described in Section 4.2.2. Two fully integrated RF MEMS switches were used to choose bands, while varactor diodes (C_{var}) were used for the continuous fine tuning of the frequency within each band.

Figure 4.51 shows a simplified schematic of the dual-band VCO. It uses a negative resistance topology created by the capacitive emitter degeneration (C_{var}). The inductance of the transmission lines at the base of the transistor pair T_1 provides a resonance and the RF MEMS switches reconfigures the inductance to change the frequency band. With the application of 28 V_{DC}, the capacitance of the switch varies from 25 fF in the OFF state to 250 fF in the ON state.

The measured response of the dual-band VCO is presented in Figure 4.52. The output power provided by the VCO is 4 dBm in the 50 GHz band and 5 dBm in the 70 GHz band, with phase noise at 1 MHz offset of −84 dBc/Hz and −86 dBc/Hz, respectively.

4.6.3   Reconfigurable Filters

A RF MEMS-based reconfigurable lowpass filter was presented in [86]. The cutoff frequency is switched from 53 GHz to 20 GHz by activating a multicontact capacitive

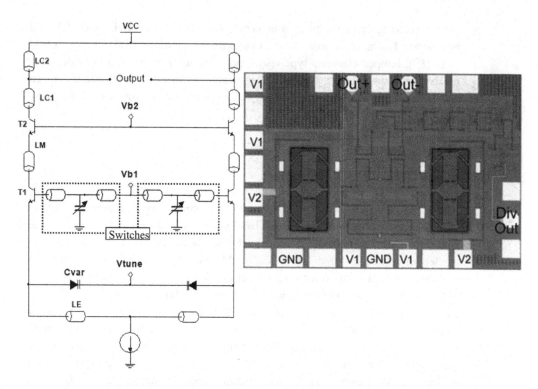

Figure 4.51 IHP's dual-band VCO schematics and optical image of the fabricated device [85].

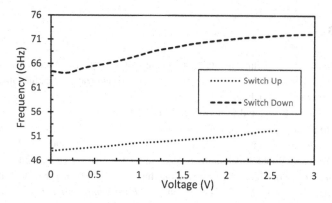

Figure 4.52 Measured response of the dual-band VCO.

MEMS switch. The switch changes the series inductance and shunt capacitance of a Tchebycheff filter. The filter uses slow-wave CPW based on the periodic structures proposed by [87].

The same principle was applied on a switchable lowpass filter in [88]. Here, the cutoff frequency is changed from 67 GHz to 28 GHz. The authors also demonstrated a bandpass filter with center frequency at 55 GHz with a 3-dB bandwidth of 32 GHz that can be reconfigured to a filter with center frequency at 20 GHz with a 12 GHz bandwidth. The insertion loss in these two states is 1.27 dB and 1.61 dB, respectively.

In [89], lumped-elements type resonators were used to create a two-pole analog tunable bandpass filters at 65 GHz, using metal–air–metal (MAM) bridge-type capacitors as tuning elements. A frequency tuning bandwidth of 10% with 3.3 ± 0.2 dB insertion loss was obtained.

4.7   Challenges and Perspectives

RF MEMS has come a long way and now can be considered as a mature technology based on the developments of research laboratories and industries of the past 30 years. Commercial devices and circuits are available and being used in large scale. The major challenges related to reliability of the MEMS switch has been resolved and these switches can be a solution for several microwave and mm-wave applications.

Switches have been demonstrated to operate for billions of cycles reliably with no contact degradation or dielectric charging [17]. Thin film encapsulation, as demonstrated by the IHP and CEA–Leti technologies (Section 4.2.2) presents an alternative for MEMS packaging, rendering them competitive cost wise. Specifically, for mm-wave application, due to the poor performance of semiconductor-based tuning devices, MEMS can become the mainstream choice for tunable circuits that can be easily integrated into CMOS technology or on an interposer.

References

1. *Analog Devices*, "The fundamentals of analog devices" revolutionary MEMS switch technology, 2019. Available at: www.analog.com/en/technical-articles/fundamentals-adi-revolutionary-mems-switch-technology.html
2. D. Bell, T. Lu, N. Fleck, and S. Spearing, "MEMS actuators and sensors: Observations on their performance and selection for purpose," *Journal of Micromechanics and Microengineering,* vol. 15, no. 7, pp. S153–S164, 2005. 10.1088/0960-1317/15/7/022.
3. M. Wietstruck, W. Winkler, A. Göritz, S. Tolunay Wipf, C. Wipf, D. Schmidt, A. Mai, and M. Kaynak, "0.13 μm BiCMOS embedded on-chip high-voltage charge pump with stacked BEOL capacitors for RF-MEMS applications," in *Proceedings of the 17th Symposium on RF-MEMS and RF-Microsystems (MEMSWAVE),* 2016.
4. D. Saias, P. Robert, S. Boret, C. Billard, G. Bouche, D. Belot, and P. Ancey, "An above IC MEMS RF switch," *IEEE Journal of Solid-State Circuits*, vol. 38, no. 12, pp. 2318–2324, 2003.
5. M. Daneshmand, S. Fouladi, R. R. Mansour, M. Lisi, and T. Stajcer, "Thermally actuated latching RF MEMS switch and its characteristics," *IEEE Transactions on Microwave Theory and Techniques,* vol. 57, no. 12, pp. 3229–3238, 2009.
6. G. Rebeiz, *RF MEMS: Theory, Design, and Technology*, Hoboken, NJ: John Wiley & Sons, 2004.

7. V. K. Varadan, J. V. Kalarickaparambil, and K. A. Jose, *RF MEMS and Their Applications*, Hoboken, NJ: John Wiley & Sons, 2003.

8. H. J. De los Santos, *RF MEMS Circuit Design for Wireless Communications*, Boston-London: Artech House, 2002.

9. S. Lucyszyn, *Advanced RF MEMS*, Cambridge: Cambridge University Press, 2010.

10. S. Timoshenko, *History of Strength of Materials: With a Brief Account of the History of Theory of Elasticity and Theory of Structures*, New York: McGraw-Hill, 1953.

11. R. Roark, W. Young, R. Budynas, and A. Sadegh, *Roark's Formulas for Stress and Strain*. New York: McGraw-Hill, 2012.

12. B. Lacroix, A. Pothier, A. Crunteanu, et al., "Sub-microsecond RF MEMS switched capacitors," *IEEE Transactions on Microwave Theory and Techniques*, vol. 55, no. 6, pp. 1314–1321, June 2007.

13. "RADAR CHIP Products - Silicon Radar GmbH," Silicon Radar GmbH, 2019. Available at: https://siliconradar.com/products/#120ghz-radar-chips

14. B. Heinemann, H. Rücker, R. Barth, et al., "SiGe HBT with fx/fmax of 505 GHz/720 GHz," in *2016 IEEE International Electron Devices Meeting (IEDM)*, San Francisco, CA, 2016, pp. 3.1.1–3.1.4.

15. M. Kaynak, M. Wietstruck, W. Zhang, et al., "MEMS module integration into SiGe BiCMOS technology for embedded system applications," *ECS Transactions* 41.7 (2011): 191.

16. Y. Lamy, O. El Bouayadi, C. Ferrandon, et al., "mmW characterization of wafer level passivation for 3D silicon interposer," in *2013 IEEE 63rd Electronic Components and Technology Conference*, Las Vegas, NV, 2013, pp. 1887–1891.

17. F. Souchon, B. Reig, C. Dieppedale, et al., "Key improvements of the MEMS switch lifetime thanks to a dielectric-free design and contact reliability investigations in hot/cold switching operations," in *2013 IEEE International Reliability Physics Symposium (IRPS)*, Anaheim, CA, 2013, pp. 6B.2.1–6B.2.8.4.

18. "FBK I MST Micro System Technology," *Mst.fbk.eu*, 2019. Available at: https://mst.fbk.eu/

19. S. Tolunay Wipf, A. Göritz, M. Wietstruck, C. Wipf, et al. "Thin film wafer level encapsulated RF-MEMS switch for D-Band applications," in *2016 11th European Microwave Integrated Circuits Conference (EuMIC)*, London, 2016, pp. 452–455.

20. S. Tolunay Wipf, A. Göritz, M. Wietstruck, C. Wipf, B. Tillack, A. Mai, and M. Kaynak, "Electromagnetic and small-signal modeling of an encapsulated RF-MEMS switch for D-band applications," *International Journal of Microwave and Wireless Technologies*, vol. 9, no. 6, pp. 1271–1278, 2017.

21. M. Kaynak, K. E. Ehwald, J. Drews, et al., "BEOL embedded RF-MEMS switch for mm-wave applications," in *2009 IEEE International Electron Devices Meeting (IEDM)*, Baltimore, MD, 2009, pp. 1–4.

22. Z. Baghchehsaraei and J. Oberhammer, "Parameter analysis of millimeter-wave waveguide switch based on a MEMS-reconfigurable surface," *IEEE Transactions on Microwave Theory and Techniques*, vol. 61, no. 12, pp. 4396–4406, Dec. 2013.

23. N. Vahabisani and M. Daneshmand, "Monolithic millimeter-wave MEMS waveguide switch," *IEEE Transactions on Microwave Theory and Techniques*, vol. 63, no. 2, pp. 340–351, Feb. 2015.

24. T. Reck, C. Jung-Kubiak, and G. Chattopadhyay, "A 700-GHz MEMS waveguide switch," *IEEE Transactions on Terahertz Science and Technology*, vol. 6, no. 4, pp. 641–643, July 2016.

25. R. E. Mihailovich, M. Kim, J. B. Hacker, et al., "MEM relay for reconfigurable RF circuits," *IEEE Microwave and Wireless Components Letters*, vol. 11, no. 2, pp. 53–55, February 2001.

26. "Radant MEMS, Inc." Available at: www.radantmems.com/

27. S. Majumder, J. Lampen, R. Morrison, and J. Maciel, "A packaged, high-lifetime ohmic MEMS RF switch," in *IEEE MTT-S International Microwave Symposium Digest, 2003*, Philadelphia, PA, 2003, vol. 3, pp. 1935–1938.

28. Guan-Leng Tan and G. M. Rebeiz, "A DC-contact MEMS shunt switch," in *IEEE Microwave and Wireless Components Letters*, vol. 12, no. 6, pp. 212–214, June 2002.

29. C. Bozler, R. Drangmeister, S. Duffy, et al., "MEMs microswitch arrays for reconfigurable distributed microwave components," *IEEE Antennas and Propagation Society International Symposium. Transmitting Waves of Progress to the Next Millennium. 2000 Digest. Held in conjunction with: USNC/URSI National Radio Science Meeting*, Salt Lake City, UT, 2000, vol. 2, pp. 587–591.

30. C. D. Patel and G. M. Rebeiz, "A high-reliability high-linearity high-power RF MEMS metal-contact switch for DC–40-GHz applications," *IEEE Transactions on Microwave Theory and Techniques*, vol. 60, no. 10, pp. 3096–3112, Oct. 2012.

31. D. Hyman, "Wideband DC-contact MEMS series switch," *Micro Nano Lett.*, vol. 3, no. 3, p. 66–69(3), Sept. 2008.

32. D. Mercier, P. L. Charvet, P. Berruyer, et al., "A DC to 100 GHz high performance ohmic shunt switch," *2004 IEEE MTT-S International Microwave Symposium Digest (IEEE Cat. No.04CH37535)*, Fort Worth, TX, 2004, vol. 3, pp. 1931–1934.

33. D. Hyman, J. Lam, B. Warneke, et al., "Surface-micromachined RF MEMs switches on GaAs substrates," *International Journal of RF Microwave and Computer Aided Engineering*, vol. 9, no. 4, pp. 348–361, July 1999.

34. C. L. Goldsmith, Zhimin Yao, S. Eshelman, and D. Denniston, "Performance of low-loss RF MEMS capacitive switches," *IEEE Microwave and Guided Wave Letters*, vol. 8, no. 8, pp. 269–271, Aug. 1998.

35. Y. Uno, K. Narise, T. Masuda, et al., "Development of SPDT-structured RF MEMS switch," *TRANSDUCERS 2009 – 2009 International Solid-State Sensors, Actuators and Microsystems Conference*, Denver, CO, 2009, pp. 541–544.

36. J. B. Muldavin and G. M. Rebeiz, "High-isolation CPW MEMS shunt switches. 1. Modeling," *IEEE Transactions on Microwave Theory and Techniques*, vol. 48, no. 6, pp. 1045–1052, June 2000.

37. D. Hyman and M. Mehregany, "Contact physics of gold microcontacts for MEMS switches," in *Electrical Contacts – 1998. Proceedings of the Forty-Fourth IEEE Holm Conference on Electrical Contacts (Cat. No.98CB36238)*, Arlington, VA, 1998, pp. 133–140.

38. R. Coutu, *Electrostatic Radio Frequency (RF) Microelectromechanical Systems (MEMS) Switches with Metal Alloy Electric Contacts*, PhD diss., Air Force Institute of Technology, 2004.

39. J. Schimkat, "Contact materials for microrelays," in *Proceedings MEMS 98. IEEE. Eleventh Annual International Workshop on Micro Electro Mechanical Systems. An Investigation of Micro Structures, Sensors, Actuators, Machines and Systems (Cat. No.98CH36176*, Heidelberg, Germany, 1998, pp. 190–194.

40. J. Y. Park, G. H. Kim, K. W. Chung, and J. U. Bu, "Electroplated rf MEMS capacitive switches," in *Proceedings IEEE Thirteenth Annual International Conference on Micro Electro Mechanical Systems (Cat. No.00CH36308)*, Miyazaki, Japan, 2000, pp. 639–644.

41. Yu Liu, T. R. Taylor, J. S. Speck, and R. A. York, "High-isolation BST-MEMS switches," in *2002 IEEE MTT-S International Microwave Symposium Digest (Cat. No.02CH37278)*, Seattle, WA, 2002, vol. 1, pp. 227–230.

42. C. H. Chang, J. Y. Qian, B. A. Cetiner, et al., "RF MEMS capacitive switches fabricated with HDICP CVD SiN/sub x/," in *2002 IEEE MTT-S International Microwave Symposium Digest (Cat. No.02CH37278)*, Seattle, WA, 2002, vol. 1, pp. 231–234.

43. P. De Silva, C. Vaughan, D. Frear, et al., "Motorola MEMS switch technology for high frequency applications," in *2001 Microelectromechanical Systems Conference (Cat. No. 01EX521)*, Berkeley, CA, 2001, pp. 22–24.

44. J. B. Muldavin and G. M. Rebeiz, "All-metal high-isolation series and series/shunt MEMS switches," *IEEE Microwave and Wireless Components Letters*, vol. 11, no. 9, pp. 373–375, Sept. 2001.

45. Jae-Hyoung Park, Sanghyo Lee, Jung-Mu Kim, Hong-Teuk Kim, Youngwoo Kwon, and Yong-Kweon Kim, "Reconfigurable millimeter-wave filters using CPW-based periodic structures with novel multiple-contact MEMS switches," *Journal of Microelectromechanical Systems*, vol. 14, no. 3, pp. 456–463, June 2005.

46. T. Purtova, *RFMEMS Periodic Structures: Modelling, Components and Circuits*," PhD diss., Universität Ulm, 2012.

47. M. Kaynak, M. Wietstruck, W. Zhang, et al., "Packaged BiCMOS embedded RF-MEMS switches with integrated inductive loads," in *2012 IEEE/MTT-S International Microwave Symposium Digest*, Montreal, QC, 2012, pp. 1–3.

48. M. Kaynak, V. Valenta, H. Schumacher, and B. Tillack, "MEMS module integration into SiGe BiCMOS technology for embedded system applications," in *2012 IEEE Bipolar/BiCMOS Circuits and Technology Meeting (BCTM)*, Portland, OR, 2012, pp. 1–7.

49. S. Tolunay Wipf, A. Göritz, M. Wietstruck, C. Wipf, B. Tillack, and M. Kaynak, "D–Band RF–MEMS SPDT Switch in a 0.13 μm SiGe BiCMOS Technology," *IEEE Microwave and Wireless Components Letters*, vol. 26, no. 12, pp. 1002–1004, Dec. 2016.

50. S. Tolunay Wipf, A. Göritz, C. Wipf, M. Wietstruck, et al., "240 GHz RF-MEMS switch in a 0.13 μm SiGe BiCMOS Technology," in *2017 IEEE Bipolar/BiCMOS Circuits and Technology Meeting (BCTM)*, Miami, FL, 2017, pp. 54–57.

51. D. Mercier, J. C. Orlianges, T. Delage, et al., "Millimeter-wave tune-all bandpass filters," *IEEE Transactions on Microwave Theory and Techniques*, vol. 52, no. 4, pp. 1175–1181, Apr. 2004.

52. Y. Lu, L. P. B. Katehi, and D. Peroulis, "High-power MEMS varactors and impedance tuners for millimeter-wave applications," *IEEE Transactions on Microwave Theory and Techniques*, vol. 53, no. 11, pp. 3672–3678, Nov. 2005.

53. S. Pu, *A Micromachined Zipping Variable Capacitor*," PhD diss., University of Southampton, 2010.

54. J. Muldavin, C. Bozler, S. Rabe, and C. Keast, "Large tuning range analog and multi-bit MEMS varactors," in *2004 IEEE MTT-S International Microwave Symposium Digest (IEEE Cat. No.04CH37535)*, Fort Worth, TX, 2004, vol. 3, pp. 1919–1922.

55. L. Dussopt and G. M. Rebeiz, "High-Q millimeter-wave MEMS varactors: extended tuning range and discrete-position designs," in *2002 IEEE MTT-S International Microwave Symposium Digest (Cat. No.02CH37278)*, Seattle, WA, 2002, vol. 2, pp. 1205–1208 .

56. G. McFeetors and M. Okoniewski, "Performance and operation of stressed dual-gap RF MEMS varactors," in *2006 European Microwave Conference*, Manchester, 2006, pp. 1064–1067.

57. Y. Lu, L. P. B. Katehi, and D. Peroulis, "High-power MEMS varactors and impedance tuners for millimeter-wave applications," *IEEE Transactions on Microwave Theory and Techniques*, vol. 53, no. 11, pp. 3672–3678, Nov. 2005.

58. D. Peroulis and L. P. B. Katehi, "Electrostatically-tunable analog RF MEMS varactors with measured capacitance range of 300%," in *IEEE MTT-S International Microwave Symposium Digest, 2003*, Philadelphia, PA, 2003, vol. 3, pp. 1793–1796.

59. G. P. Rehder, S. Mir, L. Rufer, E. Simeu, and H. N. Nguyen, "Low frequency test for RF MEMS switches," in *2010 Fifth IEEE International Symposium on Electronic Design, Test & Applications*, Ho Chi Minh City, 2010, pp. 350–354.

60. WiSpry. "Tunable Antenna Technology," 2019. Available at: http://wispry.com/

61. Cavendish-kinetics.com, "Cavendish kinetics – maximum RF performance," 2019. Available at: www.cavendish-kinetics.com

62. Hui Shen, Songbin Gong, and N. S. Barker, "DC-contact RF MEMS switches using thin-film cantilevers," in *2008 European Microwave Integrated Circuit Conference*, Amsterdam, 2008, pp. 382–385.

63. D. J. Chung, R. G. Polcawich, D. Judy, J. Pulskamp, and J. Papapolymerou, "A SP2T and a SP4T switch using low loss piezoelectric MEMS," in *2008 IEEE MTT-S International Microwave Symposium Digest*, Atlanta, GA, 2008, pp. 21–24.

64. Jaewoo Lee, Chang Han Je, Sungweon Kang, and Chang-Auck Choi, "A low-loss single-pole six-throw switch based on compact RF MEMS switches," in *IEEE Transactions on Microwave Theory and Techniques*, vol. 53, no. 11, pp. 3335–3344, Nov. 2005.

65. S. Gong, H. Shen and N. S. Barker, "A 60-GHz 2-bit switched-line phase shifter using SP4T RF-MEMS switches," *IEEE Transactions on Microwave Theory and Techniques*, vol. 59, no. 4, pp. 894–900, April 2011.

66. M. Kim, J. B. Hacker, R. E. Mihailovich, and J. F. DeNatale, "A DC-to-40 GHz four-bit RF MEMS true-time delay network," *IEEE Microwave and Wireless Components Letters*, vol. 11, no. 2, pp. 56–58, Feb. 2001.

67. Guan-Leng Tan, R. E. Mihailovich, J. B. Hacker, J. F. DeNatale, and G. M. Rebeiz, "Low-loss 2- and 4-bit TTD MEMS phase shifters based on SP4T switches," *IEEE Transactions on Microwave Theory and Techniques*, vol. 51, no. 1, pp. 297–304, Jan. 2003.

68. S. Gong, H. Shen, and N. S. Barker, "A 60-GHz 2-bit switched-line phasesshifter using SP4T RF-MEMS switches," *IEEE Transactions on Microwave Theory and Techniques*, vol. 59, no. 4, pp. 894–900, Apr. 2011.

69. Juo-Jung Hung, L. Dussopt, and G. M. Rebeiz, "Distributed 2- and 3-bit W-band MEMS phase shifters on glass substrates," *IEEE Transactions on Microwave Theory and Techniques*, vol. 52, no. 2, pp. 600–606, Feb. 2004.

70. Hong-Teuk Kim, J. H. Park, S. Lee, et al., "V-band 2-b and 4-b low-loss and low-voltage distributed MEMS digital phase shifter using metal-air-metal capacitors," in *IEEE Transactions on Microwave Theory and Techniques*, vol. 50, no. 12, pp. 2918–2923, Dec. 2002.

71. G. Rehder, T. Vo, and P. Ferrari, "Development of a slow-wave MEMS phase shifter on CMOS technology for millimeter wave frequencies," *Microelectronic Engineering*, vol. 90, pp. 19–22, 2012.

72. A. Bautista, A. L. Franc, and P. Ferrari, "An accurate parametric electrical model for slow-wave CPW," in *2015 IEEE MTT-S International Microwave Symposium*, Phoenix, AZ, 2015, pp. 1–4.

73. M. Verona, G. P. Rehder, A. L. C. Serrano, M. N. P. Carreño, and P. Ferrari, "Slow-wave distributed MEMS phase shifter in CMOS for millimeter-wave applications," in *2014 44th European Microwave Conference*, Rome, 2014, pp. 211–214.

74. C. C. Chang, Y. Chen, and S. Hsieh, "A V-band three-state phase shifter in CMOS-MEMS technology," *IEEE Microwave and Wireless Components Letters*, vol. 23, no. 5, pp. 264–266, May 2013.

75. R. De Paolis, F. Coccetti, S. Payan, M. Maglione, and G. Guegan, "Characterization of ferroelectric BST MIM capacitors up to 65 GHz for a compact phase shifter at 60 GHz," *2014 44th European Microwave Conference*, Rome, 2014, pp. 492–495.

76. H. Krishnaswamy, A. Valdes-Garcia, and J. Lai, "A silicon-based, all-passive, 60 GHz, 4-element, phased-array beamformer featuring a differential, reflection-type phase shifter," in *2010 IEEE International Symposium on Phased Array Systems and Technology*, Waltham, MA, 2010, pp. 225–232.

77. M. Tabesh, A. Arbabian, and A. Niknejad, "60GHz low-loss compact phase shifters using a transformer-based hybrid in 65nm CMOS," in *2011 IEEE Custom Integrated Circuits Conference (CICC)*, San Jose, CA, 2011, pp. 1–4.

78. S. Bulja, D. Mirshekar-Syahkal, M. Yazdanpanahi, R. James, S. E. Day, and F. A. Fernández, "Liquid crystal based phase shifters in 60 GHz band," in *The 3rd European Wireless Technology Conference*, Paris, 2010, pp. 37–40.

79. C. Fritzsch, F. Giacomozzi, O. H. Karabey, et al., "Continuously tunable W-band phase shifter based on liquid crystals and MEMS technology," in *2011 6th European Microwave Integrated Circuit Conference*, Manchester, 2011, pp. 522–525.

80. A. L. Franc, O. H. Karabey, G. Rehder, E. Pistono, R. Jakoby, and P. Ferrari, "Compact and broadband millimeter-wave electrically tunable phase shifter combining slow-wave effect with liquid crystal technology," *IEEE Transactions on Microwave Theory and Techniques*, vol. 61, no. 11, pp. 3905–3915, Nov. 2013.

81. T. Quemerais, D. Gloria, D. Golanski, and S. Bouvot, "High-Q MOS varactors for millimeter-wave applications in CMOS 28-nm FDSOI," in *IEEE Electron Device Letters*, vol. 36, no. 2, pp. 87–89, Feb. 2015.

82. G. Velu, K. Blary, L. Burgnies, et al., "A 310/spl deg//3.6-dB K-band phaseshifter using paraelectric BST thin films," *IEEE Microwave and Wireless Components Letters*, vol. 16, no. 2, pp. 87–89, Feb. 2006.

83. A. Gaebler, F. Goelden, A. Manabe, M. Goebel, S. Mueller, and R. Jakoby, "Investigation of high performance transmission line phase shifters based on liquid crystal," in *2009 European Microwave Conference (EuMC)*, Rome, 2009, pp. 594–597.

84. Ç. Ulusoy, M. Kaynak, T. Purtova, B. Tillack, and H. Schumacher, "A 60 to 77 GHz switchable LNA in an RF-MEMS embedded BiCMOS technology," *IEEE Microwave and Wireless Components Letters*, vol. 22, no. 8, pp. 430–432, Aug. 2012.

85. G. Liu, M. Kaynak, T. Purtova, A. Ç. Ulusoy, B. Tillack, and H. Schumacher, "Dual-band millimeter-wave VCO with embedded RF-MEMS switch module in BiCMOS technology," in *2012 IEEE 12th Topical Meeting on Silicon Monolithic Integrated Circuits in RF Systems*, Santa Clara, CA, 2012, pp. 175–178.

86. Sanghyo Lee, Jong-Man Kim, Jung-Mu Kim, Yong-Kweon Kim, and Youngwoo Kwon, "Millimeter-wave MEMS tunable low pass filter with reconfigurable series inductors and capacitive shunt switches," *IEEE Microwave and Wireless Components Letters*, vol. 15, no. 10, pp. 691–693, Oct. 2005.

87. J. Sor, Y. Qian, and T. Itoh, "Miniature low-loss CPW periodic structures for filter applications," *IEEE Transactions on Microwave Theory and Techniques*, vol. 49, no. 12, pp. 2336–2341, Dec. 2001.

88. Jae-Hyoung Park, Sanghyo Lee, Jung-Mu Kim, Hong-Teuk Kim, Youngwoo Kwon, and Yong-Kweon Kim, "Reconfigurable millimeter-wave filters using CPW-based periodic structures with novel multiple-contact MEMS switches," *Journal of Microelectromechanical Systems*, vol. 14, no. 3, pp. 456–463, June 2005.

89. Hong-Teuk Kim, Jae-Hyoung Park, Yong-Kweon Kim, and Youngwoo Kwon, "Low-loss and compact V-band MEMS-based analog tunable bandpass filters," *IEEE Microwave and Wireless Components Letters*, vol. 12, no. 11, pp. 432–434, Nov. 2002.

5 Microwave Liquid Crystal Technology

Rolf Jakoby, Matthias Jost, Onur Hamza Karabey, Holger Maune,
Matthias Nickel, Ersin Polat, Roland Reese, Henning Tesmer,
and Christian Weickhmann

5.1   Introduction to Microwave Liquid Crystal Technology

Chapter 1 of this book provided a first overview of recent developments in new
platforms and technologies and their implications in communications, including future
mobile traffic, the 5G vision, trends in satellite communication platforms, spectrum
allocation, key technology drivers, markets, and perspectives. For the deployment of
these platforms and services, new hardware concepts and technologies to enable smart
system functionalities are crucial for smart user devices and terminals as well as for
base stations and satellites, following to some extent the software-controlled radio
(SCR) approach.

A bottleneck of this approach are efficient hardware solutions for reconfigurable/
tunable components, circuits, and devices at the radio frequency (RF) level of a
transceiver, which are able to dynamically control frequency, bandwidth, impedance,
polarization, amplitude, and phase of electromagnetic waves by software-controlled
signal processing electronics to achieve system functionalities such as phase shifting,
beam steering and beam forming, adaptive impedance matching, and filtering.
Particularly for smart systems in the millimeter (mm)-wave range, where ordinarily
complexity, technological constraints, and cost increase, it is very challenging to find
hardware solutions that improve RF performance while simultaneously reducing the
manufacturing cost to an economical price point.

A very promising solution, particularly for mm-wave systems, is the microwave
liquid crystal (MLC) technology. Inherently, all components and circuits based on the
MLC technology provide continuous tuning, in contrast to traditional microwave
reconfigurable ones with PIN diodes, field-effect transistors (FET) or radio
frequency-microelectromechanical (RF-MEMS) switches, or monolithic microwave
integrated circuits (MMICs) [1–12] except varactor diodes, which, however, exhibit a
poor quality factor at mm-waves, as discussed in Chapter 3. This continuous tuning is
accompanied by extremely low power consumption (quasi power-less tuning), low-
frequency dispersion and very linear operation [13–15], and very high power-handling
capability. Moreover, MLC technology potentially enables low-cost, large-scale
manufacturing of devices, for example, of beam-steering antenna arrays, particularly
when standard low-temperature co-fired ceramic (LTCC) technology [16–24] or
standard liquid crystal display (LCD) processes on established LCD production lines
[25–28] can be applied for fabrication.

The MLC technology established at Technische Universität (TU) Darmstadt was initiated by Prof. Jakoby in 2002 in close collaboration with the Performance Materials Division of Merck KGaA, Darmstadt, Germany, being the largest supplier of LCs worldwide. Since then, Merck synthesizes and provides novel nematic LC mixtures with the highest material's figure of merit (FoM) over a wide frequency range from few GHz up to 8 THz [14, 23, 29–35]. Moreover, several radiation tests of these LCs for space qualification, including total dose tests, indicated no influence of radiation on LC [16, 18, 19]. Hence, they are compatible with space applications.

A first review of the potential of the MLC technology in general is given in [36] and more specifically of TU Darmstadt in [37, 38], including material properties and characterization as well as some exemplary proof-of-concept demonstrators of LC-based components, circuits, and devices. A comprehensive overview on and a comparative review of available microwave liquid crystal materials as well as techniques employed for their characterization and their key application-relevant properties by incorporating a great number of publications and their results in this specific field is presented in [39].

5.1.1   Performance Metric of Microwave LCs

Liquid crystals (LCs) are dielectric materials with highly anisotropic characteristics, which can be grouped into three different mesophases between solid/crystalline and liquid/isotropic states. These mesophases can be separated by orientation and positional order into nematic, smectic, and cholesteric [31, 40]. The nematic phase is the most commonly used phase of LCs at microwaves and is characterized by an orientation of rod-shaped molecules, which include a polar group, producing a dipole moment that can be controlled by an external electrostatic field, resulting in a highly anisotropic permittivity tensor. At macroscopic scale, the time-averaged direction of the molecules' long axis is denoted by a director $\vec{n}$.

In recent years, there have been many comprehensive and comparative studies of various high birefringence LCs, on their mesogenic and physical-chemical properties (viscosity, birefringence, permittivity, anisotropy, and elastic constants), discussing the question of how to obtain LCs with a broad range of nematic phases. A comprehensive review of high birefringence new LCs for microwave application developed at the Institute of Chemistry of the Military University of Technology in Warsaw is given in [41], which is based on their long experience in LCs for optics. In [42], dielectric properties of LC isothiocyanato-tolane derivatives with fluorine atoms at various lateral positions are analyzed from 1 kHz to 3 GHz. From the same group eutectic mixtures of isothiocyanato-tolane molecules are characterized from 26 to 40 GHz in a coplanar waveguide (CPW) with an active part made of a central cavity such as a rectangular waveguide to determine the dielectric properties and comparing them with the standard display LC K15 (5CB, cyanobiphenyl) [43].

In [44–47] systematic investigations were focused on how the dielectric anisotropy in the low frequency region up to 1 GHz and the birefringence of the nematic LC phase at optical wavelength (589 nm) influences the performance of LC mixtures at microwave frequencies. For this, the authors prepared a wide range of mixtures being nematic at room temperature and containing different classes of promising compounds. Their quasi-static and frequency-dependent dielectric properties, their birefringence and the impact of those properties on their microwave behavior were analyzed by using the cavity perturbation method [48] at 30 and 38 GHz, respectively, in a wide temperature interval from $-20°C$ to $135°C$.

Merck's LC mixtures dedicated to microwave application have been developed since 2002, where more than 400 liquid crystal components have been selected and characterized at 19 GHz [32, 34]. Table 5.1 summarizes the elastic and electromagnetic properties of some selected nematic LC mixtures at room temperature, which are regularly used for LC-tuned microwave components. In addition, K15 (5CB) is given as a benchmark, since it is well known from optics and widely used in many studies. The listed values might slightly vary, depending on the working frequency. The temperature dependency is discussed in [46, 47, 49].

All LCs in Table 5.1 offer low dielectric constants in both principal directions, typically in the range of 2.4–2.6 for the perpendicular and 2.9 to 3.5 for the parallel state, thus satisfying low permittivity requirements. The maximum loss tangents have typical values of tan $\delta \leq 0.015$. Standard display mixtures such as K15 offer only limited anisotropy and tunability, accompanied with relatively high loss. In contrast, Merck's first generation LC compounds such as the GT3 series, having long conjugation body with biphenyl or terphenyl structure, exhibit much higher anisotropy in relative permittivity,

$$\Delta\varepsilon = \varepsilon_{r,\parallel} - \varepsilon_{r,\perp}, \tag{5.1}$$

with a tunability in relative permittivity,

$$\tau_{LC} = \frac{\Delta\varepsilon}{\varepsilon_{r,\parallel}}, \tag{5.2}$$

of about 25%, accompanied with much lower dielectric losses, that is, with much higher material's FoM

$$\eta_{LC} = \frac{\tau_{LC}}{\max(\tan\delta)} \tag{5.3}$$

of about 20 when compared with K15.

Merck's second-generation LC mixtures such as the GT5 series and TUD-566 use single classes, including bistolane and other novel components with triple bonding in between. With these LCs, tunability τ_{LC} is more than doubled and the material's FoM η_{LC} is increased by a factor of 9–13 compared to standard K15, where to the authors' knowledge, η_{LC} of 46–64 are the highest values reported so far.

As stated in [32]:

It is quite easy to find its good correlation with optical birefringence, where the electron polarization also plays a major role. In contrast to that, it is not easy to find a clear relationship between molecular structure and loss tangent. Several factors, such as position

Table 5.1 Dielectric and elastic properties of different LC mixtures

LC	$\varepsilon_{r,\perp}$	$\tan\delta_\perp$	$\varepsilon_{r,\parallel}$	$\tan\delta_\parallel$	K_{11} (pN)	K_{22} (pN)	K_{33} (pN)	γ_{rot} (Pa·s)	$\Delta\varepsilon_{1kHz}$	T_c (°C)	τ_{LC} (%)	η_{LC}
K15 (5CB)	2.56	0.0264	2.94	0.0125	7.0	4.2	13.5	0.126	14.4	38.0	12.9	4.9
GT3-23001	2.46	0.0143	3.28	0.0038	24	14	34.5	0.746	4.6	173.5	25.0	17.5
GT5-26001	2.39	0.007	3.27	0.0022	12.0		41.9	1.958	1.0	146.0	26.9	38.4
GT5-28004	2.40	0.0043	3.32	0.0014	11.8		52.9	5.953	0.8	151.0	27.8	64.4
TUD-566	2.41	0.006	3.34	0.0027	13	8.0	48.0	2.100	1.0	105.5	27.8	46.4
GT7-29001	2.46	0.0116	3.54	0.0064	14.5		18.0	0.307	22.1	124	30.5	26.3

The values for K_{11}, K_{22}, and K_{33} are constants for splay, twist and bend deformations according to [50], γ_{rot} is the rotational viscosity, $\Delta\varepsilon_{1kHz}$ the anisotropy at 1 kHz, τ_{LC} and η_{LC} the material's tunability and figure-of-merit, respectively. All RF measurements are taken at 19 GHz.

and strength of permanent dipole moment, viscosity, and molecular shape, contribute to the loss mechanism. Singles having bistolane core structure are one of the best compounds having both, high tunability and low dielectric losses.

Additionally, both generations of mixtures show significantly improved voltage holding ratio, a reliability index against external heat stress of max. 24 hours under 120°C [32]. This enables use of LC-based devices outdoors, making them attractive for commercial applications. Moreover, because of the LCs' unique feature of slightly decreasing dielectric losses with frequency, these LCs enable tunable microwave devices with excellent performance, particularly at high frequencies above 10 GHz, because of lower losses, as they are flexible and continuously tunable.

However, the *tuning speed* or *response time* of Merck's second-generation LCs is much slower than for the ones of the first generation or the standard display mixtures, which will be discussed in the text that follows, after introducing the basic biasing scheme.

Hybrid Biasing Scheme To make use of the LC's anisotropic nature, the orientation of the LC molecules must be controlled by means of electrostatic fields. The simplest biasing scheme can be explained by means of a parallel-plate capacitor. Beside the pair of electrodes, an additional pair of alignment layers is required to prealign the LC molecules into their initial state in parallel to the alignment layers (perpendicular to the RF field) by means of the surface anchoring forces in analogy to LC displays. The effect of polyimide alignment layers on the LC tuning efficiency is analyzed in [14, 23, 51].

Response Time The most critical parameter for reconfigurable LC devices is *tuning speed* or *response time*, which is defined as the time interval required to reach an equilibrium state of LC devices [14, 23, 26, 40, 52–54]. This will be explained for the aforementioned LC-filled parallel-plate capacitor. When a bias voltage V_b is applied, the LC directors $\vec{n}$ start to change their orientation from the initial state (perpendicular to $\vec{E}_{RF}$ due to the alignment layers) toward the field lines of the electrostatic bias field $\vec{E}_b$, until they are nearly perpendicular between the alignment layers (nearly in parallel to $\vec{E}_{RF}$) for $V_b > V_{sat}$. The time it takes is the *rise or switch-on response time*

$$\tau_{\text{on}} \propto \frac{\gamma_{\text{rot}}}{\varepsilon_0 \Delta\varepsilon_{r,\,1kHz}\left(E_b^2 - E_{\text{th}}^2\right)} = \frac{\gamma_{\text{rot}} h_{\text{LC}}^2}{K_{11}\pi^2\left(\dfrac{V_b^2}{V_{th}^2} - 1\right)} = \frac{\tau_{\text{off}}}{\left(\dfrac{V_b^2}{V_{th}^2} - 1\right)} \qquad (5.4)$$

with the threshold voltage

$$V_{th} = E_{th} h_{LC} = \pi\sqrt{\frac{K_{11}}{\varepsilon_0 \Delta\varepsilon_{r,\,1kHz}}}. \qquad (5.5)$$

The more critical parameter for reconfigurable LC devices using this basic biasing scheme, however, is the *decay or switch-off response time*

$$\tau_{off} \propto \frac{\gamma_{rot}}{\varepsilon_0 \Delta \varepsilon_{r,1kHz} E_{th}^2} = \frac{\gamma_{rot} \cdot h_{LC}^2}{\varepsilon_0 \Delta \varepsilon_{r,1kHz} \cdot V_{th}^2} = \frac{\gamma_{rot} h_{LC}^2}{K_{11} \pi^2} \qquad (5.6)$$

when the bias voltage is released and the LC directors $\vec{n}$ (in parallel to $\vec{E}_{RF}$) are going back to the initial state (in parallel to the alignment layers and almost perpendicular to $\vec{E}_{RF}$) by means of the surface anchoring forces. Due to its asymptotic nature of the response (in orientation, capacitance or phase shift, respectively) the time for completion is infinite. Thus, according to the usual engineering approach, it is defined in practice as the 90% to 10% decay T_{10}^{90} of the change in orientation of the LCs or in the capacitance or in the phase shift for LC-based devices. Similarly, the switch-on response time is described as the 10% to 90% change T_{90}^{10} in orientation or capacitance or phase shift.

The response times depend on material's elastic properties, that is, mainly on the rotational viscosity γ_{rot} and the elastic constant K_{11} for the splay deformation, but strongly on the applied bias voltage V_b versus the threshold voltage V_{th} in case of a fully electrical biasing scheme and significantly on the effective LC layer height h_{LC}, because it follows a square law. While the elastic properties depend on the material itself, the other parameters are dependent on the device's topology and the biasing concept.

Elastic Properties In general, changing the orientation of long and stiff nematic LC molecules by an electrostatic field within a thin LC cell capacitor of 1–5 μm thickness implies relatively slow switch-off response times τ_{off} in the range of 3.2 ms up to 80 ms for GT3 and 16 ms up to 410 ms for TUD-566 according to Eq. (5.6) and Figure 5.1, compared to a fast shift of the ion from and back to the lattice center of a

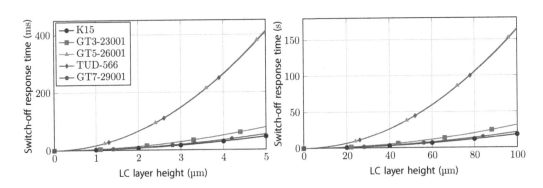

Figure 5.1 Switch-off response time of a parallel-plate capacitor cell filled with the standard display LC K15; Merck's first-generation LC GT3-23001; two Merck's second-generation LCs, TUD-566 and GT5-26001; and Merck's third generation LC GT7 versus the LC layer height (left) up to 5 μm and (right) up to 100 μm.

1 to 5 μm thick ferroelectric capacitor, which is in the nanosecond or even picosecond range. But this slowness of nematic LCs implies excellent linear behavior [13–15].

Taking into account some uncertainties of K_{11} and γ_{rot} in Table 5.1 and in the determination of the "real" LC layer height, the foregoing theoretical values for τ_{off} for GT3 agree well with experiments on single LC varactors filled with an LC of the GT3 series in [55, 56], measuring switch-off response time τ_{off} of 4 ms and 92 ms for an LC layer thickness h_{LC} of about 1 μm and 5 μm, respectively, following clearly the square law. Owing to their specific elastic properties, the switch-off response time τ_{off} is about 1.5 times longer for GT3-23001 and longer by a factor of 9 for TUD-566 or GT5-26001 compared to the common LC K15.

To enable faster relaxation of the LC molecules, special classes of LC materials have been employed such as polymer-stabilized nematic LCs [57–59], porous PTFE membranes impregnated with LC [14, 60, 61], nematic LCs doped with nanoparticles [14, 45, 62–65], ferroelectric LCs [45, 66, 67], ferroelectric LC mixtures doped with carbon nanotubes [68], $BaTiO_3$ nanoparticles suspended in a ferroelectric LC mixture [69], and dual-frequency switching LCs [15, 45, 70, 71]. This might improve the response time by one up to two orders of magnitude compared to the one with "pure" nematic LCs. Thus, for example, the phase shift response time could be reduced from more than 1 s down to 33 ms for a microstrip line phase shifter with 100 μm thick membrane impregnated LC layer [60] or from about 20 s down to less than 0.4 s in an ITO glass cell and from 80 s down to 2 s in a CPW-type phase shifter by using a polymer-stabilized nematic LC [59]. However, this improvement in response time is usually at the expense of much higher dielectric losses and/or lower anisotropy, resulting in lower tunability and much lower material's FoM, and often it requires larger driving voltages for components with the same thickness. Therefore, despite slower response time, research on nematic LCs is going on, since there is still room for improvement in the performance metric in terms of tunability, the material's FoM, and response time, where the overall performance metric can be tailored to match the requirements of a specific application.

5.1.2   Applications of Microwave LCs

First microwave applications using liquid crystals had been on waveguide and planar transmission line phase shifters [72–75] providing a phase shifter FoM $= \Delta\varphi_{max}/IL_{max} < 20°/dB$ only, where $\Delta\varphi_{max}$ is the maximum differential phase shift and IL_{max} the maximum insertion loss in all tuning states.

With the emergence and progress of newly developed LC mixtures specifically synthesized for microwaves since 2002, innovative concepts and designs with appropriate biasing schemes enabled numerous high-performance RF components and devices, which fully exploit the LC's unique properties. Thus, with Merck's first-generation LCs, an FoM of more than 110°/dB was achieved already in an early stage with a microstrip line phase shifter at 24 GHz [29, 30, 76, 77] and soon afterward, an FoM of up to 150°/dB could be reached with a fully electrical biasing and even up to 200°/dB with magnetic biasing for a partially LC-filled waveguide phase shifter around 30 GHz [78]. Since then, various planar delay line, metallic, and dielectric waveguide phase shifters at different frequencies were realized. Apart from delay line phase shifters, an increasing number of tunable high-performance RF components and devices are reported in the literature.

In contrast to traditional reconfigurable microwave components made of semiconductor materials or ferroelectrics, LC-based components might face some difficulties:

- Since it is a fluid, it must be filled in after processing via tiny filling holes into some encapsulating packages such as flat panels for phased and reflect arrays or into long closed/sealed channels with respect to the wavelengths as for delay lines or into small cavities as for filters. This might complicate the preparation of the LC-based devices, but not necessarily when standard processes are established later such as liquid crystal display (LCD) technology. During the experimental phase, it might be even beneficial, since the LC already being filled into a device can be exchanged for improvement or comparison.
- To make use of the LC's anisotropic nature, the orientation of the LC molecules must be controlled by means of electrostatic fields. This is simple for the hybrid biasing scheme, when the RF line and the ground plane can be simultaneously used as bias lines as for parallel-plate capacitors or planar microstrip delay line phase shifters, but it can be quite difficult for "thick" voluminous waveguide structures, where the prealignment forces of the polyimide layers are getting extremely weak, and hence tuning efficiency $\Delta\tau_\varphi$, which describes quantitatively how much of the material's tunability can be utilized by a proposed device layout, reduces. This requires more complex biasing schemes with multielectrode configurations, for example, processed on a thin Mylar film, which in turn degrade the RF performance to some extent by adding additional losses [23, 78–80].
- With a fully electrical biasing scheme without any alignment layer, the switch-off response time can be significantly reduced by forcing the reorientation of the LCs due to an applied voltage and not by the weak alignment layer forces, following then Eq. (5.4) for τ_{on}. Moreover, the response time can be further reduced by overshooting with higher applied voltages for a very short time [23]. The control voltages being required for tuning depend on the electrodes' distances, which might be less than 10 V for flat devices with a thickness <150 μm but could be more than 200 V for thicker components. While planar transmission line topologies are inherently flat and compact, volumetric waveguide–based structures and devices are bulky and much more complex and usually more expensive to implement. However, electrical losses of waveguide structures are inherently significantly lower than for planar structures at mm-waves, which means a much higher device performance, for example, a higher FoM of phase shifters or a much higher quality factor Q of filters, generating higher selectivity and more distinct passband properties.
- The most decisive design parameter to reduce the switch-off response time of LC-based components and devices for all biasing schemes is the LC layer height h_{LC}, because of the square law, as can be seen from Figure 5.1. It is determined by the

chosen component/device topology and concept. Hence, the choice of h_{LC} is of fundamental importance in the engineering of MLC components and devices. However, decreasing LC layer heights is limited in waveguide as well as in planar line topologies, because of increasing metallic losses and 50 Ω impedance mismatch. For example, a typical 50 Ω design of a microstrip delay line phase shifter would be of an effective LC layer height of 60–100 µm and a width of the signal electrode of 100–200 µm to avoid metallic losses, which exceeds the dielectric ones [23]. However, thicknesses of more than 100 µm lead to response times of up to several tens of seconds and more, which might be still feasible for portable and slow-moving applications but will be impractical for on-the-move applications. Thus, to reduce the response times down to a few milliseconds for pure nematic LCs, it is essential to realize LC-based devices with very thin LC layers of $h_{LC} < 5$ µm.

To overcome the response time limitations, there are some innovative approaches and concepts. For instance, a very promising approach are periodically loaded line phase shifters, for example, slot lines with periodically thin overlapping areas filled with LC [81] or CPWs periodically loaded by very thin LC varactors of $h_{LC} < 5$ µm [14, 55, 82]. The latter results in an FoM $> 60°/dB$ and switch-off response times of less than 92 ms and 340 ms for the LCs GT3-23001 and TUD-566, respectively. This concept is now employed into an LCD process for flat-panel array antennas, exhibiting switch-off response times of less than 30 ms for $h_{LC} \leq 4$ µm [28]. Other options are miniaturized and fast tunable slow-wave phase shifters with thin LC layer also of less than $h_{LC} < 5$ µm, which was demonstrated either by a 0.35 µm CMOS process [83] or by the nanowire membrane (NaM) technology [84]. Another approach might be stack-layered structures, that is, a multistack assembly of several LC sublayers separated by dielectric substrates. Early examples are given in [85] with grating patterned electrodes and thin LC layer of 100 µm thickness for measuring transmission properties at 50 GHz or in [86, 87] for LC prism cells for deflection and an LC lens with focusing properties at 94 GHz or for wavelength selector and variable phase gratings characterized from 26.5 GHz to 40 GHz in [88, 89] or for LC beam former operating from 50 GHz to 75 GHz with beam-steering angles of about ±13° [61].

5.1.3   Perspectives of Microwave LC Technology

Boosted by novel breakthroughs in microwave LC synthesis, numerous high-performance RF components and devices such as phase shifters, adaptive resonators and filters, beam steering reflectarrays, and phased array antennas are getting attractive for commercial applications.

Thus, Kymeta has recently demonstrated a flat LC metamaterial surface antenna in a satellite communication setup [90, 91]. ALCAN Systems also demonstrated a first flat beam steering phased array antenna with LC-tunable phase shifter stack [92, 93]. Both antennas have been fabricated in a partnership with display manufacturing companies.

To complement these developments in device design and manufacturing, some groups are developing LC materials of the third generation, focusing on a high tunability class, still with low losses. Thus, the systematic investigations have been updated in [94] with newly developed LC compositions with large optical anisotropy and low melting point based on synthesized quaterphenyl and quinquiphenyl LC compounds, containing lateral substituents. Some selected four-ring compounds have low melting point of less than 70°C and exhibit large anisotropy $\Delta\varepsilon \approx 1.12\text{--}1.34$, accompanied by low maximum loss tangent $\max(\tan\delta) \approx 0.002\text{--}0.006$ at 29 GHz. However, the elastic properties and switch-off response times are not given in [94].

Merck developed LC materials of the third generation [35]. These are subdivided into a low-loss material class and a high tunability class with substantially improved tunability of more than 30%. An example is GT7 in Table 5.1, with high anisotropy $\Delta\varepsilon = 1.08$ and high tunability $\tau_{LC} = 30.5\%$, but moderate material's FoM $\eta_{LC} = 26.3$, which, however, exhibit fast response time similar as K15. Beyond microwave properties, parameter improvements related to the usage of LC in a real product are aimed (1) to increase the temperature range, (2) to reduce the tuning voltage, and (3) to increase the tuning speed or switch-off response time [35].

1. Important improvements have been made to the low temperature stability of the third generation of mixture classes by lowering the crystallization temperature T_c down to below $-20°C$ and $-30°C$, respectively.
2. The dielectric anisotropy at low frequencies $\Delta\varepsilon_{1\ kHz}$ could be increased significantly to values of 10–20 and above, leading to much lower threshold voltage according to Eq. (5.5), and hence lower tuning voltage in Eq. (5.4).
3. According to Eqs. (5.4) and (5.6), response times are proportional to the rotational viscosity. This parameter has been decreased to around 0.3 Pa·s for the third-generation LCs compared to 2.1 Pa·s for second-generation LC TUD-566, thus, enabling considerable improvements in tuning speed as can be seen from Figure 5.1.

These improved performance metric of third generation LC mixtures combined with the parallel progress in processing, manufacturing and assembly technology will bring LC-based components and devices, in particular LC-based antenna technology a step closer toward becoming a part of novel satellite and 5G communication systems. As stated in [35]: "Future work will focus on reducing losses in the high tunability material class as well as further loss reduction and a tunability increase in the low loss mixture class," which in turn will open up and boost new applications of tunable LC-based components, circuits and devices for smart microwave and mm-wave systems.

## 5.2		Fundamentals of LC Material for Microwaves

LCs are a family of materials, which have mesophases in between their solid and liquid states, depending on temperature and pressure [95]. In the solid state,

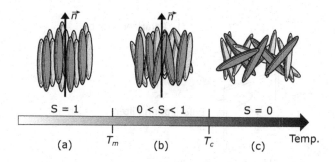

Figure 5.2 Schematic of a nematic liquid crystal material depending on the temperature. (a) Crystalline solid state with perfect alignment with the order parameter where the director $\vec{n}$ is parallel to the molecular long axis. (b) Nematic phase, where $\vec{n}$ indicates the time- and space average directions of the molecules. (c) Liquid phase with no molecular ordering resulting in isotropic material.

intermolecular forces are strong, and the molecules are held either in a regular or in an irregular order. When the molecules are bonded regularly the corresponding state is called crystalline solid state, in which the molecules feature positional and orientational order. In the liquid state, on the other hand, the intermolecular bonds are broken down, because the molecules vibrate due to an increased temperature and/or reduced pressure. Hence, they cannot hold both, positional and orientational order in this state.

LC material is constituted by anisotropic molecules in shape. As a result of this anisotropy, the intermolecular forces in the LC material are not uniform. As shown in Figure 5.2, above a *melting point* T_m, positional order largely disappears while orientational order is preserved to a certain extent [96]. When there is no positional order, the corresponding state is called as *nematic phase*, which is the most common phase for high frequency applications, and therefore, it is in the main focus of this book. LC materials feature different physical phenomena such as anisotropy in the refractive index and in the magnetic susceptibility, which will be discussed later in this chapter. Owing to molecular polarity, the molecules can be oriented by an electric or magnetic field, which makes the LC material attractive to implement dielectrically tunable components at micro and mm-wave frequencies [40]. When temperature increases further, the molecules lose their orientational order from a point T_c, the *clearing point*. The LC becomes to be isotropic in the liquid state.

LCs became well known through their application in display technology, which is posed in "The history of liquid crystal displays (LCD)" [97]. This technology allowed flat screens and small scale displays for almost any electronic device. In its easiest approach, the LC is used in a twisted nematic cell (TNC) configuration. As shown in Figure 5.3, the light is polarized through a filter, then turned into the desired polarization through a layer of LC. Eventually only, for example, the vertical part of the polarized light transmits through the second filter. When biasing the LC, the order of the twisted nematic cell is disturbed, and therefore less light is turned in its polarization.

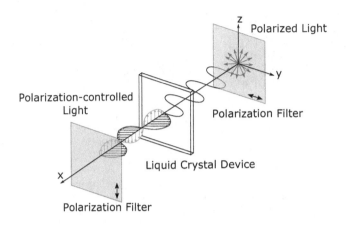

Figure 5.3 Principle of LC-based display with TNC.

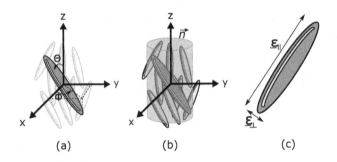

Figure 5.4 (a) LC molecules in Cartesian coordinate system with angular representation. (b) The director $\vec{n} = (0, 0, 1)^T$ of the unit volume element. (c) The permittivity values of the molecule's axes.

5.2.1   Properties of LCs

5.2.1.1   Order Parameter of a Bulk of LC

In the nematic phase, the LC molecules tend to align in parallel to each other. At macroscopic scale, the time-averaged direction of the molecules' long axis is denoted by a director $\vec{n}$. There is no preferred direction along the short axes, since the nematic liquid crystals are usually uniaxial [40]. The director $\vec{n}$ can be defined in both the solid and the nematic phase, with a different molecular ordering, but not for the liquid phase. The molecular ordering is defined quantitatively by an order parameter S [14]. This parameter measures the order of the molecules with respect to the average direction as shown in Figure 5.2. Besides the ordering used in the following chapters, S can become negative, which occurs only if the molecules are perpendicular to the vector being taken as a reference and besides that are completely unordered [40].

Figure 5.4 shows an LC molecule in a cartesian coordinate system. The deviation of the direction of a single molecule from the direction $\vec{n}$ is represented by the angle Θ. The angle Φ can be neglected, since the LC molecules are uniaxial, and therefore this deviation does not add to the order parameter S. According to [40, 98], the most useful approach is the average value of the second Legendre polynomial, considering the independency of the algebraic sign, since $\vec{n}$ is equal to $-\vec{n}$ as well. The order parameter S is defined as

$$S = \langle P_2(\cos \Theta) \rangle = \frac{\int_0^\pi f(\Theta, T) P_2(\cos \Theta) \sin \Theta \, d\Theta}{\int_0^\pi f(\Theta, T) \sin \Theta \, d\Theta}, \tag{5.7}$$

with $P_2(x) = \frac{1}{2}(3x^2 - 1)$. The function $f(\Theta, T)$ describes the probability of a molecule, having a deviation within the solid angle $d\Omega$ (i.e., $\sin \Theta \, d\Theta$) and is dependent on the temperature T. In the isotropic case, the probability for each angle is equal, and therefore, the probability $f(\Theta, T > T_c)$ is constant. The average value can either be taken from many molecules or by taking one molecule into account for a period of time. If all molecules are pointing in the same direction, they share the common direction with the director and all molecules have a deviation of $\Theta = 0$, which leads to $S = 1$. In the isotropic case $f(\Theta) = 1$, the order parameter is $|S| = 0$.

The order parameter can be described by different models. One possibility is the Landau–de Gennes theory, another is the Maier–Saupe theory. The latter one also includes the dependency of the absolute temperature T through the Boltzmann distribution [40]:

$$f(\Theta, T) = \frac{e^{-\frac{V_n(\Theta, T)}{k_B T}}}{Z(T)} = \frac{1}{Z(T)} e^{S(T)\left(\frac{1}{2}(3(\cos \Theta)^2 - 1)\right)\frac{v}{k_B T}}, \tag{5.8}$$

with

$$V_n(\Theta, T) = -v \cdot S(T) \left(\frac{1}{2} \left(3(\cos \Theta)^2 - 1 \right) \right). \tag{5.9}$$

In Eq. (5.9), the parameter v is an interaction constant, which is temperature independent, originally. The function $Z(T)$ is the single molecule separation function

$$Z = \int_0^\pi e^{vS\left(\frac{3}{2}(\cos \Theta)^2 - \frac{1}{2}\right)} \sin \Theta \, d\Theta. \tag{5.10}$$

The temperature-dependent order parameter can be calculated numerically by using the Boltzmann constant k_B, a molecule interaction constant v, and the normalized temperature $T_n = k_B T / v$ as

$$S(T_n) = \frac{\int_0^\pi \left(\frac{3}{2}(\cos \Theta)^2 - \frac{1}{2} \right) e^{S(T_n)\left(\frac{3}{2T_n}(\cos \Theta)^2 - \frac{1}{2T_n}\right)} \sin \Theta \, d\Theta}{\int_0^\pi e^{S(T_n)\left(\frac{3}{2T_n}(\cos \Theta)^2 - \frac{1}{2T_n}\right)} \sin \Theta \, d\Theta}. \tag{5.11}$$

The result of this calculation is shown in Figure 5.5. The normalized transition from the nematic to the isotropic (NI) phase is $T_{c,\mathrm{NI}} = 0.22019$. There, it can be seen, that at T_c the order from $S_c = 0.4289$ rapidly drops to 0, and therefore the LC

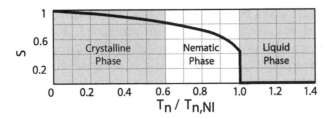

Figure 5.5 Order parameter S versus normalized temperature $T_n/T_{n,NI}$.

macroscopically becomes isotropic. This model is valid above the melting point T_m, since $S = 1$ in the crystalline phase, except from possible defects. It is shown, that theoretical values for S in the nematic phase are between 0.4 and 0.8 [23, 40]. The approach shown here is not suitable for alignment with experimental values, due to the lack of fitting parameters. Nevertheless, the qualitative behavior is explained very well.

5.2.1.2 Electromagnetic Properties of LC Materials

LC material is both, a diamagnetic and dielectric material [99]. Thus, they feature anisotropic magnetic susceptibility $\Delta \overleftrightarrow{\chi_m}$ and anisotropic electric susceptibility $\Delta \overleftrightarrow{\chi_e}$ in presence of magnetic and electric fields, respectively. Both $\Delta \overleftrightarrow{\chi_m}$ and $\Delta \overleftrightarrow{\chi_e}$ are tensorial and macroscopic quantities. They are quantities averaged over time and over space, in order to describe measurable effective electric and magnetic susceptibilities. The magnetic susceptibilities of the LC material are on the order of 10^{-5} [40, 98], and therefore, the material is not utilized as a diamagnetic material for micro- and millimeter-wave applications. Nevertheless, $\Delta \overleftrightarrow{\chi_m}$ leads to an anisotropic complex permeability $\underline{\overleftrightarrow{\mu}}_r$ of the material. Although being low, $\underline{\overleftrightarrow{\mu}}_r$ is utilized to control the orientation of the LC director field for material characterization purposes.

When designing tunable microwave components in LC technology, the dielectric properties are of utmost interest. The anisotropic complex permittivity $\underline{\overleftrightarrow{\varepsilon}}_r$ can be determined through the polarizability of the molecules [14]. When the LC molecules are exposed to an electric field, that is, in the presence of $\vec{E}_{\text{molecule}}$, dipole moments $\vec{p}$ are induced, which are determined by using the tensorial molecular polarizability vector $\overleftrightarrow{\alpha}_p$:

$$\vec{p} = \overleftrightarrow{\alpha}_p \vec{E}_{\text{molecule}}. \tag{5.12}$$

The molecular field $\vec{E}_{\text{molecule}}$ can differ from the macroscopic electric field $\vec{E}$ because of the strong dipole–dipole interactions between the molecules [40]. Hence, considering the tensorial internal field constant $\overleftrightarrow{K}$, which is $\overleftrightarrow{K} = 1/\left(1 - N\overleftrightarrow{\alpha}_p/3\varepsilon_0\right)$, where N is molecular number density and ε_0 is the permittivity of free space ($\varepsilon_0 = 8.85 \times 10^{-12}$ F/m), Eq. (5.12) can be rewritten as

$$\vec{p} = \overleftrightarrow{\alpha}_p \overleftrightarrow{K} \vec{E}. \tag{5.13}$$

This results in a macroscopic polarization $\vec{P}$ as

$$\vec{P} = N\vec{p} = \overset{\leftrightarrow}{\chi_e}\vec{E} . \tag{5.14}$$

with $\overset{\leftrightarrow}{\chi_e} = N\overset{\leftrightarrow}{\alpha}_p \overset{\leftrightarrow}{K}$. The corresponding effective susceptibility tensor is obtained by averaging $\overset{\leftrightarrow}{\chi_e}$ such that [40]

$$\langle\overset{\leftrightarrow}{\chi_e}\rangle = \frac{N}{3}\begin{pmatrix} \alpha_{p,\perp}K_\perp(2-S)+\alpha_{p,\parallel}K_\parallel(1-S) & 0 & 0 \\ 0 & \alpha_{p,\perp}K_\perp(2+S)+\alpha_{p,\parallel}K_\parallel(1-S) & 0 \\ 0 & 0 & \alpha_{p,\perp}K_\perp(2-2S)+\alpha_{p,\parallel}K_\parallel(1+2S) \end{pmatrix}, \tag{5.15}$$

where $\parallel$ and $\perp$ indicate parallel and perpendicular components, respectively, of the corresponding quantity, that is, either $\overset{\leftrightarrow}{\alpha}_p$ or $\overset{\leftrightarrow}{K}$, with respect to director $\vec{n}$.

The total displacement flux $\vec{D}$ is equal to

$$\vec{D} = \varepsilon_0\left(1+\overset{\leftrightarrow}{\chi_e}\right)\vec{E} = \overset{\leftrightarrow}{\varepsilon_r}\vec{E} \tag{5.16}$$

Thus, the components of the complex relative permittivities, which are shown in Figure 5.6, can be calculated with Eqs. (5.15) and (5.16):

$$\underline{\varepsilon}_{r,\perp} = 1 + \frac{N}{3\varepsilon_0}\left(\alpha_{p,\perp}K_\perp(2+S)+\alpha_{p,\parallel}K_\parallel(1-S)\right) \tag{5.17}$$

and

$$\underline{\varepsilon}_{r,\parallel} = 1 + \frac{N}{3\varepsilon_0}\left(\alpha_{p,\perp}K_\perp(2-2S)+\alpha_{p,\parallel}K_\parallel(1+2S)\right)$$

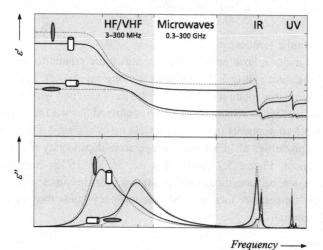

Figure 5.6 Frequency dependency of the real and imaginary part of the LC's permittivity. The permittivity of the molecules is given as dashed lines, while the permittivity of the directors is given as solid lines. The order parameter is chosen to be $S = 0.8$.

This results in the complex dielectric anisotropy

$$\Delta\underline{\varepsilon}_r = \underline{\varepsilon}_{r,\parallel} - \underline{\varepsilon}_{r,\perp} = \frac{N \cdot S}{\varepsilon_0}\left(\alpha_{p,\parallel}K_\parallel - \alpha_{p,\perp}K_\perp\right) \tag{5.19}$$

The dielectric anisotropy is proportional to the order parameter and can be either positive or negative. Nevertheless, all micro- and millimeter-wave optimized LC samples reported in literature exhibit positive dielectric anisotropy.

The relation between the director $\vec{n}$ and the corresponding relative permittivities $\varepsilon_{r,\parallel}$ and $\varepsilon_{r,\perp}$ is shown in Figure 5.6. Starting from the diagonal material tensor of an LC director parallel to one of the Cartesian axes, for example, $\vec{n} = (0,0,1)$ according to Figure 5.4, the permittivity tensor is given as

$$\overleftrightarrow{\underline{\varepsilon}}_r = \begin{pmatrix} \varepsilon_{r,\perp} & 0 & 0 \\ 0 & \varepsilon_{r,\perp} & 0 \\ 0 & 0 & \varepsilon_{r,\parallel} \end{pmatrix}. \tag{5.20}$$

5.2.1.3 Frequency and Temperature Dependency of the Permittivity

Like in almost all materials, the LC material is subject to dispersion mechanisms as well, which are reflected in both the real and the imaginary part of the complex relative permittivity $\underline{\varepsilon}_r = \varepsilon' - j\varepsilon''$. The frequency dependent components can be calculated using the Debye model, which includes the relaxation constant τ_{rel} of the material [100]

$$\varepsilon'(\omega) = \varepsilon(\infty) + \frac{\varepsilon(0) - \varepsilon(\infty)}{1 + (\omega\tau_{\text{rel}})^2}$$

$$\varepsilon''(\omega) = \omega\tau_{\text{rel}}\frac{\varepsilon(0) - \varepsilon(\infty)}{1 + (\omega\tau_{\text{rel}})^2}, \tag{5.21}$$

where ω is the angular frequency.

There are two major relaxation mechanisms, which are related to the molecules' rotation around the short axis ($\tau_{\text{rel},2}$), as well as the long axis ($\tau_{\text{rel},1}$). In the frequency range from 1 GHz on, the rotation of the molecules has failed and will not lead to any further dielectric loss. Measurement results have confirmed that the LC can reasonably be used for tunable components from about 1 GHz up to at least 8 THz [15, 33].

The dielectric properties of LCs feature temperature dependency like most materials. As an example, Figure 5.7 shows that $\Delta\varepsilon_r$ in Eq. (5.19) decreases as the temperature increases, because molecules' positional order becomes weaker and then disappears as the temperature increases. Similar tendency holds true for microwave loss of the LC material.

5.2.2 The Elastic Continuum Theory of LCs

In light of the previous part, LC materials exhibit different polarizability α_p, and therefore, different macroscopic relative permittivity $\underline{\varepsilon}_r$, depending on the angle

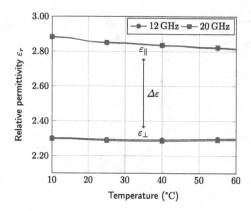

Figure 5.7 Exemplary temperature dependency of the relative permittivities of an LC mixture.

between the director $\vec{n}$ and the electric field $\vec{E}$. For instance, in operation of a tunable capacitor or tunable transmission line, the LC material is exploited as a tunable dielectric by changing the orientation of the LC directors $\vec{n}$ due to an external static field between the biasing electrodes. Hence, it is essential to discuss possible LC orientation mechanisms, which is explained by means of total free energy in an LC bulk material.

The LC material in a tunable device is modeled commonly by using the continuum theory [40, 96, 101]. In an LC bulk, the system's energy is defined based on the Gibbs free energy W_f, which is affected from the elastic deformations, external electromagnetic fields and anchoring effects. Thus, an increase in the Gibbs free energy is minimized by changing the orientation of the director, in order to reach an equilibrium state, resulting in a change in the relative permittivity.

If no electric or magnetic field is present, the LC molecules tend to align in parallel to each other, in which they minimize the Gibbs free energy. The nonparallel LC molecules are classified into certain types of elastic deformations as shown in Figure 5.8. These deformations are *splay*, *twist*, and *bend* [102]. Each of those elastic deformations leads to an increase of the Gibbs free energy, which is again released by molecule reorientation. This reorientation can be modeled using the Frank–Oseen equation of the free energy, which is given for the system per unit volume as [96, 98]

$$df_{\text{elastic}} = df_{\text{splay}} + df_{\text{twist}} + df_{\text{bend}}$$

$$= \frac{1}{2}K_{11}\left(\nabla \cdot \vec{n}\right)^2 + \frac{1}{2}K_{22}\left[\vec{n} \cdot \left(\nabla \times \vec{n}\right)\right]^2 + \frac{1}{2}K_{33}\left[\vec{n} \times \left(\nabla \times \vec{n}\right)\right]^2 \quad (5.22)$$

where K_{11}, K_{22}, and K_{33} are measurable constants for splay, twist, and bend deformations according to [50]. However, this approach assumes a continuous director field, that is, the change of director orientation is spatially much larger than the molecules length. Hence, it is not possible to model defects in the director field with an isotropic core properly. For such cases, a more accurate model would be the Landau–de Gennes model, which is quite complicated in turn.

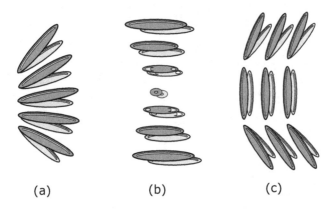

(a)   (b)   (c)

Figure 5.8 Elastic deformation of LC molecules: (a) splay, (b) twist, and (c) bend.

All types of deformation induce torque, which also affects its neighboring molecules. The minimum deformation energy is reached when all molecules are in parallel to each other, neglecting any boundary conditions. The elastic constants are dependent on the LC being used and the temperature, and therefore, the order parameter S need to be determined separately. The elastic constants are proportional to S^2 [40]. Usually it is sufficient to consider splay, twist, and bend to calculate the director orientation. In some cases, where the surfaces are large in comparison to the volume, additional surface terms need to be considered [103].

Similar to the elastic deformations, an external electromagnetic field increases the energy of the system. In the presence of an external electric field $\vec{E}$, the energy per unit volume is given by [14, 54, 98]

$$df_{\text{electric}} = -\frac{1}{2}\vec{D} \cdot \vec{E} = -\frac{1}{2}\varepsilon_0 \left(\left\langle \vec{E}^2 \right\rangle \varepsilon_{r,\perp} + \Delta\varepsilon_r \left\langle \left(\vec{n} \cdot \vec{E} \right)^2 \right\rangle \right). \qquad (5.23)$$

In analogy, in case of an external magnetic field $\vec{H}$, the energy per unit volume is given by [15, 54]

$$df_{\text{magnetic}} = -\frac{1}{2}\vec{B} \cdot \vec{H} = -\frac{1}{2}\mu_0 \left(\left\langle \vec{H} \right\rangle^2 \mu_{r,\perp} + \Delta\mu_r \left\langle \left(\vec{n} \cdot \vec{H} \right)^2 \right\rangle \right), \qquad (5.24)$$

where μ_0 is the permeability of free space and μ_r is the relative permeability, respectively.

Additionally, any anchoring forces occurring on the boundaries have impact on the system's energy. As Being a liquid substrate, LCs have to be filled into cavities, which are sealed after the cavity is filled with LC. Any physical impurities on the boundaries of the cavity, such as polished copper surfaces, result in interaction energy between the boundary and the LC molecules, reorienting the LC molecules. According to the Rapini–Papoular model, this interaction energy density is given as [26, 40]

$$df_{\text{surface}} = \frac{1}{2}W_p(\sin(\theta - \theta_0))^2 + \frac{1}{2}W_a(\sin(\phi - \phi_0))^2, \qquad (5.25)$$

where W_p and W_a are polar and azimuthal anchoring strengths, respectively; (θ, ϕ) specify the director $\vec{n}$; and (θ_0, ϕ_0) specify the preferred alignment direction for the polar and azimuthal angles, respectively, which are defined by the anchoring force [26]. Therefore, the total free energy W_f, which is minimized by reorienting the LC molecules, is the integral for the entire volume of the considered LC bulk:

$$W_f = \iiint \left[df_{\text{elastic}} + df_{\text{electric}} + df_{\text{magnetic}} + df_{\text{surface}} \right] dV. \tag{5.26}$$

5.2.3   Orientation Mechanisms of LCs and Biasing Schemes

5.2.3.1   Prealignment by Surface Anchoring

The LC molecules can be aligned either in parallel (homogeneous alignment) or perpendicular (homeotropic alignment) to the boundary [40]. As can be seen in Eq. (5.25), df_{surface} can be minimized for an equilibrium state of the molecules if the terms $\sin(\theta - \theta_0)$ and $\sin(\phi - \phi_0)$ are equal to zero. Hence, in the presence of the surface anchoring, the director $\vec{n}$ is oriented in such a way that θ_0 and ϕ_0 become equal to θ and ϕ to minimize df_{surface}.

Similar to LCDs, in planar LC-based micro- and mm-wave devices use is made of a spin-coated, cured polyimide film, for example, Nylon 6 [14], having a thickness typically between 50 nm and 100 nm, mechanically rubbed with a velvet cloth as shown in Figure 5.9.

The result of this rubbing is a microscopic grooving of the polyimide surface, which anchors the LC molecules in the vicinity of this thin layer along the rubbing direction in parallel to the surface. Accompanied with the tendency of LC molecules to align in parallel to each other, this concept can be principally used for preorientation of the LC molecules in a preferred direction, before biasing with an external electromagnetic field to change the orientation state of the LC molecules. This prealignment layer is particularly useful in planar structures with thin LC layer heights h_{LC}, typically below 150 μm, since with higher LC layers more LC molecules are no longer in parallel to the surface, because of decreasing anchoring forces, thus decreasing the orientation effectivity. Moreover, the response times of LCs are directly correlated with the LC layer heights h_{LC}, as described in the introduction. Both the

Figure 5.9 Schematic of the prealignment layer realization process. The LC molecules are anchored in the direction of the grooves, applied by rubbing the polyimide film with a velvet cloth.

thickness of the prealignment layer and the deepness of the grooves affect the device performance [14, 26].

5.2.3.2   LC Tuning by Electromagnetic Fields and Basic Biasing Concept

LC tuning, that is, the orientation of the LC molecules, can also be controlled with an external magnetic $\vec{H}_b$ or electric $\vec{E}_b$ biasing fields. In the presence of a field, the total free energy W_f in Eq. (5.26) is minimized, if the absolute value of df_{electric} or df_{magnetic} is maximized. It should be noted that both $\Delta\varepsilon_r$ and $\Delta\mu_r$ are positive for mm-wave optimized nematic LCs, and therefore both df_{electric} and df_{magnetic} are negative. To maximize their absolute values, the director is reoriented in parallel to the field lines of $\vec{H}_b$ or $\vec{E}_b$, in order to maximize the result of the dot product ($\cdot$) between $\vec{n}$ and $\vec{E}$ or $\vec{H}$ in Eq. (5.23) or (5.24), respectively.

To induce a magnetic biasing field $\vec{H}_b$, magnets can be placed as parallel plates around the LC structure or LC component. In principle, magnetic fields are usable but impractical because of the high-power consumption and large space necessary for the generation of magnetic field strengths of 0.2 T to 0.7 T, which are required, depending on the size and shape of the LC structure or component. If using permanent magnets, a mechanical reorientation of the magnets would be required to reorient the LCs within a component, which is not viable. Therefore, to tune LC-based micro- and millimeter-wave components in applications, only electrical biasing is feasible. However, magnetic biasing can be efficiently utilized either for characterization purposes or for a fast proof-of-concept and first functional tests of LC-tuned structures and components, initially without any electrical biasing network to reduce implementation efforts at the beginning.

The basic electrical biasing concept can be explained exemplary by means of a simple LC-filled parallel-plate capacitor as sketched in Figure 5.10, where the

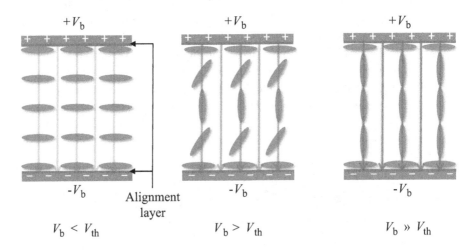

Figure 5.10 Schematic of an electrically biased parallel plate capacitor filled with LC. An alignment layer on the plates is assumed to prealign the LC molecules initially in parallel to the surface.

orientation of the LC directors $\vec{n}$ of all LC unit-volume elements within the parallel-plate capacitor is given for three snapshots of different biasing voltages V_b and where a thin, rubbed polyimide film is assumed on the inner surfaces of both plates.

Before biasing, the LC molecules are prealigned in parallel to the plates due to surface anchoring by the rubbed polyimide film as described Section 5.2.3.1. When a biasing voltage V_b (typically $\approx$ at 1 kHz) is applied to the capacitor, an electric biasing field $\vec{E_b}$ is generated inside, which is perpendicular to the initial state of the LC directors $\vec{n}$. A change in orientation of the LC directors $\vec{n}$ from parallel to the surfaces occurs only after V_b exceeds a certain threshold voltage V_{th}. In thin structures, the splay deformation is dominant, resulting in a threshold voltage of

$$V_{\text{th}} = \pi \sqrt{\frac{K_{11}}{\varepsilon_0 \Delta \varepsilon_{r,\,1\text{kHz}}}} \qquad (5.27)$$

This threshold phenomenon is well known as the *Fréedericksz transition* [14, 40, 53, 98]. It defines the voltage where the electric and elastic forces of the surface anchoring are in an equilibrium. According to Eq. (5.27), the threshold voltage for planar structures is proportional to the square root of the elastic constant for the splay deformation K_{11} over the anisotropy $\Delta \varepsilon_{r,\,1\text{kHz}}$ at the biasing frequency. It differs for the various LC mixtures, but is typically around 1 V.

If the voltage is further increased $V_b > V_{\text{th}}$, the electric torques exceed the elastic ones and the orientation of most LC directors $\vec{n}$ starts to align more and more toward the orientation of the field lines, until nearly all LC directors $\vec{n}$ are oriented along the field lines of $\vec{E_b}$. At this "maximum" biasing voltage $V_b = V_{\text{sat}}$, LC directors $\vec{n}$ are almost in parallel to the tuning field $\vec{E_b}$.

When an RF field $\vec{E}_{RF}$ is applied at the aforementioned parallel-plate capacitor, it will experience different relative permittivities ε_r, and of course, different dielectric losses $\tan \delta$, depending on the changing orientation of the LC directors $\vec{n}$ with respect to the fixed orientation of the electric field vector $\vec{E}_{RF}$, which is vertically polarized as the tuning field $\vec{E_b}$ in Figure 5.10. Initially, when no bias voltage is applied between both plates, the LC directors $\vec{n}$ are oriented in parallel to the surfaces, that is, by default perpendicularly aligned to the RF field $\vec{E}_{RF}$. Hence, the RF field experiences a relative permittivity of $\varepsilon_{r,\perp}$ and loss tangent $\tan \delta_\perp$. With increasing biasing voltage, the LC directors $\vec{n}$ start to align more and more toward the orientation of the field lines of $\vec{E_b}$, that is, in parallel to $\vec{E}_{RF}$. Hence, $\vec{E}_{RF}$ experiences a material's effective relative permittivity, which varies from $\varepsilon_{r,\perp}$ continuously toward $\varepsilon_{r,\parallel}$, accompanied by a change in the loss tangent from toward $\tan \delta_\parallel$. Beyond this saturation state $V_b > V_{\text{sat}}$, no change in permittivity can be observed anymore [14]. Thus, varying ε_r by changing the orientation of LC directors $\vec{n}$ due to the bias field $\vec{E_b}$, the capacitance of the parallel-plate capacitor can be tuned, building up a tunable LC varactor. This principle also applies to tuning the center frequency and the bandwidth of LC-based filters or to alter the propagation constant of a wave traveling along a line and hence to vary the differential phase shift of the line with a certain length, representing a simple variable delay-line phase shifter.

5.2.3.3   Electrical Biasing Schemes for MLC Devices

For the realization of LC-tunable components, the LC must be embedded into the micro- or mm-wave structure or component, for example, filled into a cavity of a resonator, transmission line or waveguide, or it can be directly integrated into an antenna element. In contrast to optical applications, where LC is rotating the polarization of the light, in RF components, the material's orientation within the LC-filled structure is changed with respect to the applied electrical RF field $\vec{E}_{RF}$ to tune its effective relative permittivity ε_r and hence to change the propagation constant. In general, two different electrical biasing schemes are proposed for LC-tunable devices.

1. A hybrid biasing scheme (classical LCD approach), which uses a prealignment layer to preorient the LC molecules in parallel to the surface by anchoring forces similar to the LCD technology, and simultaneously, an electrostatic biasing field $\vec{E}_b$ between the electrodes to tune the LC directors as for the parallel-plate capacitor in Figure 5.10 [15, 26, 54]. This basic biasing concept is feasible only for low-profile (thin, flat), planar structures with an LC layer thickness typically below 150 µm, mainly for two reasons:
    1. The tunability decreases with increasing LC layer thickness h_{LC}, since more and more LC molecules are not perfectly prealigned in parallel to the surface due to lower anchoring forces.
    2. The response times increase with thickness h_{LC}^2, which is particularly critical for the switch-off time t_{off}, where it takes much longer to reorient the LC molecules into the initial state in parallel to the surface after releasing the bias voltage.
2. If the LC layer height h_{LC} becomes too large for a reasonable utilization of a prealignment layer for the aforementioned reasons, for example, for bulky or voluminous metallic or dielectric waveguide-based components with LC layer heights of several millimeters or more, only a fully or all-electrical biasing scheme is feasible, using multiple pairs of electrodes without any prealignment layer [78, 79]. Figure 5.11 exhibits an example of three pairs of electrodes to provide a quadrupole biasing field for tuning the LC, filled into a Rexolite container, embedded inside a metallic rectangular waveguide. By varying the biasing voltages, the biasing field can be changed in both orientation and field strength.

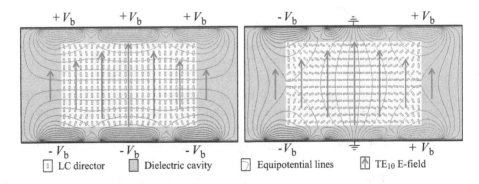

Figure 5.11 Schematic of the fully electric biasing scheme for a metallic rectangular waveguide with a Rexolite container as an LC-filled cavity.

5.3   Microwave Characterization of LCs

Linked with the fundamentals and studies on microwave LCs, there have been intensive investigations in parallel on appropriate characterization methods at different frequencies. A main principle is to determine material parameters by fitting measured S-parameters to modeled ones by using extraction algorithms either point-by-point or multipoint methods [104]. Because of its practicality, the Nicolson–Ross–Weir (NRW) [105, 106] method, which is a point-by-point extraction technique, is frequently utilized to determine initial guesses for multipoint techniques. NRW works properly except for the frequencies where the physical length of the transmission line is equal to multiple of the half wavelength. At these frequencies, extracted results diverge, because the phases of reflection coefficients cannot be accurately measured, especially for low-loss materials [104]. The multipoint techniques, on the other hand, avoid unphysical results, since they allow correlations between contiguous frequency points. They require optimization algorithms such as nonlinear least-square [104] or Newton–Raphson iteration [15]. For the characterization of LC materials, electrostatic or magnetostatic fields are utilized to align the LC molecules to a certain direction, in order to characterize the material at this orientation as an isotropic and homogeneous material. When the LC directors are parallel or orthogonal to the RF electrical field of the characterization setup, the corresponding complex relative permittivity is $\varepsilon_{r,\parallel}$ or $\varepsilon_{r,\perp}$, respectively. Some material characterization methods and setups are briefly summarized in the text that follows.

In [88], a technique is presented for quantifying the microwave permittivities of small quantities of nematic LC material filled in a 75 μm gap between two 3 mm thick aluminum plates. When a voltage is applied between the two plates, the LC realigns and the shift of the resonant peaks gives the anisotropic permittivities. In [107] a microwave dielectric resonator-based technique was introduced for LC characterization from 13 to 15 GHz to extract the permittivity values and dielectric loss tangents. In [108] dielectric properties of highly anisotropic nematic LCs have been measured by the split-post dielectric resonator technique. The measurement cell was composed of two parallel high-resistivity silicon wafers (silicon transducer) filled with LC at around 5 GHz.

In [48] use is made of the well-known cavity perturbation method for material characterization, that is, to extract the complex permittivity and permeability of novel LC samples. As a measuring setup, two rectangular waveguide resonator cavities are designed for a TE_{102}-mode at 9 GHz and 38 GHz with small holes to insert empty and LC-filled PTFE tubes, respectively. Measurements were carried out over a temperature range from $-25°C$ to $+45°C$. External magnets were used to orient the LC molecules parallel and perpendicular to the RF field. Soon thereafter, a broadband characterization method was introduced in [49], using a temperature-controlled coaxial transmission line to characterize LCs between 360 MHz and 23 GHz and in a temperature range between $+7°C$ and $+115°C$. Similar measurements of temperature-dependent properties of GT3-23001 and K15 were done with a similar broadband setup in [109, 110] from $+10°C$ to $+60°C$ at 20 GHz. Besides electromagnetic properties in the frequency range between 26–40 GHz [49] and 65–110 GHz [71], the response time of a Merck's first-generation LC has been considered for phase shifter applications. The same author presented in [111] a setup for W-band characterization of LCs by measuring the reflection parameter of the 10 mm long standard WR-10 rectangular waveguide terminated by a short circuit, using a model based on two transverse electromagnetic (TEM) transmission lines with the LC and air, respectively, to extract the LC's complex permittivity. The LC is oriented with a homogeneous magnetostatic field.

To measure the complex permittivity at microwaves with very high accuracy and perfect reproducibility, a new measurement system was designed and built up by TU Darmstadt, which is presented in detail in [23, 80, 112]. The design is based on the cavity perturbation method described in [48] with several improvements to extract the complex permittivity tensor by formulating the Maxwell equations as an eigen-susceptibility problem of the considered sample [113, 114] and by applying the variational approach to a triple-mode perturbation method [115]. This offers a direct solution of the desired material parameters by utilizing adequate and well-proven numerical techniques such as the finite element method (FEM). The unloaded Q of the cylindrical aluminum cavity, containing a small amount of LC in a quartz tube, is around 4200, enabling the measurements of very low dielectric losses of LCs. The LC orientation is controlled by means of static magnetic field of 0.35 T, generated by permanent magnets.

Other characterization setups and measurements at microwaves can be found at [58, 116–121]. Standard LCs such as K15 and E7 as well as novel highly birefringent, low-loss LCs were also characterized at terahertz frequencies from 0.1 to 0.35 THz with a free-space continuous-wave (CW) terahertz system [122], from 0.15 to 1 THz in [123], from 0.2 to 2.5 THz in [124], and 0.3 to 1.5 THz in [33], and soon thereafter, from 0.3 to 8 THz in [80] by using a terahertz time-domain spectroscopy (TDS) and a Fourier-Transform Interferometry (FITR) setup.

The following three subsections will focus only on the narrowband cavity perturbation method and the broadband coaxial transmission line method, both briefly summarized earlier, and on terahertz free-space material characterization. Finally, the last section will sumarize the characterization results of some selected microwave LC mixtures.

5.3.1   Narrowband Cavity Perturbation Method

Among a great variety of methods, narrowband (i.e., resonant) material characterization has been used for decades to obtain very precise and reliable data on permittivity $\underline{\varepsilon} = \varepsilon_0 \left(\varepsilon' - j\varepsilon'' \right)$, that is, materials' relative permittivity $\varepsilon_r = \Re \left(\underline{\varepsilon} \right)$ and loss tangent $\tan \delta = \Im \left(\underline{\varepsilon} \right) / \Re \left(\underline{\varepsilon} \right)$. A particularly common and elegant approach is the so-called cavity perturbation method (CPM), where a resonant cavity is perturbed by a small sample of material. If both the cavity's and the sample's geometry are known precisely, the shift of the resonance frequency and the change in quality factor of the response curve provide extremely sensitive and precise data.

Table 5.2 Overview of applicability of the approaches by Berk and Rumsey for different classes of materials

Material property	Berk [125]	Rumsey [126]
Loss-free isotropic	Yes	Yes
Loss-free anisotropic	Yes	Yes
Gyrotropic	Yes	No
Lossy	No/Yes	Yes

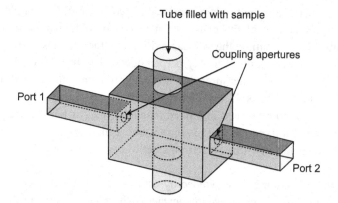

Figure 5.12 Schematic of the cavity perturbation setup.

Solutions for isotropic materials have been around since the middle of the twentieth century. In 1956 Berk introduced a very general approach: the variational principle for electromagnetic resonators and waveguides [125]. Rumsey had introduced the reaction concept in 1954 [126]. Both methods were frequently used to determine the properties of materials. The applicability of both principles is listed in Table 5.2. Since then, both methods have been extended or amended.

Another aspect of the cavity perturbation method is unintended perturbation. The simplest model for a cavity does not account for problems of sample insertion and coupling. Thus, as a general rule, all CPM approaches need to account for the frequency shift due to material properties and cavity apertures for sample insertion and coupling as shown in Figure 5.12.

In this section, the problem of characterizing liquid crystals is treated, involving lossy anisotropic material. Therefore, Berk's variational approach holds only with these modifications. Furthermore, the material is liquid, which implies an intended perturbation due to the material under test (MuT), but also a perturbation due to a solid container, usually a PTFE or quartz tube. For reasons of practicability, the sample is inserted into the cavity through holes, thus avoiding frequent opening and closing of the resonator.

Common calamitic, nematic liquid crystals have only one optical axis. Thus, if the permittivity is regarded along this axis, it shows a parallel value $\varepsilon_{r,\parallel}$. The permittivity does not depend on the orientation perpendicular to the optical axis, that is, the tensor contains two components with $\varepsilon_{r,\perp}$. Hence, the anisotropic permittivity of liquid crystals can generally be described by a permittivity tensor

$$\overset{\leftrightarrow}{\varepsilon}_r = \begin{pmatrix} \varepsilon_{r,\perp} & 0 & 0 \\ 0 & \varepsilon_{r,\perp} & 0 \\ 0 & 0 & \varepsilon_{r,\|} \end{pmatrix} \tag{5.28}$$

represented in this form when the coordinate system's z-axis is in line with the director of the material.

In a cavity where only the fundamental mode is considered, the parallel and the perpendicular component of the material tensor can be obtained by two distinct measurements: first orienting the optical axis along the mode's electrical field component and then changing it to any perpendicular direction. The advantage of this approach is that the measurement accuracy remains unchanged for both values. The main drawback is the fact that in order to change the orientation, either the cavity has to be turned in a fixed magnetic field or a magnetic yoke has to be turned around a fixed cavity [13–15, 48, 66]. Either way, mechanical action is involved to determine the electromagnetic parameters in both principal directions.

This mechanical action can be avoided by using a dual-mode or even triple-mode cavity [23, 80, 115]. If the resonance frequencies of the respective two or three modes are sufficiently far apart from each other, two or three independent measurements can be carried out at the same time without the need to turn the cavity with respect to the magnetic field. However, the accuracies of the result now depend on the respective mode.

5.3.1.1   Modeling Cavities for Material Parameter Extraction

All analytical, semianalytical or purely numerical approaches require a precise knowledge of the resonator geometry. If the material is isotropic, there is no constraint on which mode to use in a cavity apart from the fact, that in general, the Q factor of the unloaded mode should be as high as possible. For instance, a real 30 GHz setup attains an unloaded Q factor of more than 7000 [14], which can be increased further, depending on the material. The detuning due to the perturbation by the sample should be such that it is small enough so that the mode remains well separated from neighboring modes but at the same time reasonably large to be able to resolve the permittivity as finely as possible. In the anisotropic case, these constraints compete for each tensor component. The tensor in Eq. (5.28) may contain three individual components on its diagonal, and hence the problem will be to arrange three resonant modes in such a way that they have good separation, high sensitivity, and a high Q factor each.

For the sake of simplicity, a cylindrical cavity for the determination of two tensor components (uniaxial materials) in accordance to Eq. (5.28) shall be discussed here. A triple mode approach that may provide data for up to three individual components provides a good insight [23, 115].

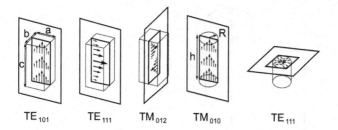

Figure 5.13 Electric fields for rectangular and cylindrical cavities.

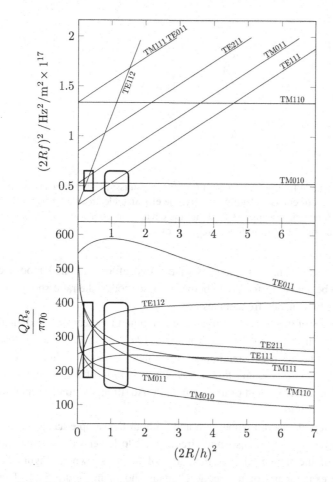

Figure 5.14 Mode chart for cylindrical cavities with varying ratios of radius to height, R/h. The top graph indicates normalized resonance frequencies and the bottom graph the normalized Q factors. The boxes indicate preferable locations for characterization applications [80].

For the design of a cavity, the mode chart is a handy tool. One can choose the two lowest, orthogonal modes TM_{010} and TE_{111}. Their electric fields for rectangular and circular cavities are shown in Figure 5.13. Figure 5.14 depicts the mode chart for the lowest 11 modes of a cylindrical cavity versus $\left(\dfrac{2R}{h}\right)^2$. At a ratio of radius to height of

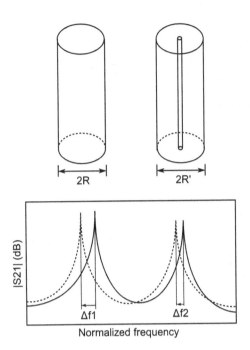

Figure 5.15 Schematic for the frequency shift of the transmission coefficients once the electrical field is disturbed: for empty cylindrical cavity, the effective electrical diameter is $2R$ (left side) and loaded by a material sample inserted into the cavity, it becomes $2R' > 2R$. As a result, the transmission coefficients shift to lower frequencies.

$R/h = 0.5$, the resonance frequencies of the two aforementioned modes coincide. Therefore, to be able to differentiate between these modes, the ratio should be adapted to get different resonance frequencies.

The Q factors of the respective modes are obtained by means of analytical expressions available in the literature, for example, [127, 128]. They are the lowest with respect to their higher-order neighbors. Nevertheless, they are chosen because they end up in the same range, and thus allow determination of the material losses with similar accuracy, independent of the orientation. Additionally, fundamental modes are comparatively easy to excite.

In a next step, the sample tube with and without the sample has to be considered. Inserting the tube shifts the resonance frequencies to lower values, since the relative permittivity of the tube is higher than that of air as shown in Figure 5.15. This modifies the exact shape of the mode chart as shown in Figure 5.14. This can be modeled, for example, using Berk's or Rumsey's approaches.

5.3.1.2   Ansatz and Numerical Extraction

Berk has presented variational expressions suited to calculate the resonance frequencies of virtually arbitrarily perturbed modes. If the cavity has perfectly conducting walls, the Eqs. (5.29) and (5.30) provide the resonance frequency for the modes:

$$\omega^2 = \frac{\iiint_V \left(\nabla \times \vec{E}^* \right) \cdot \overleftrightarrow{\mu}^{-1} \cdot \left(\nabla \times \vec{E} \right) dV}{\iiint_V \vec{E}^* \cdot \overleftrightarrow{\varepsilon} \cdot \vec{E} \ dV} \tag{5.29}$$

where $\vec{E}$ is a field satisfying Maxwell's equations in the cavity with volume V, $\vec{E}^*$ is the complex conjugate of $\vec{E}$, and $\overleftrightarrow{\varepsilon}$ and $\overleftrightarrow{\mu}$ are the Hermitian permittivity and permeability tensors, respectively. The superscript signifies that the inverse of the permeability tensor is taken into account here. If the tensors are oriented along the main axes, that is, the tensor is empty except the diagonal elements, the inverse is computed by calculating the inverse of the individual tensor diagonal entries. Another expression was pointed out, which is more general, since it allows for not perfectly conducting walls of the resonator cavity:

$$\omega = j \frac{\iiint_V \vec{E}^* \cdot \left(\nabla \times \vec{H} \right) \mp \vec{H}^* \cdot \left(\nabla \times \vec{E} \right) - \vec{E}^* \cdot \sigma \cdot \vec{E} \ dV - \oiint_{\partial V} \vec{n} \cdot \left(\vec{E} \times \vec{H}^* \right) d(\partial V)}{\iiint_V \vec{E}^* \cdot \overleftrightarrow{\varepsilon} \cdot \vec{E} - \vec{H}^* \cdot \overleftrightarrow{\mu} \cdot \vec{H} \ dV},$$

$$\tag{5.30}$$

where in addition to the aforementioned conventions, σ is the conductivity of any dielectric (a Hermitian tensor) and ∂V is the boundary of the volume V with $\vec{n}$ the inward-directed normal vector on the surface ∂V. For a lossless cavity with lossless dielectric, ω is purely real. If the cavity is nonperfectly conducting or lossy dielectrics are involved (i.e., $\sigma \neq 0$), the imaginary part of ω contains information about the Q factor of the (loaded) cavity.

Using these expressions, the mode chart mentioned earlier can now be refined to depict the real situation. In a first refinement, the sample tube can be taken into account. The expressions for the respective modes can be obtained using Maxwell's equations, cylindrical symmetry, and the continuity constraints at the dielectric boundaries:

$$\left(\frac{d^2}{d\rho^2} + \frac{1}{\rho} \frac{d}{d\rho} + \frac{1}{\rho^2} \frac{d^2}{d\varphi^2} \right) E_z + k^2 E_z = 0$$
$$\left(\frac{d^2}{d\rho^2} + \frac{1}{\rho} \frac{d}{d\rho} + \frac{1}{\rho^2} \frac{d^2}{d\varphi^2} \right) H_z + k^2 H_z = 0 \tag{5.31}$$

$$E_z(\rho, \varphi) = P_E(\rho) \cdot \Phi_E(\varphi) \text{ and } H_z(\rho, \varphi) = P_H(\rho) \cdot \Phi_H(\varphi)$$

$$\text{with } \Phi_E(\varphi) = E_0 \cdot e^{-jm\varphi} \text{ and } P_E(\rho) = A \cdot J_m(k_e \rho) + B \cdot N_m(k_e \rho)$$

$$\text{and } \Phi_H(\varphi) = H_0 \cdot e^{-jm\varphi} \text{ and } P_H(\rho) = C \cdot J_m(k_h \rho) + D \cdot N_m(k_h \rho) \tag{5.32}$$

where k_e and k_h model the influence of the permittivity with $k_e = \omega^2 \varepsilon_\perp \mu - \beta^2$ and $k_h = \left({\varepsilon_\parallel}/{\varepsilon_\perp} \right) \cdot (\omega^2 \varepsilon_\perp \mu - \beta^2)$, and J_m and N_m are the Bessel and Neumann functions of m-th order, respectively. They also contain the axial resonance condition, as resonance is obtained if the length of the resonator h is a multiple p of half the wavelength, that is, if $\beta = \pi p / h$. $P_E(\rho)$ and $P_H(\rho)$ piecewise continuous on three domains: $0 \leq \rho < R_1$ for the sample, $R_1 \leq \rho < R_2$ for the quartz tube and $R_2 \leq \rho \leq R_3$ for the remaining air-filled resonator. Their continuity is enforced by the continuity of the tangential electric field and the tangential magnetic field (permeability $\mu_r = 1$ in all three materials). At the boundaries at the radii R_1, R_2, and R_3 all six field components are subject to constraints. B and D must be zero in the inner domain due to the pole of Neumann's function at $\rho = 0$.

The two modes are now treated separately with the same prototype functions. In the case of the TM_{010} mode, the electric field is only in z, that is, extraction of the permittivity tensor's z-component is possible. If nonperturbed fields (trial fields; cf. Berk [125]) are assumed, the z-component of the permittivity tensor, $\varepsilon_{z,\text{tf}}$, for the sample volume is obtained by expressing $\varepsilon_{z,\text{tf}}$ in terms of the known parameters:

$$\varepsilon_{z,\text{tf}} = \frac{\dfrac{\lambda_1^2}{\omega_{\text{res}}^2 \mu} \cdot J_1(\lambda_1)^2 + (\varepsilon_2 - \varepsilon_2) \cdot R_1^2 \left(J_0\left(\dfrac{R_1\lambda_1}{R_3}\right)^2 + J_1\left(\dfrac{R_1\lambda_1}{R_3}\right)^2 \right) - \varepsilon_3 R_3^2 \cdot J_1(\lambda_1)^2}{R_1^2 \left(J_0\left(\dfrac{R_1\lambda_1}{R_3}\right)^2 + J_1\left(\dfrac{R_1\lambda_1}{R_3}\right)^2 \right)},$$

(5.33)

with $J_0(\lambda_1) = 0$, the first zero of Bessel's function of zeroth order.

The value obtained with Eq. (5.33) is not the exact result for the perturbed field, because the perturbation causes the field to be more "concentrated" in the sample and tube volume. For the TM_{010} mode, the true solution $\varepsilon_{z,\text{LC}}$ is confined in the interval

$$\frac{\varepsilon_{z,\text{tf}} + 1}{2} \leq \varepsilon_{z,\text{LC}} \leq \varepsilon_{z,\text{tf}}.$$

(5.34)

The foregoing closed form equation for $\varepsilon_{z,\text{tf}}$ can be obtained because the system reduces to a transcendental equation due to the fact that TM_{010} implies $\beta = 0$:

$$0 = \frac{-E \cdot J_0(k_{e,2}R_1) + F \cdot J_1(k_{e,2}R_1)}{E \cdot Y_0(k_{e,2}R_1) + F \cdot Y_1(k_{e,2}R_1)} + \frac{G \cdot J_1(k_{e,2}R_2) + H \cdot J_0(k_{e,2}R_2)}{G \cdot Y_1(k_{e,2}R_2) + H \cdot Y_0(k_{e,2}R_2)},$$

(5.35)

where

$$E = k_{e,1} \cdot (k_{h,2})^2 \cdot \varepsilon_{z1} \cdot J_1(k_{e,1}R_1)$$
$$F = k_{e,2} \cdot (k_{h,1})^2 \cdot \varepsilon_{z2} \cdot J_0(k_{e,1}R_1)$$
$$G = k_{e,2} \cdot (k_{h,3})^2 \cdot \varepsilon_{z2} \cdot (Y_0(k_{e,3}R_2) \cdot J_0(k_{e,3}R_3) - Y_0(k_{e,3}R_3) \cdot J_0(k_{e,3}R_2))$$
$$H = k_{e,3} \cdot (k_{h,2})^2 \cdot \varepsilon_{z3} \cdot (Y_0(k_{e,3}R_3) \cdot J_1(k_{e,3}R_2) - Y_1(k_{e,3}R_2) \cdot J_0(k_{e,3}R_3)).$$

(5.36)

This can then be transformed into Eq. (5.35) by using the variational expression in Eq. (5.29). Once the z-component of the permittivity tensor is calculated, the same procedure is applied to obtain the other components using the TE_{111} mode.

The influence of the aperture through which the sample tube is inserted into the cavity can be modeled by using the approach of Bethe [129]. It is not easily modeled in an analytical way because the boundaries of the domain are altered from a simple cylindrical structure. A modal expansion of the field is necessary to account for the deformation of the field due to the aperture.

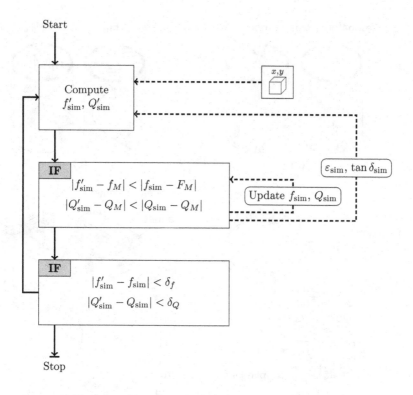

Figure 5.16 Flow chart for a precise and fast extraction algorithm.

Because it is small, it behaves like a waveguide below cutoff frequency. The field is therefore purely evanescent – no wave propagation into the filling aperture takes place. The idea is therefore to model the evanescent field like an electrostatic field using the conformal mapping technique.

At last, the perturbation induced by the coupling apertures is modeled using the formulas of Gao [130]. Their influence is small, because the electric and magnetic field in the sample volume is only very slightly deformed by the coupling apertures. The resulting fields are computed numerically in both cases (filling and coupling apertures, respectively) and the resonance frequency is obtained. A fundamental and easily implemented algorithm is shown in Figure 5.16. It does not provide best convergence properties, but previous work [23] has shown that this approach provides precise results in a short time.

5.3.1.3   Measurement Setup and Results

The result for the TM_{010} mode is analytically rigorous. Numerically it is exact within a chosen deviation in frequency δ_f and quality factor δ_Q. Apart from these numerical errors, knowledge of the geometry is essential. Typical fabrication tolerances are in the range of tens of micrometers. From a simulation and a reference measurement, the frequency shift due to material insertion can be estimated. Therefore, three

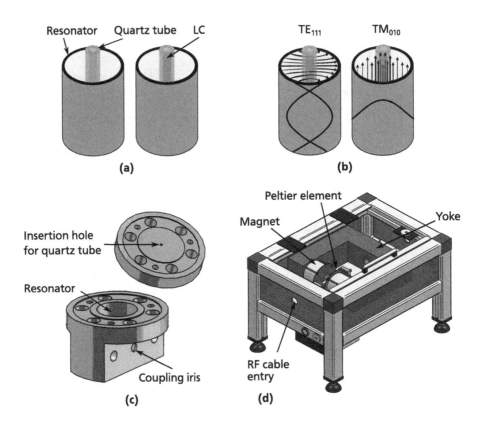

Figure 5.17 (a) Resonator with empty and filled quartz tube. (b) Used modes of the dual-mode resonator. (c) Drawing of a cylindrical dual mode resonator. (d) Temperature-controlled LC characterization setup with permanent magnets.

transmission coefficient measurements are performed: (1) the empty resonator, (2) with an inserted empty quartz tube, and (3) with the quartz tube filled with the LC material.

Cylindrical resonators have been fabricated at 12, 19, and 30 GHz [23, 80, 115]. As an example, a schematic of the 19 GHz dual mode resonator is shown in Figure 5.17. Moreover, the two modes TE_{111} and TM_{010} being used are shown in Figure 5.17b. The resonators are milled in split block technology and the most important parameters are summarized in Table 5.3. For the characterization, the resonator in Figure 5.17c is assembled and connected to a vector network analyzer (VNA) to perform the measurements with the temperature-controlled LC characterization setup depicted in Figure 5.17d. The fabricated resonator has a high empty Q factor of 8203 and 7223 for the TE_{111} and TM_{010} mode, respectively.

Figure 5.18 shows the measurement results for the triple-mode resonator operating at 30 GHz for the GT3-23001 LC mixture. The lower resonance corresponds to the TM_{010} mode, the middle to the TE_{112}, and the upper to the TM_{012} mode.

Table 5.3 Parameters of the 19 GHz dual-mode resonator

Material	Diameter resonator	Height resonator	Diameter LC	Diameter quartz tube	Q_0 TE$_{111}$ TM$_{010}$	Q_L TE$_{111}$ TM$_{010}$
Copper	12.02 mm	11.54 mm	0.5 mm	0.7 mm	8203 7223	7500 6523

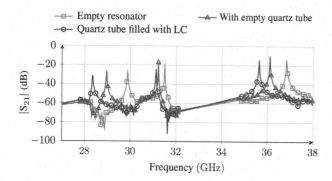

Figure 5.18 Measurement results for the characterization of the GT3-23001 mixture with the 30 GHz triple-mode resonator [80].

A summary of the extracted relative permittivities and loss tangents of the GT3-23001 and GT5-26001 LC mixtures for parallel and orthogonal orientation is given in Table 5.4.

In [23] the accuracy and stability of the measurement setup has been analyzed with the 19 GHz dual mode resonator. Therefore, 23 measurements have been performed at 20°C at different days. To increase the sensitivity of the measurement, TUD-566 has been used, since it has comparatively very low losses and reacts most sensitively to interfering influences. The results of the standard deviations are shown in Figure 5.19, where the most sensitive parameter, the material's FoM has been measured. The mean value is at 45.1, with a standard deviation of 2.05 and a maximum deviation of 3.80.

Furthermore, care needs to be taken when measuring over large temperature ranges. A variation in the radius and height of the unperturbed, closed resonator [128] results in a frequency shift of

$$f_{nml}^{TE,TM} + \delta f = \frac{c_0}{2\pi\sqrt{\mu_r \varepsilon_r}} \sqrt{\left(\frac{p_{nm}^{TE,TM}}{r_0(1+\Delta_r)}\right)^2 + \left(\frac{l\pi}{h_0(1+\Delta_h)}\right)^2} \qquad (5.37)$$

If Δ_r and Δ_h are assumed to be in the same range with $\Delta_r = \Delta_h = \Delta T \cdot \alpha$ for thermal expansion, which depends on the material temperature coefficient and temperature shift only, the expression can be drastically simplified to

$$f_{nml}^{TE,TM} + \delta f = \frac{f_{nml}^{TE,TM}}{1+\Delta} \rightarrow \therefore \delta f = -f_{nml}^{TE,TM} \cdot \frac{\Delta}{1+\Delta}. \qquad (5.38)$$

Table 5.4 Summary of the measurement results with the 30 GHz resonator [80]

Mode f_0	TM$_{010}$ 29 GHz		TE$_{112}$ 31.2 GHz		TM$_{012}$ 36.2 GHz	
LC mixture	$\varepsilon_{r,\parallel}$	$\tan\delta_\parallel$	$\varepsilon_{r,\perp}$	$\tan\delta_\perp$	$\varepsilon_{r,\parallel}$	$\tan\delta_\parallel$
GT3-23001	3.22	34.95×10^{-3}	2.57	51.06×10^{-3}	3.25	11.86×10^{-3}
GT5-26001	3.27	34.14×10^{-3}	2.51	24.15×10^{-3}	3.30	8.21×10^{-3}

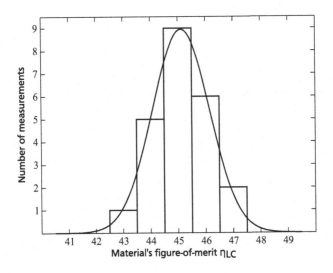

Figure 5.19 Standard deviation of the material's figure of merit; 23 measurements are performed at 19 GHz with TUD-566 [23].

For a small thermal expansion $\Delta = \Delta T \cdot \alpha$, the expansion coefficient is in the range of $10^{-6}\ldots10^{-5}K^{-1}$ and the temperature shift may in the worst case be in the range of $100K$. Using a series expansion to the first term, we obtain

$$\frac{\Delta}{1+\Delta} \approx \Delta, \tag{5.39}$$

and hence, a frequency shift due to thermal effects of

$$\delta f = -f_{nml}^{TE,TM} \cdot \Delta T \cdot \alpha. \tag{5.40}$$

For a 19 GHz aluminum resonator ($\alpha_{Al} = 23 \cdot 10^{-6}\ K^{-1}$) and a temperature shift of 100 K (typical temperature range for the characterization of liquid crystals), this results in a frequency shift of $\delta f = -46$ MHz.

Since this applies in a similar way for the perturbed mode, the thermal frequency shift does not influence the relative shift of resonance frequency due to the sample. It can, therefore, be removed by comparing the empty measurement and the sample-loaded measurement. A problem far more serious is the influence of a slightly misplaced sample tube, that is, the influence of asymmetry.

5.3.2   Broadband Coaxial Transmission Line Method

The coaxial transmission line method is widely used for broadband microwave material characterization [49, 104, 109, 110, 131]. The properties of a material under test are extracted by using the measured scattering parameters of the coaxial line. An artificial coaxial line is required for characterization of liquids, since there is no solid support for the inner conductor [15, 26]. A typical temperature-controlled artificial coaxial line setup is described here [26, 110].

A cross-sectional schematic of the setup is shown in Figure 5.20. The outer conductor and the dielectric material of a coaxial line (RG 405/U) are removed from the middle of the line with a certain length of l and without altering the inner conductor. This line is then placed into two symmetric split blocks. Once the split blocks are assembled, they form a cylindrical cavity of length l as an outer conductor and the inner conductor hangs on air without any mechanical support inside. This part is so called test cell and will be filled by LC. The test cell is excited through the semirigid cables from both ends. A mechanical bending occurs on the inner conductor since it has no mechanical support inside the test cell. However, this hardly affects the extracted material parameters [26]. The dimensions of the split blocks shown in Figure 5.20 are designed in such a way that the characteristic impedance of the empty

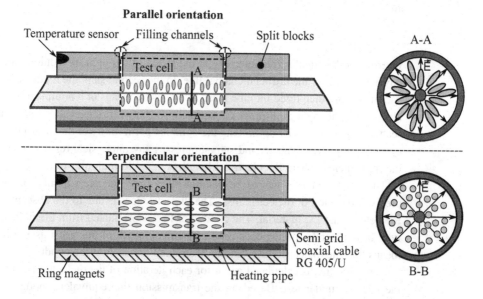

Figure 5.20 Sectional view of the temperature controlled coaxial line measurement setup for parallel and perpendicular orientation of the LC molecules on top and bottom, respectively. The test cell has a length of 35 mm [110].

test cell is 50 Ω. The liquid material under test is inserted into the test cell by means of two filling channels.

During the measurements, the LC molecules are aligned parallel and perpendicular with respect to the RF field by means of electrostatic and magnetostatic fields, respectively. As shown on the top right in Figure 5.20, the electrostatic field is generated by applying a bias voltage to the inner and outer conductors, resulting in parallel orientation of the LC molecules. For perpendicular orientation a magnetostatic field is applied by using cylindrical permanent ring magnets surrounding the split block (see bottom right in Figure 5.20).

5.3.2.1   Parameter Extraction

In [26], a point-by-point extraction method is utilized in two steps. In the first step, the Nicolson–Ross–Weir method is utilized to determine the complex relative permittivity $\underline{\varepsilon_r}$. In the second setup, these values are used as initial guesses to model the scattering parameters [104] as

$$[S]_{\text{Model}} = \frac{1}{1 - \underline{\Gamma}^2 \underline{Z}^2} \begin{pmatrix} \underline{\Gamma}(1 - \underline{Z}^2) & \underline{Z}(1 - \underline{\Gamma}^2) \\ \underline{Z}(1 - \underline{\Gamma}^2) & \underline{\Gamma}(1 - \underline{Z}^2) \end{pmatrix}, \tag{5.41}$$

with

$$\underline{\Gamma} = \frac{\sqrt{1/\underline{\varepsilon_r}} - 1}{\sqrt{1/\underline{\varepsilon_r}} + 1} \tag{5.42}$$

and

$$\underline{Z} = e^{-\underline{\gamma} l} \tag{5.43}$$

where ω is the angular frequency, c_0 is the speed of the light in vacuum, and l is the physical length of the artificial coaxial line. In the second step, the modeled scattering parameters are then fitted into the measured ones by means of a nonlinear least-square algorithm. The code minimizes the square error between the modeled and measured scattering parameters by optimizing the complex permittivity of the material under test $\underline{\varepsilon_r} = \varepsilon_r(1 - j \tan \delta)$. The second step suppresses any unphysical results determined at the frequencies, where the electrical length is equal to multiples of π.

The conductor loss has to be distinguished from the measurement results for a precise extraction of dielectric loss. The dielectric loss is due to the absorption of the RF field by the LC material. The conductor loss is originated from the finite conductivity of the metals and depends on the electrical length of the transmission line. This means that the dielectric loss can be determined precisely if the conductor loss of the transmission line is calculated again for each iteration of the dielectric permittivity of the material under test. Based on the transmission line equivalent model [128], the complex propagation constant $\underline{\gamma}$ of a coaxial line in Eq. (5.43) can be written as

$$\underline{\gamma} = l \sqrt{\omega \varepsilon_r ((K \tan \delta - L) + j(L \tan \delta + K))}, \tag{5.44}$$

where

$$K = \frac{R_s \varepsilon_0 (r_i + r_o)}{\ln (r_o/r_i)(r_o r_i)}$$ (5.45)

and

$$L = \omega \mu_o \varepsilon_0.$$ (5.46)

In Eq. (5.45), r_i and r_o are the inner and outer conductor radii, respectively and R_s is the surface resistivity. The surface resistivity can be calculated analytically as $R_s = \sqrt{\omega \mu_0 / 2\sigma}$, where σ is the conductivity of a metal [128]. However, it is preferred to measure the surface resistivity for practical reasons, for example, to include the impact of the surface roughness. Keeping this in mind, a reference measurement of scattering parameters is needed before filling the LC material into the coaxial line. Using these scattering parameters S_{ij}, the line impedance Z_0 can be calculated as [132]

$$Z_0 = 50\Omega \sqrt{\frac{(1 + S_{11})^2 - S_{21}^2}{(1 - S_{11})^2 - S_{21}^2}}.$$ (5.47)

On the other hand, the line impedance of the same coaxial line can also be calculated as

$$Z_0 = \frac{\sqrt{L - jK}}{2\pi \varepsilon_0 \sqrt{w}} \ln (r_o/r_i)$$ (5.48)

by using the transmission line equivalent circuit, where $\varepsilon_r = 1$ and $\tan \delta = 0$ since the dielectric is air. The unknown K in Eq. (5.44) can be determined for each frequency point in advance by solving the Eq. (5.48) with precalculated values of Z_0 of Eq. (5.47). Hence, the complex propagation constant γ in Eq. (5.44) can be calculated for each iteration of ε_r. Then, it is used in Eqs. (5.41) to (5.43) to model the scattering parameters of the artificial coaxial line accurately.

5.3.2.2   Measurement Setup

The measurement setup is shown in Figure 5.21. The cutoff frequency of the RG 405/U is around 64 GHz according to formulas given in [128]. Nevertheless, the operating frequency of the artificial coaxial line is limited up to be 45 GHz, since the first higher order mode appears at 48 GHz when the test cell is filled with an LC sample with $\varepsilon_r = 3.5$. Hence, K-connectors from Anritsu are installed to the feeding lines. The measured scattering parameters represents the electrical characteristic of the test cell only, since any effects caused by the connectors and the feed lines are eliminated by using a trough-reflect-line (TRL) calibration technique [133].

In [110] several LC samples were characterized from 3.5 to 30 GHz while controlling the temperature of the setup between 10 and 60°C. The measured results in tunability versus temperature are shown in Figure 5.22.

Comparing the cavity-perturbation and coaxial line characterization of similar LC materials shows that the relative difference between the material tunabilities τ_{LC} is

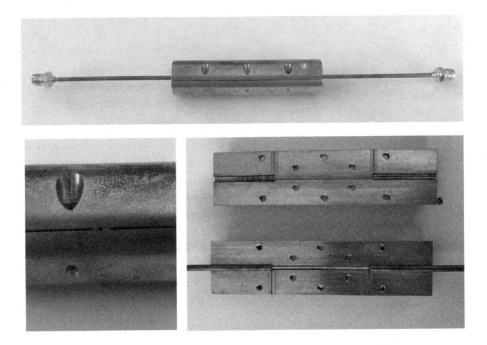

Figure 5.21 Fabricated split block for the broadband coaxial measurement setup [110].

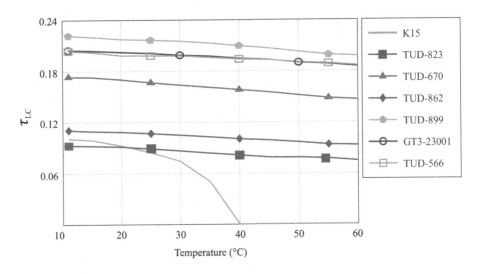

Figure 5.22 Measurement results of the broadband LC characterization [110].

about 20%. For the material's FoM η_{LC}, on the other hand, comparison of the methods is more difficult, since the extracted loss tangents tan δ are much more sensitive to any uncertainty and tolerances than ε_r. Nevertheless, the broadband systems are less accurate, especially for low-loss materials, since the uncertainties in the measured scattering parameters of low-loss materials are high [104].

5.3.3   Terahertz Material Characterization

The field of characterization of materials in the terahertz range has undergone radical changes in the past 10 years. With the emergence of terahertz VNAs and reasonably priced pulsed terahertz systems (time-domain spectroscopy, TDS) as well as continuous-wave (CW) systems with photomixing, material characterization has become easier, less expensive, and more reliable.

Typical scenarios for TDS are broadband measurements of materials with broad features, that is, spectral signatures that extend over a larger frequency range. This applies for many liquids where molecular interaction smears resonant features over often large fractions of the frequency. Typical TDS systems nowadays cover bands from several hundred GHz (200–300 GHz) at the low edge to several THz (typically up to 4 or 5 THz) at the upper, see Figure 5.23. By sampling a pulsed signal in the time domain using a variable delay, the spectrum is obtained via Fourier transformation of the pulse. For a high-resolution spectrum, many samples have to be considered, which takes a large amount of time.

Conversely, CW systems offer a smaller bandwidth (typically up to 3 THz) but they take samples at each frequency point individually. Therefore, they are very flexible: broad spectra are easily recorded by selecting a small number of frequency points while fine features are recorded by selecting the vicinity of the feature and a large number of points. Furthermore, it is possible to record transient events, for example, phase shift over time. The latter is possible only for extremely slow processes (orders of minutes or hours) in TDS systems.

5.3.3.1   Material Characterization in a Free-Space Setup

In a free-space setup it is typically not possible to determine a full set of S-parameters. This is mainly due to the fact that currently available free-space setups lack a device that separates the reflected wave from the source wave.

For the measurements, a transmission setup is used, where the magnitude and phase of the transmitted signal can be measured in one direction, while the reflection is

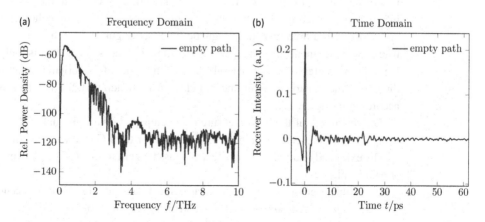

Figure 5.23 TDS spectrum of an empty path in air in (a) frequency domain and (b) time domain.

usually disregarded. This can be circumvented using two or more measurements by measuring at least with two different sample thicknesses.

The expression for the S-parameter matrix is

$$S_{\text{measured}}(f) \approx \begin{pmatrix} n.a. & S_T(f) \\ n.a. & n.a. \end{pmatrix}.$$
(5.49)

where $n.a.$ indicates a nonavailable value.

For the case of a simple slab of material, the ABCD matrix is taken as a starting point:

$$A_{\text{sample}} = \begin{pmatrix} \cos(\gamma) & jZ_0 \sin(\gamma) \\ \dfrac{j}{Z_0} \sin(\gamma) & \cos(\gamma) \end{pmatrix},$$
(5.50)

with

$$\gamma = \beta l_{\text{sample}} = \frac{\omega}{c_0} \sqrt{\varepsilon_r} l_{\text{sample}} \text{ and } Z_0 = \eta / \sqrt{\varepsilon_r}.$$
(5.51)

The expression can now be transformed into a S-parameter matrix and because the transmitted power with respect to the incident power is proportional to $|S_T|^2$, fitting is possible now. Applying the preceding expression further yields

$$S_T = -\frac{j\sqrt{\varepsilon_r}(\varepsilon_r - 1) \sin\left(\dfrac{\omega}{c_0} \sqrt{\varepsilon_r} l_{\text{sample}}\right)}{j\sqrt{\varepsilon_r}(\varepsilon_r + 1) \sin\left(\dfrac{\omega}{c_0} \sqrt{\varepsilon_r} l_{\text{sample}}\right) + 2\varepsilon_r \cos\left(\dfrac{\omega}{c_0} \sqrt{\varepsilon_r} l_{\text{sample}}\right)}.$$
(5.52)

If the sample thickness l_{sample} is known, the sample's permittivity ε_r is obtained using a fitting algorithm.

A sample holder, composed of two windows of dielectric material (e.g., HDPE, PTFE, PMP, or quartz) are separated by a spacer including a rubber ring to avoid liquid leaking from the holder. A similar structure is commercially available for infrared spectroscopy. To characterize liquid crystal, the sample holder must be embedded into a homogeneous magnetic field, see Figure 5.24. This can be done by using two permanent magnets (given the sample holder is small, that is, only several tens of millimeters on each lateral edge). Since the LC is orientated only in one direction, the polarization of the E-field must be adopted for the measurements according to Figure 5.25.

The schematic of the TDS measurement setup is shown in Figure 5.26, where the sample holder presented in Figure 5.24 is used for the LC characterization.

The extracted relative permittivities and loss tangent for the GT3-23001 mixture are shown in Figure 5.27.

Using this method, permittivity measurements are possible with an accuracy that depends mainly on the sample thickness and phase resolution of the measurement device. Loss measurements, however, are difficult to realize if the contrast between sample and empty measurement is low. In this case, a boundary can be determined as to what

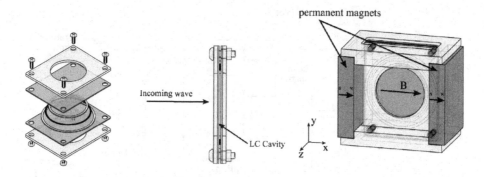

Figure 5.24 Assembly of the sample holder used for THz measurements. Two metal plates, held with four screws, hold two PP windows and distance rings that contain the LC. The assembled holder is inserted between two magnets for biasing [80].

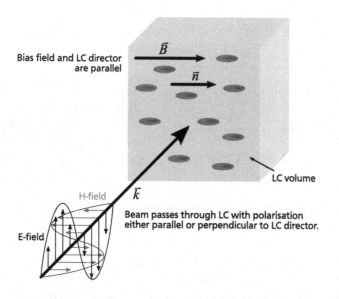

Figure 5.25 Principle of measuring $\varepsilon_{r\perp}$ and $\varepsilon_{r\parallel}$. $\varepsilon_{r\parallel}$ is obtained by aligning the E-field of the incident beam with the bias field and director direction (as depicted). The orthogonal value $\varepsilon_{r\perp}$ is obtained by turning the E-field polarization by 90° [80].

minimum loss tangent can be resolved. This depends on sample thickness, window material losses and the setup's dynamic range. A possible solution for this problem in the future is using resonant methods that resolve losses better than nonresonant methods.

5.3.4   LC Characterization Results and Further Development

Table 5.5 provides the characterization results of some selected LC mixtures at different frequencies. A more extensive table, covering measurement values of various other LCs taken from many references, can be found in [39].

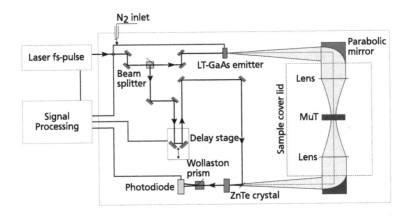

Figure 5.26 Schematic of the used TDS measurement setup.

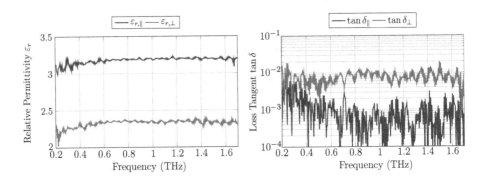

Figure 5.27 Extracted permittivity and loss tangent of the GT3-23001 mixture [80].

Figure 5.28 shows the maximum loss tangent versus tunability for the set of LCs in Table 5.5 at 19 GHz, indicating standard display mixture, Merck's first- and second-generation LC mixtures. Figure 5.29 visualizes the loss tangent versus permittivity for the two characteristic alignment states for the same set of LCs at 19 GHz, where the arrow starts at the values for perpendicular alignment and ends at the values for parallel alignment, thus indicating the difference in the loss tangent and the anisotropy. A clear development toward higher anisotropies and lower losses can be observed, accompanied by decreasing differences in the loss tangent values from perpendicular to parallel orientation. For instance, GT5-28004 indicate already a small decay in loss tangent values only from perpendicular to parallel orientation. Both Figure 5.28 and Figure 5.29 indicate the development from lossy mixtures with limited tunability toward mixtures with electromagnetic properties well suited for microwave applications. Recent developments regarding LCs of the third generation for microwave applications are well summarized in [35].

Table 5.5 Electromagnetic properties of selected LC mixtures at different frequencies

LC	f (GHz)	$\varepsilon_{r,\perp}$	$\tan\delta_{\perp}$	$\varepsilon_{r,\parallel}$	$\tan\delta_{\parallel}$	$\Delta\varepsilon$	τ_{LC}	η_{LC}	Method	Year	Reference
K15 (5CB)	19	2.7	0.0273	3.1	0.0132	0.4	0.13	4.8	CPM	2013	[32]
	30	2.58	0.0294	2.91	0.0132	0.33	0.11	3.9	CPM	2013	[26]
	60	2.86	0.0109	3.21	0.0002	0.35	0.11	10	MSL	2010	[134]
E7	19	2.53	2.2	2.98	0.009	0.45	0.15	6.9	CPM	2017	[35]
	30	2.73	4.86	3.21	0.0317	0.48	0.15	3.1	MSL	2010	[134]
	60	2.78	0.61	2.78	0.0011	0.47	0.14	2.3	MSL	2010	[134]
BL006	19	2.58	0.019	3.16	0.0069	0.58	0.18	9.7	CPM	2017	[35]
	30	2.62	0.025	3.04	0.001	0.42	0.14	5.5	CPM	2007	[135]
MDA-03-2838	9	2.25	0.03	3.1	0.005	0.85	0.51	17	CPM	2005	[48]
MDA-03-2844	9	2.4	0.02	3.4	0.007	1	0.29	14.7	CPM	2008	[136]
GT3-23001	19	2.41	0.0141	3.18	0.0037	0.77	0.24	17.2	CPM	2017	[35]
	30	2.46	0.0143	3.28	0.0038	0.82	0.25	17.5	CPM	2010	[26]
GT3-24002	19	2.5	0.0123	3.3	0.0032	0.8	0.24	19.7	NS	2013	[137]
GT5-26001	19	2.39	0.007	3.27	0.0022	0.88	0.27	38.6	NS	2018	[37]
	30	2.39	0.0055	3.27	0.0022	0.88	0.27	49	NS	2017	[22]
GT5-28004	19	2.40	0.0043	3.32	0.0014	0.92	0.28	64.4	NS	2018	[37]
TUD-566	19	2.4	0.006	3.2	0.0025	0.8	0.25	41.7	NS	2013	[79]
	30	2.32	0.0066	3.11	0.0021	0.79	0.25	38.5	CPM	2013	[26]

Characterization methods: CPM → cavity perturbation method, MSL → microstrip line, NS → not specified.

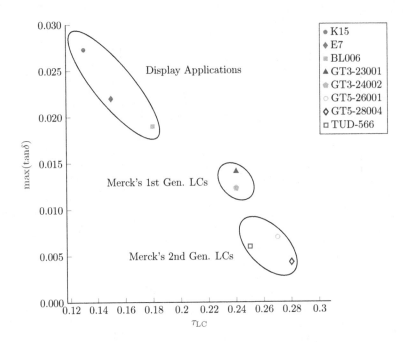

Figure 5.28 Maximum loss tangent versus tunability for selected LC mixtures at 19 GHz.

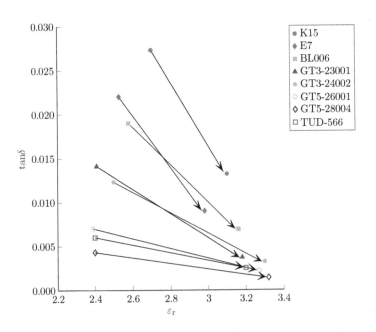

Figure 5.29 Maximum loss tangent versus permittivity of selected LC mixtures at 19 GHz.

5.4   Liquid Crystal–Based Delay Line Phase Shifter Topologies

The hardware implementation of electronically steerable antennas (ESAs) in the analog domain requires by default tunable phase shifters with a differential phase shift of at least 360° beyond each antenna element of an array (and in addition, variable gain amplifiers when beam shaping is desired) as well as a feeding network before down mixing and A/D conversion in the receive mode (and D/A conversion in the transmit mode) according to Chapter 2. These phase shifters have to be implemented within the limited area of each antenna element of the ESA then, which is about $\lambda/2 \times \lambda/2$.

Beyond ESAs, tunable phase shifters are aimed for in many micro- and mm-wave components and devices for adjustments and flexible operation, that is, to reconfigure their characteristics in operation. Examples are Wilkinson dividers, hybrid couplers, switches, or Butler matrices [54].

There are two different ways of phase shifting:

1. In non-frequency-dependent phase shifters (true phase shifters), the spectrum is shifted uniformly by a certain angle (Hilbert transformer). Each of them causes a rotation of the spectrum by 90° [138].
2. In frequency-dependent phase shifters (true delay lines), a phase shift of a specific frequency is accomplished by delaying the input signal. The phase shift becomes frequency dependent, since different frequencies have different periodic times.

Those kinds of phase shifters can be realized by delay lines or all-pass filters.

The latter are preferred for broadband systems, since they feature a linearly dependent differential phase shift performance over the bandwidth. Especially in an antenna operation, the true delay lines prevent beam squinting. All the LC-based phase shifters are inherently true delay lines. Nevertheless, they can be utilized as a true phase shifter as well, since the differential phase shift can be kept constant by controlling the bias voltage, that is, the LC material's properties.

This subchapter presents and compares LC-based phase shifters built up in different topologies, including various low-profile planar transmission delay line topologies, high-performance metallic and dielectric waveguide phase shifters with different biasing schemes, required biasing voltages, form factors, and performance metric. While planar transmission line topologies are inherently flat and compact, volumetric waveguide-based structures and devices are bulky and much more complex and usually more expensive to implement. However, electrical losses of metallic and dielectric waveguide structures, particularly at mm-waves, are significantly lower than for corresponding planar structures. But these bulky structures require complex biasing electrode configurations with higher biasing voltages than planar structures. Because of their trade-off properties between conventional planar and waveguide structures, substrate integrated waveguide (SIW) structures have attracted many researchers in recent years.

To compare the performance of different phase shifter topologies, a phase shifter FoM is defined by [76, 77]

$$\text{FoM} = \frac{\Delta\varphi_{\max}}{IL_{\max}}, \qquad (5.53)$$

where $\Delta\varphi_{\max}$ is the maximum differential phase shift and $IL_{\max}$ the maximum insertion loss in all tuning states. The insertion loss depends on the topology, material, and operational frequency.

The optimal possible FoM which can be achieved at all is [15, 23, 76]:

$$\text{FoM}_{opt} = 13.2 \cdot \frac{1 - \sqrt{1 - \tau_{\text{LC}}}}{\tan\delta_{LC,\,max}} \approx 6.6 \cdot \eta_{\text{LC}} \text{ in}^\circ/\text{dB}, \qquad (5.54)$$

where τ_{LC} is the tunability and $\tan\delta_{LC,\,max}$ the maximal loss factor of the LC and η_{LC} is the material's FoM. This equation is valid only by presuming a constant dielectric loss of the LC for all tuning states, while all other materials are assumed to be loss free. In practice, the FoM is typically in the range of 50%–80% of FoM$_{opt}$.

First microwave investigations on the birefringence of two common nematic LCs, that is, BL006 and E7, and its electro- and magneto-optical effects as well as first experimental data of a waveguide phase shifter at 30 GHz are presented in terms of phase shift in [72]. With the emergence and progress of newly developed LC mixtures specifically synthesized for microwaves since 2002, innovative concepts and designs with appropriate biasing schemes enabled numerous of high-performance RF components and devices, which fully exploit the LC's unique properties. Thus, with a Merck's first generation LCs, an FoM of more than 110°/dB was achieved already in an early stage with an microstrip line phase shifter at 24 GHz compared to 20°/dB with the standard LC K15 [29, 30, 76, 77] and soon afterward, an FoM of up to 150°/dB could be reached with a fully electrical biasing and even up to 200°/dB with magnetically biasing for a partially LC-filled waveguide phase shifter around 30 GHz [78]. Since then, various planar delay line and waveguide phase shifters at different frequencies are presented in [20, 21, 31, 34, 59, 71, 79, 80, 139–144]. Very recently, some investigations are also focused on electrically tunable dielectric delay line phase shifter, inserting an LC section inside the dielectric core [54, 37, 116, 145–150]. They offer low-loss propagation at very high frequencies and potentially low-cost fabrication by using processing technologies such as 3D printing, injection molding or milling, since no metallic component is required for it. Moreover, some approaches of monolithic integration of LC phase shifter with different technologies, including LTCC, LC-based complementary CMOS, nanowire membrane, and MEMS phase shifter, respectively. All these phase shifter topologies will be discussed in this Section.

Moreover, for the design and investigation of arbitrary LC-based structures, an in-house software tool SimLCwg is implemented at TU Darmstadt, combining the finite difference method for static fields to simulate the director dynamics and the finite difference frequency domain (FDFD) method for a full RF wave simulation [23, 151]. This software tool allows a systematic analysis of LC-based structures in terms of the tuning effectivity $\Delta\tau_\varphi$ as well as in terms of the response times τ_{on} and τ_{off}, depending on the rubbing direction of the alignment layers and the geometrical parameters of the structure [23, 152, 153]. Herein, the tuning effectivity $\Delta\tau_\varphi$ describes quantitatively how much of the material's tunability can be utilized by a proposed device layout.

5.4.1   Tunable Low-Profile Planar Transmission Line Phase Shifter

5.4.1.1   Tunable Microstrip Line

Inspired by the well-established LC display technology, several LC-based planar transmission line devices have been investigated [29, 73, 76]. Among those, the microstrip line (MSL) topology has been widely studied, because of its high performance and simple design. Figure 5.30 exhibits its assembly: the strip or signal line (top) and ground plane (bottom) are located on separated glass carrier substrates with $\varepsilon_{r,s}$. These glasses on both sides are covered by thin, rubbed polyimide films for prealignment of the LC director $\vec{n}$. Then, the substrate carrying the signal electrode is flip-chipped, introducing a cavity between both substrates, which is sealed to prevent a leakage of LC. Afterward, the LC is filled inside this cavity, having a dielectric constant $\varepsilon_{r,LC}$ which varies between $\varepsilon_{r,\perp}$ and $\varepsilon_{r,\parallel}$, depending on the bias voltage. For electrical tuning, the strip line and the ground plane are used as biasing electrodes. Hence, it uses the hybrid biasing scheme.

Initially, when no bias voltage is applied ($V_b = 0$ V) between these electrodes, the LC directors $\vec{n}$ are oriented in parallel to the alignment layer, that is, by default perpendicularly aligned to the RF field $\vec{E}_{RF}$, which is vertical polarized as the tuning field $\vec{E}_b$, since both are not decoupled. Hence, the RF field experiences a relative permittivity of $\varepsilon_{r,\perp}$ and loss tangent $\tan\delta_\perp$ during its propagation along the line. When an applied biasing voltage V_b between the strip line and ground plane exceeds the Fréedericksz threshold voltage V_{th}, the LC directors $\vec{n}$ starts to align from the initial orientation parallel to the surface toward the orientation of the field lines of the tuning field $\vec{E}_b$ [40]. If the biasing voltage V_{sat} is exceeded, nearly all LC directors $\vec{n}$ are in parallel to the RF field. Hence, $\vec{E}_{RF}$ experiences a material's effective relative permittivity, which varies from $\varepsilon_{r,\perp}$ continuously toward $\varepsilon_{r,\parallel}$, accompanied by a change in the loss tangent from $\tan\delta_\perp$ toward $\tan\delta_\parallel$. Beyond this saturation state $V_b > V_{sat}$, no change in permittivity can be observed any longer [14]. By varying ε_r, the

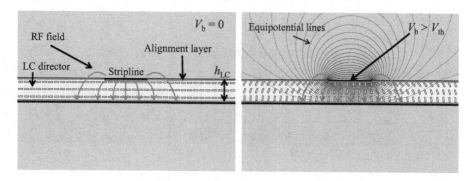

Figure 5.30 Schematic of the hybrid biasing scheme for a tunable LC filled microstrip line.

propagation constant of the wave traveling along the line alters, and hence the phase shift of the line with a certain length. Hence, it represents a simple variable delay-line phase shifter, where phase shift, however, is accompanied by a change in the wave impedance of the propagating mode, that is, some line mismatch occurs, as already discussed in Chapter 3, where the optimal state between return and insertion loss is highlighted.

A major objective in designing the microstrip transmission line is to maximize the material's tunability, that is, to achieve a high tuning efficiency $\Delta\tau_\varphi$ (maximum: $\Delta\tau_\varphi = 1$), which describes quantitatively how much of the material tunability can be utilized by a proposed device layout compared to ideal molecule alignment. For example, if a carrier substrate with a significantly higher permittivity compared to the LC is chosen ($\varepsilon_{r,s} > \varepsilon_{r,LC}$), the RF field will propagate partially in this substrate instead of in the LC section, reducing the tuning effectivity $\Delta\tau_\varphi$, and hence the differential phase shift of a transmission line phase shifter. To compensate decreasing tuning effectivity, the length of a transmission line phase shifter could be extended in practice to maintain the desired differential phase shift, for example, of 360°. However, to the expense of larger size and higher insertion loss.

Another design parameter, which affects the tuning effectivity $\Delta\tau_\varphi$ as well as the response times τ_{on} and τ_{off}, is the rubbing direction of the polyimide film, since two different preorientations are possible, as shown in Figure 5.31: (a) the preorientation along the x-direction (within the cross-sectional plane) and (b) the preorientation along the z-direction (in the propagation direction).

In the unbiased state $V_b = 0$, where only prealignment by means of surface anchoring is active, not all LC molecules are perpendicular to the RF field lines for the cross-sectional preorientation of the MSL phase shifter Figure 5.31a. Hence, the orientation efficiency is $\tau_\varphi(V_b = 0) > 0$, whereas all LC molecules are perpendicular to the RF field lines for the preorientation in propagation direction in Figure 5.31b, that is, the orientation effectivity is $\tau_\varphi(0) = 0$ (optimum). In the fully biased state of

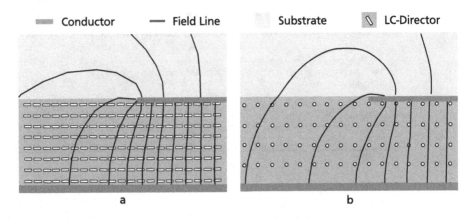

Figure 5.31 (a) Perpendicular (cross-sectional) and (b) perpendicular (propagation direction) LC alignment in an MSL phase shifter.

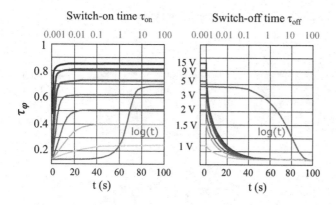

Figure 5.32 Orientation effectivity τ_φ versus response times τ_{on} and τ_{off} for the cross-sectional preorientation of an MSL with $h_{LC} = 100$ μm. The parameter is the biasing voltage V_b from 1 V to 15 V. In addition, there are logarithmically scaled curves for $V_b = 4$ V [23].

the MSL phase shifter beyond saturation ($V_b > V_{sat}$), also not all LC molecules are in parallel to the RF field according to Figure 5.30 (right), leading to an orientation effectivity $\tau_\varphi(V_{sat}) < 1$, and hence a tuning effectivity $\Delta\tau_\varphi = \tau_\varphi(V_{sat}) - \tau_\varphi(0) < 1$.

The different orientation and response time behavior for both preorientations will be investigated in more detail here for an MSL design with an effective LC layer height of $h_{LC} = 100$ μm and a width w of the signal electrode of 200 μm to achieve a line impedance of 50 Ω. The LC material used for these fundamental investigations is K15, whose material properties are given in Table 5.5. Because K15 is well known and widely used in optics and in the terahertz range, it serves here as a reference LC. Starting with a cross-sectional preorientation, Figure 5.32 shows the simulation results of the orientation efficiency $\tau_\varphi = \tau_\varphi(V_b)$ of the MSL versus response times τ_{on} and τ_{off}. The varied parameter is the biasing voltage V_b from 1 V to 15 V.

Two important aspects can be derived directly from Figure 5.32:

1. A perfectly perpendicular alignment, that is $\tau_\varphi(0) = 0$ for the unbiased state of $V_b = 0$ V as well as a perfectly parallel alignment, that is $\tau_\varphi(V_{sat}) = 1$ for $V_b = 15$ V is not possible, since not all LC molecules are perpendicular (Figure 5.31a) and parallel (Figure 5.30b) to the RF field, respectively. Actually, τ_φ is roughly between 0.15 and 0.85 from the unbiased up to the fully biased state, resulting in a tuning effectivity $\Delta\tau_\varphi$ of about 70%.
2. Once the biasing voltage is applied, the rise or switch-on time τ_{on} is much faster than the decay or switch-off time τ_{off}, which is particularly emphasized by the logarithmically scaled curves for $V_b = 4$ V. This is due to the relatively strong electromagnetic biasing field compared to the weak reorientation force, originating from the anchoring of LC molecules on the rubbed polyimide layer.

Further, as long as the biasing voltage is between the threshold voltage V_{th} and the saturation voltage V_{sat} ($V_{th} \leq V_b \leq V_{sat}$), the tuning effectivity increases with

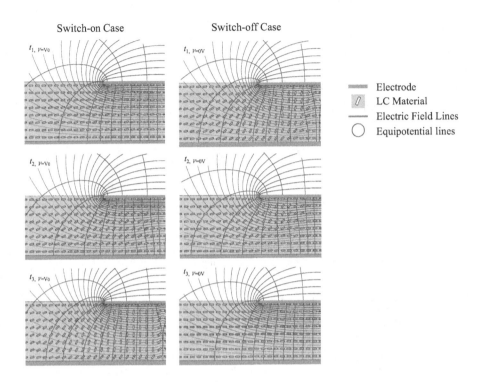

Figure 5.33 Director dynamic simulations for the (left) switch-on and (right) switch-off case of an MSL filled with LC.

increasing biasing voltage. Below the V_{th} the intermolecular forces are stronger than the force of the biasing field, and therefore no change in orientation is possible. At V_{sat}, the LC orientation is in saturation, where a further increase in V_b will generate no improvement anymore.

The results in Figure 5.32 can also be explained in more detail by the snap shots in Figure 5.33: After switching on the biasing voltage, the LC directors near the edge of the microstrip line will aligning first, since they already have an orientation of about 45° to the biasing field lines, which are highly concentrated at the edges, and thus the most effective torque is generated. Combined with the elastic force between the directors, the biasing field is aligning more and more directors even to the vicinity of the strip line, where it impinges on the area with opposite orientation. Up to this point, the change in orientation of the LC is very fast, whereas a lot of time is needed for tuning the LC molecules in the middle part under the signal electrode, where the LC molecules are directly anchored in parallel to the surfaces and which has to be aligned in parallel to the RF field, that is, requires a change in the orientation of about 90°.

Once the bias voltage is switched off, the reorientation starts fast at the beginning, due to the high elastic deformation. The directors near the surfaces will first reorient toward the preferred direction of the polyimide layer. This process slows down with

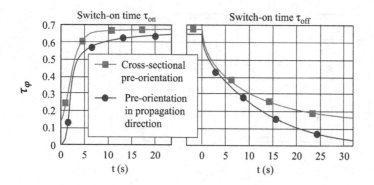

Figure 5.34 Orientation efficiencies versus response times of an MSL for two different preorientations of the LC directors [23].

increasing distance of the directors to the surfaces and with decreasing elastic deformation, since this reorientation force is based only on the elastic forces between the directors. In case of a preorientation in propagation direction as shown in Figure 5.31, only the short axis of the LC directors is effective for the RF field and the biasing field, respectively, resulting in a higher tuning effectivity than in the previous case. To compare both preorientations, Figure 5.34 exhibits the orientation effectivity τ_φ versus the response time, specifically for a biasing voltage of $V_b = 4$ V. The difference is not significant. But the MSL with preorientation in propagation direction provides a higher tuning effectivity of about $\Delta\tau_\varphi \approx 0.64$ compared to $\Delta\tau_\varphi \approx 0.53$ for the cross-sectional preorientation, and it requires more time to align the LC molecules after switching on the biasing field than with the preorientation in the cross-sectional plane. However, the drop at the beginning of the switch off process is steeper compared to the cross-sectional preorientation, which can be explained by the high amount of elastic energy in the center of the transmission line.

If new LC mixtures such as GT3-23001 and TUD-566 specifically synthesized for microwaves are used rather than K15, the aforementioned results will change, because of different LC properties according to Table 5.5. The response times τ_{on} and τ_{off} are affected by the material's parameter viscosity γ_{rot} and the elastic constant K_{11} for the splay deformation for the same effective LC layer height h_{LC}. However, these parameters effect the switch-on time not that much, because of the strong electrostatic field generated by the biasing voltage V_b. Hence, the switch-on times for the different LCs differ only slightly from each other. Moreover, the switch on process can be arbitrarily accelerated by overshooting the voltage for a short time. But the switch-off response time depends significantly on the individual parameters of the various LCs. It is the critical parameter when investigating LC-based transmission line phase shifters. This can be seen from simulations in Figure 5.35, where the orientation effectivity τ_φ is plotted versus the switch-off response time. The molecules of K15 are fastest being reoriented from the nearly parallel orientation to the RF field toward a perpendicular orientation, that is, in parallel to rubbed polyimide layer when the biasing voltage is released, whereas the new microwave LCs, GT3-23001 and TUD-566, exhibit longer

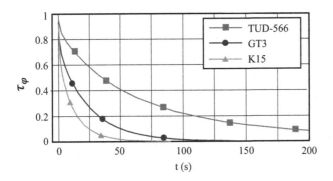

Figure 5.35 Simulated switch-off response time behavior of K15, GT3-23001, and TUD-566 filled MSL phase shifter [23].

switch-off response times compared to K15. Especially for the LC TUD-566 the molecules require a much longer time to go back to the preferred direction in parallel to the surface forced by the rubbed polyimide film. From Figure 5.35, the 90% to 10% decay T_{10}^{90} of the switch-off response times are about 25 s, 36 s, and 183 s for K15, GT3, and TUD-566, respectively.

Beside the LC parameters, the performance of the LC-based MSL phase shifter depend strongly on its geometry. Two examples for an MSL design are shown in Figure 5.36: filled with GT3-23001 (left) and TUD-566 (right). These figures exhibit curves of the same phase shifter FoM (dashed lines), orientation effectivity $\tau_\varphi(V_b)$ (full lines), and line impedances Z (thin dashed line) as a function of the strip line width w (vertical axis), thickness s (lower horizontal axis), and 90%–10% decay T_{10}^{90} of the switch-off response time (upper horizontal axis) [23, 154]. For a thin strip line of few micrometers only, the thickness between the substrate and the ground plane is about the effective LC layer height: $s \approx h_{LC}$. With these plots, LC-based MSL phase shifters can be designed with desired performance before realization. As top substrate, fused silica is chosen, having a height of $h = 300\,\mu m$, a relative permittivity of $\varepsilon_r = 3.82$, and a dielectric loss of $\tan\delta = 5 \cdot 10^{-5}$. Since changes in the geometry of the MSL cause also changes in the line impedance, they are shown in the plots of Figure 5.36 as the dashed lines nearly parallel to the values of constant τ_φ. The switch-off time τ_{off} can also be determined for given geometrical parameters.

As an example, an MSL phase shifter with an LC layer thickness of $h_{\text{LC}} = 100\ \mu m$ features a line impedance of $(50 \pm 2)\ \Omega$ when its signal electrode is $w = 200\ \mu m$ wide. The orientation effectivity is about $\tau_\varphi(V_{sat}) \approx 0.73$ for the GT3-23001 filled phase shifter, where nearly $\tau_\varphi(V_{sat}) \approx 0.75$ are reached by using TUD−566, due to its slightly higher anisotropy. Since $\tau_\varphi(0) \approx 0$, these values correspond with the tuning effectivity $\Delta\tau_\varphi = \tau_\varphi(V_{sat}) - \tau_\varphi(0)$. The FoM reaches values of up to 86°/dB and 140°/dB for GT3-23001 and TUD-566, respectively. This large difference can be explained by the drastically reduced dielectric losses of TUD−566 compared to GT3-23001 (see Table 5.5). These advantages of the TUD−566 are accompanied by the drawback of a higher viscosity, resulting in a much longer 90% to 10% decay T_{10}^{90}

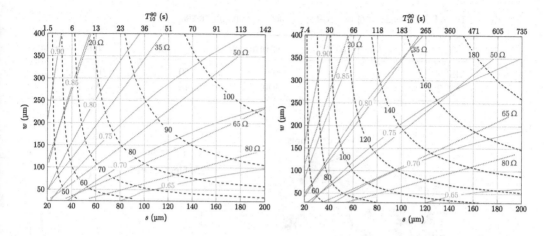

Figure 5.36 Curves of constant phase shifter FoM (dashed lines), orientation effectivity τ_φ (full lines) and line impedance Z (dotted lines), depending on the strip line width w (vertical axis), the thickness $s \approx h_{LC}$ (lower horizontal axis), and 90% to 10% decay T_{10}^{90} of the switch-off time (upper horizontal axis) for a GT3-23001–based MSL phase shifter (left) and a TUD-566–based MSL phase shifter (right), both at 30 GHz. The height h of the top substrate made of fused silica is set to 300 µm, its permittivity is $\varepsilon_r = 3.82$, and the electrical conductivity of the metal parts is set to $\sigma = 4 \cdot 10^7$ S/m. A preorientation of the LC in propagation direction was chosen [23].

of the switch-off time. While the GT3-23001-filled phase shifter already exhibits a very long decay of $T_{10}^{90} \approx 36$ s, the TUD-566-filled phase shifter is even more than five times slower $T_{10}^{90} \approx 183$ s. Hence, the response time is the critical parameter of the LC-based line phase shifter. It can be dramatically reduced by thinner LC layer thickness h_{LC}, accompanied with higher orientation effectivity $\tau_\varphi(V_b)$, but at the same time with steeply increasing impedance mismatch to 50 Ω and significantly decreasing FoM due to increasing electrical losses. Another parameter effecting the FoM to some extent is the permittivity of the carrier substrate $\varepsilon_{r,s}$, which is analyzed in [23]. At a certain frequency, the FoM decreases with higher permittivities $\varepsilon_{r,s}$. For example, for $\varepsilon_{r,s} = 2.2, 3.82, 7$, and 10, the FoM is about 92, 87, 82, and 78°/dB at 30 GHz and 110, 106, 93, and 62°/dB at 100 GHz.

In [25, 76], LC-based phase shifters have been presented in microstrip transmission line topology for K_u- and K_a-band frequencies, respectively. A schematic and the realized phase shifter are shown in Figure 5.37.

As substrate material, TMM3 from Rogers Corp., was chosen, which has a permittivity of $\varepsilon_r = 3.27$ and a dielectric loss of $\tan \delta = 2 \cdot 10^{-3}$. The ground electrode and the carrier substrate on the top side are separated by a well-defined distance from each other by using a spacer with a rectangular hole, forming the LC cavity. Since the line feeds had to be processed on this spacer, it needed to be suitable material for the microwave range, which also can withstand the pressure when screwing the ground plane to the carrier substrate. Therefore, RT/Duroid5880 from Rogers Corp. has been chosen, having $\varepsilon_r = 2.2$ and $\tan \delta = 9 \cdot 10^{-4}$. The spacer needs to protrude by a few millimeters above the carrier substrate and the ground plane in order to be able to contact the line

Table 5.6 Dimensions of two LC-based microstrip line phase shifters

	h_{LC} (μm)	w (μm)	l (mm)	Polyimide layer
MSL1	254	600	50	PI-2555
MSL2	127	300	48	AL-3046

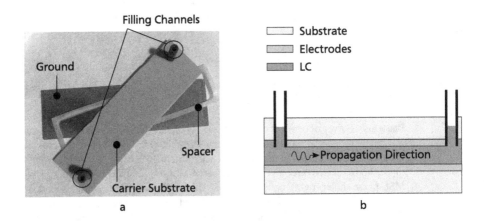

a   b

Figure 5.37 Photograph and cross-sectional schematic of the realized broadband MSL LC phase shifter (MSL1) for 1.125 GHz up to 35 GHz [15, 76].

feed. The electric field inside the phase shifter is mainly between the signal electrode and the ground plane, where the LC is filled into the cavity, affecting the electromagnetic field during its propagation by changing its permittivity, depending on the bias voltage. The length of the actual phase shifter is equivalent to the length of the rectangular hole, as LC can be filled only in this cavity. Two different phase shifters, namely MSL1 and MSL2, of this type have been realized, where their dimensions are given in Table 5.6. Both have been designed for a line impedance of 50 Ω.

For the MSL1 the LC K15 as well as MDA-03-2838 were used, whereas in MSL2 MDA-03-2844 was filled in additionally. The dielectric properties of the MDA-mixtures are given in Table 5.5. The results of the S-parameter measurements and the resulting FoM of MSL1 are shown in Figure 5.38. The maximum biasing voltage for orienting the LC from the perpendicular alignment to the parallel alignment was set to $V_{sat} = 35$ V. The frequency range of this phase shifter was set to $3.5 \text{ GHz} \leq f_{MSL_1} \leq 24.0 \text{ GHz}$.

The matching of this phase shifter is not good, since it shows reflections higher than -10 dB in most parts of the frequency range for both LCs. This also affects the transmission, where high insertion losses (*IL*) up to 6 dB for K15 and 3.5 dB for MDA-03-2838 were measured, respectively. By using MDA-03.2838 instead of K15, the *IL* could be reduced on average by at least 1 dB. While the differential phase shift at 24 GHz for K15 is only around 90°, that of MDA-03-2838 is more than 300°. This can be explained by the large difference in the materials anisotropy (Δε), which is

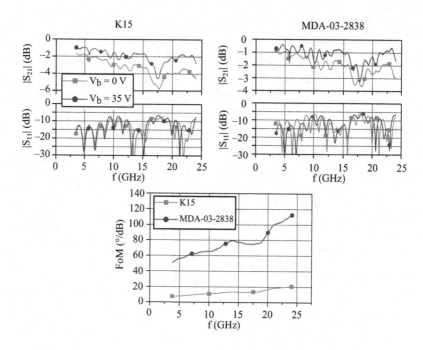

Figure 5.38 Measured S-parameters and resulting FoM for MSL1 [15].

more than three times higher for MDA-03-2838. The calculated results for the FoM are 20°/dB with K15 and 110°/dB with MDA-03-2838, that is, an increase by more than a factor of 5, just by improving LC material performance.

For the second design, MSL2, the frequency range $1.13\,\text{GHz} \leq f_{\text{MSL}_2} \leq 35.00\,\text{GHz}$ was extended compared to MSL1. The maximum biasing voltage for this phase shifter was set to $V_{\text{sat}} = 50\,\text{V}$. The measurement results of MSL2 are shown in Figure 5.39, where both measured bandwidth and bias voltage are increased compared to the previous one.

The matching of this phase shifter is much better than the former one, since reflections S_{11} are almost less than $-10\,\text{dB}$ over the wide frequency range, resulting also in much smoother curves of the transmission parameter S_{21}. However, the IL increased compared to the MSL1 approximately about 1 dB. This is due to the thinner LC layer, by which, together with the reduced signal electrode width, the ohmic losses increase according to [75]. Since MDA-03-2844 has a slightly lower maximum loss tangent, but also a lower anisotropy than MDA-03-2838, both exhibit similar FoMs, with a maximum in the range of 70°–73°/dB at 30 GHz.

By using GT3-23001 according to Table 5.5 with higher anisotropy and lower loss tangent $\tan \delta_\perp$ than MDA-03-2844 and MDA-03-2838, an MSL phase shifter with a signal electrode width of 300 µm and an LC layer thickness of 127 µm can reach an FoM of around 96°/dB at 30 GHz according to Figure 5.36 by using fused-silica instead of TMM3 from Rogers Corp. However, the transition from phase shifter to connector and connector to coaxial line are not considered in the simulations as well as fabrication tolerances.

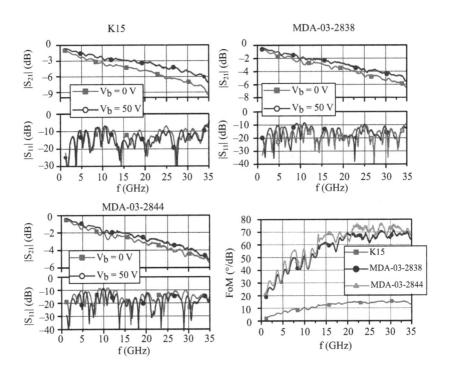

Figure 5.39 Measured S-parameter and resulting FoM for MSL2 [15].

5.4.1.2 Grounded Coplanar Waveguide Loaded with Fast Tuning LC Varactors

The coplanar waveguide (CPW) was first demonstrated for guiding microwaves by C. P. Wen in 1969 [155]. It originally consisted of three upper conductors on a substrate material. If a further ground plane is added beneath the substrate material, it is called a grounded coplanar waveguide (gCPW). Similar to the MSL, the substrate material is on the top of the gCPW, while the LC cavity is constructed with the aid of spacer. An example of an LC-filled gCPW, including the LC orientations for both extreme orientation states is presented in Figure 5.40, where preorientation in the propagation direction has been chosen as the initial state.

For a gCPW with small gap distances g compared to the distance s between the center conductor and ground plane, the orientation effectivity would be dramatically reduced for a cross-sectional preorientation, hence, reducing the tuning effectivity. Since mainly the LCs between the conductor in the center of the gCPW and the ground plane are oriented by the biasing field, the transmission line capacitance is dominated with increasing width of the centered conductor. The wider this conductor, the more the gCPW behaves like an MSL, because of a similar mode propagating through the transmission line. Nevertheless, with a grounded CPW or a CPW without additional ground plane, a tuning effectivity as high as for an MSL cannot be achieved, which is due to the specific topology of the transmission line, resulting in a higher amount of RF field travelling through the nontunable carrier substrate at the top.

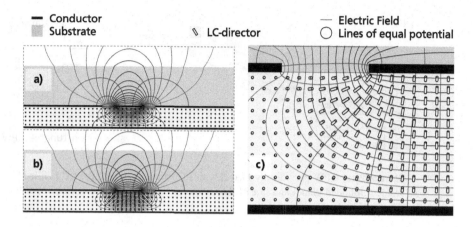

Figure 5.40 Cross section of an LC-filled CPW (a) in the initial state, (b) after switching of the biasing voltage, and (c) a magnified section of part (b). The preorientation of the LC by using a polyimide layer was set to be in the propagation direction [23].

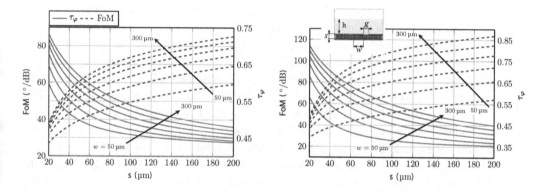

Figure 5.41 Curves of constant phase shifter FoM (dashed lines) and orientation effectivity τ_φ (full lines) at 30 GHz versus the distance s between the center conductor and the ground plane, which is about the LC-layer thickness $h_{LC} \approx s$. The varied parameter is the center conductor width w. The grounded CPW phase shifter is filled with the LC GT3-23001 (left) and TUD-566 (right). The gap between the center conductor and the adjacent conductors was chosen to be $g = (w + h_{LC})/2$. The height of the top substrate made of fused silica is set as 300 μm, its permittivity $\varepsilon_r = 3.82$ and the electrical conductivity of the metal parts is assumed to be $\sigma = 4 \cdot 10^7$ S/m. A preorientation of the LC was chosen in the propagation direction [23].

The simulated FoM and orientation effectivity $\tau_\varphi(V_b)$ of an LC GT3-23001 and TUD-566 filled grounded CPW phase shifter at 30 GHz are shown in Figure 5.41 versus the distance s between the center conductor and ground plane, which is about the LC-layer thickness $h_{LC} \approx s$. The varied parameter is the center conductor width w from 50 μm to 300 μm. Generally, the FoM, particularly above $s = 100$ μm, is lower compared to the simulations for the MSL in Figure 5.36 for the same LC, respectively. This is due to reduced tuning effectivity $\Delta\tau_\varphi = \tau_\varphi(V_{sat}) - \tau_\varphi(0)$ and

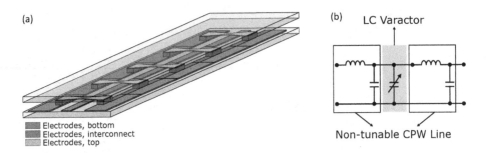

Figure 5.42 (a) CPW periodically loaded with LC varactors, where the LC is filled between the bridge (grounded plane) and line of the CPW. (b) Schematic of a unit element of the periodic structure [82].

higher metallic losses. As a result, the MSL seems to be the better choice for a planar line phase shifter.

But the grounded CPW promises decisively reduced response times compared to MSL in conjunction with the loaded line concept, using a periodic structure of a CPW segment (without ground plane) and a small grounded CPW segment, building a low bridge with a very thin LC layer height of 1–5 μm only above the line of the CPW according to Figure 5.42, which will be filled with LC. First experiments on single LC varactors filled with an LC of the GT3 series is presented in [55, 56], where tunability is around 20% and quality factor better than 40 in the frequency range of 7–7.5 GHz. The measured switch-off response time τ_{off} are 4 ms and 92 ms for an LC layer thickness h_{LC} of about 1 μm and 5 μm, respectively, following clearly the square law. It corresponds quite well with the theoretical values of 3.2 ms and 80 ms for GT3-23001 taken from Figure 5.1 with $h_{LC} = 1$ μm and 5μm, respectively, taking into account some uncertainties in the measured "real" LC layer height of the realized varactors and some uncertainties in the parameters k_{11} and γ_{rot} in Table 5.1 for GT3-23001. In [26, 82], a detailed analysis is given of the transmission line concept loaded by these thin LC varactors, its realization and proof-of-concept. Although the whole structure is actually filled with LC, the segments without additional ground plane behave as a standard CPW line, since the thin LC layer is not affecting its properties significantly.

The realized structure is given in Figure 5.43, where fused silica is used as top and bottom substrate and where the spacers with a diameter of about 5 μm determine the height of the LC cavity, which was filled with TUD-566.

The center conductor width is set to $w = 60$ μm, the gap between the center and adjacent conductors to $g = 195$ μm. The total length ℓ of the phase shifter is 12.7 mm. Additionally to the alignment layers on the substrates, a bias voltage of V_b of up to 60 V is applied via dedicated indium-tin-oxide (ITO) biasing strips to the bridges (additional ground plane) in the LC varactor section. By this, the LC can be tuned from the perpendicular preorientation to the parallel orientation, using the full anisotropy.

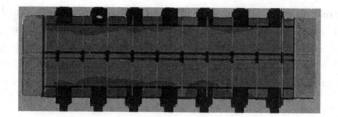

Figure 5.43 Photograph of the realized LC varactor loaded CPW line [82].

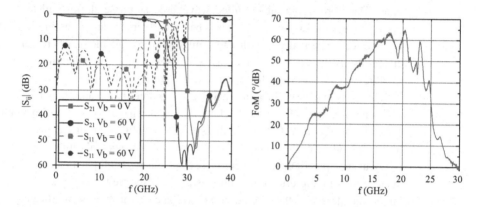

Figure 5.44 Measured S-parameters for the two extreme tuning states for $V_b = 0$ V and $V_b = 60$ V and the FoM versus frequency [82].

The results of the S-parameter measurements versus frequency are shown in Figure 5.44. For a frequency of 20 GHz, more than 90° differential phase shift $\Delta\varphi$ has been achieved, accompanied with a maximum *IL* of less than 1.5 dB, thus obtaining an FoM of more than 60°/dB. Note, since a length ℓ of this loaded line phase shifter of 12.7 mm ($\approx 0.85\lambda_0$) provides $\Delta\varphi > 90°$ at 20 GHz, a 360°-phase shifter would be $\ell \approx 3.3\lambda_0$.

Due to its small LC layer height h_{LC} of 4.5–5 µm, the measured response times are T_{90}^{10} is less than 110 ms from zero bias to fully tuned state and T_{90}^{10} is now less than 340 ms from fully tuned state back to zero, which is significantly faster than the MSL with the same aforementioned high-performance LC. The response times of max. 340 ms are somewhat slower than those obtained from the single varactors above due to the use of a more viscous high-performance LC of the second generation, where the theoretical value of τ_{off} is about 330–410 ms for 4.5–5 µm (see Figure 5.1), which suits the experimental value of 340 ms well.

As consequence, the switch-off response time T_{90}^{10} could further reduced by using varactors with a thinner LC layer and/or a nematic LCs with lower rotational viscosity γ_{rot} of Merck's third-generation LCs.

Beyond tunable microstrip and loaded CPW delay lines, more planar transmission line phase shifter topologies have been investigated in [23, 26], for example, loaded slot and microstrip lines, coplanar strip lines with and without an additional ground plane as well as an antipodal finline, showing results in orientation effectivity τ_φ, differential phase shift, insertion loss, FoM, response time, and physical length with respect to the wavelength (compactness) for different frequencies and in dependence of some design parameters as well as for various LCs.

5.4.2   Tunable High-Performance Metallic Waveguide Phase Shifter

Metallic waveguides have intrinsically lower metal losses than planar transmission lines. Therefore, using an LC-filled waveguide as phase shifter will lead to higher FoM for the same LC mixtures. This is particularly important for applications, where a high performance is decisive due to limited power available or where generally high effectivity is required by handling high power, for example, in communication satellites, and where its bulky appearance is not critical.

Because waveguides have a much larger volume than planar transmission lines at the same frequency, LC layer thicknesses h_{LC} are much greater, for example, about 3 mm in the K_a-band. Polyimide films for the prealignment of LC molecules are no longer suitable. For this kind of voluminous components, only fully electrical biasing (without any prealignment layer) is feasible, using multiple pairs of electrodes to achieve high tuning effectivity and reasonable response times. Moreover, because of larger thickness h_{LC}, the saturation bias voltage V_{sat} increases significantly.

Two opposite electrodes generate an electrostatic field in between. The issue without prealignment layer is to generate an orthogonal biasing field orientation. Unfortunately, the orthogonality cannot be fully met, since the electric field lines are always perpendicular to good conducting surfaces, that is, the electrodes and the waveguide walls. This is illustrated in Figure 5.45, where the simplest form of an electrical biasing configuration is shown with galvanically separated waveguide walls. For the fundamental TE_{10} mode, the RF field vector $\vec{E}_{RF}$ is vertical polarized with a cosine taper of its magnitude. Hence, for the biasing fields $\vec{E}_b$ in both figures, the LC directors $\vec{n}$ with respect to $\vec{E}_{RF}$ are mainly in parallel (left) and tend toward perpendicular (right), that is, $\vec{E}_{RF}$ experiences a material's effective relative permittivity of

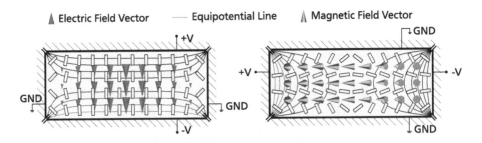

Figure 5.45 LC filled waveguide with galvanically separated walls [23]. On the left, the LC orientation is almost in parallel to the RF field vector $\vec{E}_{RF}$ (arrows) and on the right, LC orientation tends to be perpendicular to $\vec{E}_{RF}$ (arrows represent $\vec{H}_{RF}$).

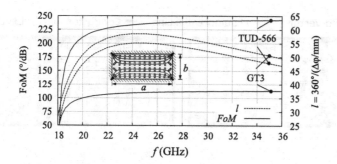

Figure 5.46 FoM and length *l* versus frequency of a GT3-23001- and TUD-566-filled metallic waveguide phase shifter to achieve a 360° differential phase shift. The width *a* and height *b* of the waveguide are set to 5.0 mm and 2.0 mm, respectively [23].

$\varepsilon_{r,\parallel}$ and nearly $\varepsilon_{r,\perp}$, accompanied by the loss tangents $\tan \delta_{\parallel}$ and $\tan \delta_{\perp}$, respectively. To ensure propagation independent of the LC director orientation, the operating frequency must be larger than the lowest cutoff frequency, that is,

$$f > \frac{1}{2b} \frac{1}{\sqrt{\mu \cdot \varepsilon_{r,\parallel}}} \qquad (5.55)$$

since $\varepsilon_{r,\parallel} > \varepsilon_{r,\perp}$.

For the configuration in Figure 5.45, the FoM and length *l* for a 360° phase shift of this waveguide phase shifter are shown in Figure 5.46 versus frequency for two LC mixtures: GT3-23001 and TUD-566. Due to intrinsically low metallic losses of waveguides, their overall losses are dominated by the dielectric losses of the LC only. Hence, despite LC-filled waveguides cannot achieve perfect perpendicular orientation of all LC directors, the FoM is significantly higher compared to LC-filled planar line topologies. For example, using TUD-566, 240°/dB could be achieved compared to about 140°/dB for a microstrip transmission line topology at 30 GHz.

As seen in Figure 5.45, galvanic isolation of the waveguide walls and the associated nonoptimal LC orientation are unfavorable, since the strength of the biasing field $\vec{E}_b$ for the perpendicular alignment of the LC directors to $\vec{E}_{RF}$ (right picture) is quite small in the middle of the waveguide, particularly where the TE_{10} mode has its highest field intensity, and hence where the LC's effective permittivity variation has its strongest impact. Moreover, smaller field strength of $\vec{E}_b$ in the center will reduce the response time.

A significant improvement can be achieved by a configuration of five electrode pairs shown in Figure 5.47, which is capable of emulating the potential curve of a parallel–plate capacitor in both directions. If the gap between two electrodes is chosen to be as wide as the electrode dimension itself and if a linear voltage gradient is applied to the electrodes, the tuning effectivity can reach values as large as 90%. The great advantage of this electrode structure can also be seen in Figure 5.47c: a reference voltage V_{ref} is applied only to the two electrodes at the edges, which are diagonally opposite to each other. It corresponds to the maximum value of the tuning voltage

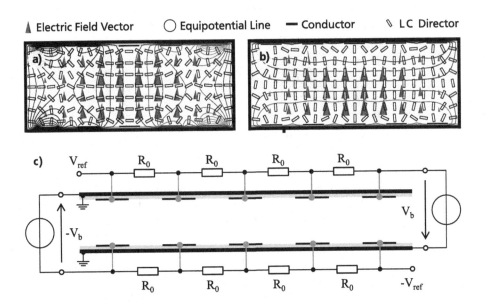

Figure 5.47 LC director orientations inside a waveguide for the two extreme tuning states parallel and perpendicular to RF field $\vec{E}_{RF}$, using a configuration of five electrode pairs [23].

$V_b = V_{sat}$. The different electrodes on the same waveguide wall are connected to each other via high-impedance resistors R_0. Hence, for continuous tuning, only an adjustable bias voltage V_b and its inverse are needed, which is shown in Figure 5.48 for four examples. If the voltage is set to $V_b = V_{ref}$, all electrodes connected to each other will have the same voltage. For example, all the electrodes at the top will have a voltage of V_{ref}, while the electrodes at the bottom will have a voltage of $-V_{ref}$. Thus, the resulting biasing field will align the LC directors almost parallel to $\vec{E}_{RF}$, experiencing $\varepsilon_{r,\parallel}$. A chosen value of V_b smaller than V_{ref} will cause a voltage gradient between the electrodes due to the resistors between them, resulting in a tilted alignment of the LC directors. With the right choice of V_b, every arbitrary intermediate state of the LC alignment is thus possible. The second extreme state will be reached when V_b is chosen to be $-V_{ref}$. In this case, a tendency of horizontal LC orientation can be observed, in particular in the center of the waveguide, that is, $\vec{E}_{RF}$ experiences a nearly perpendicular effective permittivity $\varepsilon_{r,\perp}$, which is smaller than $\varepsilon_{r,\parallel}$. In contrast to the galvanically isolated waveguide walls, the electrical biasing field $\vec{E}_b$ for this five-electrode-pair configuration is much more homogeneous over the whole LC volume. This improves the tuning effectivity, and at the same time, reduces the switch-on and switch-off response times, which are now, both depending on the biasing voltage V_b, beside the material's parameters. Generally, the more electrode pairs are available, the more homogeneous is the LC director alignment over the whole volume. But the improvement is getting smaller and smaller with increasing number of electrode pairs.

To improve the performance of the waveguide phase shifter further, it has to be filled with LC only in the central part, where the field strength of the TE_{10} mode is

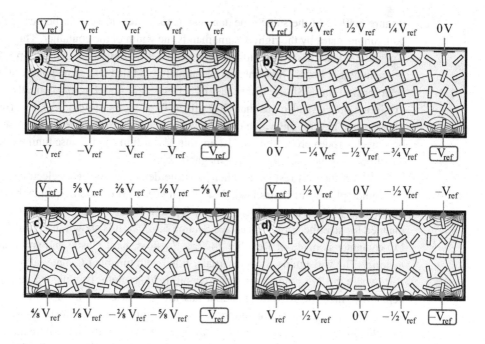

Figure 5.48 LC director orientations for four different applied bias voltages [23].

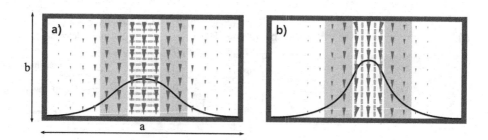

Figure 5.49 Schematic of the director distribution in a partially LC-filled waveguide phase shifter for (a) perpendicular LC orientation to $\vec{E}_{RF}$ ($\varepsilon_{r,\perp}$, $\tan \delta_\perp$) and (b) parallel LC orientation to $\vec{E}_{RF}$, where $\varepsilon_{r,\parallel} > \varepsilon_{r,\perp}$ and $\tan\delta_\parallel < \tan\delta_\perp$ [78].

strongest, as shown in Figure 5.49. In doing so, the losses of the waveguide will be kept lower in the areas where no LC is present. However, it requires an additional container to fill in the LC, which could be made of Teflon ($\varepsilon_r = 2.1$) or Rexolite ($\varepsilon_r = 2.5$). The parts outside of the container can still be air.

The two extreme tuning states are given in Figure 5.49: (a) LC directors are mainly horizontally oriented, that is, in perpendicular to $\vec{E}_{RF}$ of the TE$_{10}$ mode, which experiences $\varepsilon_{r,\perp}$ and $\tan\delta_\perp$ and (b) LC directors are mainly vertical oriented, that is, in parallel to $\vec{E}_{RF}$ experiencing $\varepsilon_{r,\parallel}$ and $\tan\delta_\parallel$.

Since $\varepsilon_{r,\parallel} > \varepsilon_{r,\perp}$, the RF field will be more concentrated in the center part for LC orientation parallel to $\vec{E}_{RF}$ (Figure 5.49b) than for LC orientation perpendicular to $\vec{E}_{RF}$

(Figure 5.49a). At the same time $\tan\delta_\parallel < \tan\delta_\perp$; thus, losses can be nearly equal in both tuning states by the right design, mainly the width of the container. This will have a positive impact in terms of lower average losses and higher differential phase shift, hence in higher FoM. This concept of a partially LC filled waveguide phase shifter has been proved in [78] by using the LC TUD-566, first by magnetically biasing, exhibiting an FoM of more than 200°/dB at frequencies around 35 GHz. However, for electrical biasing this value drops down to about 140°/dB, due to the losses of the biasing electrodes processed on a 12 μm thin Mylar film and because tuning effectivity could not fully be exploited as for magnetic biasing.

Based on the first results, another lab-scale demonstrator of an electrically biased waveguide phase shifter has been designed in [79, 143] for the frequency bands from 23 GHz to 27.5 GHz, specifically for a K_a-band beam steering horn antenna array, planned for satellite applications (see Section 5.6.7.2). It is shown in Figure 5.50. The LC container is made of Rexolite 1422, a low-loss, hard plastic and space qualified material. The waveguide is only partially filled with tunable LC material in the center part, where the TE_{10} mode has its maximum field strength. By this, the losses can be kept low, since the Rexolite material has lower losses than the LC in both tuning states. Due to the dielectric filling of the waveguide, it needs to be tapered in the metallic and in the dielectric parts to avoid reflections at the material edges. This is why the Rexolite container has a taper at each side, while the metallic taper is integrated into the transition from the feed waveguide to the phase shifter section.

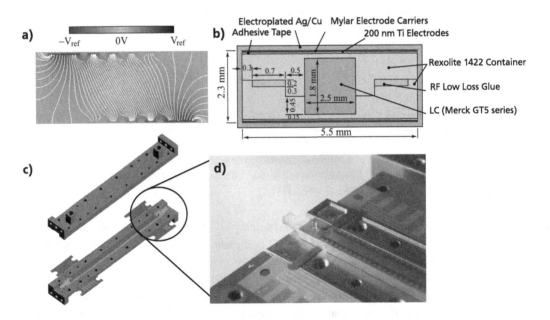

Figure 5.50 Schematics of (a) the biasing field potential, (b) the cross section of the waveguide phase shifter, and (c) waveguide phase shifter segment. (d) Photograph of one end of the waveguide phase shifter with biasing lines and electrodes as well as the dielectric taper of the Rexolite container [143].

To ensure single-mode propagation in the LC-filled waveguide, smaller dimensions with $a = 5.5$ mm and $b = 2.2$ mm are used than for regular waveguides in this frequency range. For practical reasons, a voltage range of -200 V $\leq V_b \leq +200$ V was chosen, resulting in an internal electric field of almost $E_b = 200$ V/mm. Figure 5.50 shows the electrical biasing configuration, using six electrodes on each side, where parts of the electrodes has to be equipped with $\lambda/4$-stub lines and stepped impedance low pass filters to suppress the coupling of the TE$_{10}$ mode into stripline modes between the waveguide walls and the biasing electrodes. The waveguide itself is built up in a split-block design made of brass, in which the Rexolite core is inserted. This allows a flexible testing, while the design of the container is prepared to have the waveguide walls electroplated around it. The split-block has small gaps for the biasing electrodes to be led out. They will be contacted at the outside of the waveguide to avoid additional losses due to soldering joints or bonding wires inside the waveguide.

The measurement results of the S-parameters, the resulting differential phase shift $\Delta\varphi$, and FoM of the waveguide phase shifter with LC GT5-26001 are shown in Figure 5.51 over a frequency range from 21 GHz to 35 GHz for electrical tuning only. In [143], the results for magnetic biasing are given in addition. Over the

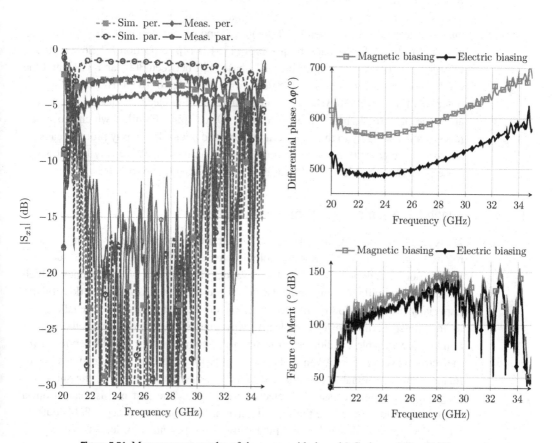

Figure 5.51 Measurement results of the waveguide-based LC phase shifter [143].

complete design range from 23 GHz to 27.5 GHz, the matching is better than -12 dB. Only below 23 GHz it goes up to more than -10 dB. The *IL* is between 2 dB and 4.0 dB in this range, which was underestimated in the simulation. This was due to the fact that the waveguide losses could not be estimated precisely, and effects of the electrode sheets could not be taken into account in the simulations. The differential phase shift is above 550° if the phase shifter is magnetically biased and above 450° if it is electrically biased with the preceding configuration. The difference between both biasing schemes is due to the fact that the LC can be oriented much better (nearly ideal) in parallel and perpendicular to $\vec{E}_{RF}$ with magnets, producing more homogeneous field inside than with electrical biasing. For the electrically biased phase shifter, an FoM can be achieved of about 120°/dB (more than 130°/dB for magnetic biasing) in the desired frequency range and 130°/dB (150°/dB for magnetic biasing) in the frequency range of 27–32 GHz. The response times of this phase shifter was estimated to be between 1 and 2 minutes at room temperature. Since the phase shifter is designed for a temperature-controlled antenna at 60°C, improvements in the order of magnitude of 10 to 12 can be expected for the response time.

Partially LC-filled waveguide phase shifters are particularly well suited for mm-wave and terahertz applications due to waveguide's inherently low metallic losses. Therefore, as a first proof-of-concept, a magnetically biased waveguide-integrated phase shifter with an LC-filled PTFE container and some tapering for better matching was demonstrated for W-band frequencies in [54, 140]. Reflection is below -10 dB, insertion loss below -3 dB, differential phase shift between 300° and 320°, and the FoM is between 116°/dB and 148°/dB from 95 GHz to 105 GHz, revealing the potential of this technology for W-band applications. For electrical biasing, electrodes are made of chromium and processed on a 20 μm thin PET film, which, however, reacts very sensitively to the metal evaporation and photolithography process, causing microscopic cracks. Hence, these components could not be electrically biased, since these processing problems could not be solved sufficiently up to now.

5.4.3   Tunable Dielectric Waveguide Phase Shifter

Designing continuously tunable phase shifter in the mm-wave or even terahertz regime suffers from increasing ohmic losses of rough metallic surfaces. Furthermore, the dimensions of metallic waveguides decrease with frequency, which makes the technological implementation of the electric biasing network difficult. Especially the fabrication of the electrodes on suitable substrates is very challenging, as mentioned earlier. Instead, dielectric waveguides, well known from the optical domain, are suitable candidates up to terahertz frequencies. A good overview over dielectric waveguides is given in [156]. Two types of dielectric waveguides are of major interest for tunable line phase shifters as shown in Figure 5.52, the subwavelength fiber and the "classical" dielectric waveguide. The first one is made from a single dielectric core, where the guided wave has decaying evanescent fields outside the dielectric core, that is, in air. Inherently, this fiber has low losses, but is very sensitive to surrounding objects, for example, metallic structures. The classical

a)   b)

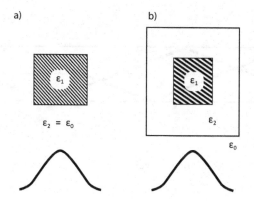

Figure 5.52 (a) Subwavelength fiber and (b) classical dielectric waveguide. For both, the typical E-field distribution is shown for $\varepsilon_1 > \varepsilon_2$.

dielectric waveguide consists of a core and a cladding material. Here, the evanescent field is decaying inside the cladding.

Inserting an LC section inside the dielectric core allows to tune the propagation constant of the wave, and hence, the differential phase by using simply two parallel pairs of electrodes around for biasing, which is easier configuration than for metallic waveguide phase shifters [79]. These parallel plates surrounding the dielectric fiber leads to a homogeneous biasing field, since no metallic boundaries are present, which disturb the electrostatic field. Like in metallic waveguides, prealignment layers are not feasible, since LC layers are too thick, causing low tuning effectivity and very long response times. Since plastics such as Teflon and polyethylene as well as processing technologies such as 3D printing, injection molding [157], or milling can be used for fabricating dielectric waveguide components, it could be very cost efficient, especially for mass production.

In [158, 159] measurements of permittivity and dissipation factor for a frequency range from 50 to 110 GHz are given for some common plastics, where PTFE and HDPE exhibit the lowest losses and both show nearly no change in their permittivity over the whole frequency range. For milling, Rexolite, a crosslinked polystyrene, shows the best manageability. Moreover, it is a space-approved material.

Figure 5.53 illustrates how an electromagnetic wave is guided inside a dielectric waveguide described by a simple geometrical approach.

A dielectric core ε_1 is surrounded by a second dielectric material or air with ε_2, where $\varepsilon_1 > \varepsilon_2$.

According to Snell's law, internal reflections occur for certain angles $\leq \theta_c$ at the interface between the two permittivities, where the upper bound is given by $\cos \theta_c = \dfrac{\varepsilon_2}{\varepsilon_1}$. Since the DW supports certain incident angles, the wave can travel inside the waveguide on different ways with different phase relations, where the phase shift must be $\Delta\varphi = 2\pi n$, with $n = 0, 1, 2, \ldots$ to achieve constructive interference. The phase fronts of two such rays with an angle of incidence θ are shown in Figure 5.53, where the phase shift of the ray with the solid line from a to b must be

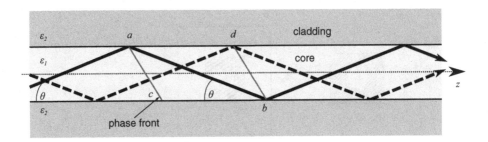

Figure 5.53 Propagation principle in a dielectric waveguide.

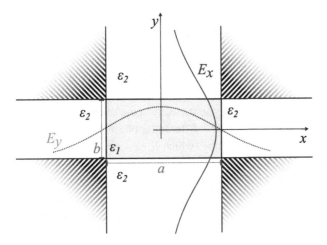

Figure 5.54 Schematic geometry of Marcatili's approximate method. The field distribution indicates the fundamental mode of a rectangular dielectric waveguide.

an integer multiple of the phase shift for the dashed lined ray from c to d. Although closed formulations of the dispersion relation of circular-shaped dielectric waveguides can be found, it does not exist for rectangular-shaped dielectric waveguides.

Problematic are the electric and magnetic fields in the corner region, since they are difficult to calculate due to the presence of two boundary conditions. An approximate method was proposed by Marcatili [160]. Figure 5.54 shows the geometry of the approach, consisting of a dielectric core ε_1 surrounded by a second dielectric ε_2. The main assumption in this approach is that the field in the shaded region can be neglected, which leads to a calculation problem of two perpendicular aligned dielectric slab waveguides. Furthermore, the fields are assumed to be sinusoidal in the core region and exponentially decaying in the outer regions. With this approach, the main task is now to match the fields at the boundaries $y = \pm\dfrac{b}{2}$ and $x = \pm\dfrac{a}{2}$.

Due to the superposition of two dielectric slabs, two independent modes exist named after the predominant polarization of the electric field. In the case of an electric field polarized in x direction, the modes are named E^x_{mn} and E^y_{mn} for a polarization in the y direction, with $m, n = 1, 2, \ldots i$, where m and n determines the number of extrema, occurring in the dielectric core. The propagation constant is given by

$$k_z = \sqrt{\varepsilon_1 k_0^2 - k_x^2 - k_y^2}, \tag{5.56}$$

using the transcendental equations

$$k_y b = m\pi - 2\tan^{-1}\left(\frac{\varepsilon_2}{\varepsilon_1}\frac{k_x}{\sqrt{k_1^2 - k_2^2 - k_y^2}}\right) \quad \text{and} \quad k_x a = m\pi - 2\tan^{-1}\left(\frac{k_x}{\sqrt{k_1^2 - k_2^2 - k_y^2}}\right). \tag{5.57}$$

This approximation method allows a first calculation of the waveguide dimensions a and b. Later, these dimensions are validated and optimized by using numerical solvers. This provides a more accurate calculation, since they allow a proper consideration of the fields in the corner region.

5.4.3.1 Tunable Subwavelength Fiber and Dielectric Waveguide

Because of its simplicity, first, a continuously tunable subwavelength waveguide with an LC segment inserted inside the fiber according to Figure 5.55 has been designed for W-band frequencies [147]. As fiber material, use was made of Rexolite with $\varepsilon_r = 2.53$

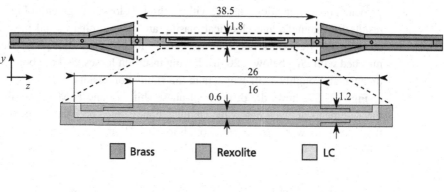

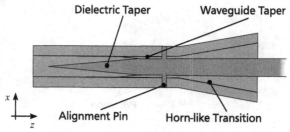

Figure 5.55 Schematic of a rectangular-shaped subwavelength fiber phase shifter with an edge length of 1.8 mm and a hollow waveguide transition [147]. All dimensions given in mm.

and $\tan \delta = 0.006$ offering good mechanical and electrical properties at the desired frequency range. The fiber has a rectangular with an edge length of 1.8 mm while the LC segment has a cylindrical shape of 0.6 mm diameter. The length of the LC segment is 26 mm.

For measurement purposes, two WR10 to subwavelength transitions are used, designed as a horn-like structure. The design principles of such a horn-like transition can be found in [156]. It is more suitable to preserve the evanescent fields of the Rexolite subwavelength fiber, whereas a dielectrically filled open-ended waveguide would cause high radiation due to the sharp edges. Furthermore, the hollow waveguide is also tapered to prevent the propagation of higher order modes due to the dielectrically filling. The fiber itself is fixed by Rexolite pins in the metallic waveguides to ensure the right position of the fiber inside the waveguide.

Since parts of the guided electromagnetic wave are decaying outside the fiber, the two electrodes around the LC segment needs to be placed in a certain distance away from the fiber. This has been realized with Rohacell ($\varepsilon_r = 1.03$) as spacer material surrounding the LC segment, where the electrodes are glued on the Rohacell with a distance of 6 mm to the fiber. The measurement setup is shown in Figure 5.56. Magnetic biasing was done by means of rare earth magnets and electric biasing was done by a voltage of about ± 550 V at the electrodes. For the measurement, the calibration plane, using a TRL calibration, was set to the WR10 connection flanges.

A comparison of the measurement and simulation with the LC GT5-26001 for perpendicular orientation, that is, with the highest losses, is given in Figure 5.57. The results for magnetic and electrical biasing are quite similar for all measured parameters. In the frequency range from 95 GHz to 110 GHz, the phase shifter is well matched with $|S_{11}|$ below -20 dB, having insertion losses $|S_{21}|$ of about 1.5 to 2.5 dB on average, roughly 1 dB worse than the simulation results.

Figure 5.57 shows the differential phase shift $\Delta \varphi$ and FoM, where the ones from measurements are well below the simulated ones, since both the magnetic and electric field strengths were not strong enough to fully align the LC, that is, less orientation effectivity τ_φ, and hence, lower tuning effectivity $\Delta \tau_\varphi$, overlapped by undesired

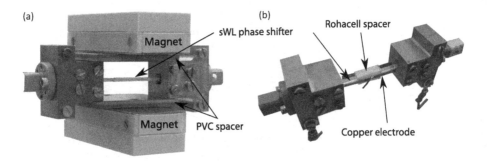

Figure 5.56 Measurement setup of the subwavelength phase shifter. (a) Magnetic biasing with rare earth magnets and (b) electric biasing with copper electrodes mounted on Rohacell blocks.

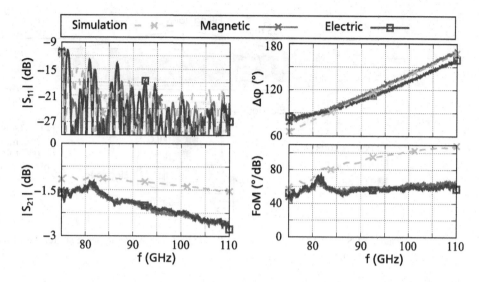

Figure 5.57 S-parameter measurements and simulation of the subwavelength fiber phase shifter with a segment filled with LC GT5-26001 in perpendicular orientation, that is, with the highest losses (left). Differential phase shift $\Delta\varphi$ and FoM (right).

additional surface anchoring effects at the cavity walls. Nevertheless, the differential phase shift is between 120° and 170° and the FoM is in a range of 55–62°/dB within the frequency range from 95 to 110 GHz for electrical biasing.

Since the electrode placement of the LC-filled subwavelength fiber has been proven as critical during the experiments owing to the exponentially decaying evanescent field outside the dielectric core, a classical dielectric waveguide, consisting of a core and a cladding, was investigated as a further step in [161]. It is expected to be less critical, since the electromagnetic wave is decaying inside the cladding. Thus, the electrodes can be placed directly on the waveguide. A design for such a dielectric waveguide phase shifter is shown in Figure 5.58.

The dielectric waveguide consists of a Rogers 3003 core ($\varepsilon_r = 3$, $\tan\delta = 0.001$) and a Rexolite cladding. In between, an LC core is inserted for tunability. For measurement purposes, again horn-like transitions are used according to Figure 5.55. To connect the Rexolite subwavelength fiber to the dielectric waveguide with the Rogers core and the Rexolite cladding, both are smoothly tapered toward the final dimensions.

Since most of the used LC mixtures have permittivity values for perpendicular orientation lower than the permittivity of the Rexolite cladding, the propagation condition for a dielectric waveguide is no longer fulfilled. To overcome this problem, air gaps are milled into the cladding material to reduce its effective permittivity. Figure 5.59 depicts the simulated field distribution at 100 GHz with electrodes attached on the cladding material. For this simulation, a permittivity of $\varepsilon_r = 2.4$ is assumed for the LC core, which is below the one of Rexolite with $\varepsilon_r = 2.53$. However, with the milled air gaps in the Rexolite cladding, the field is mainly

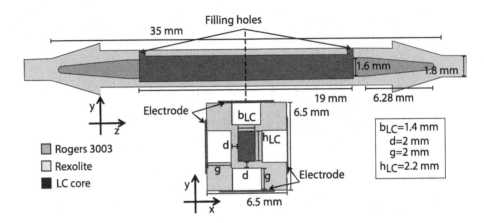

Figure 5.58 Schematic view of an LC-filled dielectric waveguide phase shifter with two horn-like transitions.

Figure 5.59 Simulated field distribution at 100 GHz of the classical dielectric waveguide with LC core and Rexolite cladding. The gray boxes indicate the milled air gaps into the cladding material.

confined inside the LC core. Furthermore, the propagating mode is not disturbed by the metallic electrodes. Figure 5.60 shows the measurements setup. The dielectric waveguide phase shifter is made from two Rexolite halves glued together with an ultraviolet glue. In between both halves, the Rogers core, which is milled out of a substrate sheet, is inserted. As a first proof-of-concept, a copper tape is stuck onto

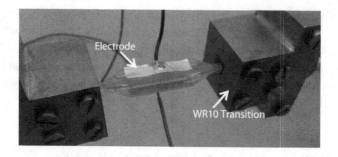

Figure 5.60 Measurement setup of the proposed dielectric waveguide phase shifter. Copper tape is used to realize the electrodes for electrical biasing.

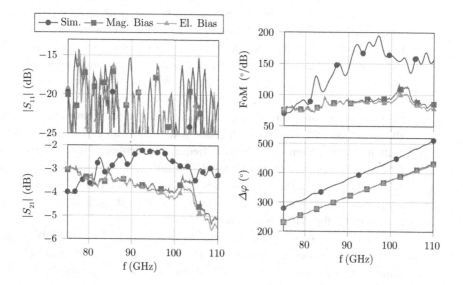

Figure 5.61 S-parameter measurements and simulation of the dielectric waveguide phase shifter with a segment filled with LC GT5-26001 in a perpendicular orientation, that is, with the highest losses (left). Differential phase shift $\Delta\varphi$ and FoM (right).

the cladding. In total four electrodes are used to obtain full LC orientation from perpendicular to parallel with a voltage difference of ±550 V. The magnetic biasing was done by means of rare earth magnets.

The measurements of the S-parameters, the derived FoM, and differential phase shift along with the simulation results are shown in Figure 5.61. Like the results from the subwavelength fiber, there are some differences between measurement and simulation results. This could be due to fabrication tolerances during the assembling and some additional losses induced by the glue. Nevertheless, the measured insertion loss is in same range of 3 to 4 dB as for the subwavelength fiber over the frequency range of 75–105 GHz, whereas matching with $|S_{11}|$ below −15 dB is better.

More decisively, also the achieved FoM with 75 up to 100°/dB with electric biasing is much better. This is also valid for the achieved differential phase shift, which is in a range of 220°–420°. Again, both, the differential phase shift and FoM from measurements are below the simulated ones. This is because magnetic and electric field strengths were not strong enough to fully align the LC, thus lowering the tuning effectivity $\Delta\tau_\varphi$, which might be further reduced by anchoring effects at the Rexolite.

5.4.3.2   Parallel-Plate Waveguide with LC-Filled Dielectric Slab

Due to the loose field confinement, some radiation occurs when the wave propagates along the aforementioned dielectric waveguide topologies, especially at discontinuities such as bends, taper, and edges. Kwan et al. [162] proposed a partially dielectric filled parallel-plate waveguide as a low-loss and quasi-radiation free waveguide topology. It consists of a dielectric slab sandwiched between two metal plates on top and bottom, forming a parallel-plate waveguide as depicted in Figure 5.62.

The fundamental mode of this parallel-plate dielectric waveguide (PPDW) is the TE_{10} mode, where the E field is perpendicular to the metal plates. The guiding mechanism of the electromagnetic wave is similar to that of the dielectric waveguide, due to internal reflection at the interface between dielectric slab and air. Figure 5.63 shows a cross section of a parallel-plate waveguide with an LC-filled dielectric slab section inside, designed for V-band frequencies between 50 GHz and 75 GHz in [150]. For the characterization of this phase shifter, it is designed for a connection to WR15 flanges, including a horn-like transition according to the Figure 5.63. For simplification, the height of the parallel plate waveguide with dielectric slab has been chosen as for the WR15 waveguide.

Figure 5.64 shows the electrode biasing configuration for the PPDW phase shifter. To avoid any undesired resonances on the electrodes, use is made of a stepped-impedance structure similar to the LC-filled metallic waveguide phase shifter, compare Section 5.4.2. In this design, a more homogeneous static biasing field is achieved

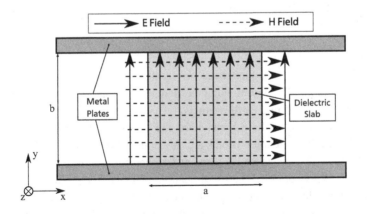

Figure 5.62 Schematic view of a parallel-plate waveguide partially filled with a dielectric slab.

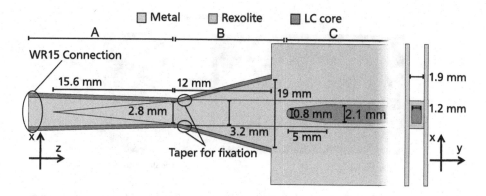

Figure 5.63 Cross section of a PPDW phase shifter with (A) a dielectric taper, (B) a horn transition from WR15 to the parallel-plate waveguide with dielectric slab, and (C) a parallel plate waveguide with an LC-filled dielectric slab inside.

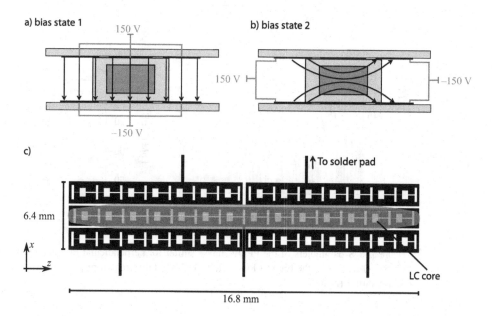

Figure 5.64 View of (a) two extreme biasing states and (b) the electrode configuration. Babinett's principle is used in the electrode design (c).

with the use of Babinett's principle, inverting the metallic structure with air as depicted in Figure 5.64. This increases the metal area but avoids also resonances on the electrodes. This figure sketches on top also the two extreme biasing states for parallel and perpendicular LC orientation.

The manufactured PPDW consists of five single parts, two WR15 transitions and two metal plates, and the dielectric slab. The WR15 transitions are made from brass in split block technology with the metal plates stuck into gaps in the brass block as

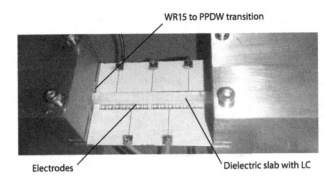

Figure 5.65 The assembled PPDW with one metal removed to facilitate a view on the dielectric slab filled with LC GT5-26001. The electrodes are glued on the metal plates.

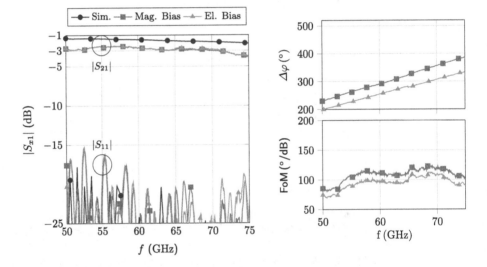

Figure 5.66 S-parameters of the PPDW phase shifter for perpendicular orientation of the LC GT5-26001, that is, the highest losses (left). FoM and differential phase shift $\Delta\varphi$ of the PPDW phase shifter (right).

shown in Figure 5.65. The dielectric slab is made from two Rexolite halves glued together with an ultraviolet glue. A maximum voltage of ± 150 V is used to tune the orientation of the used LC GT5-26001. The corresponding S-parameter measurements are shown with the simulation results in Figure 5.66. In the frequency range between 50 GHz and 75 GHz, it indicates on average -3 dB for the $|S_{21}|$ measurements, which is about 1.5 dB less than predicted by simulation due to higher losses than expected. Nevertheless, the phase shifter is well matched over the whole frequency range with $|S_{11}| < -15$ dB. The differential phase shift varies between 200° and 400° and the FoM between 75° and 120°/dB, where the values for both are better for magnetic biasing than for electric biasing as expected, because of higher tuning effectivity.

5.4.4   Monolithic Integration of LC Phase Shifter with Different Technologies

5.4.4.1   LC-Based Low-Temperature Co-fired Ceramic Phase Shifter

One potential way of realizing a hermetically sealed cavity for LC is by using the low-cost multilayer low-temperature co-fired ceramic (LTCC) technology, which is also space qualified, and hence suitable for space applications. LTCC is mechanically stable and allows a three-dimensional structuring, using conductivity, resistance, or capacitive pastes between the layers. Structures on different layers can be connected by vias and those vias can also be used to form metallic cavities inside the LTCC structure. By this, a cavity like a metallic waveguide can be built up inside the LTCC, thus, combining high performance with low profile. After processing the LTCC structure, microwave-optimized LC mixtures are filled into these cavities, thus enabling compact and tunable components such as filters and phase shifters. These phase shifters can easily be integrated into an LTCC module together with biasing lines and electronics for tuning as well as with individual antenna elements to form an array.

Two different LC-LTCC delay line phase shifter topologies have been investigated: (1) a metallic waveguide and (2) quasi-planar microstrip line, both embedded into the LTCC structure. The latter one is exemplarily shown in Figure 5.67 [23]. The different LTCC CT707 layers with $\varepsilon_{r,CT707} = 6.13$ at 30 GHz can be seen in the cross section on top of Figure 5.67, while the bottom picture illustrates the tunable microstrip line in propagation direction.

For characterization purposes with GSG-probes, CPW to MSL transitions are integrated into the LTCC structure, indicated in Figure 5.67. The CPW pads can clearly be seen also on the top view of a realized microstrip line phase shifter in LTCC technology according to Figure 5.68, having a length of 58.8 mm, a cavity height and width of about 70 μm (± 10 μm) and 300 μm, respectively. Since the MSL is on a different layer than the CPW pads, they must be connected by means of vias. The signal line of the CPW is therefore connected to the MSL, while the two ground pads are connected to the ground plane, processed on the LTCC layer. In between the two LTCC layers with the MSL and the ground plane, an additional layer is included,

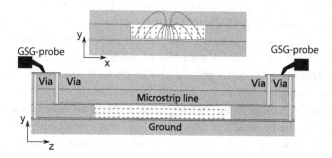

Figure 5.67 Schematics of the cross section (top) and in propagation direction (bottom) of the LC-LTCC microstrip line phase shifter. The LC orientation in the upper picture is shown for a certain biasing voltage, whereas the LC orientation in the lower picture is given for the unbiased state.

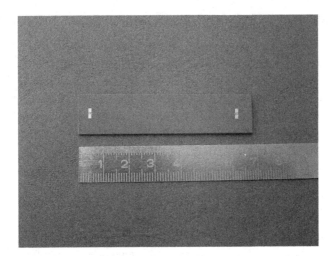

Figure 5.68 Photograph (top view) of a realized LC-LTCC microstrip line phase shifter.

having a punched−out channel in which the LC is filled in afterward. Despite a prealignment layer is not possible to implement into the LTCC cavity, the roughness of the LTCC is enough to prealign the LC mainly in the propagation direction (see lower picture in Figure 5.67). By applying a biasing voltage to the signal line, the LC molecules will change their orientations till most of them are in parallel to the biasing field lines at the saturation voltage. For a proof-of-concept and a comparison, the realized microstrip line phase shifter in LTCC technology has been filled with three different LCs: K15, TUD-188, and TUD-566.

Despite all measurements have been carried out with all three LCs in [23], the S-parameters for two biasing states are given in Figure 5.69 for the TUD-566 filled phase shifter only: matching $|S_{11}|$ is better than −13 dB over the whole frequency range from 25 GHz to 35 GHz, while the insertion loss is below 4.5 dB. Measurements of the empty phase shifter indicated that the insertion loss is dominated by metallic losses, leaving a proportion of the LC losses of just around 1.3 dB. Since this represents a minor part, a distinct increase in the electrical performance could be achieved only by a higher material tunability of the LC, assuming similar material quality factor. This allows a shorter phase shifter length for the same differential phase shift, which automatically reduces the metallic losses of the device.

The maximum differential phase shift $\Delta\varphi_{max}$ and FoM for all three LCs are given in Figure 5.69, where the phase shifter filled with TUD-566 exhibits definitely the best performance, where the differential phase shift rise from about 250° to around 350° over the whole frequency range, while the FoM keeps around 70°/dB. However, its better performance is on the expense of a much slower response time as given in [23], where the measured differential phase shift with respect to its maximum exhibit a 90% to 10% decay, that is, switch-off times T_{10}^{90} of about 13 s, 33 s, and 150 s for K15, TUD-188, and TUD-566, respectively.

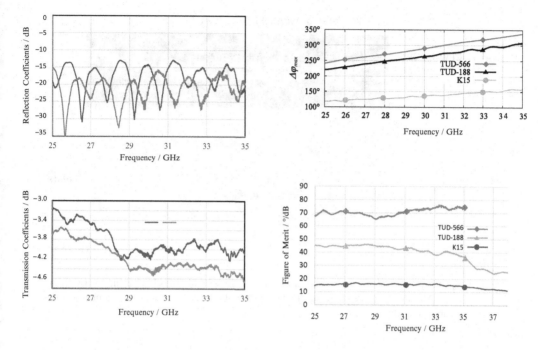

Figure 5.69 S-parameter measurements (left) and resulting FoM and differential phase shift $\Delta\varphi$ (right) of the microstrip line phase shifter in LTCC technology filled with the LC TUD-566 [23].

5.4.4.2   LC-Reflection-Type Phase Shifter in MEMS Technology

As already mentioned, important requirements for phase shifters implemented in electrically steerable planar phased array antennas is the provision of a 360° differential phase shift and an overall size not exceeding $\lambda_0/2 \times \lambda_0/2$ (λ_0 is the wavelength in free space) to fit behind a single antenna element. An option to welt a 360° phase shifter within this squared area even if its length ℓ exceeds $\lambda_0/2$ by far is meandering or spiraling it. However, each bend of a transmission line causes additional losses. Therefore, a straight delay line phase shifter is always a preferred solution.

This might be achieved by utilizing the so-called slow-wave effect, by which the phase velocity of the transmission line can be reduced to miniaturize it by factor three to five as explained in the next two sections. In this section, a 90° phase shifter with reduced length compared to a 360° phase shifter is used in combination with an MEMS switch at the end as shown in Figure 5.70, which means 0° or 180° phase shift of the reflected wave for open or short of the MEMS switch at the end. Hence, this reflection-type phase shifter requires a 90° tunable delay line only for to tune the overall phase shift continuously between 0° and 360°. To reduce the response time at the same time, this reflection-type phase shifter uses the loaded-line concept described in Section 5.4.1.2.

Thus, by using the MEMS technology apart from realizing true-time-delay phase shifters as in [2], a 360° reflection-type LC-MEMS phase shifter was proposed first

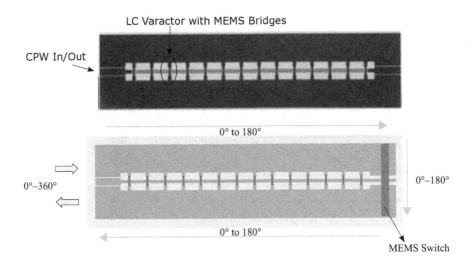

Figure 5.70 Principle of the reflection-type phase shifter with a 90°loaded-line phase shifter and an MEMS switch at the end (top). Photograph of the loaded-line LC-MEMS phase shifter (bottom) [163, 164].

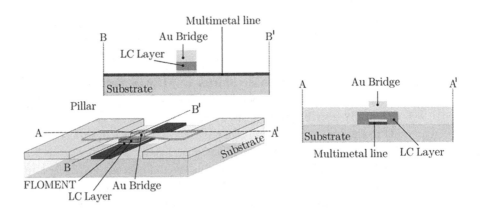

Figure 5.71 Schematics of a unit cell, the cut in the A–A′ plane and the cut in the B–B′ plane of the unit cell of the reflection-type LC-MEMS phase shifter [163, 164].

time in [163, 164], combining a 90°-coplanar waveguide phase shifter periodically loaded with LC filled underneath MEMS bridges as varactors (several unit cells) and an MEMS switch at the end for generating a 0° or 180° phase shift of the reflected wave. It was realized in W-band. Figure 5.71 exhibits a schematic of such a unit cell with an LC filled underneath an MEMS bridge. The CPW line was processed on a 300 μm thick fused silica substrate, where the ground planes are made of a 5 μm thick gold layer, while the signal line is made of a 600 nm thick multimetal Ti–TiN–Al–Ti–TiN layer. On top of the multimetal layer, a 150 nm thick gold layer is processed, creating a floating metal

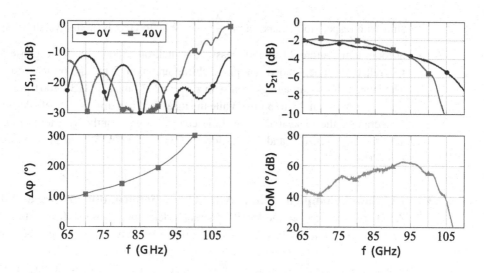

Figure 5.72 S-parameter measurement results, differential phase shift and FoM of the reflection-type LC-MEMS phase shifter filled with GT3-23001 [163, 164].

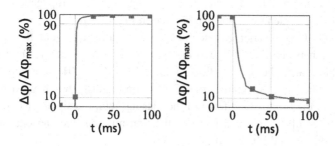

Figure 5.73 Transient measurements of the loaded-line phase shifter filled with GT3-23001 at 76 GHz and 23°C [163, 164].

bridge. To realize the moveable bridge, and thus to form the thin LC cavity, photoresist was deposited above and around this multimetal gold layer, before the gold parts such as the bridge, the pillars or the grounds have been processed by using an electroplating process. Finally, the photoresist has been removed by using oxygen plasma. The LC has been filled under the MEMS bridges by using a microinjector. The LC was then sucked under the bridges due to the capillary forces, which also kept it in there.

The results of the S-parameter measurements and the resulting FoM of this reflection-type LC-MEMS phase shifter filled with GT3-23001 are given in Figure 5.72. Matching is better than −15 dB from 65 GHz up to 95 GHz. In this frequency range, the differential phase shift varies between 100° and 240°, while the IL varies between −2 dB and −4 dB, which results in an FoM between 40°/dB and 60°/dB. The maximum biasing voltage for this measurement was set to $V_b = 30$ V. According to Figure 5.73, where the rise and decay of the measured differential phase shift with respect to its maximum $\dfrac{\Delta\varphi}{\Delta\varphi_{max}}$ in % at 76 GHz is plotted versus time t in ms at room temperature 23°C, the measured switch-on and switch-off response times of this reflection-type, loaded line phase shifter are about 2 ms and 69 ms, respectively. The latter value again agrees well with $\tau_{off} = 62\text{--}77$ ms for GT3-23001 in Figure 5.1 for h_{LC} in-between 4.5 to 5 µm. While the measured rise time is sparsely dependent on temperature, the measured decay time changes significantly with temperature: it is 136, 69, 37, 17, 11, 7 and 5.2 ms for 10, 23, 30, 40, 50, 70 and 90°C, respectively [163, 164].

5.4.4.3   LC-Based Complementary Metal-Oxide-Semiconductor Phase Shifter

Another way to reduce the length of straight delay line phase shifters is to utilize the so-called slow-wave effect, by which the phase velocity of the line can be reduced to miniaturize it by factor three to five [165, 166]. An option to decrease the phase velocity is by increasing the relative effective permittivity by adding a patterned shielding plane to a CPW as described in [100]. A first demonstrator of a tunable slow-wave phase shifter filled with LC and realized by a 0.35 µm CMOS process in the mm-wave regime has been reported in [83]. It is based on a standard CPW loaded by a patterned ground shield orthogonally arranged to the propagation direction. A schematic view of demonstrator is presented in Figure 5.74.

After the CMOS, a vapor RF-etching process was carried out to remove the dielectric layer between the CPW and the orthogonal pattern shield. The CPW was then suspended and the cavity filled with LC GT3-23001 (see top view in Figure 5.75). Then, the orientation of LC can continuously be changed by a bias voltage. Due to the electrostatic force of the biasing field, the signal electrode of the CPW deflects, resulting in an increase of the transmission line capacitance, which increases the differential phase shift.

The S-parameter measurements of this device, having a physical length of 2.1 mm, are presented for two bias voltages in Figure 5.76: Matching $|S_{11}|$ is better than

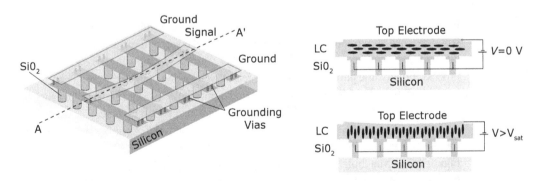

Figure 5.74 Schematic view (left) of the slow-wave CPW filled with LC. The signal electrode is deflected by an electrostatic force. Cross section (right) along the length of the signal electrode with perpendicular and parallel LC orientation, respectively [83].

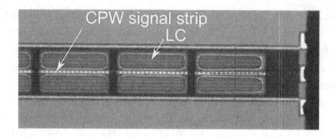

Figure 5.75 Top view photograph of the slow-wave CPW filled with LC [83].

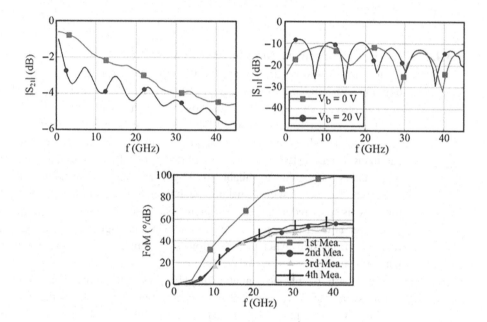

Figure 5.76 S-parameter measurements and FoM of a slow-wave CPW phase shifter realized by a 0.35 μm CMOS process and filled with GT3-23001 [83].

-10 dB over a wide frequency range and even better than -13 dB between 30 GHz and 45 GHz, while the insertion loss increases with frequency being in the range of 4–5.5 dB from 30 GHz to 45 GHz.

The differential phase shift, and hence, the FoM of the first measurement is significantly higher than for the three follow-up measurements, which is due to the deflection of the signal electrode after switching on the biasing voltage of 20 V for the first time. After switching off the biasing voltage again, the signal electrode will not fully return to its origin state. The follow-up measurements exhibit similar, that is, reliable results of the differential phase shift of about 275° at 45 GHz, resulting in an FoM of around 52°/dB. Note that FoM would increase by factor 2.1, that is, to more than 100°/dB when using LC TUD-566.

Hence, this hybrid technology enables a 360° differential phase shift with a length of the phase shifter of $\ell = 0.4\,\lambda_0$, small enough for to integrate it as a straight delay line phase shifter underneath an individual antenna element with a size of $\lambda_0/2 \times \lambda_0/2$. It demonstrates how the slow-wave effect can be utilized for miniaturization. Moreover, the LC layer height of the demonstrator could be kept as low as $h_{LC} = 1$ μm only. This would fasten up the switch-off response time τ_{off} significantly to 4 ms, due to the decrease of τ_{off} with h_{LC}^2 and the good initial anchoring of the LC to the rectangular shape of the patterned ground shield. However, the suspended CPW signal is attracted by the patterned shield when applying a bias voltage. This phenomenon is similar to MEMS behavior. Consequently, the demonstrator is difficult to control.

5.4.4.4   LC-Based Nanowire Membrane Phase Shifter

Up to now, slow-wave devices are mostly implemented in high-cost CMOS technology. In contrast to CMOS technology, the nanowire-filled membrane (NaM) technology can provide similar performance, however, with significantly reduced fabrication effort and at least 10 times reduced fabrication cost. The NaM substrate is based on alumina obtained through electrochemical oxidation of aluminum under specific anodizing voltages [167–170]. This membrane is mostly referenced as anodic alumina oxide and is commercially available from Synkera. This alumina membrane is used as a template for the growth of metallic nanowires. First, a thin copper film of 20 nm thickness is sputtered as a seed layer on the backside of the membrane. After that, the seed layer is used to grow the copper nanowires through the membrane, until they reach the front side. After this, the front side is polished to provide a smooth and flat surface.

With these membranes, nontunable [166] as well as tunable [84] slow-wave transmission lines have been realized recently. Figure 5.77 shows the schematic of a tunable slow wave microstrip line based on the combined NaM and LC technology. The nanowires are connected to the ground plane on the backside and separated from the strip line by a thin dielectric layer of few micrometers only, whereas the membrane used for the first demonstrator has a thickness of 50 μm. While the magnetic field can pass through the nanowires almost unperturbed as in a classical microstrip transmission line, the electric field is confined between the dielectric layer above the nanowires

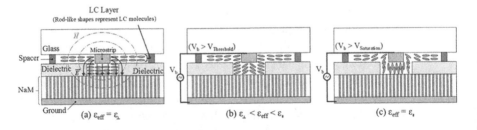

Figure 5.77 Liquid crystal filled slow-wave microstrip line for three different biasing conditions.

and the microstrip signal electrode. Hence, an increase of the equivalent capacitance C' of the transmission line is obtained while the inductance L' remains unchanged, providing a slow-wave effect characterized by a decrease of the phase velocity $v_{\text{ph}} = 1/\sqrt{L' \cdot C'}$. This slow-wave effect strongly enhances when the strip line is close to the end caps of the nanowires, that is, when the LC layer thickness h_{LC} is very thin. The NaM technology allows the realization of tunable microstrip lines for 50 Ω design with LC-layer thickness in the range of 1–3 μm, compared to 50–150 μm for conventional LC-based microstrip lines. This enables a significant reduction of the switch-off response time τ_{off}, which is inversely proportional to the square of the cavity height h_{LC}^2.

For the realization, the microstrip line has been processed on a glass substrate by means of photolithography. Additionally, the alignment layer as well as cavity walls made of SU8 have been processed on the same substrate. The SU8 cavity walls are processed in a way that the LC layer height of 1 μm is ensured. The membrane is afterward glued to the already processed substrate. LC filling is done under vacuum conditions and the parallel orientation is achieved by the application of a DC biasing voltage to the strip line itself. A photograph of the first LC-NaM phase shifter demonstrator with a total length of 1 mm is given in Figure 5.78.

The measurements for the LC GT3-23001 have been conducted in W-band and are given in Figure 5.79. Matching was poor with values of −7 dB to −8 dB from 65 GHz to 110 GHz. This is due to an improper capacitive coupling from the CPW contact pads to the NaM. The insertion loss is between 2.5 dB and 4.0 dB and a differential phase shift has been measured from 50° to 92°. This results in a phase shifter FoM above 20°/dB above 75 GHz with a maximum value of 28°/dB at 81 GHz and 96 GHz. The maximum voltage used for biasing was 20 V. Note that the FoM would increase by factor of 2.1 when using LC TUD-566.

Despite these first results not being good enough, the potential of this hybrid LC–NaM technology is supposed to be high, because it enables tunable low-profile components for high frequencies with very fast response time and low tuning voltages due to the cavity height of 1 μm only. From the measurements of the differential phase shift versus time in Figure 5.80, it can be seen that the switch-on response time of

Figure 5.78 Photograph of the LC-NaM phase shifter demonstrator.

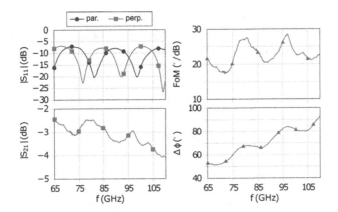

Figure 5.79 Measurement results of the straight-line LC–NaM phase shifter.

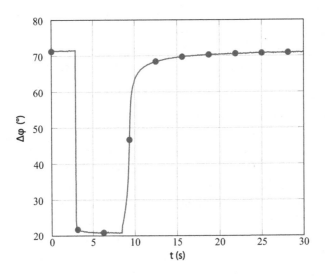

Figure 5.80 Response time measurement of the LC-NaM phase shifter demonstrator at 65 GHz.

about 25 ms is faster than the switch-off response time of around 1.9 s. However, the latter is slower as expected from simulation, mainly because the prealignment layer being processed on one substrate only seems to be harmed by the photolithography processes, which have been conducted afterward. Consequently, a fully electrical biasing is required to utilize fastest switch-off response time.

5.4.5   Comparison of LC-Based Phase Shifters

Table 5.7 summarizes some results for phase shifters from the literature in different technologies and topologies for comparison.

Table 5.7 Comparison of phase shifters in different technologies and topologies

Technology	Topology	f (GHz)	$\Delta\varphi_{max}$ (°)	IL_{max} (dB)	FoM (°/dB)	Ref.
Varactors	Loaded transmission line	6	410	6	68	[171]
	Coupled-line couplers	10	392	3.4	115	[172]
InGaAs	PIN switch diodes	28	349	7.8	45	[8]
MEMS	distributed MEMS CPW	40	84	1.8	47	[2]
	loaded transmission line (4 bit)	60	250	3	83	[173]
	Hybrid distributed RTPS	15	337.5	2.2	153	[174]
MEMS and CMOS	RTPS (CMOS 0.18 μm)	65	144	3.2	42	[10]
BST	Loaded CPW	9	405	7.6	53	[175]
	Loaded CPW	40	600	27	22	[176]
LC	Microstrip line	24	330	3	110	[76]
	Finline	40	303	4.8	62	[31]
	Ridged waveguide	94	500	7.2	69	[71]
	Loaded CPW	20	90	1.5	60	[82]
	Metallic waveguide	29	550	4.2	130	[143]
		35	110	0.55	200	[78]
		100	309	2.3	135	[140]
	Subwavelength dielectric fiber	82.5	107	1.5	75	[147]
	Classical dielectric waveguide	102.5	360	3.8	100	[161]
	Parallel-plate dielectric waveguide	68	300	3	113	[150]
LC and LTCC	SIW waveguide	28	400	9.7	41	[21]
	Microstrip line	33	320	4.2	76	[23]
LC and MEMS	Loaded line	95	240	4.0	60	[163]
LC and CMOS	CPW (CMOS 0.35 μm)	45	275	5.35	52	[83]

5.5 Tunable Resonators and Filters

Apart from delay-line phase shifters, tunable resonators and filters are key elements in communication systems, and therefore have been widely discussed in literature. One of the earliest examples of LC resonators is reported in [177], presenting a half-wavelength open-circuited stub resonator and a second-order dual behavior resonator with frequency shifts of a few percent for frequencies below 10 GHz, using two standard LCs, K15 and BL037. A first tunable terahertz filter based on a Fabry–Pérot cavity was presented in [178], formed by two metallic comb structures on top and bottom of two fused silica substrates. The fingers of the comb, with a width of 300 μm and separated by 100 μm, are oriented parallel to the electric field of the incident terahertz wave and act as highly reflective mirror for the terahertz signal. The effective lateral dimension of the filter is 35 × 40 mm² and the total height of the cavity is 200 μm, whereas the LC cell is sandwiched in-between the fused silica substrates with a thickness of 100 μm. It makes use of a dual-frequency switching LC with a crossover frequency of 10 kHz, where LC's orientation, and hence, permittivity is changed with the frequency of the control voltage. For the experimental investigation, a continuous-wave photomixing terahertz setup was used, featuring a dynamic range of 50 dB at the target frequency of 500 GHz. For tuning operation, a sinusoidal voltage with a peak-to-peak amplitude of 30 V was applied at 1 kHz and at 40 kHz to orientate the LC director parallel and perpendicular to the electrostatic field, respectively. The achieved tuning range was 16 GHz only, limited by the small anisotropy of the LC, accompanied by quite large insertion loss of the filter due to the high loss of the LC. Nevertheless, the concept was proofed and future LCs with improved properties will enhance the filter performance, suitable for spectroscopic THz measurements. In [179], a tunable three-pole bandpass filter with Chebyshev characteristics was demonstrated with a frequency shift of about 2 GHz around the center frequency of 20 GHz. It uses an LC-glass structure with a total thickness of 600 μm with an LC layer thickness of 5 μm only. The filter structure and topology are made of a series of coupled $\lambda/2$ resonators realized as periodically loaded transmission lines. [180] presented also a three-pole filter at 33 GHz with a bandwidth of about 3.3 GHz and a tunability range of 2 GHz over bias voltages 0–10 V. This concept uses tunable parallel-coupled microstrip lines with the LC E7.

A tunable resonator at 3.5 GHz was presented in [181], using two sandwiched samples, K15 and GT3-23001, which provides a tuning range of 4–8%, respectively. The same author reported a tunable LC resonators at 4.5 GHz with 6% continuous tuning range [182], and uses this kind of resonator in an array to form a tunable frequency selective surface (FSS) [183]. In a similar frequency range, [184] uses a microstrip square patch resonator with dual-mode operation as bandpass filter, tuning the frequency between 4.54 GHz and 5.19 GHz. In [185], a tunable microwave notch filter was proposed with a microstrip line, where the rejection frequency can be controlled from 4.45 GHz to 4.85 GHz.

Analogously to the different phase shifter topologies, planar filter structures are relatively easy and inexpensive to fabricate. However, resonators in planer topologies suffer from low Q factors. But because of thin LC layer heights, these filters can be tuned quickly. In contrast, metallic waveguide resonators achieve significantly higher Q factors, but at the expense of much lower tuning speed due to higher LC cavities. Furthermore, they are comparatively bulky and heavy. A trade-off between planar and waveguide resonator structures can be realized with a substrate integrated waveguide (SIW) topology with moderate Q factors and tuning speeds in between both. It allows low-cost fabrication, for example, with printed circuit board (PCB) or LTCC technology.

One example of using a substrate integrated waveguide (SIW) topology processed with the low-temperature co-fired ceramic (LTCC) technology is presented in [24, 186, 187], which focus on tunable amplitude tuner and tunable bandpass filter aimed for a satellite transceiver front end at 30 GHz. Both components made use of Merck's LC GT3-23001. With the amplitude tuner, the signal's amplitude can be controlled by using the interference concept of two signals with different phases, which exhibits a tunable attenuation from 11 dB to 30 dB at 30 GHz. Herewith, the phase of one signal path can be continuously controlled by a bias voltage, tuning an LC-based stripline phase shifter. The second component is a tunable three-pole SIW Chebyshev bandpass filter with measured quality factors in the range of 68–100 for a frequency tuning of 29.4–30.1 GHz, i. e. a tuning range of 700 MHz, accompanied by an insertion loss of 2–4 dB. It will be described in more detail in Section 5.5.1, including the filter design, simulation, realization and experimental results.

Section 5.5.2 focuses on LC-controlled high-Q waveguide filters, specifically on a center frequency tuned bandpass filter being targeted for a satellite input multiplexer section (IMUX) within the Liquida-Sky project [19]. As a first step, a cylindrical resonator is built up with a Rexolite tube cavity to fill in the LC GT3-23001 and with a pressure-release system, exhibiting Q = 1200 at 25.4 GHz [188], which is to the authors' knowledge the highest measured one for tunable filters at Ka-band. Then, three of these resonators have been used for a three-pole filter design at 20 GHz with coupling iris between the resonators. The measurement results of this lab-scale demonstrator exhibited a tuning range of 450 MHz and extracted Q factors of 484 at 20.17 GHz for parallel orientation and 327 at 19.77 GHz for perpendicular orientation of the LC [34, 188]. With a professional assembly of this promising cavity concept by Tesat-Spacecom GmbH & Co. KG, Backnang, Germany, a tunable three-pole bandpass filter achieved a tuning range of 300 MHz with a 3-dB bandwidth of 120 MHz, accompanied by an insertion loss of around 1.3 dB only, resulting in a quality factor of more than 1000. It enables applications for tunable high-Q filters at mm-waves, however, with limited tuning range due to limited tunability of the LCs.

A completely different topology is finally presented in more detail in Section 5.5.3, an LC continuously tunable nonradiative dielectric waveguide three-pole bandpass filter at 60 GHz [189]. For the orientation of the LC GT3-23001, an electrode network with parasitic mode suppressive structure is realized, tuning the center frequency from about 58.9 GHz to 60.4 GHz with a bias voltage of 200 V, that is, a tunability of 2.5%. The maximum fractional bandwidth was 1% and the filter's insertion loss between 4.9 and 6.2 dB over the tuning range, resulting in extracted unloaded quality factors from 96 to 141. Section 5.5.4 summarizes and compares the results of selected state-of-the-art tunable filters at higher frequencies.

5.5.1   SIW Bandpass Filter

An SIW consists of a dielectric substrate with top and bottom metallization, which are connected with each other by several metallic vias on the left and right side. To support the TE_{01} mode, the vias' diameter d_{via} and separation p_{via}, as shown in Figure 5.81, must be chosen in such a way that the following conditions are fulfilled [190]:

$$d_{via} < p_{via}, \; \frac{p_{via}}{\lambda_g} < 0.25, \; \frac{p_{via}}{\lambda_g} > 0.05 \; \text{ and } \; \frac{\alpha_l}{k_0} < 10^{-4}, \tag{5.58}$$

where λ_g is the wavelength of the guided wave and α_l is the total loss of the waveguide. If all conditions are fulfilled, the TE_{01}-mode can be operated without any bandgap in the operating frequency range and leakage losses are negligible. Hence, it is equivalent to the TE_{01} mode of a standard rectangular waveguide.

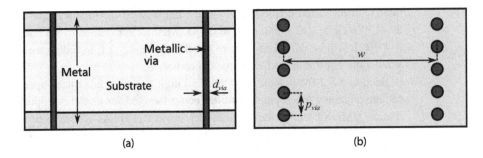

Figure 5.81 (a) Cross section and (b) top view of a SIW transmission line.

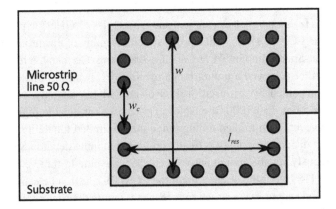

Figure 5.82 Top view of a SIW resonator excited with 50 Ω microstrip lines as input and output coupling.

The effective width w_{eff} of a rectangular waveguide can be directly mapped to the dimensions of the SIW according to the following equation [186]:

$$w_{eff} = w - 1.08 \cdot \frac{d_{via}^2}{p_{via}} + 0.1 \cdot \frac{d_{via}^2}{w}, \tag{5.59}$$

for $p_{via}/d_{via} < 3$ and $d_{via}/w < 0.2$. The SIW has the same cutoff frequency as its equivalent rectangular waveguide. The SIW can be used to build up a resonator by shorting the two ends with vias. By placing coupling slots at both ends, the fundamental resonant TE_{101} mode can be excited with the resonance frequency [186]:

$$f_{res, 101} = \frac{c}{2\sqrt{\varepsilon_r}} \sqrt{\left(\frac{1}{w_{eff}}\right)^2 + \left(\frac{1}{l_{res, eff}}\right)^2} \tag{5.60}$$

where $l_{res, eff}$ is the effective length of the resonator. By knowing the desired resonant frequency, the effective length of the resonator can be calculated and mapped back to the SIW resonator length l_{res}.

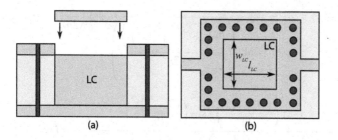

Figure 5.83 Cross section (a) and top view (b) of an LC SIW resonator.

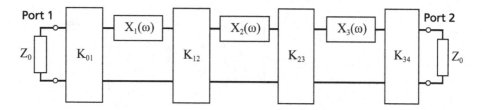

Figure 5.84 Bandpass filter model with three resonators using K-impedance inverters.

A SIW resonator can be excited, for example, by a 50 Ω microstrip line through an inductive iris with the width w_c as shown in Figure 5.82. By removing a section of the substrate and top metallization of the resonator, an LC cavity with the length l_{LC} and width w_{LC} is obtained according to Figure 5.83. After the LC is filled inside the resonator, the top metallization is reattached, to seal the LC cavity. The size of the LC cavity has an impact on the tunability and the losses. For a larger LC cavity, tunability, but also losses increase, since the LC usually has higher losses than the substrate material. Hence, there is a trade-off between tunability and loss, which must be considered in the design process.

For a design of a three-pole Chebyshev filter as a lab-scale demonstrator, Rogers RT/duroid RT5880 is used as a nontunable substrate and 50 Ω microstrip lines as feed. It is designed for 22 GHz for the lowest permittivity of the LC, that is, for perpendicular orientation, aiming for a bandwidth of 600 MHz and a return loss of at least −20 dB. The corresponding filter model with K-impedance inverters is shown in Figure 5.84, where $Xi(\omega)$ is the reactance of the ith resonator.

As a first step, a single resonator is design by using full-wave simulations for a resonance frequency of 22 GHz, providing the following geometrical dimensions: $w = 7.8$ mm, $l_{res} = 6.0$ mm, $w_{LC} = 1.0$ mm, and $l_{LC} = 4.0$ mm. Then, to determine the input and output coupling as well as the inter-resonator couplings, the element values $g_0, g_1, \ldots g_n$ of the lowpass prototype of the K-impedance inverters has to be calculated, depending on the filter order n and the ripple level in the passband L_{Ar} according to following formulas [186]:

$$g_0 = 1 \qquad (5.61)$$

$$g_1 = \frac{2}{\sinh\left(\dfrac{\ln\left(\coth\left(\dfrac{L_{Ar}}{17.37}\right)\right)}{2n}\right)} \sin\left(\frac{\pi}{2n}\right) \tag{5.62}$$

$$g_i = \frac{1}{g_{i-1}} \frac{4\sin\left(\dfrac{(2i-1)\pi}{2n}\right)\cdot\sin\left(\dfrac{(2i-3)\pi}{2n}\right)}{\sinh^2\left(\dfrac{\ln\left(\coth\left(\dfrac{L_{Ar}}{17.37}\right)\right)}{2n}\right) + \sin^2\left(\dfrac{(i-1)\pi}{n}\right)} \quad for\ i = 2,3,\ldots,n \tag{5.63}$$

$$g_{n+1} = \begin{cases} 1, & n\ odd \\ \coth\left(\dfrac{\ln\left(\coth\left(\dfrac{L_{Ar}}{17.37}\right)\right)}{4}\right), & n\ even. \end{cases} \tag{5.64}$$

Using the calculated element values, the K-impedance inverter parameters are given by [191]

$$K_{01} = \sqrt{\frac{Z_0 x_1 FBW}{g_0 g_1 \omega_1}} \tag{5.65}$$

$$K_{j,j+1} = \frac{FBW}{\omega_1}\sqrt{\frac{x_j x_{j+1}}{g_j g_{j+1}}} \quad for\ j = 1,\ldots,n-1 \tag{5.66}$$

$$K_{n,n+1} = \sqrt{\frac{Z_0 x_n FBW}{g_n g_{n+1}\omega_1}}, \tag{5.67}$$

where x_i is the reactance slope parameter of the ith resonator and ω_1 is the lower cutoff frequency.

One option to realize these desired impedance inverter parameters is by coupling two resonators with an inductive iris, where two resonance frequencies occur as shown in Figure 5.85.

For the inter-resonator couplings, the coupling coefficient k can be determined with these two resonance frequencies f_1 and f_2 according to

$$k = \frac{f_2^2 - f_1^2}{f_2^2 + f_1^2}. \tag{5.68}$$

The coupling coefficient k depends on the iris width w_c (see Figure 5.82). Therefore, full-wave simulations are performed by varying the iris widths. First, for the input and output coupling parameters a single resonator is simulated, which has a 50 Ω microstrip line feed on the one side and is shorted on the other side. The simulated external quality factor Q_e is obtained from the reflection parameter's phase according to [192]

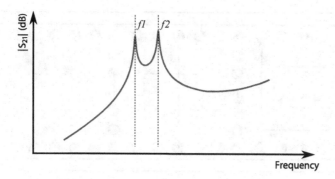

Figure 5.85 Response of two coupled resonators.

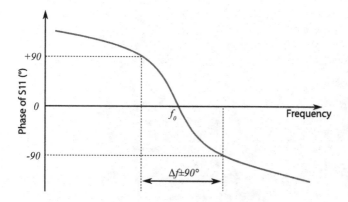

Figure 5.86 Phase response of a short-circuited resonator.

$$Q_e = \frac{f_0}{\Delta f_{\pm 90°}}, \qquad (5.69)$$

where $\Delta f_{\pm 90°}$ is the bandwidth in which the phase is shifted $\pm 90°$ from the phase at resonance, which is zero in the ideal case as depicted in Figure 5.86.

The input and output coupling parameters are related to Q_e by [191]:

$$K_{01} = \sqrt{\frac{x_1}{\frac{Q_{e,input}}{Z_0}}} \quad \text{and} \quad K_{n,n+1} = \sqrt{\frac{x_n}{\frac{Q_{e,output}}{Z_0}}}. \qquad (5.70)$$

Since the Q_e depends on the iris width, the right dimension can be obtained by the full-wave simulation. After the resonator and coupling structures are determined, the whole three-pole Chebyshev filter is simulated by using the schematic in Figure 5.87, providing the filter design parameters: $w_{c,01} = w_{c,12} = 4.3$ mm, $w_{c,12} = w_{c,23} = 3.1$ mm, $w_1 = w_3 = 6.6$ mm, $w_2 = 6.9$ mm, $l_{LC} = 4.0$ mm, $w_{LC} = 1.0$ mm, and $l_{res} = 6.0$ mm.

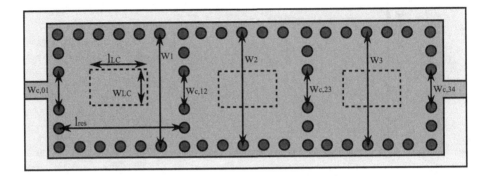

Figure 5.87 Schematic of the tunable three-pole LC bandpass filter.

Figure 5.88 Fabricated SIW LC bandpass filter [186].

Using these design parameters, the lab-scale filter demonstrator is manufactured, where each of the LC cavities are milled inside the substrate and sealed with a copper plate, which is glued on the top metallization; see Figure 5.88. This plate has two holes for the LC filling. After that, via holes are drilled and filled with LPKF ProConduct Paste, which is hardened in an oven at 90°C. Thereafter, the LC is injected through the holes of the sealing plate. For the measurements, the LC mixture GT5-26001 is utilized and two rare earth magnets are used for the orientation for a first proof-of-concept.

In the simulation, a frequency shift of 1100 MHz (dashed lines) with an insertion loss of less than 3 dB is predicted as shown in Figure 5.89. However, the measured tuning range (solid lines) decreased to 600 MHz from 21.4 to 22.5 GHz. Furthermore, the losses are about 3 dB and the measured return loss around 10 dB higher. A reason for the lower tuning range has been an imperfect LC filling with some air gaps, reducing the tuning effectivity. Simulation with an air gap of 0.02 mm revealed similar results as the measurements [186]. However, this air gap also reduces the losses. The higher measured loss is caused by the sealing of the LC cavity, where glue is used to

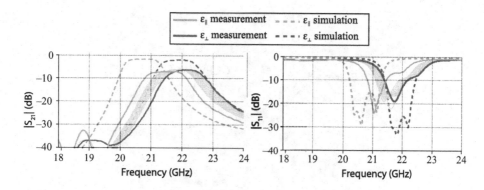

Figure 5.89 Simulation and measurement results of the designed filter [186].

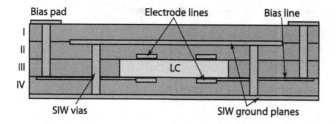

Figure 5.90 Cross section of an SIW LTCC-LC resonator with four layers and electrode lines. Thin bias lines are connected to the electrode lines in between the SIW vias [24].

attach the metal plate on top of the resonators, and where a small amount of this glue flowed inside the LC cavity, increasing the losses drastically and additionally decreasing the tunability.

To design an electrically tunable SIW LC filter, an electrode network must be integrated inside the resonators, which can be realized by means of multilayer technology. Therefore, LTCC technology is utilized for the second design with electrical tunability, since the fabrication is comparatively easy. The cross section of such a SIW LTCC-LC resonator is shown in Figure 5.90 [24].

Four layers of LTCC are necessary for the design, where the electrode lines are placed between the layers II–III and III–IV. These lines are feed by thin bias lines, which are led to the outside between the vias. Moreover, they are connected by through vias with bias pads on layer I. A fabricated resonator is shown in Figure 5.91.

The designed three-pole filter has a similar structure like the magnetically tunable filter shown in Figure 5.88, except that the dimensions changed, since this filter is designed for 30 GHz. $w_{c,01} = w_{c,12} = 1.9$ mm, $w_{c,12} = w_{c,23} = 1.6$ mm, $w_1 = w_3 = w_2 = 3$ mm, $l_{LC} = 2.0$ mm, $w_{LC} = 2.1$ mm, and $l_{res} = 2.9$ mm.

The measured S-parameters are shown in Figure 5.92 [24]. A tuning range from 29.4 GHz to 30.1 GHz was measured with an FBW of 11.6% and 11.2% for orthogonal and parallel orientation, respectively. The insertion losses range between 2 and 4 dB.

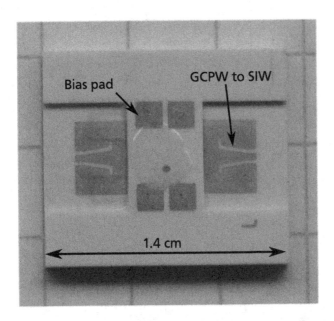

Figure 5.91 Fabricated SIW resonator in LTCC-LC technology at 30 GHz with GCPW to SIW transitions [24].

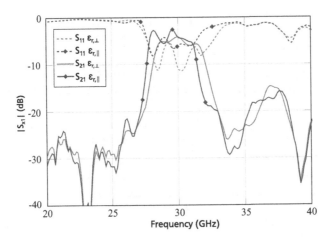

Figure 5.92 S-parameter measurement results of the electrically tunable three-pole LTCC-LC SIW filter [24].

5.5.2   Waveguide Filter

The design process of a waveguide filter is similar to the one of a SIW filter. However, the waveguide filter presented in this section was specifically designed for space applications, where long-term reliability is very important [188]. Furthermore, the filter should work for a large temperature range and it should resist vibrations and mechanical shocks without damage. Therefore, the LC container must be sealed in such a way that no leakage occurs. Moreover, Rexolite as a radiation-hard material being space approved has been used for the LC cavity. Furthermore, its relative permittivity $\varepsilon_r = 2.53$ is in the same range of the LC. In [188] two different LC cavities for LC waveguide filters are presented and tested for space applications. The metallic resonator is cylindrically shaped and has coupling irises on both sides as shown in Figure 5.93. One way to bring in LC into the resonator without leakage, is to use a cylindrical Rexolite container of the same diameter as the resonator; see Figure 5.94. After the LC is injected into the cylindrical Rexolite container, it can be sealed by gluing the top plate on this Rexolite LC cavity and inserted into the resonator afterward. A disadvantage of this design is that the space-approved Rexolite glue for sealing is very lossy. Another drawback is that the Rexolite container filled with LC can burst at higher temperatures due to the LC's expansion; that is, this design is not suitable for large temperature changes.

Therefore, a modified version has been investigated hereinafter: a thin cylindrical Rexolite tube is inserted into the resonator through an opening as depicted in

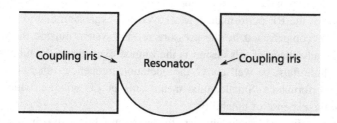

Figure 5.93 Cross section of a cylindrical metallic resonator coupling via irises with the rectangular waveguides on both sides.

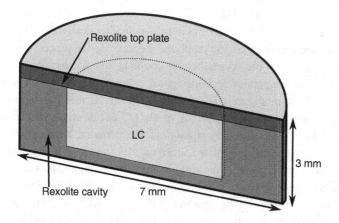

Figure 5.94 Cross section of the Rexolite LC container being inserted into the resonator. The container has the same diameter as the resonator.

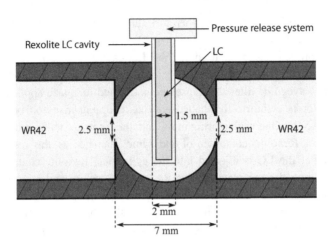

Figure 5.95 Cross section of the resonator with the inserted cylindrical Rexolite LC tube.

Figure 5.95. After the tube is filled with LC, it is sealed with a pressure release system, which is glued on top of the tube. Since the glued joint is outside of the resonator, it no longer influences the RF performance. Moreover, LC expansion due to temperature changes is now compensated by the pressure release system outside the resonator. However, a disadvantage of this design is the limited diameter of the tube, since the cutoff of the hole must be well above the operation frequency; otherwise it would decrease the performance. Smaller tube means smaller LC volume, which reduces losses, but at the expense of tunability.

Nevertheless, a first lab-scale demonstrator has been fabricated, using brass which is gold-platted afterward shown in Figure 5.96. The most important dimensions are given in Figure 5.95. The resonator is closed by screwing two polished copper plates on it. A Q factor of 1200 was measured at 25.42 GHz with the Rexolite LC tube cavity, compared to around 450 for the cylindrical Rexolite LC cavity, both filled with the LC mixture GT3-23001 [188].

Then, this resonator has been used for a three-pole filter design at 20 GHz with 1% FBW and a desired tunability of 3%. Applying the same design process as for the SIW filter, the dimensions of the resonator and coupling iris widths are obtained. The measurement results for parallel and orthogonal orientation are shown in comparison with the simulations in Figure 5.97. It can be observed that the tuning range is 450 MHz only, due to small air bubbles in the LC cavity and glue protuberances, instead of 600 MHz as predicted by simulation. Furthermore, the measured losses of around 6.5 dB are higher than simulated, since the loss tangent of glue is unknown and could not be taken into account in the simulations. This additional loss can be avoided by using the thin cylindrical Rexolite tubes, which have their glue joint outside the resonator. This promising cavity concept has been further investigated by Tesat Spacecom GmbH & Co. KG, Backnang, Germany, and a tunable three-pole bandpass filter has been designed at 20 GHz, see Figure 5.98. The measurement results are shown in Figure 5.99.

Figure 5.96 Fabricated resonator with the Rexolite tube cavity inside.

This filter has a tuning range of 300 MHz with a 3 dB bandwidth of 120 MHz. Furthermore, the insertion losses are around 1.3 dB, resulting in a quality factor of better than 1000.

5.5.3   NRD Filter

The above presented filters are operating either in the K- or K_a-band. For higher frequencies, the electrode design and implementation become even more challenging. Therefore, an alternative waveguide topology is investigated in this subsection.

Dielectric waveguides that are used for phase shifter designs are not suitable for filter design because of high radiation losses at discontinuities. However, the non-radiative dielectric (NRD) waveguide, introduced by Yoneyama et al. in 1981 [193], makes a filter design feasible. Similar to the PPDW, it consists of a dielectric core sandwiched in between two parallel metal plates as depicted in Figure 5.100. Hence, the electrode implementation is comparatively easy, even at higher frequencies.

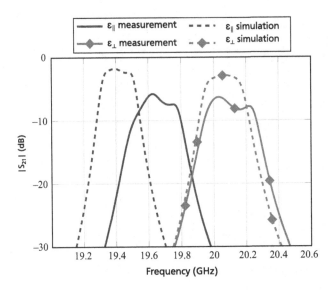

Figure 5.97 Measurement and simulation results of the three-pole waveguide LC bandpass filter.

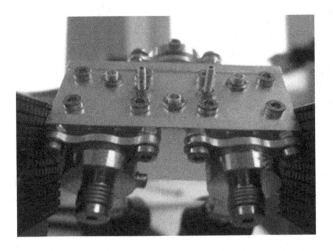

Figure 5.98 Fabricated tunable three-pole LC filter.
Courtesy of Tesat Spacecom GmbH & Co. KG

Due to the metal plates, it is nonradiative and, at the same time, preserving the natural low loss characteristics of dielectric waveguides, since there are no metal walls orthogonal to the E-field of the operating mode, which would have increased the conduction loss. By setting the metal plate separation $a < \lambda_0/2$, no propagation is possible between the plates in air. To eliminate the cutoff behavior, a dielectric with a proper relative permittivity must be inserted between the metal plates. Depending on its dimensions a and b as well as on its relative permittivity ε_r, propagation is feasible

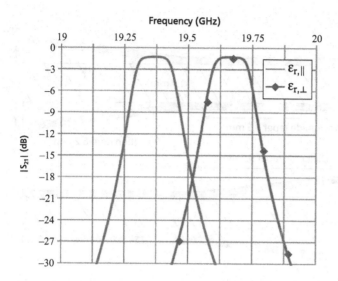

Figure 5.99 Measured transmission coefficients for parallel and orthogonal orientation. Courtesy of Tesat Spacecom GmbH & Co. KG

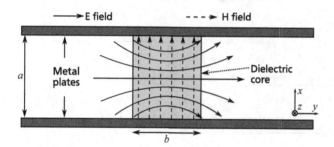

Figure 5.100 Cross section of the NRD waveguide, including the E-field lines of the operating LSM_{01} mode.

in and along the dielectric core only for a certain bandwidth. To prevent higher-order modes, the following requirement must be fulfilled:

$$\frac{\lambda_{g0}}{2} < a < \lambda_{g0}, \tag{5.71}$$

where λ_{g0} is the guide wavelength of the LSM_{01} mode.

For the design, Rexolite is chosen as dielectric material, which has a relative permittivity of $\varepsilon_r = 2.53$ The plate separation is set to $a = 2.3$ mm and the width of the dielectric is $b = 1.9$ mm. With these dimensions, the non-radiative characteristic of the NRD is guaranteed in the V-band. To excite the LSM_{01} mode, a rectangular waveguide is used followed by a transition from the TE_{10} to the LSM_{01} mode. By inserting Rexolite into the waveguide, higher-order modes can propagate in the

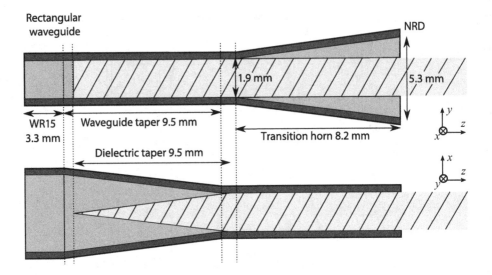

Figure 5.101 Design of the transition from the WR15 rectangular waveguide to the NRD waveguide.

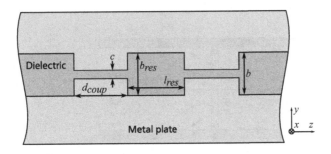

Figure 5.102 Cross section of an NRD resonator coupled with thin coupling lines.

dielectric-filled waveguide; hence, the width of the waveguide is tapered to guarantee single mode operation, see Figure 5.101. The end of the Rexolite core is tapered to achieve a good impedance matching. Furthermore, to obtain also a good matching between the dielectric-filled waveguide and the NRD waveguide, a transition horn in the E-plane was designed. All dimensions are optimized by using full-wave simulation.

NRD waveguide resonators can be realized in different ways. A straightforward design is the rectangular resonator, where the resonator is excited by thin coupling lines below cutoff with width c as depicted in Figure 5.102. Furthermore, the coupling lines are keeping the resonator in position. To keep the design simple, the resonator's width b_{res} is equal to b.

Since, the resonator's material and width are fixed, the resonance frequency depends only on its length. After the length is determined by using full wave simulations, LC is inserted into the resonator to obtain tunability of the center frequency as shown in Figure 5.103. The size of the LC cavity is restricted by the

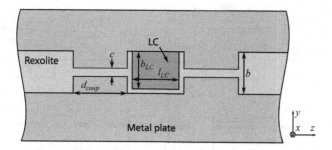

Figure 5.103 Cross section of the LC-filled resonator.

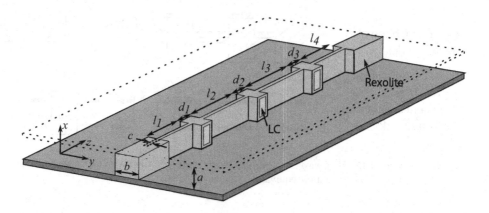

Figure 5.104 Schematic of a tunable NRD bandpass filter.

manufacturing process and the material itself. If the wall thickness is too small, the cavity can break during the assembly. Therefore, the wall thickness is set to 200 μm to guarantee mechanical stability.

Then, a tunable three-pole Chebyshev bandpass filter is designed at 60 GHz with a passband ripple level of 0.2 dB and 1% FBW. The coupling elements and resonators are formed by varying the width of the NRD's Rexolite core, as shown in Figure 5.104 [189].

To obtain the dimensions of the resonators (d_1, d_2, and d_3) and coupling lines (l_1, l_2, l_3, and l_4), a design method is applied, which is based on a stepped-impedance half-wave microstrip filter. A stepped-impedance filter consists of alternating high- and low-impedance half-wave sections; see Figure 5.105. This filter can be described with an equivalent K-impedance inverter network. Each impedance step is replaced with an impedance inverter according to [194]

$$K_{i-1,i} = \sqrt{\frac{Z_i}{Z_{i-1}}}.$$ (5.72)

However, to achieve the same impedance steps with the NRD, the dimensions get too large. Therefore, the resonators are connected with thin below cutoff coupling lines and are not consecutively lined up. Hence, each impedance step of the

Stepped-impedance
microstrip filter

Equivalent K-impedance
inverter network

NRD filter (shape of
the dielectric)

Equivalent S-paramter
matrix network

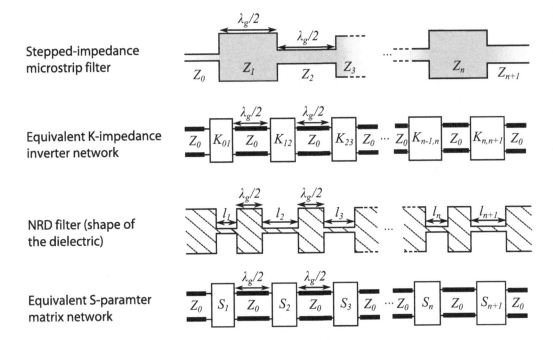

Figure 5.105 Half-wave stepped-impedance microstrip filter with the corresponding structure of the NRD core and the equivalent S-parameter matrix network.

stepped-impedance filter results in a coupling line in the NRD filter design. The length of a coupling line depends on the ratio of the impedance step. To determine the coupling lines' lengths an equivalent network with S-parameter matrices is used, as shown in Figure 5.105, where the K-impedance inverters are replaced by S-parameter matrices. By assuming lossless junctions between resonators and coupling lines, it is sufficient to know only one element of an S-parameter matrix to determine all matrix elements; hence, only the reflection coefficient S_{11} is considered in the following. The relation between S_{11} and the impedance inverter is given by [194]

$$(S_{11})_{i-1,i} = \frac{K_{i-1,i}^2 - 1}{K_{i-1,i}^1 + 1}.$$ (5.73)

Hence, for the calculation of the S_{11} values, the K-impedance inverters must be determined for the desired filter parameters [194]

$$K_{i,i+1} = \frac{\sqrt{1 + \left(\frac{\sin\left(\frac{i\pi}{n}\right)}{y}\right)^2}}{\sqrt{Z_i Z_{i+1}}}, \quad for\ i = 0, 1, \cdots, n,$$ (5.74)

with

$$y = \sinh\left(\frac{1}{n}\sinh^{-1}\left(\frac{1}{L_r}\right)\right),\tag{5.75}$$

where L_r is the passband ripple level, and with

$$Z_i = \begin{cases} 1, & for\, i = 0\, and\, n+1 \\[2ex] \dfrac{2\sin\left(\dfrac{(2i-1)\pi}{2n}\right)}{y\,FBW} - \\[3ex] \dfrac{FBW}{4y}\left(\dfrac{y^2+\sin^2\left(\dfrac{i\pi}{n}\right)}{\sin\left(\dfrac{(2i+1)\pi}{2n}\right)} + \dfrac{y^2+\sin^2\left(\dfrac{(i-1)\pi}{2n}\right)}{\sin\left(\dfrac{(2i-3)\pi}{2n}\right)}\right), & for\, i = 1, 2, \cdots, n. \end{cases}$$

$$\tag{5.76}$$

After the required S_{11} values are calculated, the corresponding coupling lengths must be determined. Therefore, full-wave simulations are performed for varying coupling lengths to obtain its influence on the reflection coefficient. The used simulation model is shown in Figure 5.106.

Without the coupling line the magnitude is 0 (linear), for increasing coupling lengths the magnitude is increasing, where the results are summarized in Figure 5.107.

By knowing the calculated magnitude of S_{11}, the required coupling length can be read from the plot. Hence, the coupling lengths can be derived for the defined filter parameters. Since lossless junctions are assumed, the reflection coefficient must be real at the junction, and hence, its phase must be either 0 or π. For the resulting coupling lengths, the phase deviations are obtained from Figure 5.107 and compensated according to

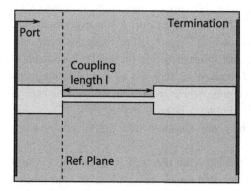

Figure 5.106 One-port simulation model of the coupling length.

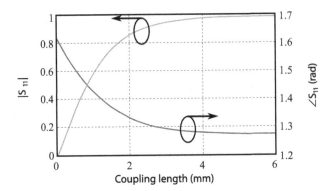

Figure 5.107 Simulation results of the reflection coefficient for varying coupling lengths.

$$\Theta_n = \frac{\angle S_{11}(l_n)}{2},\tag{5.77}$$

where $\angle S_{11}(l_n)$ is the simulated junction phase for the n-th coupling line. Since half-wave resonators are used, the electrical length of each resonator must be π. That is why the physical length of the resonators must be adapted in consideration of the phase compensations with

$$d_n = \frac{\lambda_g}{2\pi}(\Theta_n + \Theta_{n+1}),\tag{5.78}$$

where λ_g is the resonator's guide wavelength, varying with the LC's orientation inside the resonator. For the aforementioned resonator dimensions, the center of the guide wavelength range is at 6.65 mm and the resulting lengths of the resonators and coupling lines for the given filter parameters are [189] $d_1 = d_3 = 1.37$ mm, $d_2 = 1.35$ mm, $l_1 = l_4 = 2.38$ mm, and $l_2 = l_3 = 4.88$ mm.

With an electrode network an electrical bias field can be adjusted to align the LC molecules between parallel and orthogonal alignment to the RF field. By using three pairs of electrodes and adjusting the bias configuration, continuous tuning of the center frequency is obtained. The bias scheme of a single resonator for orthogonal and parallel orientation are shown in Figure 5.108. Since the electrode network is realized on a thin substrate and inserted inside the NRD, its influence on the RF performance is minimized. Therefore, to suppress unwanted mode excitation in the substrate, a parasitic mode suppressive electrode structure is designed. The resulting electrode network is shown in Figure 5.109.

The slots, tapers, and stubs are designed and optimized by using full-wave simulations.

For the measurements of the lab-scale demonstrator, the NRD is milled together with the WR15 transitions out of a single brass piece in split-block technology. Afterward, the Rexolite structure with the open LC cavities is milled. The electrode network is etched on a 12 µm thin flexible Pyralux substrate with an 18 µm thin

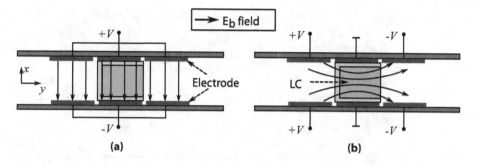

Figure 5.108 Bias schemes for a single NRD LC resonator for orthogonal (a) and parallel (b) orientation. All resonators are equally biased.

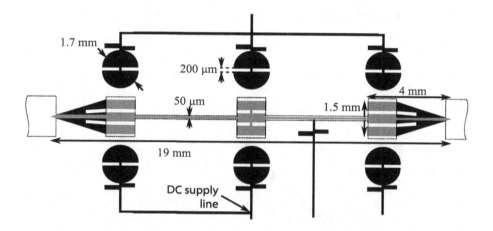

Figure 5.109 Designed electrode network with parasitic mode suppressing structure.

copper cladding. Finally, the whole filter is assembled and LC is injected into the cavities; see Figure 5.110.

The measurement and simulation results are shown in Figure 5.111. The deviation of the center frequency and FBW between the initial simulations and measurements are too large. Therefore, the fabrication tolerances of the Rexolite structure has been analyzed in more detail, which revealed that the tolerances especially at the thin places are too large. A reason for this is that Rexolite becomes flexible at a certain thinness and precise milling is not possible. These tolerances are measured as precisely as possible and were adapted to the simulation model. The new simulation results then fit very well with the measurement results. For the measurements, a bias voltage of ± 200 V was applied. A tunability of 2.5% was obtained with insertion losses ranging between 4.9 and 6.2 dB. The measured FBW is maximal 1% over the tuning range [189]. However, the ripple characteristic in the passband vanished, due to the high losses.

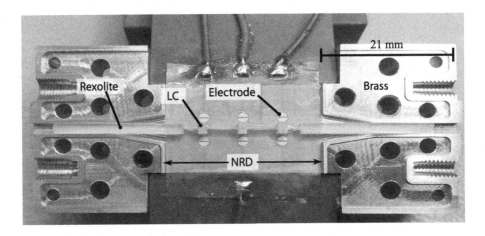

Figure 5.110 Photograph of the partially assembled tunable NRD filter.

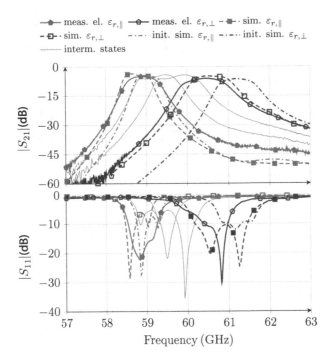

Figure 5.111 Measured *S*-parameters with electrical biasing.

5.5.4   Comparison of Tunable Filters

For comparison, a selection of state-of-the-art tunable filters is given in Table 5.8. Tunable MEMS filter achieves lower insertion losses and a higher tunability. Whereas, the LC-tuned waveguide filter exhibits a high Q factor, but its tunability is

Table 5.8 Selection of state-of-the-art tunable filters

Technology	Topology	f (GHz)	FBW (%)	τ (%)	IL (dB)	Q	Work
MEMS	Waveguide	65	8.3	6.5	1.5-5	68	[195]
		95	4.3	2.7	2.4	400	[196]
		30	4.7	66.6	0.8-2.9	264	[197]
BST	CPW	35	9	6.9	4.9-7.6	N/A	[198]
	CPW	31.5	12.2	17	2.5-6.9	N/A	[199]
	Microstrip	9	18.8	25	4-8	N/A	[200]
	Microstrip	13	13	20	6-10	N/A	[200]
LC	SIW	22	3.54	2.7	6	102	[186]
		30	11.6	2.3	2-4	68	[24]
	Waveguide	20	1	2.3	5-7	N/A	[188]
		20	0.6	1.5	1.3	>1.000	Tesat Spacom
	NRD	60	1	2.5	4.2-6.2	96	[189]
	Microstrip	50	18	10.3	3.8	6	[201]
		85	16.5	10.4	7.6	6	[201]
		33	10	6.1	4.5	N/A	[180]

very limited. Therefore, the new microwave LC mixtures of the third generation with higher anisotropy would allow higher tunability, that is, a larger tuning range of the center frequency.

5.6 Electronically Steerable Antennas

Electronically steerable or reconfigurable antennas (ESAs) are widely discussed in recent years, because of perspective markets and attractive commercial applications in the new satellite and 5G communication hardware platforms described in Chapter 1, operating at mm-waves where still large frequency resources are available. However, to overcome path losses at these high frequencies, large, highly directive, high-gain antennas are required to focus the narrow beam toward the desired hub (satellite, access point, relay or base station) or mobile terminal. When the hub and/or mobile terminal are moving, then these antennas have to be to some extent "smart": to steer the beam dynamically toward a desired direction, or even, creating dynamically a desired antenna pattern, maybe accompanied with a polarization-agile capability, or being frequency agile within a desired tuning range. For mm-waves, where complexity, technological constrains and cost increase, there are different activities worldwide to enable cost-efficient ESAs such as frequency-agile antennas, polarizer and polarization-agile antennas, deflecting gratings, transmit arrays and lens antennas, mixed beam-switching and beam-steering antennas, beam-steering reflectarrays, beam-steering leaky-wave antennas, and beam-steering phased array antennas, active and passive. These categories will be first briefly summarized in the text that follows, before focusing on the last category only, presenting three different concepts of phased array antennas based on different topologies.

5.6.1   Frequency-Agile Antennas

One of the initially reported frequency-agile antennas, where LC is placed inside its intermediate foam substrate, underneath the patch is given in [202]. By applying an external bias voltage to this structure, a frequency shift of 140 MHz was obtained between 4.74 GHz and 4.6 GHz. A comparable rectangular patch demonstrated a 5.5% tunability with respect to 4.5 GHz [203]. For these two antennas, standard commercial K15 and BL037 liquid crystal samples were used. Similarly, a sandwich structure antenna operating at a frequency of 5 GHz was presented in [204], using three layers of Taconic glass–reinforced PTFE substrate placed on top of each other with E7 liquid crystal in its central layer. Measurements demonstrated a tuning range of 4%. [205, 206] proposed a design for a frequency-agile antenna, where the structure consists of a microstrip patch and an LC cell which serves as substrate. It is shown that the resonant frequency of the antenna can be altered between 5.45 and 5.84 GHz with a bias voltage between 0 and 10 V. Fritzsch et al. [207] present a frequency-tunable patch antenna with a Merck's first-generation LC as a tunable dielectric layer between a ground plane and an inset feed microstrip patch in combination with a liquid crystal polymer (LCP) substrate, whose permittivity fits quite well to the permittivity of the tunable LC. Due to its high chemical and mechanical stability, LCP can form a leak-proof cavity for the tunable LC. The measured resonance frequency of the LCP patch antenna was tuned continuously from 34.1 GHz to 37.7 GHz with external bias voltages between 0 and 90 V, which corresponds to a frequency tuning range of about 10%. As a perspective of the author: the design may be adapted to a three-layer PCB process and planar LC-based phase shifter discussed above might be realized in the same technology, giving the opportunity to realize a frequency-tunable phased array. Zhao et al. [208] simulated a 1×4 element series-fed patch array which can dynamically tune the operation frequency and achieve beam steering. First, the antenna element is composed of the microstrip patch array and LC-based microstrip line (MSL) phase shifter. Second, an LC substrate is used to support the patch array for controlling its resonance frequency, while the LC-based MSL phase shifter is adopted to tune the transmission phase by changing the effective permittivity of the LC. Simulation results predicts a scanning range of $+20°$ to $-20°$, while its operation frequency can be tuned between 14.5 and 16.4 GHz.

5.6.2   Polarizer and Polarization-Agile Antennas

Some recent publications presented tunable polarizer and polarization-agile antennas using liquid crystal mixtures. The polarizer introduced in [22, 209] continuously changed the linear polarization state of a TE_{11} mode in a circular waveguide in the range from $+90°$ to $-90°$ with a relatively low insertion loss between 4 dB and 5 dB at 50 GHz. In [210], an electrically tunable LC polarization selector LC is built up and tested in combination with a microwave radiometry line scanner for polarimetric measurements of cars. In [211], an electronically tunable reflection polarizer has been designed, fabricated and measured in a frequency band centered at 130 GHz. The phase-agile polarizing mirror converts an incident slant 45° signal upon reflection to right-hand circular, orthogonal linear ($-45°$) or left-hand circular polarization, depending on the low-frequency bias voltage of 0 V, 40 V, and 89 V, respectively, across the cavity containing the LC material. A polarization reconfigurable microstrip patch antenna element at 13.75 GHz is presented in [212]. It uses an LC-based tunable coupled line by which the antenna polarization is tuned continuously between linear and circular polarizations by a bias voltage from 0 to 40 V to the coupled line. The prototype is fabricated in a similar fashion to LC displays, resulting in a low-cost and simple fabrication compared to other tunable device fabrication technologies. Based on this work, a polarization-agile 2×2 dual-fed microstrip patch array with two separate feeding networks is presented at the same frequency [26, 213]. The polarization state of the antenna can be controlled continuously between dual linear and dual circular polarizations, depending on a differential phase shift between the antenna feedings. The feeding networks are implemented in microstrip transmission line topology with the LC material as a tunable dielectric to obtain a desired differential phase shift between the feeding networks. Owing to the continuous tuning of the LC material, any polarization state between the circular and linear ones are achievable. This polarization-agile antenna can be fabricated with more radiating elements and can be efficiently scaled for higher operating frequencies, because of decreasing losses of the LC, it is well suited for larger arrays at the K_a- or W-band.

5.6.3   Deflecting Gratings and Lens Antennas

Early examples of grating patterned electrodes with thin LC layers for LC prism cells for deflection or LC lens with focusing properties at 94 GHz are published in [86, 87]. Similar structures were used in [61] for LC beam former operating from 50 GHz to 75 GHz with beam steering angles of about $\pm 13°$. In a recent publication [214], an electrically controlled LC-based phase gratings for manipulating terahertz waves is presented, which can be used as a tunable beam splitter with a beam splitting ratio of the zeroth-order diffraction to the first-order diffraction from 10:1 to 3:5. The insertion loss is lower than 2.5 dB for the one with the smaller base. However, the rise and decay response times of the grating are approximately 23 s and 290 s, respectively, due to the thick LC layer. Based on the grating structure, the paper demonstrated an electrically tunable grating-structured phase shifter array. Using a designed voltage gradient biasing on the grating structure, broadband terahertz signals below 0.5 THz can be steered by 8.5° with respect to the incident beam. In a more recent paper [215], an LC-filled lens antenna is investigated for beam steering for the first time, requiring neither phase shifters nor switching networks. The LC's anisotropy is applied inside a semicircular shaped parallel plate waveguide to achieve a deflection of the propagating electromagnetic wave at the interface of lower and higher permittivity. Using an electrode network, the direction of the sector of higher permittivity can be adjusted, and therefore, also the beam direction. Operating from 50 to 75 GHz, the first lab-scale demonstrator proves the antenna concept by reconfiguring the beam in predefined directions of $-30°$, $0°$, and $+30°$ with a simple electrode biasing network. The input reflection is always below -10 dB and the measured gain is between 13 and 15 dBi over the whole frequency and scanning range, while the SLL varies between -8 and -13 dB for the unbiased and between -4.5 and -6 dB for the biased state.

5.6.4   Mixed Beam-Switching and Beam-Steering Antennas

A radiation pattern reconfigurable 18 element waveguide slot array antenna by using liquid crystal was proposed in [216]. Together with the waveguide slot, the designed complementary electric field coupled resonator functions like a switch for the antenna element to radiate or not by varying the permittivity of the LC. Thus, the array factor and radiation pattern can be controlled to some extent. Its radiation direction can be reconfigured to $46°$ or $0°$ at about 15 GHz. [137] presents a Rotman lens, forming fixed beams at $0°$, $\pm15°$, and $\pm30°$. This is augmented by using voltage-controlled phase shifters with Merck's LC GT3-24002 to simultaneously steer each beam of up to $\pm7.5°$. Hence, it can be exploited to provide continuously scanned beams with full coverage over an angular range of $\pm37.5°$, and with operation over the frequency band 6–10 GHz. The phase shifter insertion losses are between 1.2 and 2.7 dB, and the measured beam squint is at worst $4.5°$ at 6 GHz. Alternative to LC-based beam steering reflectarrays and phased arrays, [54] focused on an LC-based network with mixed discrete beam switching and continuous beam-steering capability between the switching states for high-gain antennas at W-band. His approach consists of a Butler matrix as the core component for beam switching combined with continuously tunable phase shifters at the output and an interference based RF switch at the input to select the different input ports of the Butler matrix, and hence, to generate a single antenna beam toward different discrete directions, depending on the selected input port. Moreover, the interference principle in combination with phase shifting sections of his new type of switch allows a continuously adjustable power splitting ratio [147, 217], and therefore, the selection of more than one input port of the Butler matrix, and hence, the generation of multiple beams in discrete directions at the same time which can be used to cover different areas simultaneously. Additionally, the tunability allows the operational frequency to be shifted too. By using, in addition, tunable phase shifters at the output ports of the Butler matrix, a continuous beam steering between two switching states is possible. The Butler matrix itself, consisting of several hybrid couplers and fixed phase shifters, was realized in-plane based on a novel waveguide crossover with a dielectric insert allowing a very simple and very compact design [218]. For the hardware implementation of the various components of this complex mixed beam switching and beam steering network, different metallic and dielectric waveguide topologies are used with comparatively low losses at W-band.

5.6.5   *Beam-Steering Reflectarrays*

Apart from phase shifters, most publications on microwave liquid crystal technology seem to be related to reflectarrays, which have been in the focus of research already in an early stage. The first reflectarray with LC tuning is reported in [219–221], being realized and characterized at 35 GHz. It consists of 16×16 unit cells placed at a distance of 0.55λ apart. Each unit cell was filled with a standard LC BL006 to steer the main beam continuously between $+20°$ and $-20°$ controlled by a bias voltage of up to 15 V. Studies on a tunable unit cell for reflectarrays was reported in [222], where numerical and measured results demonstrated $180°$ of tunable phase shift at X-band using K15. A year later, the same author presented in [223] a monopulse reflectarray antenna at the X-band to electronically switch from a sum to a difference radiation pattern. This is achieved by applying a bias voltage of 20 V to one half of the aperture, which is constructed above a 500 µm cavity containing BL006.

Due to the decreasing loss tangent of LCs with increasing frequency, it was very attractive to design reflectarrays at higher frequencies such as 77 GHz as in [220, 224, 225]. A tunable phase range of almost $300°$ was achieved, good enough for the realization of large reflectarrays with phase errors less than $30°$. A lab-scale demonstrator with 16×16 unit cells being 0.55λ apart illustrated beam steering of $\pm25°$ with an LC of the GT3 series.

In [226], a phase range greater than $360°$ within a 7.4% bandwidth and losses lower than 4.7 dB was achieved by a multiresonant unit-cell made up of three parallel dipoles in the same cell and a tunable liquid crystal as substrate, operating in a frequency range from 30 to 40 GHz. Hence, it showed a significant improvement in bandwidth, range of phase and losses compared with that of a single-resonant element. This multiresonant unit-cell was scaled up in the frequency band 117–130 GHz [227], where the simulation results show a tunable phase-shift in a range larger than $300°$ for a 10% bandwidth and low sensitivity to the angle of incidence. Then, a reconfigurable reflectarray was designed to operate in the frequency range from 96 to 104 GHz [228], where the geometrical parameters of the unit cells have been adjusted to simultaneously improve the bandwidth, maximize the tunable phase-range and reduce the sensitivity to the angle of incidence. This time, the performance of the LC-based unit cells was experimentally evaluated by measuring the reflection amplitude and phase of a reflectarray, consisting of 52×54 identical cells. GT3-23001 was used to demonstrate the potential of the proposed reflectarray for beam scanning in the F-band. A similar approach was reported in [229], but with lower driving voltage of 10 V only instead of 40 V above. Numerical simulations in [230] are used to study the electromagnetic scattering from phase-agile microstrip reflectarray cells, where two arrays of equal size elements on top of a 15 µm thick tunable LC layer were designed in the computer model to operate at 102 GHz and 130 GHz, respectively. Micromachining processes based on the metallization of quartz/silicon wafers and an industry compatible LCD packaging technique were employed to fabricate the grounded periodic structures. A $165°$ and $130°$ phase shift with losses between 4.5–6.4 dB and 4.3–7 dB at 102 GHz and at 130 GHz, respectively, was obtained by applying a low frequency AC bias voltage of 10 V. A year later, the design, construction, and measured performance were described for a 94 GHz dual-reflector antenna configuration with an offset parabolic main reflector and a tunable 28×28 element patch reflectarray subreflector to tilt the focused beam $5°$ from the boresight direction in the azimuth plane [231]. The measured 5° tilted radiation pattern exhibited a maximum gain reduction of only 0.15 dB. This design concept exploits successfully recent advances in tunable reflectarray technology based on liquid crystals to create an electronically scanned dual reflector antenna.

A similar concept is used in [232] for a reconfigurable folded reflectarray antenna comprised of a planar lower reflector with an incorporated feed at its center and a polarizing grid on top as an upper reflector. The lower reflector is utilized to collimate the beam and to twist the polarization. The polarizing grid selects the polarization for the transmission and reflects the orthogonally polarized waves toward the lower reflector. By using LC GT3-23001 as a tunable substrate in the upper reflector with 16×16 patch elements allow additional phase adjustment for beam steering with a bias voltage, which is proofed by steering the main beam within ±6° at 78 GHz. The measured gain was 25.1 dB and the cross-polarization suppression measured at 0° angular direction is 29.7 dB at 78 GHz. Fabrication technology of the proposed antenna combines liquid crystal display and printed circuit board technologies, allowing low-cost automated manufacturing technique. Such antenna can be utilized for applications such as 77 GHz automotive radar or for 60 GHz wireless communication services. An antenna pattern synthesis method based on particle swarm optimization, which includes characterization measurements of the upper reflector and the feed, may be used to improve the radiation patterns [233, 234]. Hence, the proposed antenna configuration is a promising candidate for reconfigurable, high-gain, low-profile, and low-cost antennas.

5.6.6   Beam-Steering Leaky-Wave Antennas

While leaky-wave antennas provide low profile and high directivity, metamaterials offer design flexibility by means of dispersion engineering. Additionally, liquid crystal is employed as tunable material to achieve voltage-tunable beam scanning. One of the earliest reported leaky-wave antennas able to steer its main beam direction at a fixed frequency is presented in [235, 236]. It is based on composite right/left-handed waveguide at 7.6 GHz, providing a frequency-independent tunability by applying either an electro- or magnetostatic field to the embedded LC section. Simulations as well as measurements of a built prototype are shown for scattering parameters and far field patterns. The maximum obtained beam tilt amounts to ±10° around broadside. A similar approach of a continuously electrically tunable composite right/left-handed leaky-wave antenna based on liquid crystal for K_a-band applications is presented in [237–240]. In these publications, simulations, vector network analysis, and far field measurements of lab-scale demonstrators are presented together with the detailed fabrication process. For instance, for a 32-unit cell leaky-wave antenna with voltage-tunable interdigital capacitors (measured tunability of 9%) loaded with LC TUD-649 with material's tunability of 24.5% [237, 240], the measured radiation pattern can be scanned from −5° for the unbiased state to +9° for 60 V biasing with respect to broadside at a fixed operation frequency of 26.7 GHz, accompanied with a maximum gain reduction of 2 dB while scanning. [241] proposes a novel electrically

LC controllable composite right/left-handed leaky wave antenna with a simulated scanning range from $-21°$ to $+23°$ at 12.4 GHz by tuning the permittivity of LC. The simulated bandwidth is from 11.78 GHz to 13.09 GHz.

Based on the metamaterial surface antenna technology, Kymeta developed flat-panel antennas going back to the fundamental research on metamaterials of Prof. Smith and his team at Duke University [242, 243] and Dr. Kundtz's pioneering use of metamaterials technologies in electronic beamforming applications. It ultimately led to the spin-out of Kymeta Corporation in August 2012.

5.6.7 Beam-Steering Phased Arrays

Phased arrays are very common; the first phased array transmission was originally shown in 1905 by Nobel laureate Karl Ferdinand Braun, who demonstrated enhanced transmission of radio waves in one direction. Since then, phased arrays have been studied intensively for maritime, aviation, and military applications. Those antennas feature electronically beam steering capability. According to Section 2.2, phased arrays create high gain by many small antennas or radiating elements, for example, 64×64, each with a ($\frac{1}{2}\lambda_0 \times \frac{1}{2}\lambda_0$) dimension. Beneath each radiating element, a phase shifter is integrated, capable of electronically changing its relative phase of up to $360°$. By adjusting the right phase shift of each radiating element, all their small signals are summed up properly in the feeding network to receive maximum signal, which is being further proceed in the receiver. Because of reciprocity, the same array can be used in the transmit mode, too. Then the signal from the transmitter is first distributed equally by the feeding network to the phase shifter. By proper electronically change of the relative phase of the small signal that each antenna element transmits, their overall contribution in the far field creates a larger focused beam in a particular direction. This beam can be directed instantaneously in any direction by fully electronic control, and hence track the movement, for example, of any satellite, no matter how or where you move, without the need for any mechanical moving parts.

A key element of the phased array approach is the phase shifter with the challenge to miniaturize it, in order to integrate it beneath any individual radiating element and to improve the performance while reducing the manufacturing cost to an economical price point. Discrete and continuous phase shifters can be realized in different technologies:

- In semiconductor technologies (CMOS, BiCMOS, and GaAs) [1, 5, 7–9, 244]
- By using RF microelectromechanical systems (RF-MEMS) [2, 3, 6, 10–12, 173, 174, 195, 245–249]
- With ferrites [250–253]
- With ferroelectrics, mainly barium–strontium–titanate (BST) varactors in thin or thick-film technology [175, 176, 254–262]
- With the microwave liquid crystal (MLC) technology [25, 30, 31, 33, 38, 70, 73, 75, 78, 82, 83, 140, 143, 150, 263–265]

Advantages and drawbacks of the different technologies for higher frequency bands have been summarized in Section 2.4. Because of slightly decreasing dielectric losses of microwave LCs with increasing frequency in contrast to all other technologies, it was very attractive at the beginning to build up antenna arrays at higher frequencies. Hence, [266–269] presents first time reconfigurable Vivaldi antenna array concepts for the W-band, based on the fundamental work in [31] on an LC finline phase shifter. Using K15, BL111, and GT3-23001, a maximum differential phase shift larger than 60°, 100°, and 150° could be achieved at 100 GHz with 25 V biasing voltage, respectively. [270, 271] present an LC-based 4 × 1 slotted patch antenna array for the 60 GHz band with a compact planar LC-based meander line phase shifter with a maximum differential phase shift of 47°, enabling to steer the beam in the E-plane by a maximum of 14°. The slotted patch antenna elements are coupled to a microstrip line passing below the array element. In [272], the maximum differential phase shift could be improved up to 79.4° at the center frequency of 63.5 GHz, providing a scanning of up to 27° in the E-plane. A similar approach is done in [273] for an electrically steerable planar array antenna based on LC, featuring a configuration of in series phase shifters and power dividers and in parallel patches. The meandered LC phase shifter achieves 146° phase shift, leading to a scanning range of about 30° at 12.5 GHz for bias voltages between 0 and 20 V and a reflection coefficient less than −10 dB over a bandwidth of about 0.8 GHz. A hybrid integrated 4 × 4 phased array antenna demonstrator at 30 GHz, using TUD-566-filled microstrip line phase shifters in LTCC technology for space application is described in [17, 274, 275] . The scanning range of the measured antenna pattern at 30 GHz is ±30° in the elevation plane with a side lobe level lower than −13 dB. However, all these demonstrators with MLC-based phase shifters face two critical characteristics as discussed in Section 5.4:

1. The length for 360° phase shift usually exceeds the size of $\lambda_0/2$ of an individual radiating element. This is due to the low permittivity and low tunability of microwave LCs. But a tunable delay line could be spiraled or meandered to integrate it within the restricted area beneath the radiating elements of about $(\lambda_0/2)^2$ as in [272, 273, 275], however, with increasing radiation losses per bend. This could be avoided by using clever concepts of straight line phase shifter configurations of few wavelengths beneath the antenna array or by using reflection-type phase shifters with MEMS switch, LC-based CMOS-phase shifters, or LC-based nanowire membrane phase shifters, which are already in the range of less than $\lambda_0/2$.
2. The response time of LC-based devices is usually slow, for example, in the range of several seconds or more, depending on the transmission line topology and LC layer heights required by it. But there are hardware solutions for flat-panel antennas such as delay lines periodically loaded with LC varactors of less than 5 μm height only or by using the nanowire membrane technology also with an effective LC layer height of less than 5 μm. Both approaches allow a response time T_{10}^{90} of less than 30 ms as required for high beam-steering rates in mobile applications [163].

Other important characteristics of ESA are the scanning range and the antenna gain-to-noise-temperature G/T, which means for the phase shifters a differential phase shift of at least 360°, accompanied with the lowest insertion losses possible, that is, a high FoM. This is particularly important for high power-handling in the transmit mode of many communication systems, where antennas must manage tens up to hundreds of watts, for example, in satellites, to avoid massive power losses and power dissipation in the antenna. In addition, for systems with limited power supply, power consumption to steer these antennas must be as low as possible too, to operate the ESA terminals, particularly for vehicles, ships, mobile terminals, and autarkic M2M systems, with battery, renewable energy, off-grid power solutions. Moreover, ESAs must be designed to withstand the vibration and thermal environment on the aircrafts, vehicles, trains and ships . Hence, a robust and reliable technology must be used for manufacturing. A key figure will be the cost per terminal to enable wide-scale applications.

As mentioned in Chapter 1, there are currently different approaches by some spin-off companies for developing, producing, and commercializing cost-efficient large-scale ESA for satellite systems and for future 5G mm-wave systems. All of them aim for low-profile and light ESA with modular concept to be scaled to any requirement and a thickness of few centimeters only. Moreover, all work on the form factor of their ESA, for compact stand-alone terminals with aesthetic appearance and to integrate them smartly or nearly seamless into the structure or body of a carrier object such as into the skin of an airplane or into the rooftop of a vehicle.

One of the promising approaches of a flat-panel beam-steering antenna array for on-the-move applications will be presented in Section 5.6.7.1. It goes back to a first small 2D antenna demonstrator of a 2×2 microstrip patch array presented in [25, 26] at 17.5 GHz, adapting the well-established LCD technology. Soon later, an 8×8 microstrip patch antenna module was processed by using an existing liquid crystal display (LCD) production capacity to significantly reduce manufacturing costs. This ultimately led to the spin-off of TU Darmstadt, ALCAN Systems GmbH, Darmstadt, Germany, in June 2016. By using the loaded-line concept with periodic tunable LC varactors of about 4 μm thickness and a Merck LC of the third generation, the phase shifter stack of ALCAN's flat-panel antenna achieved response times of less than 30 ms. This represents a technology breakthrough in satellite and cellular communications, addressing high-performance, future-proof data connectivity solution for any location to meet the needs of a range of markets including maritime, aero, land mobility, consumer broadband, and enterprise, where data might be provided via satellites in any orbital altitude or via 5G mm-wave communication links.

Section 5.6.7.2 describes a different concept of a lightweight Tx/Rx electronic beam steering horn antenna array at K_a-band for a relay satellite in geostationary orbit with fully integrated LC-tuned rectangular waveguide phase shifters into each branch of the feeding network, all built in an electroplating process for minimum weight. For biasing the LC-phase shifter, custom-built electronics provide up to 513 voltage channels with voltages between +164 V and −164 V. While differential phases of up to 450° and phase change rates between 5.1°/s and 45.4°/s meet the expectations, the measured insertion loss of 4–6 dB for different biasing voltages and in the frequency range from 23 GHz to 27.5 GHz missed the goal of 3 dB.

More recently, a fully dielectric 1×4 antenna array is presented in [149] for the first time, consisting of four tapered dielectric rod antenna elements with rectangular shape fed by a single, four-port multimode interference power divider, operating in the frequency range from 85 to 105 GHz. It is made from a single sheet of cross-linked polystyrene Rexolite, enabling a compact and light-weight antenna array at low cost. The measured half-power beam width is 12° in the H-plane and the gain 16.9 dBi at 95 GHz, accompanied with a sidelobe level of around −11 dB. The return loss of the array is better than 15 dB. This concept is extended to a fully dielectric 4 × 4 antenna array in [276] with a gain of up to 22.5 dBi. Because of its high potential, a future step will be to embed LC cavities within the dielectric rod antenna elements for beam steering. Concept, design and first experimental results of this fully dielectric antenna array will be introduced in Section 5.6.7.3.

5.6.7.1   Flat-Panel Beam-Steering Antenna Array

This section will focus on flat-panel beam-steering antenna arrays, combining state-of-the-art LCD technology with the introduced microwave liquid crystal technology, featuring very low power consumption and enabling low-cost, large-scale manufacturing of large arrays with a great number of antenna elements, since the LC phase shifter stack is fabricated in standard LCD processes and established LCD production lines. Moreover, those smart antennas can provide combinations of frequency, polarization, and pattern agility.

A first small lab-scale demonstrator was already presented in [25, 26] for a proof-of-concept at 17.5 GHz, which is depicted in Figure 5.112. It is a 2 × 2 microstrip patch array with an element spacing of $0.65\lambda_0 \times 0.65\lambda_0$. The overall thickness is 1.5 mm only, where the four spiraled 360°-delay line phase shifters are on the backside together with an SMA connector for the RF signal and a four-pin DC biasing connector for to control the four delay line phase shifters. These microstrip line phase shifters must be spiraled, in order to implement them within the limited area beneath the radiating element, since their total length ℓ is about $4.7\lambda_0$ for a 360° phase shift. They have been investigated in detail in [26], together with meander-like ones. Their FoM is assumed to be around 105°/dB without noticeable coupling at the bends. The insertion loss is in a range between 3.5 dB and 4.25 dB.

The assembly of this lab-scale demonstrator consists of three stacked substrates according to Figure 5.113, where the LC material layer of 100 μm thickness is sandwiched between two 700 μm thick Borofloat glass (front and back) substrates with relative permittivity and dielectric loss tangent of 4.65 and 0.008, respectively. Both sides of a large glass substrate are processed one by one by using the same lithography processes, which are chromium/gold evaporation, photolithography, gold plating, and wet etching. When the lithography is completed, the substrate is diced precisely within ±5 μm accuracy into two pieces [25].

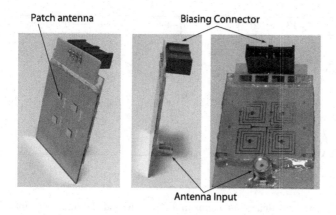

Figure 5.112 Photograph of realized two-dimensional electronic steerable antenna array using a MLC technology similar to LCD technology [25, 26]. (Left) Front side with the four microstrip patches. (Middle) Side view of the array with an overall thickness of 1.5 mm only. (Right) Back side with the four meandered microstrip lines, the feeding network, and four thin bias lines (not visible) on the substrate as well as an SMA connector for the RF signal and a four-pin DC connector to control the four delay-line phase shifters.
Courtesy of ALCAN Systems GmbH

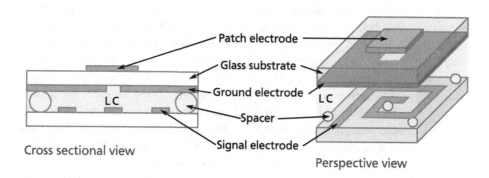

Figure 5.113 Assembly of the LCD-like phased array antenna, consisting of three stacked substrates (left) which are multilayer structures as shown on the right side for a unit cell.

On the front glass substrate, 2×2 microstrip patches made of gold are processed on top, and on the backside, a ground plane coated with a thin polyimide film for prealignment of the LC (see details in Figure 5.113 for one unit cell of $0.65\lambda_0 \times 0.65\lambda_0$). On the back substrate, first the spiraled microstrip lines for 360° phase shifting within the limited area, the RF feeding network accomplished by using standard $\lambda/4$-impedance transformers and four thin highly resistive Cr bias lines for to control each phase shifter individually are processed, again coated with a thin polyimide film for prealigning the LC molecules on top. The spacers (micro pearls with 100 µm diameter) are used to maintain a cavity of constant thickness to fill in the LC after gluing both glass substrates together and sealing them. The signals of the four microstrip delay lines are coupled through the first polyimide film, the LC layer, the

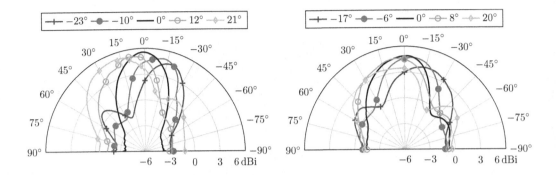

Figure 5.114 Measured antenna gain patterns in the E- (left) and H- (right) principal planes [25, 26].

second polyimide film, through the four slots in the ground plane, though the front glass to the radiating elements.

Test measurements with this aperture-coupled array have been carried out by using the LC TUD-566. The antenna matching at 17.5 GHz is around -20 dB for broadside (unbiased) and goes down to -15 dB when it is steered [25, 26]. Figure 5.114 exhibits the far-field radiation pattern, measured in the E-and H-principal planes in an anechoic chamber. In order to demonstrate its two-dimensional steering capability, the antenna is mounted on top of a turn table and electronically steered within $\pm 25°$ in both planes. As being a small demonstrator with 2×2 elements only and with a relatively large element spacing of $0.65\lambda_0$, the antenna gain reduces and the side lobe level (SLL) increases at wide-scanning angles Θ_m. Thus, for $\pm 25°$, the gain loss is about 2 dB and the SLL is -4 dB. Both are confirmed by the simulations [25, 26]. When the main beam is steered to $\Theta_m = -45°$, a grating lobe occurs at $+45°$. The measured beam steering speed is about 150 s to tilt the main beam from $\Theta_m = 0°$ to $\Theta_m = 45°$ and vice versa. This empirical result verifies the fact that the beam steering rate is lower than the response time of the delay line, which is about 180 s (see Figure 5.35) for the LC TUD-566 [26].

According to simulations in Figure 5.115, a larger size of the proposed phased-array antenna provides higher gain, and a smaller spacing between the radiating elements a wider scanning range, where the antenna can be steered with SLL less than -11 dB.

Based on these scientific results above, ALCAN's antenna technology made use of a modular concept, where several basic modules of 8×8 radiating elements build up the ESA with an appropriate size to meet the application's specific requirements. The feasibility of this approach was demonstrated in 2018 with a flat-panel antenna array in Figure 5.116 that allowed steering the antenna beam between different K_u-band satellites in geostationary orbit for television reception. This flat-panel antenna array can also be conformed to fit a curved surface.

ALCAN's antenna technology combines state of the art LCD technology with microwave LCs, appropriate phase shifter and antenna design, where each module is

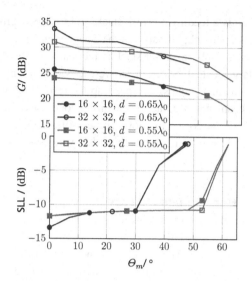

Figure 5.115 Antenna gain G and side lobe level SLL versus main beam direction Θ_m of a 16×16 and 32×32 array of the proposed topology with two spacing $d = 0.65\lambda_0$ and $d = 0.55\lambda_0$ between the radiating elements at 17.5GHz [26].

Figure 5.116 K_u-band TV reception demonstrator using ALCAN's antenna technology to steer the beam between different GEO satellites.

built up in different stacks, mainly a feeding network stack, an LC-phase shifter stack, and a radiator stack. Using this approach, all parts can be designed independently and in a modular fashion. The core LC-phase shifter stack is fabricated in standard LCD processes. This allows for large-scale fabrication as well as virtually any aperture size – including segments or antenna groups. To speed up the response times of the phase shifters, and hence the beam steering rate to few milliseconds, use is made of a new specific LC with much lower viscosity and 360° phase shifters using a topology similar to the grounded CPW loaded with 4 μm thick LC varactors according to Section 5.4.1.2.

5.6.7.2   Electronic Beam-Steering Horn Antenna Array

For a relay satellite in geostationary orbit, establishing a link between a low-Earth orbit satellite and a ground station (see Figure 5.117), a lightweight Tx/Rx electronic beam steering horn antenna array was designed for K_a-band within LISA-ES (lightweight intersatellite antenna – electrical steering), a project funded by the German Aerospace Center (DLR).

The frequency bands of interest for down- and uplink are 23–23.5 GHz and 27–27.5 GHz. The desired scanning range is +11° to −11°, aiming for a phase change rate of at least 5°/s to track the LEO satellite. For the final application, a 16 × 16 horn antenna array with two polarizations, that is, with 512 phase shifters in total, is aimed to implement into the satellite. It consists of a three-dimensional feeding network in rectangular waveguide topology because of its intrinsically low loss, LC-tuned rectangular waveguide phase shifters, intended to be fully integrated into each branch of the feeding network for robustness and reliability, and the horn antenna array, all built in an electroplating process for minimum weight. The feeding network is split into two identical feeding networks for two polarizations, which fit

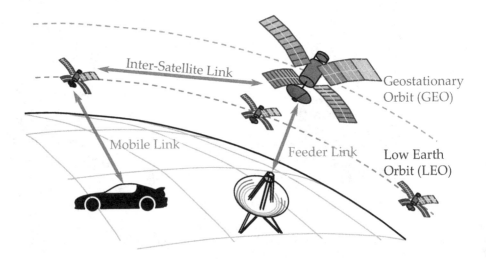

Figure 5.117 System concept for LISA-ES 5.6.7.2 electronic beam steering horn antenna array.

Figure 5.118 RF-performance demonstrator: 4 x 4 horn antenna array with feeding network, but without phase shifters. The copper structure is fabricated in a lightweight manner.
Photo courtesy TU Munich/LRT

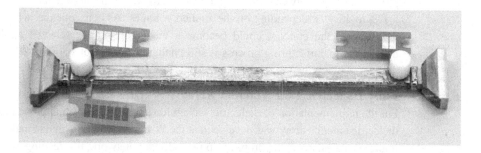

Figure 5.119 Electrically biased LC waveguide phase shifter with electroplated waveguide walls and full biasing capability [80].

into each other such that they can be connected to the horns by polarization separating dividers [80, 142, 277].

For the RF-performance tests, however, only a 4 × 4 horn antenna array and a plug-in-solution of the LC-based phase shifters for individual measurements are used as shown in Figures 5.118 and 5.119, respectively [80, 142]. Choosing the MLC technology implies continuous tunability with an excellent FoM due to low insertion losses and high power-handling capability, including high linearity. Moreover, the LC mixtures are space approved.

For biasing the LC-phase shifter, custom-built electronics provide up to 513 voltage channels. One provides a fixed voltage pair of ±164 V for all phase shifters, the other channels are tunable between ±0 V and ±164 V. In this way, each phase shifter can be individually tuned to a desired differential phase shift. The pointing direction of the array is set by applying voltages to each phase shifter individually resulting in a delay on each individual horn [80, 142].

Based on the experience with the heavy waveguide phase shifters in split-block design and made of brass according to Section 5.4.2, a lightweight design has been proposed for space applications [80]. It consists of an LC-filled Rexolite core, including the biasing electrodes, which is enclosed by a galvanically deposited silver-primed copper cladding of 0.5 mm thickness. Due to the higher electrical conductivity of silver and copper compared to brass, the wall losses will decrease slightly. More importantly, the electroplated wall fits more closely and regularly to the biasing electrode sheets, which reduces the insertion losses. A first prototype phase shifter with a length of 120 mm and additional aluminum flanges with screws is projected to weigh 12 g only.

In [142], various design iterations are realized for environmental and RF-performance tests to prove the viability of the selected design. The outcome of the tests: While differential phases and phase change rates meet the expectations, the measured insertion loss missed the goal of 3 dB. Thus, depending on the applied biasing voltages, the transmission varies between -4 dB and -6 dB in the frequency range from 23 GHz to 27.5 GHz, the reflection is below -13.7 dB to -18.7 dB in the lower band and below -11.1 dB to -11.7 dB in the upper band and the measured maximum differential phase shift is about $450°$. While the minimum time for a $360°$ phase change is between 16.1 s and 138 s, the maximum phase change rate reaches $5.1°$/s to $45.4°$/s, depending on the applied voltages. As consequence, when loss can be reduced, the concept could become a viable option for selected applications, despite the manufacturing process is still challenging and not stable [142].

5.6.7.3 Fully Dielectric Rod Antenna Array

For the implementation of dielectric waveguide phase shifters, an approach for a fully dielectric antenna array was developed at the W-band [161]. The main challenge is the implementation of an appropriate power divider network for feeding the antenna elements. This is especially challenging for dielectric waveguides, because discontinuities cause radiation and the cascading of multiple dividers leads to high losses and requires large space. Furthermore, power dividers such as Magic T-junctions, which are well known for metallic waveguides are not applicable for dielectric waveguides. Therefore, multimode interference is adopted from the optics to design a single dielectric power divider, which achieves power division for any output number in one single step. The used multimode interference (MMI) phenomenon generates several field maxima due to a higher order mode propagation. Assuming an electromagnetic wave propagating in the z-direction, the E-field distribution inside the multimode waveguide at distance z is given by

$$\sum_{i=}^{i-1} c_i \psi_i e^{j(\beta_0 - \beta_i)z}, \tag{5.79}$$

with the excitation coefficients c_i, the modal field distributions ψ_i, and the propagation constant β_i of each ith mode [278]. Due to the spacing of the propagation constants given by

$$\beta_i - \beta_0 \approx \frac{\pi n(n+2)}{3L_\pi}, \tag{5.80}$$

(a) (b) (c)

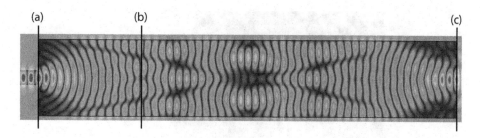

Figure 5.120 Typical interference pattern of a multimode dielectric waveguide. (a) Single-mode input. (b) Multiple field maxima. (c) Repetition of the single-mode input field distribution.

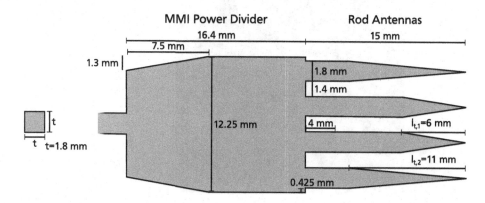

Figure 5.121 Fully dielectric 1 × 4 rod antenna array. On the left, the cross section of the single-mode waveguide is shown.

a spatial interference pattern occurs inside the multimode waveguide. Figure 5.120 shows typical interference pattern along the multimode dielectric waveguide being excited by a symmetrically positioned single mode input at cross section (a). It can be seen, that after a certain distance, multiple field maxima occur such as at cross section (b) and that the input field distribution is being repeated after a longer distance at cross section (c).

For the above example, power division can be achieved if four single-mode dielectric output waveguides are exactly placed at the four field maxima at cross section (b). This was utilized for a fully dielectric 1 × 4 rod antenna array shown in Figure 5.121, consisting of the single-mode dielectric waveguide, the MMI power divider, and four rod antennas.

As mentioned in [149], a phase shift occurs between the two middle and the two outer output waveguides. Therefore, to compensate the phase difference, two different taper lengths of the rod antennas, $l_{t,1}$ and $l_{t,2}$, are required to achieve accurate broadside radiation. Milled out of Rexolite, the antenna weighs less than 1 g. Figure 5.122 depicts the measurement setup for the far field characterization.

The H- and E-plane antenna patterns are shown in Figures 5.123 and 5.124, respectively. There it can be seen that the measurements for different frequencies

Figure 5.122 Far field measurement setup of the fully dielectric antenna array.

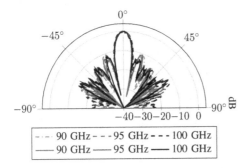

Figure 5.123 Individually normalized H-plane measurements for different frequencies. Simulations are given in dashed lines.

are in good agreement to the corresponding simulations. The achieved antenna gain is about 16.5 dBi at 95 GHz.

Beyond this planar 1D-array, a 2D-array was already demonstrated with a single step two-dimensional MMI power divider in [276]. Future work aims to integrate subwavelength LC-phase shifters between the MMI power divider and the rod antenna

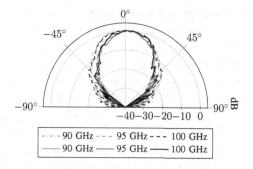

Figure 5.124 Individually normalized E-plane measurements for different frequencies. Simulations are given in dashed lines.

elements, which is a straightforward approach. Challenging will be the integration of the biasing electrodes, as disturbance on the electromagnetic fields must be avoided to reduce unwanted parasitic radiation. However, decoupling between the output waveguides could even be positively influenced by a metallic electrode mesh.

5.7   LC-Based Tunable Power Divider, Metamaterials, and Surfaces

Beyond phase shifters, resonators and filters as well as the various types of beam-steering antennas, there are many other tunable or reconfigurable devices presented in literature for different applications, which are based on the MLC technology. Exemplarily, tunable power divider and RF switches as well as tunable metamaterials, frequency-selective and high-impedance surfaces will be briefly summarized in this final section.

5.7.1   Tunable Power Divider and RF Switches

Tunable power dividers are a key component for many applications such as for adaptive arrays employing full-beam synthesis. [279] presents an LC tunable microwave bandpass filter integrated power divider with the relevant simulated and measured scattering parameters for biasing voltages from 0 V to 20 V. At the center frequency of 6.2 GHz, the measured return loss is higher than 14 dB, but the insertion losses are up to 6.5 dB. Nevertheless, the proposed power divider achieves equal power dividing performance within the working frequency range. In [210], an LC-based tunable reflection-type power divider is presented at 12 GHz, which utilizes the edge slope of a tunable Butterworth filter with LC GT3-23001. The measurement proofs the functionality of the concept, achieving a dynamic range of 9 dB for both output branches.

At high frequencies, single-pole double-throw (SPDT) RF switches are attractive in radiometers for switching between signal path and calibration load for power calibration or to select different input ports of the Butler matrix network. In [147, 217, 263], new LC-based SPDTs with continuously adjustable power splitting ratio are investigated in different topologies and technologies at two frequency bands. [263] presents the design and characterization of a tunable SPDT at 30 GHz in low-temperature co-fired ceramic (LTCC) technology with a fully embedded LC GT3-23001 filled phase shifter in strip line topology as key component. The SPDT shows a good matching with less than -10 dB over a bandwidth of 18% and a good isolation between 22 dB and 32 dB but accompanied with high insertion loss of 9–11 dB, which arises from a surface roughness and decreased conductivity of the gold metallization. The continuous tunability not only allows the selection of each preferred power splitting ratio at the output ports, but also a tuning of the center frequency. [147, 217] presents continuously tunable, interference-based SPDTs at W-band frequencies in rectangular waveguide and fully implemented in subwavelength dielectric fiber topology, respectively. Key components are the phase shifters in the two branches of the SPDT, which are realized as line sections filled with Merck's LC GT5-26001. They allow not only to adjust the signal ratios at the two output ports, but also to tune the operating frequency between 90 GHz and 110 GHz with isolation in the range of 10–12 dB for the waveguide and between 93 GHz and 110 GHz with isolation better than 25 dB (maximum 40 dB) for the fiber topology, respectively. For the first, matching is better than -12 dB and the insertion loss is less than 3 dB between 89 GHz and 105 GHz [217]. For the second, matching is better than -20 dB and the insertion loss at the thru-port is between 3.5 dB and 5 dB in the frequency range from 90 GHz to 110 GHz without WR10 to fiber transitions [147].

Another dielectric waveguide switch made from Rexolite is proposed in [149] at the W-band. For the first time, it uses the multimode interference pattern, also known as self-imaging in the optical domain, which occurs due to a higher-order mode propagation in a multimode waveguide. By exciting the multimode waveguide asymmetrically, a phase shift difference arises between two occurring field maxima. Placing LC GT3-23001 in the region of one field maximum, this phase shift can be adjusted by means of the tunable permittivity. This in turn allows control of the interference pattern, and therefore, the position of the field maximum at the output of the multimode waveguide being connected to single-mode dielectric waveguides. Around 94 GHz, the measured port isolation is better than 40 dB and 10 dB for perpendicular and parallel orientation of the LC, respectively.

5.7.2   Tunable Metamaterials, Frequency-Selective and High-Impedance Surfaces

Metamaterials are artificial structures with macroscopic electromagnetic properties beyond those of classical, natural materials. By proper design, one can engineer their permittivity and permeability values, which can lead to negative refractive index in a small frequency range and associated physical phenomena, e. g. negative refraction, antiparallel phase velocity, or subwavelength focusing. Frequently, such metamaterials are built of periodic structures, of so-called unit cells with dimensions well below the operating wavelength. This approach is not new since already Kock in

1948 investigated the realization of "artificial" dielectric lenses. Owing to their novel electromagnetic properties, negative index metamaterials have received a lot of attention in the last two decades from optical down to microwave frequencies, being envisaged as the core element in a wide range of applications. In combination with LCs it offers dynamically tuning of their properties by a local electro- or magnetostatic field, enabling the design of functional components for electromagnetic processing such as tunable phase shifters and modulators, additive resonators and filter, tunable frequency-selective and high-impedance surfaces, tunable absorbers as well as beam-switching and beam-steering gratings and arrays.

Thus, [280, 281] developed a metamaterial that has a tunable negative-zero-positive index of refraction in the optical frequency range. Many realizations employ periodic structures such as photonic crystals and nano-patterned noble metal particles or holes and bulk metamaterials with random distributions of nanoparticles. For instance, reconfigurability of the index of refraction was suggested by using LC as the host medium, containing a random distribution of coated dielectric nanospheres in [280]. The reconfigurability to metamaterials is extended in [282] by cladding LC as substrate and superstrate layers onto a conventional negative-index metamaterial. Tuning the permittivity of the LC layers controls the value of the effective permittivity in an averaging manner. Combined with proper magnetic resonances provided by a pair of silver strips separated by a thin layer of alumina, the effective index of refraction can be reconfigured between negative, zero, and positive values at a fixed wavelength. The ability to vary the permittivity of the LC layers also provides a method to control the bandwidth of the negative index behavior. In [283], anisotropic LCs were also used for tunable optical negative-index metamaterials for near-IR operation, particularly to tune the response of the magnetic resonator rather than changing the electric properties. In this article, a hybrid finite element-boundary integral technique has been developed to rigorously compute the scattering from periodic structures composed of inhomogeneous, anisotropic and dispersive materials of arbitrary shape.

Another class are the LC-tuned fishnet metamaterials [284, 285]. A typical configuration of fishnet metamaterials consists of a dielectric slab cavity formed between two identical periodic arrays of interconnected metallic patches. Fishnet structures provide ease of fabrication and scalability of their properties in a wide spectral range, which has been demonstrated from the microwave to the near-infrared and visible spectrum [285]. For instance, [286] demonstrates a terahertz tunable fishnet metamaterial by using an electrically controlled polymer-dispersed LC matrix, that is, a thin 1.5 μm polyimide "skin layer" to form a uniform surface for metal electrodes while minimizing the Fabry–Perot effect of the skin layer on the measurements. The tunability was verified by measuring the frequency shift of about 10 GHz and phase shift of 70° at 280 V, respectively, in the reflection coefficient, observing a minimum negative refractive index of −15 at 0.55 THz. In [287–290], various concepts of a voltage-tunable gradient-index fishnet metamaterial are investigated at microwaves. The unit cells of the fishnet are loaded with LC material, providing a continuous tuning of each column of the array by introducing a voltage gradient over the width of the fishnet. Hence, the phase distribution over the aperture can be tuned, thus, controlling the radiation direction. With a phase tuning of about 180° for two unit-cell layers and using GT3-23001, the measured scanning range of a realized 8 × 8 array at 27.5 GHz was ±5° with a control voltage of up to 20 V only. By stacking five fishnet unit-cell layers, a phase tuning of more than 360° would be possible, allowing arbitrary scanning angles. Moreover, the fishnet structure and properties are scalable to higher frequencies as simulation exhibits for about 330 GHz [287, 290], since LC properties even slightly increase with higher frequencies.

Another area which have recently widely published are frequency-selective surface (FSS), which exploits the dielectric anisotropy of LCs to generate an electronically tunable filter response by an external bias voltage. In these structures, an LC layer is generally sandwiched in between two identical parallel structures, a transmitting and a reflecting structure, consisting of printed element patterns. A FSS at higher frequencies of 110–170 GHz has been presented in [291], using a 130 μm thick LC BL037 layer, sandwiched between two printed arrays of slot elements, illustrating a 3% tunability by applying a control voltage of 10 V. Measured insertion loss increases from 3.7 dB to 10.4 dB at resonance (134 GHz), thus demonstrating the potential to create a FSS.

Another example is given in [292], using a magnetically tunable negative permeability metamaterial. This structure consists of an array of broadside coupled split-ring resonators (SRR) filled with LC, where a resonant frequency shift of 0.3 GHz was obtained at the X-band. [293] utilize LC cells to design a metamaterial structure, whose index of refraction is tunable from negative, through zero, to positive values. It uses randomly doped LC substrate with coated dielectric (non-magnetic) spheres. Metallodielectric FSSs and an all-dielectric FSS exhibit broadband tunable filter characteristics at mid-infrared (mid-IR) wavelengths and could be tuned over a wide range of frequencies by an LC superstrate. The authors believe these LC tunable FSS structures can be used to develop a new class of infrared/optical switches for terahertz applications.

A specific application is a dynamically adaptive Salisbury screen absorber using liquid crystals [294, 295]. In [295], the structure is backed by a high-impedance surface of a 250 μm thick LC substrate, varying its anisotropic properties by a control voltage from 0 V to 20 V, which causes a shift in the resonant frequency of the radar absorber. Simulated results for Merck's standard LC BL037 show that the reflection minimum is tunable over a 2.3% bandwidth and the reflectivity is below −10 dB in the frequency range between 8.8 and 10.05 GHz (13%). The measured reflectivity of the structure in [294] is less than −38 dB with a 10 dB bandwidth of 200 MHz at 10.19 GHz when a 4 V bias voltage is applied and it shows a 34 dB reduction in signal loss when the bias voltage is increased to 20 V.

The design of an electronically reconfigurable ground plane composed of a high-impedance surface based on GT3-23001 liquid crystals is illustrated in [296]. It is used as a backing structure for a two-arm Archimedean spiral antenna, generating either sum or difference beam for a monopulse radar application by changing the permittivity, and hence, reflection phase of the high-impedance surface reflector when a bias voltage is applied between the periodic Jerusalem cross slot array and ground plane. Simulation results at 6.5 GHz exhibit a deep null for the difference pattern and also high gain in the boresight direction for the sum pattern, when changing the permittivity of the substrate. Measurements are not reported.

References

1. J. L. Vorhaus, R. A. Pucel, and Y. Tajima, "Monolithic dual-gate GaAs FET digital phase shifter," *IEEE Transactions on Microwave Theory and Techniques,* vol. 30, no. 7, pp. 982–992, 1982. DOI: http://10.1109/TMTT.1982.1131187

2. N. S. Barker and G. M. Rebeiz, "Optimization of distributed MEMS phase shifters," in *IEEE MTT-S International Microwave Symposium,* 1999, vol. 1, pp. 299–302.

3. S. Barker and G. M. Rebeiz, "Distributed MEMS true-time delay phase shifters and wide-band switches," *IEEE Transactions on Microwave Theory and Techniques,* vol. 46, no. 11, pp. 1881–1890, 1998. DOI: http://10.1109/22.734503

4. H.-T. Kim et al., "V-band 2-b and 4-b low-loss and low-voltage distributed MEMS digital phase shifter using metal-air-metal capacitors," *IEEE Transactions on Microwave Theory and Techniques,* vol. 50, no. 12, pp. 2918–2923, 2002.

5. S. E. Shih et al., "A W-band 4-bit phase shifter in multilayer scalable array systems," in *2007 IEEE Compound Semiconductor Integrated Circuits Symposium,* pp. 1–4, 2007.

6. A. Stehle et al., "RF-MEMS switch and phase shifter optimized for W-band," in *European Microwave Conference (EuMC),* 2008/October, pp. 104–107.

7. D.-W. Kang, J.-G. Kim, B.-W. Min, and G. M. Rebeiz, "Single and four-element Ka-band transmit/receive phased-array silicon RFICs with 5-bit amplitude and phase control," *IEEE Transactions on Microwave Theory and Techniques,* vol. 57, no. 12, pp. 3534–3543, 2009. DOI: http://10.1109/tmtt.2009.2033302

8. J. G. Yang and K. Yang, "Ka-band 5-bit MMIC phase shifter using InGaAs PIN switching diodes," *IEEE Microwave and Wireless Components Letters,* vol. 21, no. 3, pp. 151–153, 2011. DOI: http://10.1109/LMWC.2010.2104314

9. N. Chen, J. Zhen, and Q. Pang, "A millimeter-wave GaAs 5–bit MMIC digital phase shifter," in *2013 International Workshop on Microwave and Millimeter Wave Circuits and System Technology,* 2013/October: IEEE.

10. C. Chang, Y. Chen, and S. Hsieh, "A V-band three-state phase shifter in CMOS-MEMS technology," *IEEE Microwave and Wireless Components Letters,* vol. 23, no. 5, pp. 264–266, 2013. DOI: http://10.1109/LMWC.2013.2253309

11. U. Shah et al., "Submillimeter-wave 3.3-bit RF MEMS phase shifter integrated in micro-machined waveguide," *IEEE Transactions on Terahertz Science and Technology,* vol. 6, no. 5, pp. 706–715, 2016. DOI: http://10.1109/TTHZ.2016.2584924

12. A. Chakraborty and B. Gupta, "Paradigm phase shift: RF MEMS phase shifters: An overview," *IEEE Microwave Magazine,* vol. 18, no. 1, pp. 22–41, 2017. DOI: http://10.1109/mmm.2016.2616155

13. F. Goelden, S. Mueller, P. Scheele, M. Wittek, and R. Jakoby, "IP3 measurements of liquid crystals at microwave frequencies," in *European Microwave Conference (EuMC),* 2006/September: IEEE.

14. F. Goelden, "Liquid crystal based microwave components with fast response times: Material, technology, power handling capability," 2010. http://tuprints.ulb.tu-darmstadt.de/2203/

15. S. Mueller, *Grundlegende Untersuchungen steuerbarer passiver Flüssigkristall Komponenten für die Mikrowellentechnik.* 2007.

16. A. Gaebler et al., "Liquid crystal-reconfigurable antenna concepts for space applications at microwave and millimeter waves," *International Journal of Antennas and Propagation,* vol. 2009, pp. 1–7, 2009. DOI: http://10.1155/2009/876989

17. I. Wolff, "Integrated beam steerable antennas in LTCC-technology," in *Proceedings of the International Workshop on Antenna Technology (iWAT),* 2010/March, pp. 1–4.

18. A. Heunisch, B. Schulz, T. Rabe, S. Strunck, R. Follmann, and A. Manabe, "LTCC antenna array with integrated liquid crystal phase shifter for satellite communication," *Additional Conferences (Device Packaging, HiTEC, HiTEN, & CICMT),* vol. 2012, no. CICMT, pp. 000097–000102, 2012. DOI: http://10.4071/cicmt-2012–tp15

19. R. Follmann et al., "Liquida-Sky: A tunable liquid crystal filter for space applications," in *2013 IEEE-APS Topical Conference on Antennas and Propagation in Wireless Communications (APWC),* 2013/September: IEEE.

20. M. Jost et al., "Continuously tuneable liquid crystal based stripline phase shifter realised in LTCC technology," in *European Microwave Conference (EuMC),* 2015/September: IEEE.

21. S. Strunck et al., "Reliability study of a tunable Ka-band SIW-phase shifter based on liquid crystal in LTCC-technology," *International Journal of Microwave and Wireless Technologies,* vol. 7, no. 5, pp. 521–527, 2014. DOI: http://10.1017/S175907871400083X

22. S. Strunck, *Flüssigkristall-basierte und LTCC-integrierte elektrisch steuerbare Mikrowellenphasenschieber und -polarisatoren.* Herzogenrath, Germany: Shaker, 2015.

23. A. Gaebler, "Synthese steuerbarer Hochfrequenzschaltungen und Analyse Flüssigkristall-basierter Leitungsphasenschieber in Gruppenantennen für Satellitenanwendungen im Ka-Band," *ETIT,* 2015. http://tuprints.ulb.tu-darmstadt.de/4691/.

24. A. E. Prasetiadi et al., "Liquid-crystal-based amplitude tuner and tunable SIW filter fabricated in LTCC technology," *International Journal of Microwave and Wireless Technologies,* vol. 10, no. 5–6, pp. 674–681, 2018. DOI: http://10.1017/S1759078718000600 www.cambridge.org/core/article/liquidcrystalbased-amplitude-tuner-and-tunable-siw-filter-fabricated-in-ltcc-technology/73D699A1F8CF43C44D3AEB4E128F5F5D

25. O. H. Karabey, A. Gaebler, S. Strunck, and R. Jakoby, "A 2-D electronically steered phased-array antenna with 2 x 2 elements in LC display technology," *IEEE Transactions on Microwave Theory and Techniques,* vol. 60, no. 5, pp. 1297–1306, 2012. DOI: http://10.1109/TMTT.2012.2187919

26. O. H. Karabey, *Electronic Beam Steering and Polarization Agile Planar Antennas in Liquid Crystal Technology.* Cham, Switzerland: Springer International Publishing, 2013.

27. C. Weickhmann, A. Mehmood, A. B. Olcen, Y. Sun, and R. Jakoby, "A low-cost, flat, electronically steerable array antenna for new massive NGEO constellations ground terminals and future 5G," in *ESA Antenna Workshop,* 2018.

28. C. Weickhmann, A. Mehmood, A. B. Olcen, Y. Sun, and R. Jakoby, "A low-cost, flat, electronically steerable array antenna for new massive NGEO constellations ground terminals and future 5G," in *European Conference on Antennas and Propagation (EuCAP),* 2019.

29. C. Weil, G. Luessem, and R. Jakoby, "Tunable inverted-microstrip phase shifter device using nematic liquid crystals," in *IEEE MTT-S International Microwave Symposium*, 2002, vol. 1, pp. 367–371.

30. C. Weil, S. Muller, P. Scheele, P. Best, G. Lussem, and R. Jakoby, "Highly-anisotropic liquid-crystal mixtures for tunable microwave devices," *Electronics Letters*, vol. 39, no. 24, pp. 1732–1734, 2003. DOI: http://10.1049/el:20031150

31. S. Mueller, C. Felber, P. Scheele, M. Wittek, C. Hock, and R. Jakoby, "Passive tunable liquid crystal finline phase shifter for millimeter waves," in *European Microwave Conference (EuMC)*, 2005.

32. A. Manabe, "Liquid crystals for microwave applications." In Proceedings of SPIE 8642, Emerging Liquid Crystal Technologies VIII, 86420S. DOI: http://10.1117/12 .2016578

33. C. Weickhmann, R. Jakoby, E. Constable, and R. A. Lewis, "Time-domain spectroscopy of novel nematic liquid crystals in the terahertz range," in *38th International Conference on Infrared, Millimeter, and Terahertz Waves (IRMMW-THz)*, 2013, pp. 1–2. DOI: http:// 10.1109/IRMMW-THz.2013.6665423

34. M. Jost et al., "Evolution of microwave nematic liquid crystal mixtures and development of continuously tuneable micro- and millimetre wave components," *Molecular Crystals and Liquid Crystals*, vol. 610, no. 1, pp. 173–186, 2015. DOI: http://10.1080/15421406 .2015.1025645

35. C. Fritzsch and M. Wittek, "Recent developments in liquid crystals for microwave applications," in *2017 IEEE International Symposium on Antennas and Propagation & USNC/URSI National Radio Science Meeting*, 2017/July: IEEE.

36. P. Yaghmaee, O. H. Karabey, B. Bates, C. Fumeaux, and R. Jakoby, "Electrically tuned microwave devices using liquid crystal technology," *International Journal of Antennas and Propagation*, vol. 2013, pp. 1–9, 2013. DOI: http://10.1155/2013/824214

37. H. Maune, M. Jost, R. Reese, E. Polat, M. Nickel, and R. Jakoby, "Microwave liquid crystal technology," *Crystals*, vol. 8, no. 9, 2018. DOI: http://10.3390/cryst8090355

38. H. Maune et al., "Tunable microwave component technologies for SatCom-platforms," *Frequenz*, vol. 71, p. 129, 2017.

39. D. C. Zografopoulos, A. Ferraro, and R. Beccherelli, "Liquid-crystal high-frequency microwave technology: Materials and characterization," *Advanced Materials Technologies*, pp. 1800447–1800447, 2018. DOI: http://10.1002/admt.201800447

40. D.-K. Yang and S.-T. Wu, *Fundamentals of Liquid Crystal Devices*, 2nd ed. Hoboken, NJ: John Wiley & Sons, 2014.

41. R. Dąbrowski, P. Kula, and J. Herman, "High Birefringence Liquid Crystals," *Crystals*, vol. 3, no. 3, pp. 443–482, 2013. DOI: http://10.3390/cryst3030443

42. J. Czub, S. Urban, R. Dąbrowski, and B. Gestblom, "Dielectric properties of liquid crystalline isothiocyanato-tolane derivatives with fluorine atom at various lateral positions," *Acta Physica Polonica A*, vol. 107, no. 6, pp. 947–958, 2005. DOI: http://10 .12693/aphyspola.107.947

43. F. Dubois et al., "Large microwave birefringence liquid-crystal characterization for phase-shifter applications," *Japanese Journal of Applied Physics*, vol. 47, no. 5, pp. 3564–3567, 2008. DOI: http://10.1143/jjap.47.3564

44. F. Goelden et al., "Tunable microwave phase shifter using thin layer ferroelectric liquid crystals," 2007/January. http://tubiblio.ulb.tu-darmstadt.de/29060/

45. A. Lapanik, "Liquid crystal systems for microwave applications: Single compounds and mixtures for microwave applications; Dielectric, microwave studies on selected systems," PhD thesis, *Technische Universität Darmstadt*, 2009.

46. A. Lapanik et al., "Nematic LCs mixtures with high birefringence in the microwave region," *Frequenz*, vol. 65, no. 1–2, pp. 15–19, 2011.

47. A. Lapanik, W. Haase, F. Golden, S. Muller, and R. Jakoby, "Highly birefringent nematic mixtures at room temperature for microwave applications," *Optical Engineering*, vol. 50, no. 8, pp. 081208–081208, 2011.

48. A. Penirschke et al., "Cavity perturbation method for characterization of liquid crystals up to 35 GHz," in *European Microwave Conference (EuMC)*, 2004/October, vol. 2, pp. 545–548.

49. S. Mueller et al., "Broad-band microwave characterization of liquid crystals using a temperature-controlled coaxial transmission line," *IEEE Transactions on Microwave Theory and Techniques*, vol. 53, no. 6, pp. 1937–1945, 2005. DOI: http://10.1109/TMTT.2005.848842

50. H. L. Ong, "Measurement of nematic liquid crystal splay and bend elastic constants with obliquely incident light," *Journal of Applied Physics*, vol. 70, no. 4, pp. 2023–2030, 1991. DOI: http://10.1063/1.349461

51. P. Yaghmaee, T. Kaufmann, B. Bates, and C. Fumeaux, "Effect of polyimide layers on the permittivity tuning range of liquid crystals," in *European Conference on Antennas and Propagation (EUCAP)*, 2012, pp. 3579–3582.

52. K. Tarumi, U. Finkenzeller, and B. Schuler, "Dynamic behaviour of twisted nematic liquid crystals," *Japanese Journal of Applied Physics*, vol. 31, no. Part 1, No. 9A, pp. 2829–2836, 1992. DOI: http://10.1143/jjap.31.2829

53. I. W. Stewart, *The Static and Dynamic Continuum Theory of Liquid Crystals: A Mathematical Introduction*. London: Taylor & Francis, 2004.

54. M. Jost, *Liquid Crystal Mixed Beam-Switching and Beam-Steering Network in Hybrid Metallic and Dielectric Waveguide Technology*. PhD thesis, Teschnische Universität of Darmstadt, 2018.

55. F. Goelden, A. Gaebler, S. Mueller, A. Lapanik, W. Haase, and R. Jakoby, "Liquid-crystal varactors with fast switching times for microwave applications," *Electronics Letters*, vol. 44, no. 7, pp. 480–480, 2008. DOI: http://10.1049/el:20080161

56. F. Goelden, A. Lapanik, A. Gaebler, S. Mueller, W. Haase, and R. Jakoby, "Characterization and application of liquid crystals at microwave frequencies," *Frequenz*, vol. 62, no. 3–4, 2008. DOI: http://10.1515/freq.2008.62.3–4.57

57. H. Fujikake, T. Kuki, T. Nomoto, Y. Tsuchiya, and Y. Utsumi, "Thick polymer-stabilized liquid crystal films for microwave phase control," *Journal of Applied Physics*, vol. 89, no. 10, pp. 5295–5298, 2001. DOI: http://10.1063/1.1365081

58. Y. Utsumi, T. Kamei, K. Saito, and H. Moritake, "Increasing the speed of microstrip line-type PDLC devices," in *IEEE MTT-S International Microwave Symposium*, 2005/June.

59. T. Nguyen, S. Umeno, H. Higuchi, H. Kikuchi, and H. Moritake, "Improvement of decay time in nematic-liquid-crystal-loaded coplanar-waveguide-type microwave phase shifter by polymer stabilizing method," *Japanese Journal of Applied Physics*, vol. 53, no. 1S, pp. 01AE08–01AE08, 2013.

60. T. Kuki, H. Fujikake, H. Kamoda, and T. Nomoto, "Microwave variable delay line using a membrane impregnated with liquid crystal," in *IEEE MTT-S International Microwave Symposium*, 2002: IEEE.

61. H. Kamoda, T. Kuki, H. Fujikake, and T. Nomoto, "Millimeter-wave beam former using liquid crystal," *Electronics and Communications in Japan (Part II: Electronics),* vol. 88, no. 8, pp. 10–18, 2005. DOI: http://10.1002/ecjb.20173

62. S. N. Paul, R. Dhar, R. Verma, S. Sharma, and R. Dabrowski, "Change in dielectric and electro-optical properties of a nematic material (6CHBT) due to the dispersion of $BaTiO_3$ nanoparticles," *Molecular Crystals and Liquid Crystals,* vol. 545, no. 1, pp. 1051329–1111335, 2011. DOI: http://10.1080/15421406.2011.571961

63. A. V. Ryzhkova, F. V. Podgornov, A. Gaebler, R. Jakoby, and W. Haase, "Measurements of the electrokinetic forces on dielectric microparticles in nematic liquid crystals using optical trapping," *Journal of Applied Physics,* vol. 113, no. 24, pp. 244902–244902, 2013. DOI: http://10.1063/1.4809976

64. O. H. Karabey, "Microwave material properties of nanoparticle-doped nematic liquid crystals," *Frequenz,* vol. 69, no. 3–4, 2015. DOI: http://10.1515/freq-2014–0169

65. Y. Garbovskiy and A. Glushchenko, "Ferroelectric nanoparticles in liquid crystals: Recent progress and current challenges," *Nanomaterials,* vol. 7, no. 11, pp. 361–361, 2017. DOI: http://10.3390/nano7110361

66. F. Goelden, A. Lapanik, S. Mueller, A. Gaebler, W. Haase, and R. Jakoby, "Investigations on the behavior of ferroelectric liquid crystals at microwave frequencies," in *European Microwave Conference (EuMC),* 2007: IEEE.

67. H. Moritake, J. Kim, K. Toda, and K. Yoshino, "Dynamic viscosity change measurement of liquid and liquid crystal using propagation velocity change of shear horizontal wave," in *IEEE International Conference on Dielectric Liquids, 2005. ICDL 2005,* June 26–July 1, 2005, pp. 257–260. DOI: http://10.1109/ICDL.2005.1490075

68. P. Arora, A. Mikulko, F. Podgornov, and W. Haase, "Dielectric and electro-optic properties of new ferroelectric liquid crystalline mixture doped with carbon nanotubes," *Molecular Crystals and Liquid Crystals,* vol. 502, no. 1, pp. 1–8, 2009. DOI: http://10 .1080/15421400902813592

69. A. Mikułko, P. Arora, A. Glushchenko, A. Lapanik, and W. Haase, "Complementary studies of $BaTiO_3$ nanoparticles suspended in a ferroelectric liquid-crystalline mixture," *EPL (Europhysics Letters),* vol. 87, no. 2, pp. 27009–27009, 2009. DOI: http://10.1209/ 0295–5075/87/27009

70. T. Kuki, H. Fujikake, and T. Nomoto, "Microwave variable delay line using dual-frequency switching-mode liquid crystal," *IEEE Transactions on Microwave Theory and Techniques,* vol. 50, no. 11, pp. 2604–2609, 2002. DOI: http://10.1109/tmtt.2002 .804510

71. S. Mueller, F. Goelden, P. Scheele, M. Wittek, C. Hock, and R. Jakoby, "Passive phase shifter for W-band applications using liquid crystals," in *European Microwave Conference (EuMC),* 2006, pp. 306–309.

72. K. C. Lim, J. D. Margerum, and A. M. Lackner, "Liquid crystal millimeter wave electronic phase shifter," *Applied Physics Letters,* vol. 62, no. 10, pp. 1065–1067, 1993.

73. D. Dolfi, M. Labeyrie, P. Joffre, and J. P. Huignard, "Liquid crystal microwave phase shifter," *Electronics Letters,* vol. 29, no. 10, pp. 926–928, 1993. DOI: http://10.1049/ el:19930618

74. N. Martin, P. Laurent, G. Prigent, P. Gelin, and F. Huret, "Technological evolution and performances improvements of a tunable phase-shifter using liquid crystal," *Microwave and Optical Technology Letters,* vol. 43, no. 4, pp. 338–341, 2004. DOI: http://10.1002/ mop.20463

75. T. Kuki, H. Fujikake, T. Nomoto, and Y. Utsumi, Design of a microwave variable delay line using liquid crystal, and a study of its insertion loss. *Electronics and Communications in Japan*, Pt. II, 85, pp. 36–42, 2002. DOI: http://10.1002/ecjb.1091

76. S. Mueller, P. Scheele, C. Weil, M. Wittek, C. Hock, and R. Jakoby, "Tunable passive phase shifter for microwave applications using highly anisotropic liquid crystals," in IEEE MTT-S International Microwave Symposium, 2004, vol. 2, pp. 1153–1156.

77. R. Jakoby, P. Scheele, S. Muller, and C. Weil, "Nonlinear dielectrics for tunable microwave components," in 15th International Conference on Microwaves, Radar and Wireless Communications (IEEE Cat. No.04EX824), May 17–19, vol. 2, pp. 369–378, 2004. DOI: http://10.1109/MIKON.2004.1357043

78. A. Gaebler, F. Goelden, A. Manabe, M. Goebel, S. Mueller, and R. Jakoby, "Investigation of high performance transmission line phase shifters based on liquid crystal," in *European Microwave Conference (EuMC)*, 2009, pp. 594–597.

79. C. Weickhmann, A. Gaebler, M. Jost, R. Gehring, N. Nathrath, and R. Jakoby, "Recent measurements of compact electronically tunable liquid crystal phase shifter in rectangular waveguide topology," *Electronics Letters*, vol. 49, no. 21, pp. 1345–1347, 2013. DOI: http://10.1049/el.2013.2281

80. C. Weickhmann, "Liquid crystals towards terahertz: Characterisation and tunable waveguide phase shifters for millimetre-wave and terahertz beamsteering antennas," PhD thesis, Technische Universität Darmstadt, 2017.

81. O. H. Karabey, F. Goelden, A. Gaebler, S. Strunck, and R. Jakoby, "Tunable loaded line phase shifters for microwave applications," in *IEEE MTT-S International Microwave Symposium*, 2011/June: IEEE.

82. F. Goelden, A. Gaebler, M. Goebel, A. Manabe, S. Mueller, and R. Jakoby, "Tunable liquid crystal phase shifter for microwave frequencies," *Electronics Letters*, vol. 45, no. 13, pp. 686–687, 2009. DOI: http://10.1049/el.2009.1168

83. A. Franc, O. H. Karabey, G. Rehder, E. Pistono, R. Jakoby, and P. Ferrari, "Compact and broadband millimeter-wave electrically tunable phase shifter combining slow-wave effect with liquid crystal technology," *IEEE Transactions on Microwave Theory and Techniques*, vol. 61, no. 11, pp. 3905–3915, 2013. DOI: http://10.1109/TMTT.2013.2282288

84. M. Jost et al., "Miniaturized liquid crystal slow wave phase shifter based on nanowire filled membranes," *IEEE Microwave and Wireless Components Letters*, vol. 28, no. 8, pp. 681–683, 2018. DOI: http://10.1109/lmwc.2018.2845938

85. M. Tanaka, T. Nose, and S. Sato, "Millimeter-wave transmission properties of nematic liquid-crystal cells with a grating-patterned electrode structure," *Japanese Journal of Applied Physics*, vol. 39, no. Part 1, no. 11, pp. 6393–6396, 2000. DOI: http://10.1143/jjap.39.6393

86. M. Tanaka and S. Sato, "Millimeter-wave deflection properties of liquid crystal prism cells with stack-layered structure," *Japanese Journal of Applied Physics*, vol. 40, no. Part 2, no. 10B, pp. L1123–L1125, 2001. DOI: http://10.1143/jjap.40.l1123

87. M. Tanaka and S. Sato, "Electrically controlled millimeter-wave focusing properties of liquid crystal lens," *Japanese Journal of Applied Physics*, vol. 41, no. Part 1, No. 8, pp. 5332–5333, 2002. DOI: http://10.1143/jjap.41.5332

88. F. Yang and J. R. Sambles, "Determination of the microwave permittivities of nematic liquid crystals using a single-metallic slit technique," *Applied Physics Letters*, vol. 81, no. 11, pp. 2047–2049, 2002. DOI: http://10.1063/1.1507615

89. F. Yang and J. R. Sambles, "Microwave liquid-crystal variable phase grating," *Applied Physics Letters,* vol. 85, no. 11, pp. 2041–2043, 2004. DOI: http://10.1063/1.1787898

90. N. Kundtz, "Next generation communications for next generation satellites," *Microwave Journal,* vol. 57, 2014.

91. Kymeta, "Kymeta delivers sustained industry-first performance levels," press release, 2016. www.kymetacorp.com/news/kymeta-delivers-sustained-industry-first-performance-levels/

92. Alcan Systems, "SES and ALCAN, a German smart antenna company, are working together to develop a new flat panel antenna for SESs O3b mPOWER system," press release, 2018. www.alcansystems.com/de/press-release-ses-and-alcan-a-german-smart-antenna-company-are-working-together-to-develop-a-new-flat-panel-antenna-for-sess-o3b-mpower-system/

93. A. Systems, "ALCAN successfully completes world's first liquid crystal based phased array antenna field test for satellite communication," 2018.

94. V. Lapanik, G. Sasnouski, S. Timofeev, E. Shepeleva, G. Evtyushkin, and W. Haase, "New highly anisotropic liquid crystal materials for high-frequency applications," *Liquid Crystals,* vol. 45, no. 8, pp. 1242–1249, 2018/06/21 2018. DOI: http://10.1080/02678292 .2018.1427810

95. P. Gilles Gennes de and J. Prost, *The Physics of Liquid Crystal.* Oxford: Clarendon Press, 1993.

96. J. W. Goodby, P. J. Collings, T. Kato, C. Tschierske, H. Gleeson, and P. Raynes, *Handbook of Liquid Crystals.* Hoboken, NJ: Wiley-VCH, 1998.

97. H. Kawamoto, "The history of liquid-crystal displays," *Proceedings of the IEEE,* vol. 90, no. 4, pp. 460–500, 2002. DOI: http://10.1109/JPROC.2002.1002521

98. P. J. Collings and M. Hird, *Introduction to Liquid Crystals: Chemistry and Physics.* Boca Raton, FL: CRC Press, 2017.

99. L. M. Blinov and V. G. Chigrinov, *Electrooptic Effects in Liquid Crystal Materials.* New York: Springer-Verlag, 1994.

100. T. S. D. Cheung and J. R. Long, "Shielded passive devices for silicon-based monolithic microwave and millimeter-wave integrated circuits," *IEEE Journal of Solid-State Circuits,* vol. 41, no. 5, pp. 1183–1200, 2006. DOI: http://10.1109/JSSC.2006.872737

101. C. W. Oseen, "The theory of liquid crystals," *Transactions of the Faraday Society,* vol. 29, no. 140, pp. 883–899, 1933. DOI: http://10.1039/TF9332900883

102. F. C. Frank, "I. Liquid crystals. On the theory of liquid crystals," *Discussions of the Faraday Society,* vol. 25, no. 0, pp. 19–28, 1958. DOI: http://10.1039/DF9582500019

103. M. Kleman and O. D. Laverntovich, *Soft Matter Physics: An Introduction.* New York: Springer-Verlag, 2004.

104. J. R. Baker-Jarvis, M. D. Janezic, J. H. Grosvenor Jr, and R. G. Geyer, "Transmission/ reflection and short-circuit line methods for measuring permittivity and permeability| NIST," *Technical Note (NIST TN)-1355,* vol. 1355, no. Technical Note (NIST TN)-1355, 1992.

105. A. M. Nicolson and G. F. Ross, "Measurement of the intrinsic properties of materials by time-domain techniques," *IEEE Transactions on Instrumentation and Measurement,* vol. 19, no. 4, pp. 377–382, 1970. DOI: http://10.1109/TIM.1970.4313932

106. W. B. Weir, "Automatic measurement of complex dielectric constant and permeability at microwave frequencies," *Proceedings of the IEEE,* vol. 62, no. 1, pp. 33–36, 1974. DOI: http://10.1109/PROC.1974.9382

107. J. Parka, J. Krupka, R. Dąbrowski, and J. Wosik, "Measurements of anisotropic complex permittivity of liquid crystals at microwave frequencies," *Journal of the European Ceramic Society,* vol. 27, no. 8–9, pp. 2903–2905, 2007. DOI: http://10.1016/j .jeurceramsoc.2006.11.015

108. R. Kowerdziej et al., "Dielectric properties of highly anisotropic nematic liquid crystals for tunable microwave components," *Applied Physics Letters,* vol. 103, no. 17, p. 172902, 2013/10/21 2013. DOI: http://10.1063/1.4826504.

109. O. H. Karabey, F. Goelden, A. Gaebler, and R. Jakoby, "Precise broadband microwave material characterization of liquids," in *European Microwave Conference (EuMC),* 2010, pp. 1591–1594.

110. W. Hu, O. H. Karabey, A. E. Prasetiadi, M. Jost, and R. Jakoby, "Temperature controlled artificial coaxial line for microwave characterization of liquid crystal," *GeMiC 2014; German Microwave Conference,* pp. 1–4, 2014.

111. S. Mueller et al., "W-band characterization of anisotropic liquid crystals at room temperature," in *European Microwave Conference (EuMC),* 2008, pp. 119–122.

112. C. Weickhmann et al., "Measuring liquid crystal permittivity with high accuracy," in *Annual Condensed Matter and Materials Meeting,* 2015.

113. A. Gaebler, F. Goelden, O. H. Karabey, and R. Jakoby, "A FDFD based eigen-dielectric formulation of the Maxwell equations for material characterization in arbitrary waveguide structures," in *IEEE MTT-S International Microwave Symposium,* 2010/May: IEEE.

114. A. Gaebler, F. Goelden, S. Mueller, A. Penirschke, and R. Jakoby, "Direct simulation of material permittivities by using an eigen-susceptibility formulation of the vector variational approach," in *2009 IEEE Instrumentation and Measurement Technology Conference,* 2009/May: IEEE.

115. A. Gaebler, F. Goelden, S. Mueller, and R. Jakoby, "Triple-mode cavity perturbation method for the characterization of anisotropic media," in *European Microwave Conference (EuMC),* 2008, pp. 909–912.

116. L. Cai, H. Xu, J. Li, and D. Chu, "High figure-of-merit compact phase shifters based on liquid crystal material for 1–10 GHz applications," *Japanese Journal of Applied Physics,* vol. 56, no. 1, pp. 011701–011701, 2017. DOI: http://10.7567/jjap.56.011701

117. P. Deo, D. Mirshekar-Syahkal, L. Seddon, S. E. Day, and F. A. Fernandez, "Microstrip device for broadband (15–65 GHz) measurement of dielectric properties of nematic liquid crystals," *IEEE Transactions on Microwave Theory and Techniques,* vol. 63, no. 4, pp. 1388–1398, 2015. DOI: http://10.1109/tmtt.2015.2407328

118. R. James, F. A. Fernandez, S. E. Day, S. Bulja, and D. Mirshekar-Syahkal, "Accurate modeling for wideband characterization of nematic liquid crystals for microwave applications," *IEEE Transactions on Microwave Theory and Techniques,* vol. 57, no. 12, pp. 3293–3297, 2009. DOI: http://10.1109/tmtt.2009.2033864

119. J. Krupka, K. Derzakowski, B. Riddle, and J. Baker-Jarvis, "A dielectric resonator for measurements of complex permittivity of low loss dielectric materials as a function of temperature," *Measurement Science and Technology,* vol. 9, no. 10, pp. 1751–1756, 1998. DOI: http://10.1088/0957–0233/9/10/015

120. D. E. Schaub and D. R. Oliver, "A circular patch resonator for the measurement of microwave permittivity of nematic liquid crystal," *IEEE Transactions on Microwave Theory and Techniques,* vol. 59, no. 7, pp. 1855–1862, 2011. DOI: http://10.1109/ TMTT.2011.2142190

121. M. Yazdanpanahi, S. Bulja, D. Mirshekar-Syahkal, R. James, S. E. Day, and F. A. Fernandez, "Measurement of dielectric constants of nematic liquid crystals at mm-wave frequencies using patch resonator," *IEEE Transactions on Instrumentation and Measurement*, vol. 59, no. 12, pp. 3079–3085, 2010. DOI: http://10.1109/TIM.2010.2062910

122. M. Koeberle et al., "Material characterization of liquid crystals at THz-frequencies using a free space measurement setup," in *German Microwave Conference*, 2008/March, pp. 1–4.

123. E. M. Pogson, R. A. Lewis, M. Koeberle, and R. Jacoby, "Terahertz time-domain spectroscopy of nematic liquid crystals," in *Nonlinear Optics and Applications IV*, B. J. Eggleton, A. L. Gaeta, and N. G. R. Broderick, Eds., 2010/April: SPIE.

124. M. Reuter et al., "Highly birefringent, low-loss liquid crystals for terahertz applications," *APL Materials,* vol. 1, no. 1, pp. 012107–012107, 2013. DOI: http://10.1063/1.4808244

125. A. Berk, "Variational principles for electromagnetic resonators and waveguides," *IRE Transactions on Antennas and Propagation*, vol. 4, no. 2, pp. 104–111, 1956. DOI: http://10.1109/TAP.1956.1144365

126. V. H. Rumsey, "Reaction concept in electromagnetic theory," *Physical Review*, 1954. DOI: http://10.1103/PhysRev.94.1483

127. R. E. Collin, *Foundations for Microwave Engineering, 2nd ed.* Hoboken, NJ: Wiley-IEEE Press, 2001.

128. D. M. Pozar, *Microwave Engineering*. Hoboken, NJ: John Wiley & Sons, 2012.

129. H. A. Bethe, "Theory of diffraction by small holes," *Biophysical Reviews,* vol. 66, no. 7–8, pp. 163–182, 1944. DOI: http://10.1103/PhysRev.66.163

130. J. Gao, "Analytical formulas for the resonant frequency changes due to opening apertures on cavity walls," *Nuclear Instruments and Methods in Physics Research Section A: Accelerators, Spectrometers, Detectors and Associated Equipment,* vol. 311, no. 3, pp. 437–443, 1992. DOI: http://10.1016/0168–9002(92)90638–K. www.sciencedirect .com/science/article/pii/016890029290638K.

131. L. F. Chen, C. K. Ong, C. P. Neo, V. V. Varadan, and V. K. Varadan, *Microwave Electronics: Measurement and Materials Characterization*. Hoboken, NJ: John Wiley & Sons, 2005.

132. W. R. Eisenstadt and Y. Eo, "S-parameter-based IC interconnect transmission line characterization," *IEEE Transactions on Components, Hybrids, and Manufacturing Technology,* vol. 15, no. 4, pp. 483–490, 1992. DOI: http://10.1109/33.159877

133. R. Collier and D. Skinner, *Microwave Measurements*, 3rd ed. London: The Institution of Engineering and Technology, 2007.

134. S. Bulja, D. Mirshekar-Syahkal, R. James, S. E. Day, and F. A. b. Fernandez, "Measurement of dielectric properties of nematic liquid crystals at millimeter wavelength," *IEEE Transactions on Microwave Theory and Techniques*, 2010. DOI: http://10 .1109/tmtt.2010.2054332

135. S. Mueller et al., "Liquid crystals: Microwave characterization and tunable devices," *Frequenz,* vol. 61, no. 9–10, 2007. DOI: http://10.1515/freq.2007.61.9–10.217

136. A. Mössinger et al., "Electronically reconfigurable LC-reflectarray with 2D scanning capability and SU-8 structured cavity," *Frequenz,* vol. 62, no. 3–4, 2008. DOI: http://10 .1515/freq.2008.62.3–4.62

137. S. Christie, R. Cahill, N. Mitchell, A. Manabe, and Y. Munro, "Electronically scanned Rotman lens antenna with liquid crystal phase shifters," *Electronics Letters*vol. 49, no. 7, pp. 445–447, 2013. DOI: http://10.1049/el.2013.0020

138. H. W. Schüßler and P. Steffen, "Halfband filters and Hilbert transformers," *Circuits, Systems, and Signal Processing,* vol. 17, no. 2, pp. 137–164, 1998. DOI: http://10.1007/bf01202851

139. Y. Garbovskiy et al., "Liquid crystal phase shifters at millimeter wave frequencies," *Journal of Applied Physics,* vol. 111, no. 5, pp. 054504–054504, 2012. DOI: http://10.1063/1.3691202

140. M. Jost et al., "Liquid crystal based low-loss phase shifter for W-band frequencies," *Electronics Letters,* vol. 49, no. 23, pp. 1460–1462, 2013. DOI: http://10.1049/el.2013.2830

141. S. Strunck, O. H. Karabey, C. Weickhmann, A. Gaebler, and R. Jakoby, "Continuously tunable phase shifters for phased arrays based on liquid crystal technology," in *Proceedings of the IEEE International Symposium on Phased Array Systems and Technology,* 2013/October, pp. 82–88.

142. M. Tebbe, A. Hoehn, N. Nathrath, and C. Weickhmann, "Manufacturing and testing of liquid crystal phase shifters for an electronically steerable array," in *Proceedings of the IEEE Aerospace Conference,* 2017/March, pp. 1–12.

143. C. Weickhmann, N. Nathrath, R. Gehring, A. Gaebler, M. Jost, and R. Jakoby, "A lightweight tunable liquid crystal phase shifter for an efficient phased array antenna," in *European Microwave Conference (EuMC),* 2013, pp. 428–431.

144. F. Sahbani, N. Tentillier, A. Gharsallah, A. Gharbi, and C. Legrand, "New tunable coplanar microwave phase shifter with nematic crystal liquid," in *2008 3rd International Design and Test Workshop,* December 20–22 2008, pp. 78–81. DOI: http://10.1109/IDT.2008.4802470

145. V. B. Bui, Y. Inoue, and H. Moritake, "NRD waveguide-type terahertz phase shifter using nematic liquid crystal," *Japanese Journal of Applied Physics,* vol. 58, no. 2, pp. 022001–022001, 2019. DOI: http://10.7567/1347-4065/aaf282

146. S. Bulja and D. Mirshekar-Syahkal, "Meander line millimetre-wave liquid crystal based phase shifter," *Electronics Letters,* vol. 46, no. 11, pp. 769–769, 2010. DOI: http://10.1049/el.2010.3513

147. M. Jost, R. Reese, M. Nickel, H. Maune, and R. Jakoby, "Fully dielectric interference-based SPDT with liquid crystal phase shifters," *IET Microwaves, Antennas & Propagation,* vol. 12, no. 6, pp. 850–857, 2018. DOI: http://10.1049/iet-map.2017.0695

148. M. Jost et al., "Tunable dielectric delay line phase shifter based on liquid crystal technology for a SPDT in a radiometer calibration scheme at 100 GHz," in *IEEE MTT-S International Microwave Symposium,* 2016, pp. 1–4.

149. R. Reese, M. Jost, M. Nickel, E. Polat, R. Jakoby, and H. Maune, "A fully dielectric lightweight antenna array using a multimode interference power divider at W-band," *IEEE Antennas and Wireless Propagation Letters,* vol. 16, pp. 3236–3239, 2017. DOI: http://10.1109/LAWP.2017.2771385

150. R. Reese, E. Polat, M. Jost, M. Nickel, R. Jakoby, and H. Maune, "Liquid crystal based phase shifter in a parallel-plate dielectric waveguide topology at V-band," in *European Microwave Integrated Circuit Conference (EuMIC),* 2017, pp. 353–356.

151. A. Gaebler, F. Goelden, S. Mueller, and R. Jakoby, "Modeling of electrically tunable transmission line phase shifter based on liquid crystal," in *2008 IEEE Antennas and Propagation Society International Symposium,* 2008/July: IEEE.

152. A. Gaebler, F. Goelden, S. Mueller, and R. Jakoby, "Efficiency considerations of tuneable liquid crystal microwave devices," in *German Microwave Conference,* 2008, pp. 1–4.

153. O. H. Karabey, B. G. Saavedra, C. Fritzsch, S. Strunck, A. Gaebler, and R. Jakoby, "Methods for improving the tuning efficiency of liquid crystal based tunable phase shifters," in *European Microwave Integrated Circuit Conference (EuMIC)*, 2011, pp. 494–497.

154. A. Gaebler, F. Goelden, S. Mueller, and R. Jakoby, "Multiphysics simulations for tunability efficiency evaluation of liquid crystal based RF," *Frequenz*, vol. 62, 2008, pp. 240–240. DOI: http://10.1515/FREQ.2008.62.9-10.240

155. C. P. Wen, "Coplanar waveguide, a surface strip transmission line suitable for nonreciprocal gyromagnetic device applications," in *1969 G-MTT International Microwave Symposium*, pp. 110–115, 1969. DOI: http://10.1109/GMTT.1969.1122668

156. C. Yeh and F. Shimabukuro, *The Essence of Dielectric Waveguides*. New York: Springer Science+Business Media, 2008.

157. M. Weidenbach et al., "3D printed dielectric rectangular waveguides, splitters and couplers for 120 GHz," *Optics Express*, vol. 24, no. 25, pp. 28968–28976, 2016. DOI: http://10.1364/OE.24.028968

158. G. L. Friedsam and E. M. Biebl, "Precision free-space measurements of complex permittivity of polymers in the W-band," in *IEEE MTT-S International Microwave Symposium*, 1997, vol. 3, pp. 1351–1354.

159. J. Krupka, "Measurements of the complex permittivity of low loss polymers at frequency range from 5 GHz to 50 GHz," *IEEE Microwave and Wireless Components Letters*, vol. 26, no. 6, pp. 464–466, 2016. DOI: http://10.1109/LMWC.2016.2562640

160. E. A. J. Marcatili, "Dielectric rectangular waveguide and directional coupler for integrated optics," *The Bell System Technical Journal*, vol. 48, no. 7, pp. 2071–2102, 1969.

161. R. Reese, M. Jost, H. Maune, and R. Jakoby, "Design of a continuously tunable W-band phase shifter in dielectric waveguide topology," in *IEEE MTT-S International Microwave Symposium*, 2017, pp. 180–183.

162. G. K. C. Kwan and N. K. Das, "Excitation of a parallel-plate dielectric waveguide using a coaxial probe-basic characteristics and experiments," *IEEE Transactions on Microwave Theory and Techniques*, vol. 50, no. 6, pp. 1609–1620, 2002. DOI: http://10.1109/TMTT.2002.1006423

163. C. Fritzsch et al., "Continuously tunable W-band phase shifter based on liquid crystals and MEMS technology," in *European Microwave Integrated Circuit Conference (EuMIC)*, 2011, pp. 522–525.

164. C. Fritzsch, *Flüssigkristallbasierte elektronisch steuerbare Gruppenantennen Technologie, Konzepte und Komponenten*. PhD thesis, Technische Universität Darmstadt, 2015.

165. A. Franc, F. Podevin, L. Cagnon, P. Ferrari, A. Serrano, and G. Rehder, "Metallic nanowire filled membrane for slow wave microstrip transmission lines," in *2012 International Semiconductor Conference Dresden-Grenoble (ISCDG)*, pp. 191–194, 2012. DOI: http://10.1109/ISCDG.2012.6360022

166. A. L. C. Serrano et al., "Modeling and characterization of slow-wave microstrip lines on metallic-nanowire-filled-membrane substrate," *IEEE Transactions on Microwave Theory and Techniques*, vol. 62, no. 12, pp. 3249–3254, 2014. DOI: http://10.1109/TMTT.2014.2366108

167. H. Masuda and K. Fukuda, "Ordered metal nanohole arrays made by a two-step replication of honeycomb structures of anodic alumina," *Science*, vol. 268, no. 5216, pp. 1466–1466, 1995. DOI: http://10.1126/science.268.5216.1466

168. H. Masuda, F. Hasegwa, and S. Ono, "Self-ordering of cell arrangement of anodic porous alumina formed in sulfuric acid solution," *Journal of The Electrochemical Society,* vol. 144, no. 5, pp. L127–L130, 1997. DOI: http://10.1149/1.1837634

169. H. Masuda, K. Yada, and A. Osaka, "Self-ordering of cell configuration of anodic porous alumina with large-size pores in phosphoric acid solution," *Japanese Journal of Applied Physics,* vol. 37, no. Part 2, No. 11A, pp. L1340–L1342, 1998. DOI: http://10.1143/jjap .37.l1340

170. A. P. Li, F. Müller, A. Birner, K. Nielsch, and U. Gösele, "Hexagonal pore arrays with a 50–420 nm interpore distance formed by self-organization in anodic alumina," *Journal of Applied Physics,* vol. 84, no. 11, pp. 6023–6026, 1998. DOI: http://10.1063/1.368911

171. F. Ellinger, H. Jackel, and W. Bachtold, "Varactor-loaded transmission-line phase shifter at C-band using lumped elements," *IEEE Transactions on Microwave Theory and Techniques,* vol. 51, no. 4, pp. 1135–1140, 2003. DOI: http://10.1109/TMTT.2003 .809670

172. J. J. P. Venter, T. Stander, and P. Ferrari, "X-Band reflection-type phase shifters using coupled-line couplers on single-layer RF PCB," *IEEE Microwave and Wireless Components Letters,* vol. 28, no. 9, pp. 807–809, 2018. DOI: http://10.1109/LMWC .2018.2853562

173. K. Hong-Teuk et al., "V-band 2-b and 4-b low-loss and low-voltage distributed MEMS digital phase shifter using metal-air-metal capacitors," *IEEE Transactions on Microwave Theory and Techniques,* vol. 50, no. 12, pp. 2918–2923, 2002. DOI: http://10.1109/TMTT .2002.805285

174. B. Pillans, L. Coryell, A. Malczewski, C. Moody, F. Morris, and A. Brown, "Advances in RF MEMS phase shifters from 15 GHz to 35 GHz," in *IEEE MTT-S International Microwave Symposium,* 2012, pp. 1–3.

175. M. Sazegar et al., "Low-cost phased-array antenna using compact tunable phase shifters based on ferroelectric ceramics," *IEEE Transactions on Microwave Theory and Techniques,* vol. 59, no. 5, pp. 1265–1273, 2011. DOI: http://10.1109/TMTT.2010 .2103092

176. G. Velu et al., "A 360°BST phase shifter with moderate bias voltage at 30 GHz," *IEEE Transactions on Microwave Theory and Techniques,* vol. 55, no. 2, pp. 438–444, 2007. DOI: http://10.1109/TMTT.2006.889319

177. J. F. Bernigaud et al., "Liquid crystal tunable filter based on DBR topology," in *European Microwave Conference (EuMC),* 2006/September 2006: IEEE.

178. T. Gobel, P. Meissner, A. Gaebler, M. Koeberle, S. Mueller, and R. Jakoby, "Dual-frequency switching liquid crystal based tunable THz filter," in *2009 Conference on Lasers and Electro-Optics and 2009 Conference on Quantum electronics and Laser Science Conference,* June 2–4, 2009 pp. 1–2. DOI: http://10.1364/CLEO.2009.CThFF4.

179. F. Goelden, A. Gaebler, O. Karabey, M. Goebel, A. Manabe, and R. Jakoby, "Tunable band-pass filter based on liquid crystal," in *German Microwave Conference Digest of Papers,* 2010/March, pp. 98–101.

180. M. Yazdanpanahi and D. Mirshekar-Syahkal, "Millimeter-wave liquid-crystal-based tunable bandpass filter," in *2012 IEEE Radio and Wireless Symposium,* 2012/January: IEEE.

181. P. Yaghmaee, C. Fumeaux, B. Bates, A. Manabe, O. H. Karabey, and R. Jakoby, "Frequency tunable S-band resonator using nematic liquid crystal," *Electronics Letters,* vol. 48, no. 13, pp. 798–800, 2012. DOI: http://10.1049/el.2012.1366

182. P. Yaghmaee, W. Withayachumnankul, A. K. Horestani, A. Ebrahimi, B. Bates, and C. Fumeaux, "Tunable electric-LC resonators using liquid crystal," in *Proceedings of the IEEE Antennas and Propagation Society International Symposium (APSURSI)*, 2013/July, pp. 382–383.

183. S. N. Novin, S. Jarchi, and P. Yaghmaee, "Tunable frequency selective surface based on IDC-loaded electric-LC resonator incorporated with liquid crystal," in *Proceedings of the Conference on Microwave Techniques (COMITE)*, 2017/April, pp. 1–4.

184. J. Torrecilla, C. Marcos, V. Urruchi, and J. M. Sánchez-Pena, "Tunable dual-mode bandpass filter based on liquid crystal technology," in *European Microwave Conference (EuMC)*, 2013/October, pp. 806–809.

185. V. Urruchi, C. Marcos, J. Torrecilla, J. M. Sánchez-Pena, and K. Garbat, "Note: Tunable notch filter based on liquid crystal technology for microwave applications," *Review of Scientific Instruments,* vol. 84, no. 2, p. 026102, 2013/02/01 2013. DOI: http://10.1063/1.4790555

186. A. E. Prasetiadi, "Tunable substrate integrated waveguide bandpass filter and amplitude tuner based on microwave liquid crystal technology," PhD thesis, Technische Universität Darmstadt, 2017.

187. A. E. Prasetiadi et al., "Continuously tunable substrate integrated waveguide bandpass filter in liquid crystal technology with magnetic biasing," *Electronics Letters,* vol. 51, no. 20, pp. 1584–1585, 2015. DOI: http://10.1049/el.2015.2494

188. T. Franke, A. Gaebler, A. E. Prasetiadi, and R. Jakoby, "Tunable Ka-band waveguide resonators and a small band band-pass filter based on liquid crystals," in *European Microwave Conference (EuMC)*, 2014, pp. 339–342.

189. E. Polat et al., "Tunable liquid crystal filter in nonradiative dielectric waveguide technology at 60 GHz," *IEEE Microwave and Wireless Components Letters,* vol. 29, no. 1, pp. 44–46, 2019. DOI: http://10.1109/LMWC.2018.2884152

190. X. Chen and K. Wu, "Substrate integrated waveguide filter: Basic design rules and fundamental structure features," *IEEE Microwave Magazine,* vol. 15, no. 5, pp. 108–116, 2014. DOI: http://10.1109/MMM.2014.2321263

191. G. L. Matthaei, L. Young, and E. M. T. Jones, *Microwave Filter, Impedance Matching Networks and Coupling Structures*. Norwood, MA: Artech House, 1963.

192. J.-S. Hong, M. J. Lancaster, and J. S. L. M. J. Hong, *Microstrip Filters for RF/Microwave Applications*. Hoboken, NJ: John Wiley & Sons, 2001.

193. T. Yoneyama and S. Nishida, "Nonradiative dielectric waveguide for millimeter-wave integrated circuits," *IEEE Transactions on Microwave Theory and Techniques,* vol. 29, no. 11, pp. 1188–1192, 1981. DOI: http://10.1109/TMTT.1981.1130529

194. S. Yi-Chi, "Design of waveguide E-plane filters with all-metal inserts," *IEEE Transactions on Microwave Theory and Techniques,* vol. 32, no. 7, pp. 695–704, 1984. DOI: http://10.1109/TMTT.1984.1132756

195. D. Psychogiou, D. Peroulis, Y. Li, and C. Hafner, "V-band bandpass filter with continuously variable centre frequency," *Antennas Propagation IET Microwaves,* vol. 7, no. 8, pp. 701–707, 2013. DOI: http://10.1049/iet-map.2012.0722

196. F. Sammoura and L. Lin, "Micromachined W-band polymeric tunable iris filter," *Microsystem Technologies,* vol. 17, no. 3, pp. 411–416, 2011. DOI: http://10.1007/s00542–010–1184–8

197. Z. Yang, D. Psychogiou, and D. Peroulis, "Design and optimization of tunable silicon-integrated evanescent-mode bandpass filters," *IEEE Transactions on Microwave Theory*

and Techniques, vol. 66, no. 4, pp. 1790–1803, 2018. DOI: http://10.1109/TMTT.2018 .2799575

198. H. Jiang, B. Lacroix, K. Choi, Y. Wang, A. T. Hunt, and J. Papapolymerou, "Ka- and U-band tunable bandpass filters using ferroelectric capacitors," *IEEE Transactions on Microwave Theory and Techniques,* vol. 59, no. 12, pp. 3068–3075, 2011. DOI: http:// 10.1109/tmtt.2011.2170088

199. S. Courreges et al., "A Ka-band electronically tunable ferroelectric filter," *IEEE Microwave and Wireless Components Letters,* vol. 19, no. 6, pp. 356–358, 2009. DOI: http://10.1109/lmwc.2009.2020012

200. J. Sigman, C. D. Nordquist, P. G. Clem, G. M. Kraus, and P. S. Finnegan, "Voltage-controlled Ku-band and X-band tunable combline filters using barium-strontium-titanate," *IEEE Microwave and Wireless Components Letters,* vol. 18, no. 9, pp. 593–595, 2008. DOI: http://10.1109/LMWC.2008.2002453

201. E. e. Economou et al., "Electrically tunable open-stub bandpass filters based on nematic liquid crystals," *Physical Review Applied,* vol. 8, no. 6, 2017. DOI: http://10.1103/ physrevapplied.8.064012

202. N. Martin, P. Laurent, C. Person, P. Gelin, and F. Huret, "Patch antenna adjustable in frequency using liquid crystal," in *European Microwave Conference (EuMC),* 2003/ October, pp. 699–702.

203. N. Martin, P. Laurent, C. Person, P. Gelin, and F. Huret, "Size reduction of a liquid crystal-based, frequency-adjustable patch antenna," in *European Microwave Conference (EuMC),* 2004/October, vol. 2, pp. 825–828.

204. L. Liu and R. J. Langley, "Liquid crystal tunable microstrip patch antenna," *Electronics Letters,* vol. 44, no. 20, pp. 1179–1180, 2008.

205. M. A. Christou, N. C. Papanicolaou, and A. C. Polycarpou, "A nematic liquid crystal tunable patch antenna," in *European Conference on Antennas and Propagation (EUCAP),* 2014: IEEE.

206. N. C. Papanicolaou, M. A. Christou, and A. C. Polycarpou, "Frequency-agile microstrip patch antenna on a biased liquid crystal substrate," *Electronics Letters,* vol. 51, no. 3, pp. 202–204, 2015. DOI: http://10.1049/el.2014.3856

207. C. Fritzsch, S. Bildik, and R. Jakoby, "Ka-band frequency tunable patch antenna," in *Proceedings of the 2012 IEEE International Symposium on Antennas and Propagation,* 2012/July: IEEE.

208. Y. Zhao, C. Huang, A. Qing, and X. Luo, "A frequency and pattern reconfigurable antenna array based on liquid crystal technology," *IEEE Photonics Journal,* vol. 9, no. 3, pp. 1–7, 2017. DOI: http://10.1109/JPHOT.2017.2700042

209. S. Strunck, O. H. Karabey, A. Gaebler, and R. Jakoby, "Reconfigurable waveguide polariser based on liquid crystal for continuous tuning of linear polarisation," *Electronics Letters,* vol. 48, no. 8, pp. 441–443, 2012. DOI: http://10.1049/el.2012.0259

210. M. Nickel et al., "Liquid crystal based tunable reflection-Type Power Divider," in *European Microwave Conference (EuMC),* 2018/September, pp. 45–48.

211. E. Doumanis et al., "Electronically reconfigurable liquid crystal based mm-wave polarization converter," *IEEE Transactions on Antennas and Propagation,* vol. 62, no. 4, pp. 2302–2307, 2014. DOI: http://10.1109/tap.2014.2302844

212. O. H. Karabey, S. Bausch, S. Bildik, S. Strunck, A. Gaebler, and R. Jakoby, "Design and application of a liquid crystal varactor based tunable coupled line for polarization agile antennas," in *European Microwave Conference (EuMC),* 2012, pp. 739–742.

213. O. H. Karabey, S. Bildik, S. Bausch, S. Strunck, A. Gaebler, and R. Jakoby, "Continuously polarization agile antenna by using liquid crystal-based tunable variable delay lines," *IEEE Transactions on Antennas and Propagation,* vol. 61, no. 1, pp. 70–76, 2013. DOI: http://10.1109/tap.2012.2213232

214. C.-L. Pan, C.-J. Lin, C.-S. Yang, W.-T. Wu, and R.-P. Pan, "Liquid-crystal-based phase gratings and beam steerers for terahertz waves," in *Liquid Crystals: Recent Advancements in Fundamental and Device Technologies*: InTech, 2018.

215. R. Reese et al., "A millimeter wave beam steering lens antenna with reconfigurable aperture using liquid crystal," *IEEE Transactions on Antennas and Propagation,* pp. 1–1, 2019. DOI: http://10.1109/TAP.2019.2918474.

216. H. Shi, J. Li, S. Zhu, A. Zhang, and Z. Xu, "Radiation pattern reconfigurable waveguide slot array antenna using liquid crystal," *International Journal of Antennas and Propagation,* vol. 2018, pp. 1–9, 2018. DOI: http://10.1155/2018/2164065

217. M. Jost, R. Reese, M. Nickel, S. Schmidt, H. Maune, and R. Jakoby, "Interference based W-band single-pole double-throw with tunable liquid crystal based waveguide phase shifters," in *IEEE MTT-S International Microwave Symposium,* 2017/ June: IEEE.

218. M. Jost, R. Reese, H. Maune, and R. Jakoby, "In-plane hollow waveguide crossover based on dielectric insets for millimeter-wave applications," in *2017 IEEE MTT-S International Microwave Symposium (IMS),* June 4–9, 2017, pp. 188–191. DOI: http://10.1109/ MWSYM.2017.8059015

219. R. Marin, A. Mossinger, J. Freese, A. Manabe, and R. Jakoby, "Realization of 35 GHz steerable reflectarray using highly anisotropic liquid crystal," in *2006 IEEE Antennas and Propagation Society International Symposium,* 2006: IEEE.

220. R. Marin, A. Mossinger, J. Freese, S. Muller, and R. Jakoby, "Basic investigations of 35 GHz reflectarrays and tunable unit-cells for beamsteering applications," in *European Radar Conference (EURAD),* 2005: IEEE.

221. A. Moessinger, R. Marin, S. Mueller, J. Freese, and R. Jakoby, "Electronically reconfigurable reflectarrays with nematic liquid crystals," *Electronics Letters,* vol. 42, no. 16, pp. 899–900, 2006. DOI: http://10.1049/el:20061541

222. W. Hu et al., "Tunable liquid crystal reflectarray patch element," *Electronics Letters,* vol. 42, no. 9, pp. 509–509, 2006. DOI: http://10.1049/el:20060571

223. W. Hu et al., "Liquid-crystal-based reflectarray antenna with electronically switchable monopulse patterns," *Electronics Letters,* vol. 43, no. 14, pp. 744–744, 2007. DOI: http:// 10.1049/el:20071098

224. A. Moessinger, R. Marin, J. Freese, S. Mueller, A. Manabe, and R. Jakoby, "Investigations on 77 GHz tunable reflectarray unit cells with liquid crystal," in *European Conference on Antennas and Propagation (EUCAP),* 2006, pp. 1–4.

225. R. Marin, "Investigations on liquid crystal reconfigurable unit cells for mm-wave reflectarrays," *Fachgebiet Mikrowellentechnik,* 2008.

226. G. Perez-Palomino, J. A. Encinar, M. Barba, and E. Carrasco, "Design and evaluation of multi-resonant unit cells based on liquid crystals for reconfigurable reflectarrays," *Antennas Propagation IET Microwaves,* vol. 6, no. 3, pp. 348–354, 2012. DOI: http:// 10.1049/iet-map.2011.0234

227. G. Perez-Palomino et al., "Wideband unit-cell based on liquid crystals for reconfigurable reflectarray antennas in f-band," in *Proceedings of the IEEE International Symposium on Antennas and Propagation,* 2012/July, pp. 1–2.

228. G. Perez-Palomino et al., "Design and experimental validation of liquid crystal-based reconfigurable reflectarray elements with improved bandwidth in F-band," *IEEE Transactions on Antennas and Propagation*, vol. 61, no. 4, pp. 1704–1713, 2013. DOI: http://10.1109/TAP.2013.2242833

229. S. Gao et al., "Tunable liquid crystal based phase shifter with a slot unit cell for reconfigurable reflectarrays in F-band," *Applied Sciences*, vol. 8, no. 12, pp. 2528–2528, 2018. DOI: http://10.3390/app8122528

230. W. Hu et al., "Design and measurement of reconfigurable millimeter wave reflectarray cells with nematic liquid crystal," *IEEE Transactions on Antennas and Propagation*, vol. 56, no. 10, pp. 3112–3117, 2008. DOI: http://10.1109/tap.2008.929460

231. W. Hu et al., "94 GHz dual-reflector antenna with reflectarray subreflector," *IEEE Transactions on Antennas and Propagation*, vol. 57, no. 10, pp. 3043–3050, 2009. DOI: http://10.1109/tap.2009.2029275

232. S. Bildik, S. Dieter, C. Fritzsch, W. Menzel, and R. Jakoby, "Reconfigurable folded reflectarray antenna based upon liquid crystal technology," *IEEE Transactions on Antennas and Propagation*, vol. 63, no. 1, pp. 122–132, 2015. DOI: http://10.1109/tap.2014.2367491

233. S. Dieter, *Charakterisierung und Optimierung von quasiplanaren Millimeterwellenantennen bezüglich Rekonfigurierbarkeit.* Göttingen, Germany: Cuvillier Verlag, 2015.

234. S. Dieter, A. Moessinger, S. Mueller, R. Jakoby, and W. Menzel, "Characterization of reconfigurable LC-reflectarrays using near-field measurements," in *2009 German Microwave Conference*, 2009/March: IEEE.

235. C. Damm, M. Maasch, R. Gonzalo, and R. Jakoby, "Tunable composite right/left-handed leaky wave antenna based on a rectangular waveguide using liquid crystals," in *IEEE MTT-S International Microwave Symposium*, 2010/May: IEEE.

236. C. Damm, *Artificial Transmission Line Structures for Tunable Microwave Components and Microwave Sensors.* Düren, Germany: Shaker Verlag, 2011.

237. M. Roig, "Tunable metamaterial leaky wave antenna based on Microwave Liquid Crystal Technology," PhD thesis, Technische Universität Darmstadt, 2015.

238. M. Roig, M. Maasch, C. Damm, and R. Jakoby, "Dynamic beam steering properties of an electrically tuned liquid crystal based CRLH leaky wave antenna," in *Proceedings of the 8th International Congress on Advanced Electromagnetic Materials in Microwaves and Optics*, 2014/August, pp. 253–255.

239. M. Roig, M. Maasch, C. Damm, O. H. Karabey, and R. Jakoby, "Liquid crystal based tunable composite right/left-handed leaky-wave antenna for Ka-Band applications," in *European Microwave Conference (EuMC)*, 2013/October, pp. 759–762.

240. M. a. Roig, M. Maasch, C. Damm, and R. Jakoby, "Investigation and application of a liquid crystal loaded varactor in a voltage tunable CRLH leaky-wave antenna at Ka-band," *International Journal of Microwave and Wireless Technologies*, vol. 7, no. 3–4, pp. 361–367, 2015. DOI: http://10.1017/s1759078715000367

241. B.-J. Che, F.-Y. Meng, Y.-L. Lyu, and Q. Wu, "A novel liquid crystal based leaky wave antenna," in *IEEE MTT-S International Microwave Workshop Series on Advanced Materials and Processes (IMWS-AMP)*, 2016/July: IEEE.

242. D. R. Smith, W. J. Padilla, D. C. Vier, S. C. Nemat-Nasser, and S. Schultz, "Composite medium with simultaneously negative permeability and permittivity," *Physical Review Letters*, vol. 84, no. 18, pp. 4184–4187, 2000. DOI: http://10.1103/physrevlett.84.4184

243. R. A. Shelby, "Experimental verification of a negative index of refraction," *Science,* vol. 292, no. 5514, pp. 77–79, 2001. DOI: http://10.1126/science.1058847

244. X. Deng, Z. He, S. Yuan, Z. Shao, and L. Liu, "W-band high bit passive phase shifter for automotive radar applications in BiCMOS," in *2011 International Conference on Computational Problem-Solving (ICCP),* October 21–23, 2011, pp. 115–119. DOI: http://10.1109/ICCPS.2011.6092220

245. E. Ozturk, M. H. Nemati, M. Kaynak, B. Tillack, and I. Tekin, "SiGe process integrated full-360° microelectromechanical systems-based active phase shifter for W-band automotive radar," *IET Microwaves, Antennas & Propagation,* vol. 8, no. 11, pp. 835–841, 2014. DOI: http://10.1049/iet-map.2013.0594

246. S. Reyaz, C. Samuelsson, R. Malmqvist, M. Kaynak, and A. Rydberg, "Millimeter-wave RF-MEMS SPDT switch networks in a SiGe BiCMOS process technology," in *7th European Microwave Integrated Circuit Conference,* October 29–30, 2012, pp. 691–694.

247. M. C. Scardelletti, G. E. Ponchak, and N. C. Varaljay, "MEMS, Ka-band single-pole double-throw (SPDT) switch for switched line phase shifters," in *IEEE Antennas and Propagation Society International Symposium (IEEE Cat. No.02CH37313),* June 16–21, 2002, vol. 2, pp. 2–5 vol.2, DOI: http://10.1109/APS.2002.1016014

248. N. Somjit, G. Stemme, and J. Oberhammer, "Performance optimization of multi-stage MEMS W-band dielectric-block phase-shifters," in *7th European Microwave Integrated Circuit Conference,* October 29–30, 2012, pp. 433–436.

249. A. H. Zahr et al., "A DC-30 GHz high performance packaged RF MEMS SPDT switch," in *2015 European Microwave Conference (EuMC),* September 7–10, 2015, pp. 1015–1017. DOI: http://10.1109/EuMC.2015.7345938

250. S. K. Koul and B. Bhat, *Microwave and Millimeter Wave Phase Shifters.* Norwood, MA: Artech House, 1991.

251. C. E. Patton, "Hexagonal ferrite materials for phase shifter applications at millimeter wave frequencies," *IEEE Transactions on Magnetics,* vol. 24, no. 3, pp. 2024–2028, 1988. DOI: http://10.1109/20.3395

252. A. Nafe and A. Shamim, "An integrable SIW phase shifter in a partially magnetized ferrite LTCC package," *IEEE Transactions on Microwave Theory and Techniques,* vol. 63, no. 7, pp. 2264–2274, 2015. DOI: http://10.1109/TMTT.2015.2436921

253. Z. Wang et al., "Millimeter wave phase shifter based on ferromagnetic resonance in a hexagonal barium ferrite thin film," *Applied Physics Letters,* vol. 97, no. 7, p. 072509, 2010/08/16 2010. DOI: http://10.1063/1.3481086

254. K. Choi, S. Courreges, Z. Zhao, J. Papapolymerou, and A. Hunt, "X-band and Ka-band tunable devices using low-loss BST ferroelectric capacitors," in *18th IEEE International Symposium on the Applications of Ferroelectrics,* August 23–27, 2009, pp. 1–6. DOI: http://10.1109/ISAF.2009.5307566

255. A. B. Kozyrev, A. V. Ivanov, O. I. Soldatenkov, A. V. Tumarkin, S. V. Razumov, and S. Y. Aigunova, "Ferroelectric $(Ba,Sr)TiO_3$ thin-film 60-GHz phase shifter," *Technical Physics Letters,* vol. 27, no. 12, pp. 1032–1034, December 1, 2001. DOI: http://10.1134/1.1432340

256. S. Gevorgian, *Ferroelectrics in Microwave Devices, Circuits and Systems: Physics, Modeling, Fabrication and Measurements.* New York: Springer Science & Business Media, 2009.

257. M. Nikfalazar et al., "Fully printed tunable phase shifter for L/S-band phased array application," in *2014 IEEE MTT-S International Microwave Symposium(IMS2014),* June 1–6, 2014, pp. 1–4. DOI: http://10.1109/MWSYM.2014.6848295

258. R. D. Paolis, F. Coccetti, S. Payan, M. Maglione, and G. Guegan, "Characterization of ferroelectric BST MIM capacitors up to 65 GHz for a compact phase shifter at 60 GHz," in *2014 44th European Microwave Conference*, October 6–9, 2014, pp. 492–495. DOI: http://10.1109/EuMC.2014.6986478

259. R. D. Paolis, S. Payan, M. Maglione, G. Guegan, and F. Coccetti, "High-tunability and high-Q-factor integrated ferroelectric circuits up to millimeter waves," *IEEE Transactions on Microwave Theory and Techniques,* vol. 63, no. 8, pp. 2570–2578, 2015. DOI: http://10.1109/TMTT.2015.2441073

260. M. Sazegar, Y. Zheng, H. Maune, C. Damm, X. Zhou, and R. Jakoby, "Compact tunable phase shifters on screen-printed BST for balanced phased arrays," *IEEE Transactions on Microwave Theory and Techniques,* vol. 59, no. 12, pp. 3331–3337, 2011. DOI: http://10.1109/TMTT.2011.2171985.

261. S. Shi, J. Purden, J. Lin, and R. A. York, "A 24 GHz wafer scale electronically scanned antenna using BST phase shifters for collision avoidance systems," in *2005 IEEE Antennas and Propagation Society International Symposium*, July 3–8, 2005, vol. 1B, pp. 84–87 vol. 1B. DOI: http://10.1109/APS.2005.1551489

262. M. Zhang, M. Liu, S. Ling, P. Chen, X. Zhu, and X. Yu, "K-band tunable phase shifter with microstrip line structure using BST technology," in *2015 Asia-Pacific Microwave Conference (APMC)*, December 6–9, 2015, vol. 2, pp. 1–3. DOI: http://10.1109/APMC.2015.7413118

263. M. Jost et al., "Liquid crystal based SPDT with adjustable power splitting ratio in LTCC technology," in *European Microwave Conference (EuMC)*, 2018/September: IEEE.

264. M. Jost et al., "Electrically biased W-band phase shifter based on liquid crystal," in *39th International Conference on Infrared, Millimeter, and Terahertz Waves (IRMMW-THz)*, September 14–19, 2014, pp. 1–2. DOI: http://10.1109/IRMMW-THz.2014.6956435

265. C. Weickhmann, M. Jost, D. Laemmle, and R. Jakoby, "Design and fabrication considerations for a 250 GHz liquid crystal phase shifter," in *39th International Conference on Infrared, Millimeter, and Terahertz Waves (IRMMW-THz)*, September 14–19, 2014, pp. 1–2. DOI: http://10.1109/IRMMW-THz.2014.6956330

266. M. Hoefle, M. Koeberle, M. Chen, A. Penirschke, and R. Jakoby, "Reconfigurable Vivaldi antenna array with integrated antipodal finline phase shifter with liquid crystal for W-Band applications," in *35th International Conference on Infrared, Millimeter, and Terahertz Waves*, 2010/September: IEEE.

267. M. Hoefle, M. Koeberle, A. Penirschke, and R. Jakoby, "Improved millimeter wave Vivaldi antenna array element with high performance liquid crystals," in *2011 International Conference on Infrared, Millimeter, and Terahertz Waves*, 2011/October: IEEE.

268. M. Hoefle, M. Koeberle, A. Penirschke, and R. Jakoby, "Millimeterwave Vivaldi antenna with liquid crystal phase shifter for electronic beam steering," presented at the 6th ESA Workshop on Millimetre-Wave Technology and Applications, 2011. http://tubiblio.ulb.tu-darmstadt.de/56414/

269. M. Koeberle, M. Hoefle, C. Mo, A. Penirschke, and R. Jakoby, "Electrically tunable liquid crystal phase shifter in antipodal finline technology for reconfigurable W-Band Vivaldi antenna array concepts," in *Proceedings of the 5th European Conference on Antennas and Propagation (EUCAP)*, April 11–15, 2011, pp. 1536–1539.

270. P. Deo, D. Mirshekar-Syahkal, L. Seddon, S. E. Day, and F. A. Fernandez, "Liquid crystal based patch antenna array for 60 GHz applications," in *IEEE Radio and Wireless Symposium*, 2013/January: IEEE.

271. P. Deo, M. Yazdanpanahi, and D. Mirshekar-Syahkal, "Effect of test device thickness on liquid crystal characterisation," *2012 Asia Pacific Microwave Conference Proceedings*, pp. 1280–1282, 2012. DOI: http://10.1109/APMC.2012.6421895

272. P. Deo, D. Mirshekar-Syahkal, L. Seddon, S. E. Day, and F. A. b. Fernández, "60 GHz liquid crystal phased array using reflection-type phase shifter," in *European Conference on Antennas and Propagation (EUCAP)*, 2013, pp. 927–929.

273. S. Ma et al., "Compact planar array antenna with electrically beam steering from backfire to endfire based on liquid crystal," *IET Microwaves, Antennas & Propagation*, vol. 12, no. 7, pp. 1140–1146, 2018. DOI: http://10.1049/iet-map.2017.1070

274. B. Sanadgol, S. Holzwarth, and J. Kassner, "30 GHz liquid crystal phased array," in *2009 Loughborough Antennas & Propagation Conference*, 2009/November: IEEE.

275. O. H. Karabey et al., "Liquid crystal based reconfigurable antenna arrays," 2010/October. http://tubiblio.ulb.tu-darmstadt.de/56398/

276. R. Reese et al., "A compact two-dimensional power divider for a dielectric rod antenna array based on multimode interference," *Journal of Infrared, Millimeter, and Terahertz Waves*, vol. 39, no. 12, pp. 1185–1202, 2018. DOI: http://10.1007/s10762-018-0535-x

277. M. Tebbe, A. Hoehn, N. Nathrath, and C. Weickhmann, "Simulation of an electronically steerable horn antenna array with liquid crystal phase shifters," in *Proceedings of the IEEE Aerospace Conference*, 2016/March, pp. 1–15.

278. L. B. Soldano and E. C. M. Pennings, "Optical multi-mode interference devices based on self-imaging: principles and applications," *Journal of Lightwave Technology*, vol. 13, no. 4, pp. 615–627, 1995. DOI: http://10.1109/50.372474

279. Y. Liu et al., "Tunable microwave bandpass filter integrated power divider based on the high anisotropy electro-optic nematic liquid crystal," *Review of Scientific Instruments*, vol. 87, no. 7, pp. 074709–074709, 2016.

280. I. C. Khoo, D. H. Werner, X. Liang, A. Diaz, and B. Weiner, "Nanosphere dispersed liquid crystals for tunable negative-zero-positive index of refraction in the optical and terahertz regimes," *Optics Letters*, vol. 31, no. 17, pp. 2592–2592, 2006. DOI: http://10.1364/ol.31.002592

281. D. H. Werner, D.-H. Kwon, I.-C. Khoo, A. V. Kildishev, and V. M. Shalaev, "Liquid crystal clad near-infrared metamaterials with tunable negative-zero-positive refractive indices," *Optics Express*, vol. 15, no. 6, pp. 3342–3347, March 19, 2007. DOI: http://10.1364/OE.15.003342

282. D.-H. Kwon, D. H. Werner, I.-C. Khoo, A. V. Kildishev, and V. M. Shalaev, "Liquid crystal clad metamaterial with a tunable negative-zero-positive index of refraction," in *IEEE Antennas and Propagation Society International Symposium*, 2007, pp. 2881–2884.

283. W. Xiande, K. Do-Hoon, D. H. Werner, and K. Iam-Choon, "Anisotropic liquid crystals for tunable optical negative-index metamaterials," in *IEEE Antennas and Propagation Society International Symposium*, July 5–11, 2008, pp. 1–4. DOI: http://10.1109/APS.2008.4619734

284. A. Minovich et al., "Liquid crystal based nonlinear fishnet metamaterials," *Applied Physics Letters*, vol. 100, no. 12, pp. 121113–121113, 2012. DOI: http://10.1063/1.3695165

285. D. C. Zografopoulos and R. Beccherelli, "Tunable terahertz fishnet metamaterials based on thin nematic liquid crystal layers for fast switching," *Scientific Reports*, vol. 5, no. 1, 2015. DOI: http://10.1038/srep13137

286. C.-L. Chang, W.-C. Wang, H.-R. Lin, F. Ju Hsieh, Y.-B. Pun, and C.-H. Chan, "Tunable terahertz fishnet metamaterial," *Applied Physics Letters,* vol. 102, no. 15, p. 151903, 2013/04/15 2013. DOI: http://10.1063/1.4801648

287. M. Maasch, "Tunable microwave metamaterial structures," 2016. www.ebook.de/de/product/25351590/matthias_maasch_tunable_microwave_metamaterial_structures.html

288. M. Maasch, A. Groudas, O. Karabey, C. Damm, and R. Jakoby, "Electrically tunable open split-ring resonators based on liquid crystal material," 2012/January.

289. M. Maasch, M. Roig, C. Damm, and R. Jakoby, "Voltage-tunable artificial gradient-index lens based on a liquid crystal loaded fishnet metamaterial," *IEEE Antennas and Wireless Propagation Letters,* vol. 13, pp. 1581–1584, 2014. DOI: http://10.1109/LAWP.2014.2345841

290. M. Maasch, M. Roig, C. Damm, and R. Jakoby, "Realization of a voltage tunable gradient-index fishnet loaded with liquid crystal," in *Proceedings of the 8th International Congress on Advanced Electromagnetic Materials in Microwaves and Optics,* 2014/August, pp. 196–198.

291. W. Hu et al., "Liquid crystal tunable mm wave frequency selective surface," *IEEE Microwave and Wireless Components Letters,* vol. 17, no. 9, pp. 667–669, 2007. DOI: http://10.1109/lmwc.2007.903455

292. F. Zhang et al., "Magnetic control of negative permeability metamaterials based on liquid crystals," in *European Microwave Conference (EuMC),* 2008/October, pp. 801–804.

293. J. A. Bossard et al., "Tunable frequency selective surfaces and negative-zero-positive index metamaterials based on liquid crystals," *IEEE Transactions on Antennas and Propagation,* vol. 56, no. 5, pp. 1308–1320, 2008. DOI: http://10.1109/tap.2008.922174

294. V. F. Fusco, R. Cahill, W. Hu, and S. Simms, "Ultra-thin tunable microwave absorber using liquid crystals," *Electronics Letters,* vol. 44, no. 1, pp. 37–37, 2008. DOI: http://10.1049/el:20082191

295. F. C. Seman, R. Cahill, and V. F. Fusco, "Electronically tunable liquid crystal based Salisbury screen microwave absorber," in *Proceedings of the Loughborough Antennas Propagation Conference,* 2009/November, pp. 93–96.

296. S. Mohamad and R. Cahill, "Spiral antenna with reconfigurable HIS using liquid crystals for monopulse radar application," in *Proceedings of the IEEE Conference on Antenna Measurements Applications (CAMA),* 2017/December, pp. 55–58.

Appendix 1 Satellite History, Orbits, and Classification

A1.1 Early History of Satellites

The following is a brief but uncomplete history of satellites, with an emphasis the early major milestones up to the year 2000 only [1, 2]:

1957	The first man-made artificial satellite, Sputnik 1, was launched by former Soviet Union on October 4, 1957.
1958	A few months later, the United States launched its first satellite, Explorer 1, into space on January 31, 1958. The world's first communications satellite in space was SCORE (Signal Communications by Orbiting Relay Equipment), launched aboard an American Atlas rocket on December 18, 1958. It broadcast a taped Christmas message from President Eisenhower and lasted 35 days in a low Earth orbit (LEO).
1960	The first passive satellite, Echo I, was also placed in a low Earth orbit. Echo was a metalized balloon satellite acting as a passive reflector of microwave signals, which were bounced off from one point on Earth to another.
1962/63	The first real-time active communication satellites from AT&T were Telstar 1 and 2, launched in 1962 and 1963, respectively, and placed in a medium Earth orbit (MEO). Both successfully relayed the first television pictures and telephone calls.
1961/64	In 1961, NASA started its program Syncom (for "synchronous communication satellite") and launched Syncom 1 and 2 in February and July 1963, respectively. The first one failed, but Syncom 2 was able to conduct voice, teletype, and facsimile tests as well as a number of television transmission tests. Syncom 3, launched in August 1964, was the world's first geostationary satellite. In addition to voice and facsimile, it provided a wideband channel for television used to broadcast the 1964 Summer Olympics in Tokyo to the United States.
1964	A milestone in the satellite history was the creation of the International Telecommunication Satellite Organization (INTELSAT), operating Intelsat I (nicknamed Early Bird), the first commercial communications satellite to be placed in geosynchronous orbit in April 1965, providing

television, telephone, and telefacsimile transmissions between Europe and North America.

1968	First European satellite ESRO 2B
1970	First Japanese satellite: Ohsumi
	First Chinese satellite: Dong Fang Hong 01
1971	ITU-WARC for Space Telecommunications
	INTELSAT IV launched
	INTERSPUTNIK – Soviet Union equivalent of INTELSAT formed
1974	First direct broadcasting satellite: ATS 6
1976	MARISAT – Three satellites for first civil maritime communication.
1977	EUTELSAT – European regional satellite.
	ITU-WARC for Space Telecommunications in the Satellite Service.
1979	International Mobile Satellite Organization (Inmarsat) established.
1980	INTELSAT V launched – 3 axis stabilized satellite built by Ford Aerospace.
1982	First mobile satellite telephone system INMARSAT-A.
1983	ECS (EUTELSAT 1) launched – built by European consortium supervised by ESA.
1984	First direct-to-home broadcast system (Japan)
1988	First satellite system for mobile phones and data communication INMARSAT-C.
1988	First SES (Luxemburg) satellite Astra 1A.
1992	OLYMPUS finally launched – large European development satellite with K_a-band, DBTV and K_u-band SS/TDMA payloads – fails within 3 years.
1993	INMARSAT II, 39 dBW EIRP, global beam mobile satellite built by Hughes/British Aerospace.
1994	INTELSAT VIII – first INTELSAT satellite built to a contractor's design. DirecTV begins Direct Broadcast to Home.
1996	INMARSAT III – first of the multibeam mobile satellites (built by GE/Marconi). EchoStar begins Direct Broadcast Service.
1998	Global satellite systems for small mobile phones.
1999	AceS launch first of the L-band MSS Super-GSOs – built by Lockheed Martin.
	Iridium bankruptcy – the first major failure?
2000	Globalstar begins service.

A1.2 Satellite Orbits

Satellites are moving around Earth in well-defined orbits. In a first step, they are classified into *Geostationary Earth Orbit* (*GEO*) and *Non-GeoStationary Orbit*

(*NGSO*) satellite systems. The NGSO systems are further subdivided into *Low Earth Orbit (LEO)*, *Medium Earth Orbit (MEO)*, and *Highly Elliptical Orbit (HEO)* systems, depending on their altitudes or the shapes of their orbits. The LEO systems are subdivided into big LEO and little LEO systems; big LEOs generally target real-time near-toll-quality voice as well as data, paging, facsimile, and radiodetermination satellite service, while little LEOs target low-data-rate (in the order of kbits/s) data messaging and radionavigation satellite services using frequencies below 1 GHz.

GEO satellites are in orbit 35 786 km above the Earth surface along the equator. Objects in the geostationary orbit revolve around Earth at the same speed as Earth rotates. This means GEO satellites remain in the same position relative to an Earth terminal, thus having a 24-hour view of a particular area, which is ideally suited for broadcast and other multipoint services. GEO systems have relatively simple configurations of both space segments and Earth stations; require simple space segment control systems; and have time-invariant elevation angles to the satellites and an extremely wide footprint on the ground, that is, a large coverage area of up to a fourth of Earth's surface. Thus, to achieve nearly global coverage, only three GEO satellites are required. However, GEO satellites above the equator have extremely low elevation angles in high-latitude countries and popular regions and some difficulties in providing services to near polar regions.

LEO and MEO systems are much closer to Earth, ranging 200–2000 km for LEOs and 2000–20 000 km for MEO, usually positioned below or between the outer and inner van Allen radiation belts at 1500–5000 km and 13 000–20 000 km, which are a major source of potentially damaging ionizing radiation. These MEO/LEO systems feature global service capability, including even high-latitude areas, and their satellites are easier to launch. However, to achieve a nearly global coverage on Earth, LEO systems especially need a quite large network of satellites for continuous communications, which can be costly, and because of their relative movements with respect to a fixed Earth terminal, they are visible only for a period in the range of 10–20 minutes each pass, which means they need more frequent handoffs and more complex onboard control systems and have to compensate for Doppler shifts. Because of higher altitudes, fewer satellites are needed in MEO networks. Moreover, they have larger coverage areas and are longer visible than LEO satellites in the range of 2–8 hours each pass, but at the cost of longer time delay and larger propagation losses.

The HEO systems allow a flexible system design and feature comparable propagation delays and losses at the apogee altitude to those of GEO systems, high elevation angles at high-latitude countries such as European countries and Canada, and no eclipsing within service areas. However, they require large, tracking onboard antennas, and have larger Doppler shifts, high fuel consumption for satellite altitude controls, and a shorter satellite lifetime as a result of periodically passing through the van Allen radiation belts.

Table A1.1 indicates the orbits and corresponding typical systems with their related main characteristics, respectively. More details about key concepts related to orbits are given in [3].

Table A1.1 Main characteristics of GEO, LEO, MEO and HEO orbits

	GEO	**LEO**	**MEO**	**HEO**
Orbit type	Circular	Circular	Circular	Elliptical
Altitude/apogee	>35 786 km	200–2000 km	2000–20 000 km	40 000 km
perigee	24 hours	86–127 minutes	2–8 hours	500 km
period				12 hours
Typical system	**Inmarsat**	**Globalstar**	**O3b**	**Molniya (USSR)**
Number of orbits	1	8	4	4
Altitude/apogee	36 000 km	1414 km	8062 km	40 000 km
perigee	24 hours	114 minutes	288 minutes	500 km
period				12 hours
Satellite weight	About 1500 kg	About 700 kg	About 700 kg	12 (3/orbit)
Numbers	3	48 (6/orbit)	12 (3/orbit)	80 degree
Min. elev. angle	5 degree	10 degree	45 minutes	8 hours
Visible time	24 hours	16.4 minutes		

A1.3   Effective Aperture, Gain, and Half-Power Beam Width of Antennas

Antennas collect all the power, flowing through the effective aperture A_e, which is related with the gain of the antenna [4]:

$$A_e = \frac{\lambda^2}{4\pi} G_0 \quad \text{with } A_e = \eta_{ap} A_{phy} \tag{A1.1}$$

for all antennas where λ is the wavelength, G_0 the gain, η_{ap} the aperture efficiency, and A_{phy} the physical aperture. The aperture efficiency $\eta_{ap} = \eta_{rad} \cdot \eta_t$ of the antenna includes the antenna radiation efficiency and the taper efficiency, which is a measure of how efficiently the physical aperture is being used. In general, η_{ap} ranges from 30% to 80%. Optimum gain pyramidal horns have an aperture efficiency near 50%. Parabolic reflector antennas have an efficiency of 55% or larger and a Cassegrain antenna could achieve 80% [4]. Hence, the gain of the antenna can be simplified to

$$A_e = \frac{\lambda^2}{4\pi} G_0 = \frac{4\pi}{\lambda^2} \eta_{ap} A_{phy} \quad \text{with } \eta_{ap} = 0.3 \cdots 0.8. \tag{A1.2}$$

Assuming a rectangular and circular aperture with an area of $A_{phy}^{\square} = A \cdot B$ and $A_{phy}^{\circ} = \pi (D/2)^2$, the gain is given by

$$G_0^{\square} = \frac{4\pi}{\lambda^2} \eta_{ap} (A \cdot B) \tag{A1.3}$$

and

$$G_0^{\circ} = \frac{4\pi}{\lambda^2} \eta_{ap} \left(\pi \left(\frac{D}{2} \right)^2 \right) = \eta_{ap} \left(\frac{\pi D}{\lambda} \right)^2. \tag{A1.4}$$

In parallel, gain is related to directivity by [4]:

$$G_0 = \eta_{rad} \cdot D_0 = \eta_{ap} \cdot D_{0,u} \text{ with } \eta_{ap} = \eta_{rad}\eta_t \quad (A1.5)$$

where $D_{0,u}$ is the directivity for uniform field distribution across the aperture.
Hence, the gain for a rectangular aperture is

$$G_0 = \eta_{ap} \cdot \pi \cdot D_{x,u} D_{y,u} = \eta_{ap} \cdot \pi \cdot \frac{2A}{\lambda}\frac{2B}{\lambda} = \eta_{ap} \cdot \pi \cdot \frac{2k_x}{k_x\frac{\lambda}{A}}\frac{2k_y}{k_y\frac{\lambda}{B}}$$

$$= \eta_{ap} \cdot 4\pi \cdot \frac{k_x}{\text{HPBW}_x}\frac{k_y}{\text{HPBW}_y} \quad (A1.6)$$

where HPBW_x and HPBW_y are the half-power beam width in x- and y-directions and where the factors k_x and k_y depend on the field distribution along both directions, with values of 0.886, 1.19, and 1.55 for a uniform, cos and $\cos^2$ distribution. Note, that the field distributions along both directions are related to the corresponding far-field pattern via Fourier transform.

With $k = k_x = k_y$ and $\text{HPBW} = \text{HPBW}_x = \text{HPBW}_y$ for a circular aperture, its gain reduces to

$$G_0 = \eta_{ap} \cdot 4\pi \cdot \frac{k^2}{\text{HPBW}^2}. \quad (A1.7)$$

Using both equations for the gain of a circular aperture, the half-power beam width is

$$\text{HPBW}^\circ = k \cdot \sqrt{4\pi\frac{\eta_{ap}}{G_0}} = k \cdot \sqrt{4\pi\frac{\eta_{ap}}{\eta_{ap} \cdot \left(\frac{\pi D}{\lambda}\right)^2}} = k \cdot \sqrt{\frac{4}{\pi}} \cdot \frac{\lambda}{D} \quad (A1.8)$$

For a parabolic antenna with circular aperture and a diameter of $D = 50\lambda$, the HPBW is 1.15°, 1.54°, and 2° for the three aforementioned distributions. Figure A1.1 shows the gain and HPBW for a circular aperture versus its diameter D for $\eta_{ap} = 0.6$ and a cos distribution ($k = 1.19$) for various frequencies.

A1.4 Comparison of Satellite Systems in Different Orbits

For a first, simple and rough comparison of the major physical characteristics of GEO and LEO/MEO satellites, related to their orbits, we assume a GEO satellite in a distance above Earth of $d_{GEO} = 40\ 000$ km and a MEO/LEO satellite in a distance $d_{MEO} = 8000$ km and $d_{LEO} = 1500$ km respectively. All satellites are assumed to operate at 20 GHz downlink.

A1.4.1 Covered Area

Assuming a high-gain, narrow beam, parabolic antenna with a circular aperture of diameter D and hence, a HPBW according to Eq. (A1.8), then, the diameter a of the nearly circular covered area on Earth, neglecting the curvature of Earth, is roughly

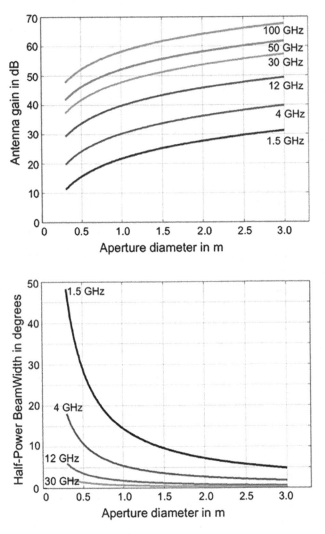

Figure A1.1 Gain and HPBW for a circular aperture versus its diameter D for $\eta_{ap} = 0.6$ and a cos distribution ($k = 1.19$) for various frequencies.

$$a \approx d \cdot \tan(\text{HPBW}) \approx d \cdot \text{HPBW} \qquad (A1.9)$$

for small HPBW, that is, large aperture of diameter D. Figure A1.2 illustrates the covered area on Earth's surface for a satellite antenna.

Figure A1.3 shows the footprint diameter a versus the distance d between satellite and Earth for various diameters D of the dish, where a increases approximately linearly with the distance d and with the HPBW of the satellite dish, which is inversely proportional to the diameter D of the dish, according to the above equation. Accordingly, assuming the same antenna, operating at the same frequency onboard

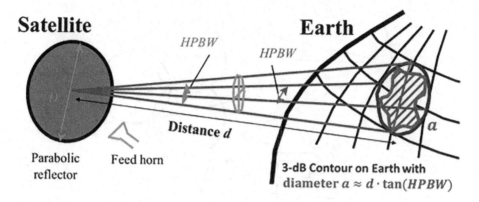

Figure A1.2 Covered area on Earth's surface of a satellite dish antenna.

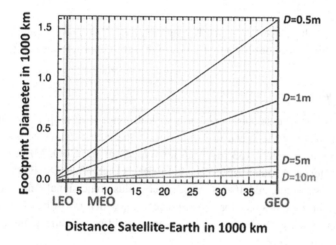

Figure A1.3 Footprint diameter a versus the distance d between satellite and Earth for various diameter D of the dish, operating at 20 GHz.

a satellite in different orbits, then the GEO satellite with $d_{GEO} = 40\,000$ km would cause a much wider footprint on Earth than a MEO and LEO satellite with $d_{MEO} = 8000$ km and $d_{LEO} = 1500$ km, which is roughly 5 and 26.6 times larger, respectively.

Generally, for a certain distance d and frequency f, the footprint diameter a decreases significantly with larger antennas of diameter D. By using multiple feeds with different frequencies and/or polarizations, illuminating the same reflector as shown in Figure A1.4, different footprints or cells would be generated on Earth, similar to the concept in terrestrial mobile networks. The cell size could be in the range of 100–250 km for GEO satellites with relatively large multibeam antennas. Repeating these cell clusters allows frequency reuse, increasing the overall throughput

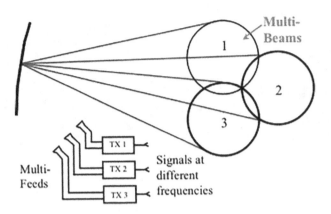

Figure A1.4 Multibeam antennas.

of the satellite with multi-beam capability significantly. This concept was decisive for the development of the fourth generation of high-throughput satellites (HTSs).

A1.4.2   Signal Delay

Signal delay τ between satellite and Earth with distance d is simply determined by $\tau = d/c_0$, where c_0 is the speed of light. Hence, the signal delays of a satellite in different orbits are roughly 133 ms, 26.6 ms, and 5 ms for GEO. MEO, and LEO, respectively. Hence, a round-trip propagation (RTP) Earth–Satellite–Earth would be roughly 260 ms, 50 ms, and 10 ms, respectively, plus some signal delay in satellite and Earth station modules and for signal processing, which might be in the range of another 30 to 100 ms. Duplex-or multiplex-hop would cause correspondingly more signal delay, respectively.

A1.4.3   Received Power for a Satellite–Earth Path

The received power for a satellite–Earth path can be calculated by

$$L_R = L_T + G_T + G_R - 20 \log \left(4\pi \frac{d}{\lambda} \right), \tag{A1.10}$$

where L_T is the transmitting power of the satellite, G_T and G_R the gain of the transmitting and receiving antenna, respectively, d the distance satellite–Earth station, and λ the wavelength. The last term represents the free-space propagation loss a_{fs}. Again, assuming the same power L_T, the same antennas, and the same wavelength, the free-space propagation losses for MEO and LEO are 14 dB and 28.5 dB lower, respectively, than for GEO.

To summarize, owing to their lower altitudes, MEO and LEO have much smaller footprints and more favorable for higher capacity, and face much lower propagation delay and definitely much lower path attenuation than GEO. Thus, in particular LEO constellations are potentially able to use relatively small, low-power (e.g., less than 0.5 W) and much lower cost handheld user terminals, which are more suitable for interactive multimedia services. But, to achieve a nearly global illumination on Earth, it is necessary to place many more satellites in low-earth orbits than in geostationary orbit, that is, more complex systems. In turn, that means, that GEO satellite systems for interactive multimedia services have to generate many small "spot beams" because of the capacity requirements, for example, multibeam antennas and Earth station antennas in the customer terminal with an extreme high gain. Nevertheless, the problem remains to generate sufficient transmitting power for small customer stations and at least to reduce any delay in the components and signal processing. Compared to GEO, MEO and in particular LEO satellites are much smaller, weigh less, and are explicitly cheaper, with lower costs to locate them in their orbits than GEO satellites. But they last only about 5–8 years because of atmospheric contamination and radiation from the van Allen belts, while GEO satellites last about 12–15 years, sometimes even 20 years (limited only conditional on fuel). Furthermore, atmospheric drag effects can cause gradual orbital deterioration for LEO/MEO satellites.

Hence, selection of the satellite orbit is a tradeoff between many factors, including satellite cost, satellite antenna size, battery power, weight, lifetime, the number of satellites needed for the network and the launch flexibility and cost, minimum elevation angles within service areas, uplink and downlink frequencies, effect of van Allen radiation belts, handset power, propagation delay, system reliability, and required service quality.

A1.5   Examples of Satellite Systems in Different Orbits

A typical example of LEO satellite systems is iridium, which failed, because it was not competitive to terrestrial-based systems at that time when satellites were launched. It was planned to use 66 satellites in circular low Earth orbits of 780 km in altitude.

The Globalstar second-generation constellation consists of 24 LEO satellites at an altitude of 1414 km, covering more than 80% of Earth's surface, everywhere outside the extreme polar regions and some mid-ocean regions, providing customers with satellite voice and data services. The satellites weight is approximately 700 kg and is designed with a life expectancy of 15 years. Each satellite consists of a communications system of S- and L-band antennas. The satellites utilize "bent-pipe" architecture. On any given call, several satellites transmit a caller's signal via CDMA technology to a satellite dish at the appropriate gateway where the call is then routed locally through the terrestrial telecommunications system [5, 6].

ORBCOMM is an American company that operates a global network of 44 LEO satellites and accompanying ground infrastructure, including 16 gateway Earth stations (GESs) around the world, offering industrial IoT and M2M communications solutions designed to track, monitor, and control fixed and mobile assets in markets including transportation, heavy equipment, maritime, oil and gas, utilities, and government. The company provides hardware devices, modems, web applications, and data services delivered over multiple satellite and cellular networks. Next-generation OG2 satellites were launched aboard a SpaceX Falcon 9 rocket. Orbcomm satellites operate at frequencies of 137–153 MHz. The OG2 satellites operate from a circular orbit of 750 km at an inclination of 52 degrees with the satellites in different planes to achieve timely global coverage. Compared to OG1 satellites, ORBCOMM's OG2 satellites are designed for faster message delivery, larger message sizes, and better coverage at higher latitudes, while increasing network capacity [7].

An example of a MEO-HTS system is the O3b satellite constellation, which is described in Chapter 1, Section 1.2 of the book.

A typical HEO system is the Molniya, which has operated since 1965 in the former USSR mainly for domestic communications. The apogee and perigee of Molniya's orbit are 40 000 km and 500 km, respectively. Molniya was initially the name of a satellite, but more recently Molniya has given its name to the elliptical orbit first used by the Molniya system.

Another example is Sirius Satellite Radio, which uses HEO orbits to keep two satellites positioned above North America while another satellite quickly sweeps through the southern part of its 24-hour orbit.

A1.6   Classification of Satellite Services

Depending on the purpose, satellite communications can be classified into several services [3]. These services, terminology, and allocated frequency bands have been defined by the Radio Regulations (RR) of the International Telecommunication Union (ITU).

Fixed Satellite Service (FSS) FSS provides services between fixed Earth or ground stations through geostationary satellite links, typically for the transmission of video, voice, and IP data. An FSS system consists of a satellite and gateway Earth stations, which are typically quite large and expensive, requiring complex facilities with large antennas. It might contain transmission from one point on the globe either to a single point (point-to-point) or to multiple points/receivers (point-to-multipoint). FSS may include satellite-to-satellite links or feeder links for other satellite services such as the Mobile Satellite Service or the Broadcast Satellite Service [3]. A typical example of FSS is the INTELSAT system. The first generation of INTELSAT systems operates in the C-band (6/4 GHz). At present, many domestic systems operate in the K_u-band (14/12 GHz). A few systems in the world are using the K_a-band (30/20 GHz).

Broadcasting Satellite Service (BSS) BSS broadcasts TV and radio programs via geostationary satellites to Earth stations. Present broadcasting satellite systems usually operate at 12 GHz. They are designed for community reception (fixed terminals with large antennas). If a satellite has enough power to transmit signals to be received by small antennas suitable for individual reception (fixed terminal with small antennas), the system is called a Direct Broadcasting Satellite (DBS) system. For this service, only a smaller number of geostationary satellites are required, because of global coverage. Although present systems are designed for fixed terminals and not for mobiles, some mobiles such as large ships, trains, and buses have received TV programs from direct broadcasting satellites (such as the BS satellite in Japan) while moving. Some advanced broadcasting system are using the K_a band, for example, the COMETS program in Japan.

Mobile Satellite Service (MSS) MSSs are categorized into three services: maritime, aeronautical, and land mobile communications. MSS is usually provided by a network of satellites, which can provide wireless communication to any point on the globe.

While first-generation *mobile satellite communication systems* provide radio communication services between mobiles such as ships, aircraft, and land vehicles and a gateway Earth station through (typically GEO) satellite links, the second-generation *personal mobile satellite communication systems* provide services between very small handheld terminals by using direct access to one or more LEO/MEO satellites, without any connection to a gateway Earth station.

In first-generation MSS systems, direct communication channels between mobiles cannot be serviced mainly because of a lack of satellite capabilities such as transmission power and channel switching functions. A typical example is the INMARSAT-3/-4 system, characterized by global beam features of geostationary satellites and relatively large user terminals. It operates in the L-band (1.6/1.5 GHz) and provides worldwide commercial services for ships, aircraft, and land mobiles [8].

In the second-generation of MSS systems, the satellites may have sophisticated functions onboard such as channel switching, networking, and signal processing. It could provide mobile services wherever the terrestrial-based systems do not exist, for example, because of exceeding infrastructure costs in undeveloped countries, or are not competitive because of low traffic density in rural and remote areas and to ease the terrestrial system during times of congestion. A typical example is Globalstar.

Radionavigation Satellite Service (RNSS) Typical examples of radio navigation systems are the Navy Navigation Satellite System (NNSS), sometimes better known as the TRANSIT, and the Navigation System with Time and Ranging/Global Positioning System, (NAVSTAR/GPS). The NNSS was the first radionavigation system in the world. It was originally developed as a US Navy military satellite system but has been open to civilian use since 1967. The Global Positioning System (GPS) started in 1978 and uses an array of satellites at an altitude of 20 230 km, operating at two frequencies of 1.6 GHz and 1.2 GHz. It is a second-generation radionavigation system and is the most widely used radionavigation system in the world. Tsikada and GLONASS have been used in Russia, which are equivalent to NNSS and GPS. In addition, Europe implements a navigation system, called Galileo, aiming for 24 plus satellites transmitting the highly accurate signals necessary to deliver navigation services across a wide range of activities.

Intersatellite Service (ISS) There are two types of ISS from the standpoint of satellite orbits. The first is to establish links between GEO–GEO satellites, and the second is to establish links between GEO–LEO satellites, using the K_a band, millimeter waves, and laser to carry out intersatellite communications.

Appendix 2 Multiphysical Modeling of Nematic Liquid Crystals

So far, the application area of liquid crystal technology has been limited to the optical area in form of the well-known liquid crystal displays (LCDs). Opening up a new area of application by introducing LC technology in microwave or millimeter-wave systems requires a different design approach. In contrast to optical systems, the dimensions of components are usually in the range of a few wavelengths or even smaller.

The tuning performance of a microwave or millimeter-wave LC device, e.g. differential phase shift or response times of an LC phase shifter, depends on the viscosity and tunability of the LC material itself as well as the utilization of the LC. The LC utilization in turn depends on the electromagnetic field distribution and is connected to the physical waveguide and biasing electrode structure. Hence, it is of interest for the optimization of LC based microwave components to be able to evaluate those performance parameters for different waveguide structures and electrode configurations. Owing to the vast increase in computing power, numerical modeling of LC has been established for the optimization of the more advanced display applications. Commercial software packages available today are DIMOS, LCD-Master, TecWiz-LCD, and Mouse/Pol-LCD. However, the optimization criteria in optical applications are based on free space wave propagation and involve Jones calculus and ray tracing as well as wide-angle vector beam methods. In contrast to microwave applications, optical characteristics such as transmission and reflection are evaluated. The method proposed by A. Gaebler [9] adopts the numerical modelling of LC for microwave applications. In optical as well as microwave applications, the director alignment is performed utilizing the aforementioned preorientation methods and/or electrostatic fields. Magnetostatic fields are used only in experimental setups due to the small susceptibilities of LC materials, which necessitates large field strengths and hence leads to rather bulky setups. Thus, the computation method for the LC director dynamic calculation can be adopted from the optical application. The connection of the permittivity tensors components with the director leads to a mutual coupling of the director field and the electrostatic field. Hence, the computation of the LC director dynamic involves the joint solution of the respective differential equation and the Laplace equation for the electrostatic field. The computation is performed in discrete time steps until a steady-state director configuration is achieved. Based on the results, the waveguides dominant mode can be determined by solving the wave equation. Hence, the problem can be divided in three steps: director dynamic computation, solution of the Laplace equation, and solution of the wave equation, where the first two steps interlock due to the mutual coupling. These steps are described in further detail in the following sections.

A2.1   Director Dynamics

As already mentioned in Section 5.2.2, applying a static electric field will induce a torque on the permanent dipole moments of the LC molecules. In turn, this will lead to an alignment of the LC directors with respect to the electric field to minimize Gibbs free energy. Counteracting forces due to dipole–dipole interactions and surface anchoring effects (preorientation, for example) will prevent a perfect alignment and lead to a certain elastic distortion of the director field. Mathematically, such a system can be described by energy considerations, keeping in mind that a physical system always aims to minimize its free energy. The corresponding torque can be gained from this energy consideration with the help of the Lagrangian mechanics. In conjunction with the Rayleigh dissipation function and the rotational viscosity of the LC, the equation of motion is obtained. Subsequently, the resulting differential equation is solved by means of the finite difference method to calculate the director configuration.

The equation of motion is gained from Lagrangian mechanics with the Lagrange equation of motion and the Rayleigh dissipation term $\partial F_d / \partial \dot{n}_i$

$$\frac{d}{dt}\left(\frac{\partial L}{\partial \dot{n}_i}\right) - \frac{\partial L}{\partial \dot{n}_i} + \frac{\partial F_d}{\partial \dot{n}_i} = 0, \tag{A2.1}$$

where $\dot{n}_i$ represents one coordinate of the director field. The Lagrangian mechanics is based on the idea of variational calculus and aims for minimization of the Lagrangian function L:

$$L = K - U = -f + \frac{1}{2}\lambda\left(n_x^2 + n_y^2 + n_z^2\right). \tag{A2.2}$$

In Eq. (A2.2), K represents the kinetic energy density, which is zero in this case, and U is the potential energy density given by the total free energy density of the system

$$f = f_{\text{elastic}} + f_{\text{electric}}, \tag{A2.3}$$

where f_{elastic} and f_{electric} are given in Eq. (5.23) and (5.24), respectively. The factor λ is called Lagrange multiplier and serves to meet the optimization constraint, which is to maintain the unit length of the director. Additionally, dissipative forces due to the viscosity of the LC are taken into account by the dissipation function

$$F_d = \frac{1}{2}\gamma_{\text{rot}}\left(\dot{n}_x^2 + \dot{n}_y^2 + \dot{n}_z^2\right) \tag{A2.4}$$

with the material dependent rotational viscosity γ_{rot}. These equations inhere some simplifications, since the effects of surfaces, hydrodynamic flow coupling, and local influence of the directors due to electrostatic or mechanical fields are neglected.

By combining Eqs. (A2.1), (A2.2), and (A2.4), the update equation of the director field is written as

$$\gamma_{\text{rot}}\frac{\mathrm{d}n_i}{\mathrm{d}t} = -\left\{\frac{\partial f}{\partial n_i} - \frac{\mathrm{d}}{\mathrm{d}x}\left[\frac{\partial f}{\partial(\mathrm{d}n_i/\mathrm{d}x)}\right] - \frac{\mathrm{d}}{\mathrm{d}y}\left[\frac{\partial f}{\partial(\mathrm{d}n_i/\mathrm{d}y)}\right] - \frac{\mathrm{d}}{\mathrm{d}z}\left[\frac{\partial f}{\partial(\mathrm{d}n_i/\mathrm{d}z)}\right]\right\} + \lambda n_i$$

(A2.5)

In practice, the Lagrange multiplier term λn_i is neglected during the recalculation of the director field. Instead, the directors are normalized after each iteration step. For numerical computation, Eq. (A2.5) is discretized in time and space using the finite difference method. The initial director configuration has to be specified at the beginning and spatial limits can be accounted with Dirichlet, Neumann, or open boundary conditions. Such a problem statement is called the initial boundary condition problem. Preorientation layers, such as polyimide (PI) films, can be modelled with this technique. The involved material-dependent parameters are the elastic constants K_{11}, K_{22}, and K_{33}, the rotational viscosity λ_{rot} and the permittivity of the LCs long and short axis $\varepsilon_{\parallel}$ and $\varepsilon_{\perp}$, respectively.

For the spatial computation, the calculation area is split into small pieces using the grid depicted in Figure A2.1, where an LC-based microstrip line is shown. To estimate the spatial derivatives in Eq. (A2.5), the components of the director field are modeled locally with the help of a multivariate Taylor approximation

$$n_i\left(\vec{r} + \Delta\vec{r}\right) = \sum_{v=0}^{\infty}\frac{1}{v!}\left(\Delta\vec{r}\cdot\nabla\right)^v n_i\bigg|_{\vec{r}},$$

(A2.6)

where $\vec{r}$ is the center of the series expansion and $\Delta\vec{r}$ the corresponding difference vector pointing to the evaluated coordinate.

To reduce the problem's complexity, the waveguide structures are assumed to be of infinite extent, assuming constant cross-section along direction of propagation (z). This will set the derivatives in Z direction to zero. Furthermore, only terms up to an order of 2 are considered in the series expansion. This leads to the following expression containing eight derivative terms:

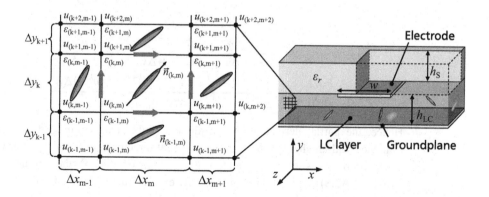

Figure A2.1 Placement of the LC directors and DC potentials for the LC-based microstrip line.

$$n_i(x + \Delta x, y + \Delta y)$$

$$\approx n_i(x + y) + \left\{ \Delta x \frac{\partial n_i}{\partial x}\bigg|_{x,y} + \Delta y \frac{\partial n_i}{\partial y}\bigg|_{x,y} \right\}$$

$$+ \frac{1}{2} \left\{ \Delta x^2 \frac{\partial^2 n_i}{\partial x^2}\bigg|_{x,y} + 2\Delta x \Delta y \frac{\partial^2 n_i}{\partial x \partial y}\bigg|_{x,y} + \Delta y^2 \frac{\partial^2 n_i}{\partial y^2}\bigg|_{x,y} \right\}$$

$$+ \frac{1}{2} \left\{ \Delta x^2 \Delta y \frac{\partial^3 n_i}{\partial x^2 \partial y}\bigg|_{x,y} + \Delta x \Delta y^2 \frac{\partial^3 n_i}{\partial x \partial y^2}\bigg|_{x,y} \right\}$$

$$+ \frac{1}{4} \left\{ \Delta x^2 \Delta y^2 \frac{\partial^4 n_i}{\partial x^2 \partial y^2}\bigg|_{x,y} \right\}. \tag{A2.7}$$

With the help of this relation, the estimation of the partial derivatives is possible. Considering $n_i(x, y)$ and its partial derivatives in Eq. (A2.7) as unknown variables, it is evident that the evaluation of the director components at nine grid points is sufficient to solve the system of equations. The solution is stated in the following for the relevant partial derivatives.

$$\frac{\partial n_i}{\partial x}\bigg|_{x,y} \approx \frac{\left(n_i^{k,m} - n_i^{k,m-1}\right)\Delta x_m^2 - \left(n_i^{k,m} - n_i^{k,m+1}\right)\Delta x_{m-1}^2}{\Delta x_{m-1} \Delta x_m (\Delta x_{m-1} + \Delta x_m)}$$

$$\frac{\partial n_i}{\partial y}\bigg|_{x,y} \approx \frac{\left(n_i^{k,m} - n_i^{k-1,m}\right)\Delta y_k^2 - \left(n_i^{k,m} - n_i^{k+1,m}\right)\Delta y_{k-1}^2}{\Delta y_{m-1} \Delta y_m (\Delta y_{m-1} + \Delta y_m)}$$

$$\frac{\partial^2 n_i}{\partial x^2}\bigg|_{x,y} \approx 2 \frac{\left(n_i^{k,m-1} - n_i^{k,m}\right)\Delta x_m + \left(n_i^{k,m+1} - n_i^{k,m}\right)\Delta x_{m-1}}{\Delta x_{m-1} \Delta x_m (\Delta x_{m-1} + \Delta x_m)}$$

$$\frac{\partial^2 n_i}{\partial x \partial y}\bigg|_{x,y} \approx \frac{\Delta x_{m-1} \Delta y_{k-1} \left(n_i^{k,m} + n_i^{k+1,m+1} - n_i^{k+1,m} - n_i^{k,m+1}\right)}{\Delta x_m \Delta y_k (\Delta x_{m-1} + \Delta x_m)(\Delta y_{k-1} + \Delta y_k)}$$

$$+ \frac{\Delta x_m \Delta y_{k-1} \left(n_i^{k+1,m} + n_i^{k,m-1} - n_i^{k-1,m-1} - n_i^{k,m}\right)}{\Delta x_{m-1} \Delta y_k (\Delta x_{m-1} + \Delta x_m)(\Delta y_{k-1} + \Delta y_k)}$$

$$+ \frac{\Delta x_{m-1} \Delta y_k \left(n_i^{k,m+1} + n_i^{k-1,m} - n_i^{k1,m} - n_i^{k-1,m+1}\right)}{\Delta x_m \Delta y_{k-1} (\Delta x_{m-1} + \Delta x_m)(\Delta y_{k-1} + \Delta y_k)}$$

$$+ \frac{\Delta x_m \Delta y_k \left(n_i^{k,m} + n_i^{k-1,m-1} - n_i^{k,m-1} - n_i^{k-1,m}\right)}{\Delta x_{m-1} \Delta y_{k-1} (\Delta x_{m-1} + \Delta x_m)(\Delta y_{k-1} + \Delta y_k)}$$

$$\frac{\partial^2 n_i}{\partial y^2}\bigg|_{x,y} \approx 2 \frac{\left(n_i^{k-1,m} - n_i^{k,m}\right)\Delta y_k + \left(n_i^{k+1,m} - n_i^{k,m}\right)\Delta y_{k-1}}{\Delta y_{k-1} \Delta y_k (\Delta y_{k-1} + \Delta y_k)} \tag{A2.8}$$

Hence, the partial derivatives of the director field at a certain grid point can be estimated with the help of the surrounding eight director field samples. After working out the derivatives in the director update equation Eq. (A2.5), the estimated derivatives Eq. (A2.8) can be used to numerically evaluate the problem statement. This is a quite laborious task and results in very long terms which are not stated here. Computer–algebra systems, such as Mathematica or MuPAD, can be used to obtain the numerical update expressions. A solution is given by James E. Anderson et al. in [10].

On the whole, Eq. (A2.5) can be summarized in an equation that relates the director field at a considered grid point and its surrounding samples to a previous director field due to the time derivative in Eq. (A2.5). Evaluating this equation for every grid point in the whole computational domain will lead to a linear system of equations that has to be solved for each time step. At the boundaries of the computational domain, the surrounding samples are determined by the respective boundary conditions. The time derivative can be approximated using the Euler method. Hence, an initial director configuration has to be known in addition. As mentioned earlier, this originates from the initial boundary condition problem statement from Eq. (A2.5).

However, the contribution of the electric free energy also depends on the director configuration due to the permittivity tensor. This leads to a coupling of director configuration and electric field. Usually, this coupling is performed explicitly, that is, the electric field is computed based on a given, previous, director configuration. In turn, these results are used to determine the acting torque and compute the new director configuration. An update of the electric field is performed at a later point in time. Implicit coupling, instead, considers previous and current director configurations and electric fields in the computation process. For this purpose, a corresponding coupled system of equations has to be solved each time step. This method possesses advantages concerning the stability and accuracy. Hence, it is possible to achieve the same accuracy as with the explicit method but with less frequent recalculations. However, solving the coupled equation every time step is out of all proportion to the gain of reduced number of time samples. This is due to the slow orientation of conventional LC molecules which, in turn, leads to a negligible change of the electric field per time step. Achieving the same accuracy, the explicit method reaches the steady state much faster than the implicit method. Thus, only a recalculation of the electric field at the start and at certain points in time with the explicit method is usually preferred.

A2.2   Laplace Equation of Anisotropic Continuous Materials

For the computation of the electric field, the Laplace equation for inhomogeneous, anisotropic material, which is given as

$$\nabla \cdot \left(\overleftrightarrow{\varepsilon} \cdot \nabla u \right) = 0, \tag{A2.9}$$

has to be solved. In Eq. (A2.9), the permittivity $\overleftrightarrow{\varepsilon}$ is assumed to be a symmetric/hermitian tensor. Working out the Nabla operators (∇) results in

$$\varepsilon_{xx}\frac{\partial^2 u}{\partial x^2} + \varepsilon_{yy}\frac{\partial^2 u}{\partial y^2} + \varepsilon_{zz}\frac{\partial^2 u}{\partial z^2} + 2\left(\varepsilon_{xy}\frac{\partial^2 u}{\partial x\partial y} + \varepsilon_{xz}\frac{\partial^2 u}{\partial x\partial z} + \varepsilon_{yz}\frac{\partial^2 u}{\partial y\partial z}\right)$$

$$+\frac{\partial u}{\partial x}\left(\frac{\partial \varepsilon_{xx}}{\partial x} + \frac{\partial \varepsilon_{xy}}{\partial y} + \frac{\partial \varepsilon_{xz}}{\partial z}\right) + \frac{\partial u}{\partial y}\left(\frac{\partial \varepsilon_{xy}}{\partial x} + \frac{\partial \varepsilon_{yy}}{\partial y} + \frac{\partial \varepsilon_{yz}}{\partial z}\right)$$

$$+\frac{\partial u}{\partial z}\left(\frac{\partial \varepsilon_{xz}}{\partial x} + \frac{\partial \varepsilon_{yz}}{\partial y} + \frac{\partial \varepsilon_{zz}}{\partial z}\right) = 0. \tag{A2.10}$$

In contrast to the Laplace equation for isotropic material, also mixed derivatives occur due to the tensor components ε_{xy}, ε_{xz}, and ε_{yz}. Additionally, derivatives of the tensor components are included due to the inhomogeneous property of the LC. This will lead to a dense system of equations with eight instead of four extra-diagonals. Similar to the computation of the director field, the calculation area is subdivided using the grid depicted in Figure 5.9. Also, the derivatives in z direction are neglected since an infinite extent of the structure is assumed. Modelling the electric potential u with a multivariate Taylor series similar to Eq. (A2.6) leads to the following approximations of the derivatives.

$$\varepsilon_{xx}\frac{\partial^2 u}{\partial x^2}\bigg|_{x,y} \approx 2\varepsilon_{xx}\bigg|^{k,m}\frac{u_{k,m+1}\Delta x_{m-1} + u_{k,m-1}\Delta x_m + u_{k,m}(\Delta x_{m-1} - \Delta x_m)}{\Delta x_{m-1}\Delta x_m(\Delta x_{m-1} - \Delta x_m)}$$

$$\varepsilon_{yy}\frac{\partial^2 u}{\partial y^2}\bigg|_{x,y} \approx 2\varepsilon_{yy}\bigg|^{k,m}\frac{u_{k+1,m}\Delta y_{k-1} + u_{k-1,m}\Delta y_k + u_{k,m}(\Delta y_{k-1} - \Delta y_k)}{\Delta y_{k-1}\Delta y_k(\Delta y_{k-1} + \Delta y_k)}$$

$$2\varepsilon_{xy}\frac{\partial^2 u}{\partial x\partial x}\bigg|_{x,y} \approx 2\varepsilon_{xy}\bigg|^{k,m}\left(\frac{\Delta x_m\Delta y_{k-1}(u_{k+1,m} - u_{k,m} + u_{k,m-1} - u_{k+1,m-1})}{\Delta y_k\Delta x_{m-1}(\Delta x_m + \Delta x_{m-1})(\Delta y_k + \Delta y_{k-1})} + \cdots\right)$$

$$\tag{A2.11}$$

which are derived in a similar manner as in the previous section by truncating the series expansion at terms including powers greater than two. Inserting these approximations in the Laplace equation, Eq. (A2.7), yields an equation relating the potentials at eight outer grid points to the potential at the center grid point

$$\sum_{\substack{i=n|^{k+1,m-1}\\i\neq n|^{k,m}}}^{n|^{k-1,m+1}} A_{n|^{k,m},i}u_i = \left(\sum_{\substack{i=n|^{k+1,m-1}\\i\neq n|^{k,m}}}^{n|^{k-1,m+1}} A_{n|^{k,m},i}\right)u_{n|^{k,m}} = A_{n|^{k,m},n^{k,m}}u_{n|^{k,m}}, \tag{A2.12}$$

where $n|^{k,m}$ denotes the number of the node at the grid point k,m. Evaluating this equation for all N grid points leads to the following system of equations.

$$\mathbf{A}\cdot \vec{u} = \vec{b}, \tag{A2.13}$$

where the vector $\vec{u}$ is the solution of the system of equations containing the potentials of the corresponding grid points. $\vec{b}$ incorporates the boundary conditions and the matrix A contains the coefficients of the potentials which are determined by Eq. (A2.7) as follows:

$$A_{n|^{k,m}, n|^{k+1,m}} = \frac{2\varepsilon_{xy}\Delta x_h(\Delta y_k - \Delta y_{k-1})}{\Delta x_m \Delta y_k \Delta y_{k-1}} + \frac{2\varepsilon_{xx} + \Delta x_{m-1}\left(\dfrac{\partial \varepsilon_{xx}}{\partial x} + \dfrac{\partial \varepsilon_{xy}}{\partial y}\right)}{\Delta x_m(\Delta x_m + \Delta x_{m-1})}$$

$$A_{n|^{k,m}, n|^{k+1,m+1}} = \frac{2\varepsilon_{xy}\Delta x_{m-1}\Delta y_{k-1}}{\Delta x_m \Delta y_k(\Delta x_m + \Delta x_{k-1})(\Delta y_k + \Delta y_{k-1})}$$

$$A_{n|^{k,m}, n|^{k,m}} = A_{n|^{k,m}, n|^{k-1,m}} + A_{n|^{k+1,m}, n|^{k,m+1}} + \ldots + A_{n|^{k,m}, n|^{k-1,m}} + A_{n|^{k,m}, n|^{k,m-1}} + A_{n|^{k,m}, n|^{k-1,m+1}}$$

$$(A2.14)$$

The inhomogeneity of the permittivity tensors components is modelled with a linear function according to the Frank–Oseen continuum theory [11]. Hence, the permittivity at a certain potential grid point is determined by averaging the surrounding tensor components as

$$\varepsilon|^{k,m} \approx \frac{\varepsilon_{k,m}\Delta x_{m-1}\Delta y_{k-1} + \varepsilon_{k,m-1}\Delta x_m \Delta y_{k-1} + \varepsilon_{k-1,m-1}\Delta x_m \Delta y_k + \varepsilon_{k-1,m}\Delta x_{m-1}\Delta y_k}{(\Delta x_m + \Delta x_{m-1})(\Delta y_k + \Delta y_{k-1})}.$$

$$(A2.15)$$

The corresponding derivatives of the permittivities are given by

$$\frac{\partial \varepsilon_{xx}}{\partial x} \approx 2\frac{\left(\varepsilon_{xx,(k-1,m)} - \varepsilon_{xx,(k-1,m-1)}\right)\Delta y_k + \left(\varepsilon_{xx,(k,m)} - \varepsilon_{xx,(k,m-1)}\right)\Delta y_{k-1}}{(\Delta x_m + \Delta x_{m-1})(\Delta y_k + \Delta y_{k-1})}$$

$$\frac{\partial \varepsilon_{yy}}{\partial x} \approx 2\frac{\left(\varepsilon_{yy,(k,m-1)} - \varepsilon_{yy,(k-1,m-1)}\right)\Delta x_m + \left(\varepsilon_{yy,(k,m)} - \varepsilon_{yy,(k-1,m)}\right)\Delta y_{k-1}}{(\Delta x_m + \Delta x_{m-1})(\Delta y_k + \Delta y_{k-1})}$$

$$(A2.16)$$

Thus, all components in **A** are determined by the grid spacing and the material properties, that is, the permittivity tensors components. Every change in the director configuration will also change the permittivity tensor's base and hence, the tensor's components. This will also lead to a change in the electric potential and field. In the presented method, the point in time for the recalculation of the electric field is determined from a given number of field computations N_{FU}. This number of computations is spread over the computation time with the help of a reference error that is determined from the residual of the updated system of equations evaluated with the old distribution of potential:

$$F_{\text{Ref}} = \left| \mathbf{A}_{\text{current}}\vec{u}_{\text{previous}} - \vec{b}_{\text{current}} \right|.$$

$$(A2.17)$$

A2.3   Computation of Waveguide Modes

After the director dynamic simulation has achieved the steady state, the waveguide modes can be computed to evaluate the parameters of interest. For this purpose, the Maxwell equations are evaluated in the frequency domain. The derivatives are approximated with finite differences on the Yee grid depicted in Figure A2.2.

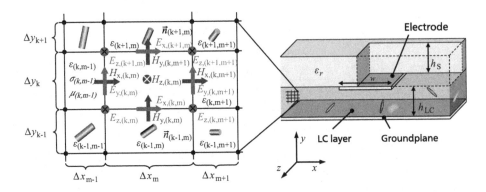

Figure A2.2 Representation of the two-dimensional Yee gridding in Cartesian coordinates.

Assuming a harmonic excitation $(\partial/\partial t \to -\mathrm{j}\omega)$ propagating in z-direction $(\partial/\partial z \to \gamma)$, results in

$$
\begin{pmatrix}
\gamma E_{y,(k,m)} + \dfrac{\left(E_{z,(k+1,m)} - E_{z,(k,m)}\right)}{\Delta y_k} \\[3mm]
-\gamma E_{x,(k,m)} - \dfrac{\left(E_{z,(k,m+1)} - E_{z,(k,m)}\right)}{\Delta x_m} \\[3mm]
\left(E_{y,(k,m+1)} - E_{y,(k,m)}\right)/\Delta x_m - \dfrac{\left(E_{x,(k+1,m)} - E_{x,(k,m)}\right)}{\Delta y_k}
\end{pmatrix}
= -\mathrm{j}\omega\mu
\begin{pmatrix}
H_{x,(k,m)} \\
H_{y,(k,m)} \\
H_{z,(k,m)}
\end{pmatrix}
\tag{A2.18}
$$

for the Faraday's law of induction and

$$
\begin{pmatrix}
\gamma H_{y,(k,m)} + \dfrac{\left(H_{z,(k+1,m)} - H_{z,(k,m)}\right)}{\Delta y_k} \\[3mm]
-\gamma H_{x,(k,m)} - \dfrac{\left(H_{z,(k,m+1)} - H_{z,(k,m)}\right)}{\Delta x_m} \\[3mm]
\left(H_{y,(k,m+1)} - H_{y,(k,m)}\right)/\Delta x_m - \dfrac{\left(H_{x,(k+1,m)} - H_{x,(k,m)}\right)}{\Delta y_k}
\end{pmatrix}
= \left(\sigma + \mathrm{j}\omega\, \overleftrightarrow{\varepsilon}\right)
\begin{pmatrix}
E_{x,(k,m)} \\
E_{y,(k,m)} \\
E_{z,(k,m)}
\end{pmatrix}
\tag{A2.19}
$$

for the Maxwell–Ampère equation respectively. The material parameters conductivity σ, $\overleftrightarrow{\varepsilon}$ and μ are averaged from the surrounding mesh cells of the evaluated grid point.

In case of all directors being aligned perpendicular to the direction of propagation, that is, only in the cross-sectional plane, the tensor components ε_{xz} and ε_{yz} are zero. This frequently analyzed special case is called "In Plane Switching" (IPS) and leads to a simplification of the above stated problem. After substitution of the longitudinal components in Eq. (A2.18) and Eq. (A2.19), the transversal magnetic components can be summarized in the two coupled systems of equations

$$
H_{x,(k,m)}\left(A_{x0} - \gamma^2\right) + H_{x,(k,m-1)}A_{x1} + \cdots + H_{y,(k,m)}A_{x17} = 0
\tag{A2.20}
$$

and

$$H_{y,(k,m)}\left(A_{y0} - \gamma^2\right) + H_{y,(k,m-1)}A_{y1} + \cdots + H_{x,(k,m)}A_{y17} = 0. \tag{A2.21}$$

where the coefficients A_{xn} and A_{yn} are determined by the material and grid dependent parameters. With the help of Gauss's law, that is, the divergence-free property of the magnetic flux, Eq. (A2.20) and Eq. (A2.21) can be expressed as an eigenvalue problem with 16 extradiagonals. This can be solved with common numerical program libraries. However, an arbitrary LC director configuration will lead to equation systems involving all four transversal field components. This, in turn, will result in a quadratic eigenvalue problem of shape

$$\left(\gamma^2 \mathbf{A} + \gamma \mathbf{B} + \mathbf{C}\right)\cdot \vec{x} = \vec{0}. \tag{A2.22}$$

Substituting both longitudinal components directly from Eqs. (A2.20) and (A2.21) without the material equations will lead instead to a simple eigenvalue problem. With the identity matrix $\mathbf{I}$, Eq. (A2.18) and Eq. (A2.19) will be reduced to

$$\left(\gamma \mathbf{I} - \mathbf{A}\right)\cdot \vec{x} = \vec{0}, \tag{A2.23}$$

which has been used in the presented method to determine the modes of propagation.

References

1. B. R. Elbert, *Introduction to Satellite Communication*, 3rd ed. The Artech House space applications series. Norwood, MA: Artech House.
2. L. J. Ippolito, *Satellite Communications Systems Engineering: Atmospheric Effects, Satellite Link Design and System Performance*, 2nd ed. Hoboken, NJ: John Wiley & Sons.
3. D. Minoli, *Innovations in Satellite Communication and Satellite Technology*. Hoboken, NJ: John Wiley & Sons.
4. W. L. Stutzman and G. A. Thiele, *Antenna Theory and Design*, 2nd ed. Hoboken, NJ: John Wiley & Sons.
5. Globalstar. Second-Generation Satellite Constellation. Available at: www.globalstar.com/en/index.php?cid=8300
6. Big LEO tables. Available at: http://personal.ee.surrey.ac.uk/Personal/L.Wood/constellations/tables/tables.html
7. Orbcomm Second Generation. Available at: http://spaceflight101.com/spacecraft/orbcomm-g2/
8. G. D. Krebs. Inmarsat-4 F1, 2, 3. Available at: http://space.skyrocket.de/doc_sdat/inmarsat-4.htm
9. A. Gaebler, "Synthese steuerbarer Hochfrequenzschaltungen und Analyse Flüssigkristall-basierter Leitungsphasenschieber in Gruppenantennen für Satellitenanwendungen im Ka-Band," *ETIT*, 2015.
10. J. E. Anderson, P. J. Bos, and P. E. Watson, *LC3D: Liquid Crystal Display 3-D Director Simulator: Software and Technology Guide*. Artech House, 2001.
11. F. C. Frank, "I. Liquid crystals. On the theory of liquid crystals," *Discussions of the Faraday Society*, vol. 25, no. 0, pp. 19–28, 1958, doi: 10.1039/DF9582500019.

Index

Printed in the United States
by Baker & Taylor Publisher Services